Springer Complexity

Springer Complexity is an interdisciplinary program publishing the best research and academic-level teaching on both fundamental and applied aspects of complex systems – cutting across all traditional disciplines of the natural and life sciences, engineering, economics, medicine, neuroscience, social and computer science.

Complex Systems are systems that comprise many interacting parts with the ability to generate a new quality of macroscopic collective behavior the manifestations of which are the spontaneous formation of distinctive temporal, spatial or functional structures. Models of such systems can be successfully mapped onto quite diverse "real-life" situations like the climate, the coherent emission of light from lasers, chemical reaction-diffusion systems, biological cellular networks, the dynamics of stock markets and of the internet, earthquake statistics and prediction, freeway traffic, the human brain, or the formation of opinions in social systems, to name just some of the popular applications.

Although their scope and methodologies overlap somewhat, one can distinguish the following main concepts and tools: self-organization, nonlinear dynamics, synergetics, turbulence, dynamical systems, catastrophes, instabilities, stochastic processes, chaos, graphs and networks, cellular automata, adaptive systems, genetic algorithms and computational intelligence.

The two major book publication platforms of the Springer Complexity program are the monograph series "Understanding Complex Systems" focusing on the various applications of complexity, and the "Springer Series in Synergetics", which is devoted to the quantitative theoretical and methodological foundations. In addition to the books in these two core series, the program also incorporates individual titles ranging from textbooks to major reference works.

Editorial and Programme Advisory Board

Understanding Complex Systems

Founding Editor: J.A. Scott Kelso

Future scientific and technological developments in many fields will necessarily depend upon coming to grips with complex systems. Such systems are complex in both their composition – typically many different kinds of components interacting simultaneously and nonlinearly with each other and their environments on multiple levels – and in the rich diversity of behavior of which they are capable.

The Springer Series in Understanding Complex Systems series (UCS) promotes new strategies and paradigms for understanding and realizing applications of complex systems research in a wide variety of fields and endeavors. UCS is explicitly transdisciplinary. It has three main goals: First, to elaborate the concepts, methods and tools of complex systems at all levels of description and in all scientific fields, especially newly emerging areas within the life, social, behavioral, economic, neuro- and cognitive sciences (and derivatives thereof); second, to encourage novel applications of these ideas in various fields of engineering and computation such as robotics, nano-technology and informatics; third, to provide a single forum within which commonalities and differences in the workings of complex systems may be discerned, hence leading to deeper insight and understanding.

UCS will publish monographs, lecture notes and selected edited contributions aimed at communicating new findings to a large multidisciplinary audience.

J. Awrejcewicz · V. A. Krysko

Chaos in Structural Mechanics

With 195 Figures and 45 Tables

Springer

Jan Awrejcewicz
Technical University Lodz
Dept. Automatics
and Biomechanics
ul. Stefanowskiego 1/15
90-924 Lodz
Poland
awrejcew@p.lodz.pl
kparmesw@telcordia.com

Vadim Anatolevich Krysko
Saratov State University
Dept. Mathematics/Mechanics
Polyteshnycheskaya ul. 77
Saratov
Russia 410005
tak@san.tu

ISBN: 978-3-540-77675-8 e-ISBN: 978-3-540-77676-5

DOI 10.1007/978-3-540-77676-5

Understanding Complex Systems ISSN: 1860-0832

Library of Congress Control Number: 2008922887

Cover design: WMXDesign Gmbh

Printed on acid-free paper

9 8 7 6 5 4 3 2 1

springer.com

Preface

This monograph is devoted to construction of novel theoretical approaches of modeling non-homogeneous structural members as well as to development of new and economically efficient (simultaneously keeping the required high engineering accuracy) computational algorithms of nonlinear dynamics (statics) of strongly nonlinear behavior of either purely continuous mechanical objects (beams, plates, shells) or hybrid continuous/lumped interacting mechanical systems.

In general, the results presented in this monograph cannot be found in the existing literature even with the published papers of the authors and their coauthors. We take a challenging and originally developed approach based on the integrated mathematical–numerical treatment of various continuous and lumped/continuous mechanical structural members, putting emphasis on mathematical and physical modeling as well as on the carefully prepared and applied novel numerical algorithms used to solve the derived nonlinear partial differential equations (PDEs) mainly via Bubnov-Galerkin type approaches.

The presented material draws on the fields of bifurcation, chaos, control, and stability of the objects governed by strongly nonlinear PDEs and ordinary differential equations (ODEs), and may have a positive impact on interdisciplinary fields of nonlinear mechanics, physics, and applied mathematics. We show, for the first time in a book, the complexity and fascinating nonlinear behavior of continual mechanical objects, which cannot be found in widely reported bifurcational and chaotic dynamics of lumped mechanical systems, i.e., those governed by nonlinear ODEs. On the other hand, we illustrate also how the strongly nonlinear PDEs modeled by chains of many (infinitely many) coupled nonlinear oscillators may also exhibit nonlinear behavior well known and studied via simple systems, usually in one or two degrees-of-freedom mechanical objects.

In the Introduction an overview of both Ritz and Bubnov-Galerkin methods with emphasis on achievements of Eastern countries is given. A historical origin of both the Bubnov-Galerkin and Galerkin procedures is described and their status versus that of other projection approaches is addressed.

In Chap. 1 a general theory of non-homogeneous shells is introduced. First, fundamental relations and assumptions are given, shells' non-homogeneities are introduced, and then the governing variational equations and equations of motion are derived. After the boundary and initial conditions are introduced, the equations are cast to non-dimensional form and the so-called variable parameters of a shell stiffness are defined. In addition, a flexural stiffness coefficient of a shell element is formally introduced. The section devoted to generalized functions completes this chapter.

Chapter 2 deals with the static instability problems of rectangular plates. First, fundamental concepts of the theory of elastic stability are illustrated and discussed. Second, two fundamental formulas of the energy-based criterion of bifurcational stability loss of an elastic continuous mechanical system are derived. In addition, advantages and disadvantages of today's stability investigation approaches are critically revisited, placing emphasis on problems not yet satisfactorily solved. In the next section various methods devoted to stability investigations are briefly addressed, exhibiting their strong and weak points regarding applications with particular attention to computational advantages of the Galerkin methods. In Sect. 2.4 the Bubnov-Galerkin method of high-order approximations and the associated numerical algorithm are presented. The rectangular form additions of other materials to a shell are described in Sect. 2.5, whereas the next section deals with static shell stability problems. Finally, the central square element, cross addition element, and perforation-type non-homogeneities are introduced.

In Chap. 3 vibrations of rectangular shells are studied. Linear and weakly nonlinear vibrations are revisited in Sect. 3.1, and then natural vibrations of non-homogeneous shells applying the Bubnov-Galerkin method of higher order approximations are analyzed. Section 3.3 is devoted to investigation of free nonlinear vibrations of homogeneous plates and shells regarding any choice of control parameters. The relatively extensive Sect. 3.4 addresses the spectral analysis of a stress–strain problem of any plate/shell approximated by systems of n degrees of freedom. A harmonic process convergence and spectral analysis of free nonlinear vibrations are illustrated and discussed in Sects. 3.5 and 3.6, respectively.

In Chap. 4 dynamic stability loss of rectangular shells is addressed. A background containing types of dynamic buckling exhibited by perfect constructions, as well as the concept of finite-time stability, is given in Sects. 4.1–4.3. Mathematical modeling of dynamical systems, problems of synchronization, chaos, and quasi-periodicity are also briefly revisited. Sections 4.6–4.10 refer to both static and dynamic bifurcations and their numerical estimations. Stability loss of homogeneous shells subjected to an action of transversal loads is rigorously studied in Sect. 4.11.

Chapter 5 concerns stability of closed cylindrical shells subjected to an axially non-symmetrical load action. In the beginning (Sect. 5.1), equations of motion are derived, and then the influence of imperfections on shell stability is studied. Both static and dynamic problems of buckling with the use of the Bubnov-Galerkin method of higher approximations are analyzed and many computational results are reported.

Composite shells are studied in Chap. 6. First, equations governing behavior of composite shells are derived, and then both static and dynamic problems of stability loss of composite shells are addressed.

In Chap. 7 the problem of interaction between flexible construction and a moving lumped body is reduced to that of essentially simpler ones, i.e., that of vibrations subject to moving force P_0 and that of displacement in the domain of the moving masses under the action of the mentioned force. Advantages of the proposed method are illustrated and discussed. It is characteristic for the modeling of interaction between a construction and moving bodies that the impact on the construction is expressed by the weight and inertial forces of the objects moving on the studied construction. This crucial feature of the applied approach constitutes also the essential difficulty in the mathematical analysis of the problem. Let us note that most publications in this field and related to the analysis of the interaction between a shell and a moving mass are based on the application of a geometrically linear model or on the assumption that a movable mass does not tear off the leading construction. Apart from that, calculations were usually conducted with the use of approximated methods with only a few first approximations taken into account. This chapter deals with this problem in a complex and detailed way and it provides the methods leading to the proper and accepted solution in engineering.

Chapter 8 deals with a novel approach to study chaotic vibrations of deterministic mechanical systems represented by shallow sector-type spherical shells. Scales of vibration character of such shells being transversally and harmonically excited vs. control parameters are constructed. Scenarios to chaos are illustrated and discussed. Control of chaotic state applying synchronous action of harmonic loading torque is proposed. Investigations are carried out using the qualitative theory of differential equations and nonlinear dynamics.

In Chap. 9 scenarios of transition from harmonic to chaotic motions are illustrated and discussed. First, a historical background of the problem is described. The Landau-Hopf scenario; the Ruelle, Takens, and Newhouse scenario; the Feigenbaum scenario; and the Pomeau-Manneville scenario are addressed, among others.

In Chap. 10 complex vibrations of closed cylindrical shells of infinite length and circular cross section subjected to transversal local load in the frame of the classical nonlinear theory are studied. A transition from PDEs to ODEs is carried out using the higher order Bubnov-Galerkin approach and Fourier representation. On the other hand, the Cauchy problem is solved using the fourth-order Runge-Kutta method. In the first part of this work static problems of the theory of closed cylindrical shells are studied. Reliability of the obtained results is verified by comparing them with the results taken from the literature. The second part is devoted to the analysis of stability, bifurcation, and chaos of closed cylindrical shells. In particular, an influence of sign-changeable external pressure and the control parameters such as magnitude of pressure measured by φ_0, relative linear shell dimension $\lambda = \frac{L}{R}$, frequency ω_p, and amplitude q_0 of external transversal load on the shell's nonlinear dynamics are studied.

Chapter 11 is devoted to control of temporal-spatial chaos exhibited by cylindrical shells. Process of controlling chaos is understood as the transformation of chaotic dynamics into regular, or the other chaotic ones, but of different characteristics, with the use of small external periodic input functions and by the influence of transverse load applied in anti-phase.

In Chap. 12 chaotic vibrations of flexible rectangular shells forced by transversal harmonic load are analyzed via application of the qualitative theory of differential equations and nonlinear dynamics. An infinitely dimensional problem is reduced to a finitely dimensional one with the application of the Bubnov-Galerkin method with higher approximations and the method of finite differences with approximation $O(\Delta^2)$. An initial problem is solved with the use of Runge-Kutta method of the fourth order. It is shown that within the range of harmonic vibrations, the results obtained from both methods are fully convergent, whereas in the range of chaos such convergence can be obtained only in relation to the character of vibrations, i.e., in relation to the frequency spectra. The increase in the number of element partitions in the method of finite differences and the number of approximations in the Bubnov-Galerkin method leads to better results, but there is some threshold value beyond which further calculations are impossible.

Chapter 13 deals with both regular and chaotic vibrations, various bifurcations, and scenarios exhibited by three-layered nonlinear uncoupled beams with constraints. The finite difference approximation is applied and the reliability of the numerical results is first rigorously discussed. New scenarios of transition to chaos and synchronization phenomena are reported, and the essential influence of four boundary conditions on various nonlinear behaviors is outlined.

In Chap. 14 dynamics of physically dissipative nonlinear multi-layer sandwich of three beams is analyzed. The boundary conditions are arbitrary. The transversal load can be applied either simultaneously to all beams or separately to each of the beams. The finite difference method is used to solve the governing equations. Different types of beam material are considered: ideally elastic-plastic, elastic-plastic with linear straightness, and pure aluminum. Some new bifurcation and chaotic phenomena of the system are reported and discussed.

Chapter 15 deals with complex vibrations of a flexible Euler-Bernoulli type beam driven by a dynamic load, and the various type of inputs on its edge are studied. The governing equations include damping terms with damping coefficients ε_1, ε_2 associated with deflection w and displacement u, respectively. Damping coefficients ε_1, ε_2 and the transversal load coefficients (q_0, ω_p) serve as control parameters. The formulated infinite dimensional problem is reduced to that of finite dimension applying the finite difference method with approximation $O(h^2)$ with regard to spatial coordinates and it is solved via the fourth-order Runge-Kutta method. This approach enabled identification of damping coefficients ε_1 and ε_2, as well as investigation of elastic waves generated by an impact introduced through a mass (lumped body) moving at constant velocity. The introduced analysis is supported by applied achievements of the qualitative theory of differential equations and nonlinear dynamics.

Finally, this book is accessible to readers with a fundamental knowledge of applied mathematics, differential equations, and modern theory of nonlinear dynamical systems. One of the co-authors (J. Awrejcewicz) acknowledges the financial suppport by the Polish Ministry of Science and Higher Education for years 2005–2008 (grant No. 4 07A 03128) regarding the book part devoted to impact phenomena.

We deeply thank P. Olejnik, M. Kaźmierczak, and A. Dębska for their help in the preparation of this book.

Łódź, Poland Jan Awrejcewicz
Saratov, Russia Vadim A. Krysko

Contents

Introduction

An overview of both Ritz and Bubnov-Galerkin methods, with an emphasis on the achievements of Eastern European countries, is provided here. The historical origin of the Bubnov-Galerkin procedure is described and its approach versus that of other projection approaches is addressed.

We begin our considerations with a definition of non-homogeneity. There exist many different definitions concerning non-homogeneity but in this book the term is understood as a dependence of Young modulus $E(x, y, z)$ and the density $\rho(x, y, z)$ of the analyzed constructions on the three spatial coordinates x, y, z.

The Ritz (MR) and Bubnov-Galerkin (MBG) methods belong to the class of projection methods. The origin of the Ritz method is exhibited mainly via singularity of a functional whose extremes are achieved at some stationary points. The main idea of both MR and MBG methods relies on the basis of a state-space approximation of the investigated system by a subspace belonging to the space at hand, in which any stationary points of the considered functional should be determined.

In such a situation the transition from a system of input partial differential equations (PDEs) to the system of algebraic equations (static problem) or ordinary differential equations (ODEs; dynamic problem) is realized.

Note that the MBG projection method can be formally applied to the solution to differential equations not necessarily belonging to the class of Euler equations and for the arbitrarily chosen functional (an existing functional can be indefinite as well).

In the last case, an application of the MBG method consists of the projection of output equations on a subspace of the analyzed functional space where a solution to the problem is included. Assuming that an approximated solution is satisfied in the subspace, the searching process can be realized by introduction of some important relations.

The MR and MBG methods began with the pioneering works of Ritz [246] and Bubnov [62]. Development and application of MR and MBG methods in the problems concerning mechanics were introduced in works of Galerkin [102], Leibenson [184, 185], Papkovich [239, 240], Galimov [104], and many others.

A deep mathematical background of MR for self-coupled linear problems has been given in a series of fundamental works by Krylov [153, 155, 156, 158]. In

J. Awrejcewicz, V.A. Krysko, *Chaos in Structural Mechanics*,
DOI: 10.1007/978-3-540-77676-5_1, © Springer-Verlag Berlin Heidelberg 2008

Krylov's work [154] the MR method was applied to the analysis of Theodore von Kármán's equation; in addition, thanks to the application of a uniform norm, a direction of error estimation has been provided by the use of this method. In the work of Kieldysh [134], MBG convergence in linear problems for the even-order ODEs and for a class of partial equations of elliptic type has been demonstrated as well.

Mikhlin [210] extended Kieldysh's results to the linear case of operational equation with the non-self-coupled operator and generalized the approach to the wide class of boundary problems for elliptic-type equations. In Krasnoselskiy's work [150] some basic theorems referred to the MBG convergence for nonlinear operational equations were given. Furthermore, by the best possible approximation of the function sought through linear combinations of coordinate elements an error estimation for that method has been proposed.

In works of Liashko [189, 190] and Martyniuk [199, 200] (linear equations with a non-self-coupled operator), a quadratic functional construction was introduced. Analogously, the functional has been used in the Ritz method for solving self-coupled problems and the convergence of the variational method has been proved. In Liashko [191] and Nashed's [227] works the obtained results have been generalized to the form of a nonlinear equation with a non-potential operator.

Contrary to MR, the MBG can be effectively applied to the solution of not only any stationary problems (boundary problems for the elliptic-type equations), but also to the problems of evolution (initial-boundary problems for the parabolic- and hyperbolic-type equations). Since for the first time in 1949 MBG was applied in analysis of non-stationary problems by Faedo, in the mathematical literature the term *Faedo-Galerkin method* [107] is often used.

However, in 1931, in the work by Krylov-Bogolubov [157], MBG was used in a broad investigation of Cauchy-Dirichlet problems for the second-order hyperbolic equation, and the scheme considered is very close to the one used today. It seems that, in spite of this, Faedo's impact in the field of MBG application to evolutional problems is not as strong as it had been thought, though there is no doubt that in the field [113] belongs to the first ones publications even if it is not necessarily a fundamental one.

From the earlier works devoted to the problem it is appropriate to enumerate publications of Hopf [122], in which MBG was adequately applied to parabolic linear equations and to the nonlinear Navier-Stokes equations.

The level of development and utilization of MR and MBG methods in nonlinear problems of the theory of plates and shells were sufficiently well illustrated in 1956 in the monographs of Volmir [303] and Mushtari and Galimov [219]. Taking into consideration only first and second approximations, the mentioned methods were successfully used there.

Successively the development of computational numerical techniques has allowed the application of both of the discussed methods by means of higher order approximations. Let us enumerate at this stage some of the work devoted to the direction of investigations. Some interesting results were given in Kornishin's monograph [143] devoted to analysis of large rectangular deflections of shells using third-order approximation. Kantor [130] applied MR to the solution of problems of the theory of axially symmetric shallow shells regarding geometrical and physical nonlinearities (sufficiency of application of fourth- or fifth-order approximations has been proved). In a survey work by Vorovich and Minakova [309], the wide spectrum of literature devoted to MBG applications with higher order approximations is extensively presented. In works of Amielchenko and Krysko [10], Krysko [159], and Potash [242], as many as 36 parameters subjected to some variation have been taken into consideration during approximation in a search for a solution.

Some results of various MR and MBG applications based on the theory of shell stability were presented in Grigoliuk and Kabanov's monograph [114]. During determination of the form of stability loss in systems with high non-homogeneity many series terms are required (using information a priori). Furthermore, an appropriate selection of a number of base functions has to be indicated also.

A particular MBG presentation for the geometrically nonlinear dynamical problems devoted to the theory of plates and shells was reported in monographs [304, 305], where stability of constructions together with any gas–fluid interactions have been particularly investigated.

A wide class of problems of the theory of shallow shells with geometrical and physical nonlinearities, as well as the cyclic load problems (in particular, some problems such as an estimation of influence of impulse action on the stability of shells, or influence of damping on the value of dynamic critical loading, etc.) can be solved using MBG as shown in Krysko's monograph [159]. The work indicates a purposefulness of MBG application in comparison to the traditional finite differences method (mainly owing to the shorter time for numerical computations).

The Ph.D. thesis of Kutsemako [171] focused on MBG applications in computations of dynamic problems of the theory of shells within a framework of both Kirchhoff-Love and Timoshenko models.

Computations of the dynamic stability loss of a cylindrical shell under a band loading and with the application of the MBG method were carried out in [140].

The paper by Kirichenko et al. [135] presents an interesting application of MBG to the analysis of dynamical stability of a stiffly clamped shallow shell with as well as without any coupling between both temperature and deformation fields.

Miscellaneous MBG problems devoted to numerical schemes of theory of shells have been given in the following works: Svirsky [286], Krotkova [151], Bacinov [49], Yakushev [322], Tcherniak [287], Mukhopadhyay [218], Mioduhovsky et al. [213], Chen and Hwang [71], Gelos et al. [110], and others.

In the beginning, the Bubnov-Galerkin method was established as a typical computational method, but today it is one of the strong tools for investigations of solvability problems of a wide class of either stationary and evolutional or linear and nonlinear problems of mathematical physics.

The significant development of mathematically exact schemes of the proofs of the existence of solutions based on both MBG and the fundamental theorems of mathematical analysis became possible after Sobolev's [279] discovery of the generalized derivatives of the Lebesque-class measurable functions. In the next stage, an approach to the generalized solution of mathematical problems of physics could be successfully formulated.

Mikhlin [206, 207, 209] and Ladyzhenskaya's work [175], in which a history of the problem is presented, are devoted to convergence investigations of MR and MBG for linear problems including plate and shell analysis, and to the formal exhibition of solvability of problems of that type in Sobolev space.

The principal results regarding the mathematical validation of application of various approximated methods (including MBG) are presented in monographs [149, 194].

Regarding the solvability of nonlinear problems of plates and shells, it is worth mentioning Vorovich's work [310], which often served as an inspiration for other fruitful ideas in this field. Some fundamental results in the field of mechanics were also obtained by Morozow, and presented in [214, 215] devoted to nonlinear problems of the theory of thin plates. Note that [214] focuses on vibrations of a prismatic rod, where by means of the Sobolev theorem as well as because of curiosity regarding the considered problem, Morozov managed to prove not only the existence, but also the uniqueness of the solution obtained.

Using MBG Lebedev [182, 183], Skrypnik [276] and other authors investigated problems related to solvability of miscellaneous mechanical tasks of thin-walled constructions. The comprehensive basics of MBG and MR focused on the static problems of nonlinear theory of the shallow shells were introduced in Vorovich's monograph [308].

The existence and uniqueness of solutions to the linear and nonlinear thermal-elasticity problems of plates and shells within a framework of Kirchhoff-Love's and Timoshenko-type kinematic theories, and with the application of three-dimensional heat transfer equation were widely investigated by Kirichenko and Krysko [136]. The solvability of the problems formulated is confirmed on the basis of MBG method in Sobolev space. In Wenk's work [311], with assumption of the temperature linear law distribution along the thickness of a plate, the nonlinear problem of thermal-elasticity of plates was examined.

In the capacity of boundary conditions some sub-differential inclusions leading to problem generalization in the form of variational inequalities have been considered. Applying MBG, it was possible to define a large number of theorems concerning existence, uniqueness, regularity, and the continuous solution dependence on the data assumed. ChrzAszczyk obtained some results related to uniqueness and smoothness of the problem solution, whose solvability was proved in the previously mentioned work [136].

In a series of works of Kowalski and Piskorek [148], Gawinecki et al. [106, 107, 108, 109], Kowalski et al. [147], but with the use of MBG method a number of theorems devoted to the existence, uniqueness, and regularity of solutions to many spatial linear problems of the theory of thermal-elasticity for both isotropic and anisotropic solid bodies, including heat transfer equation of both parabolic and hyperbolic type, were formulated and proved as well.

In a large majority of the previously specified works the convergence of MBG and MR methods appears as a by-product formed after the existence of a solution for a considered problem has been proved. In the case of problems of statics, the classical result (selection of a base is made arbitrarily) is then exposed as the convergence in the sense of energetic norm and the weak convergence in some "energetic" spaces of dynamical problems. When a proper rational selection of the basic system is done, the convergence results can be gained. Of course, some effective velocity estimations of the convergence rate should be formulated as well.

The direction of MBG development brings another fundamental achievement of Mikhlin [211], who introduced a concept of convergent operators. Furthermore, if as the base a system of eigenvalues for an auxiliary operator convergent to the operator of the equation at hand is chosen adequately, then the estimation of MR error convergence to the zero value is possible to obtain almost exactly. In [206] the result obtained was corrected, and additional inequality conditions (the so-called acute angle conditions) were imposed on both the auxiliary and output operator.

The principal results related to the convergent operators satisfying the type of acute angle inequalities have been given in Sobolevskiy [280] and Ladyzhenskaya's [176] works. On the basis of those results and for a series of specific cases, Mikhlin [206] derived some special fundamentals guaranteeing convergence of the MR error to zero.

Mikhlin's results for the MBG case have been propagated by Bogarian [60]. Dzishkariani's publication [89] takes into consideration some linear problems of error estimation of MR in a form of energetic norm. He used eigenvalues of the auxiliary operator that was convergent to the operator of an output problem.

Vainikko [302] and Dzishkariani's [89] works are essentially close to each other, showing similar results of MBG applications. In Vainikko's work [301] in the case of the auxiliary operator the spectrum of which does not purely consist of a point, the MBG error estimation for stationary problems has been achieved.

Dzishkariani's work [88] extends the error estimation of MR to the case of a quasi-linear equation containing as a nonlinear operator the continuous potential operator possessing a positive Frechet differential.

Some general velocity convergence estimations of the Galerkin-Pietrov method for linear and quasi-linear stationary problems have been proposed. Continuing, they result in error estimations of MBG when a system of elements of the self-coupled convergent operator is chosen as the base and, in particular, when the operator under investigation makes an acute angle with the main part operator of the problem analyzed. A wide overview of works devoted to the theory of numerical methods including MBG and for the stationary problems has been presented in Mikhlin's monograph [208].

Among the works related to evolutional problems of convergence investigations of MBG it is necessary to mention Sobolevskiy's publication [281], in which for a quasi-linear parabolic equation and for an arbitrarily chosen base some theorems related to the strong convergence of approximated solutions to the exact one and other zero convergence estimations have been given. Zarubin and Tiunchik [329] applied MBG to the nonlinear operator parabolic equation. It was assumed there that a system of eigenelements of the auxiliary operator satisfying the acute angle type inequality is taken as a base. Conditions of the zero error convergence of MBG as well as a quite different one constituting a strong convergence of a series of approximated solutions adequate to their exact one were provided.

In a series of Zarubin's works [327, 328] with reference to the notions of eigenvalues of the convergent operator making an acute angle with the self-coupled output operator being placed in the original problem formulation, some various kinds of estimation of MBG error for parabolic equations equipped with an appended operator have been provided.

A problem associated with MBG rate convergence for some hyperbolic equations and the coupled problems of thermoelasticity-type equations have been analyzed by Zhelozovskiy [333, 334].

Modern analysis of the theory of plates and shells as well as convergence investigations of MBG have led to the following observations. MBG is one of the most effective and widely used numerical methods for the solution of static and dynamic problems in the theory of plates and shells. Present capabilities of computational techniques and the well-established mathematical knowledge concerning MBG allow for finding of solutions to many complex problems of the theory of non-homogeneous shells after their finite deflections.

The monograph does not aim for the full and multilateral presentation of the problem under investigation, but it does attempt to show the simplicity of realization and effectiveness of MBG applications to the theoretical problems of plates and shells.

Beams belong to structural members widely described in the recent bibliography from the aspect of their chaotic dynamics.

Pezeshki and Dowell [245] determined fractal basin boundaries for two types of basins of attractions of a buckled beam dynamics using the forced Holmes-Duffing equation. Shaw [265] studied the transverse vibrations of a slender beam exhibiting free rotation around its longitudinal axis. It was shown using the Melnikov-type approach that chaotic motions exist for both first mode approximation and the full beam equation.

Higuchi and Dowell [121] investigated chaotic vibrations of a buckled beam under sinusoidally varying and static constant transverse external forces, and onset of chaos was predicted. Moorthy et al. [216] applied the finite element method to solve the chaotic vibration problem of beams with nonlinear boundary conditions. It was shown that a temporally discrete solution of the spatially discrete model could capture the phenomenon of chaos.

The Galerkin approach was applied to convert PDEs to ODEs during investigation of bifurcational behavior of a pre-buckled beam in [263]. Both pre-buckled amplitude and external loading were used as control parameters in the bifurcational plane, and the existence of a chaotic attractor was shown.

Yagasaki [320] investigated nonlinear vibrations of a straight beam clamped at both ends and forced with two frequencies near the first beam mode frequency. The Galerkin approximation and the averaging and Melnikov techniques were applied to predict chaos occurrence. Numerical simulations and experiments were compared with the theoretical prediction.

Lenci and Tarantino [186] studied chaotic dynamics by means of the Melnikov method. Qualitatively different chaotic zone diagrams were reported, and the physical interpretation of the applied mathematical tools was also addressed.

The generalized Melnikov method for predicting the onset of chaos of buckled beams was applied in references [277, 278]. Direct computation of stable and unstable perturbed manifolds allowed prediction of a chaotic threshold. Necessary and sufficient conditions for steady-state chaos were obtained.

Moorthy et al. [217] showed that the direct integration and mode superposition schemes are efficient for problems of chaos in structural mechanics systems. The solution scheme was built in algorithms for equilibrium interaction of the nonlinear forces of the temporal solution trajectory.

In reference [119] a magneto-elastic beam governed by the Duffing equation with delay was analyzed. A necessary and sufficient condition for the solution stabilization regarding control parameters and initial constants was formulated. A control of chaos was proposed.

Dwivedy and Kar [85] studied the second-order differential equation of a slender beam with an attached mass at an arbitrary position under vertical base excitation. The normal forms method was applied and bifurcation set, mixed-mode vibration, period-doubling, quasi-periodic orbits, and different routes to chaos were addressed.

A T-shaped beam-mass structure was studied experimentally and theoretically for the case of one-to-two internal and parametric resonances of the lower mode in [129]. Multiple scales method yielded four first-order amplitude- and phase-modulation equations, where the Melnikov approach allowed prediction of a Smale horseshoe type of chaos.

Yagasaki [321] studied an infinite degrees-of-freedom Hamiltonian system representing a mathematical model for an undamped buckled beam. Orbits transversally homoclinic to periodic orbits and orbits transversally homoclinic and heteroclinic to invariant tori consisting of quasi-periodic orbits were detected. The occurrence of chaotic dynamics via and beyond the Smale-Birkhoff homoclinic theorem was addressed.

Tunnel-shaped chaotic orbits, fractal orbits, cascade of period-doubling, torus doubling, and intermittency routes to chaos were reported by Dwivedy and Kar [86] with reference to dynamics of a slender beam carrying a lumped mass under principal parametric resonance with three-mode interactions. This study was extended in reference [87], where also two distinct zones of trivial stability, blue sky catastrophe phenomena, jump down phenomena, and boundary and attractor merging crises were illustrated and discussed.

A rotating viscoelastic beam with variable pitch angle was investigated by Abolghasemi and Jalali [1]. Both perturbation technique and Galerkin projection were used to obtain a non-autonomous ODE, and a bifurcation analysis was carried out.

The existence of chaotic responses for certain weakly damped linear beam equations with slowly periodic perturbations resting on weakly nonlinear elastic bearings was illustrated by Battelli and Fečkan [51].

Bifurcation and chaos in transverse motion of axially accelerating viscoelastic beams were studied in reference [72]. The Galerkin method was applied to reduce the PDE to a set of ODEs. Various bifurcation diagrams were reported.

Bifurcations and possible chaotic motions of a cantilever system excited by a periodic force with impacts were studied in reference [193]. One- and two-sided impact models were considered.

Optimization of the electromechanical control of beams dynamics was investigated in reference [174]. The snap-through instability and horseshoe chaos was controlled and the analytical investigation was validated by numerical simulation.

El-Bassiouny [90] proposed a nonlinear control law to suppress vibrations of the first mode of a cantilever beam subjected to both primary and principal parametric excitations. Numerical simulations confirmed the analytical predictions of chaos and unbounded motions.

Control with delay of an undamped buckled beam subjected to parametric excitations to suppress chaos was applied in [230]. In addition, the threshold condition for the inhibition of Smale horseshoes chaos was reported.

It should be emphasized that chaotic dynamics of plates is investigated less than chaotic dynamics of beams.

Paidoussis and Li [237] studied chaotic dynamics of heat exchanger tubes impacting on the generally loose baffle plates. It was shown that for a sufficient flow velocity chaos was born, and a negatively damped impact oscillator was used to understand the system behavior.

Yang and Sethna [324] studied nonlinear flexural vibrations of a nearly square plate periodically excited with forces normal to the mid-plane of the plate. Dynamics consisting of anti-symmetric or mixed modes occurred. The Hopf bifurcation yielded the amplitude-modulated traveling waves with jerky motions. Global chaotic phenomena were observed.

A homogeneous fully clamped rectangular plate subject to spatially thermal loads and narrow-band acoustic excitation was studied in reference [221]. Time series power spectra, autocorrelation functions, spatial dimension, and temporal complexity were applied to characterize the occurred chaotic orbits.

Subharmonic resonance of a rectangular plate with uniform stretching when two distinct linear modes are near one-to-one internal resonance was studied in reference [68]. It was shown via an averaging procedure that the plate can exhibit harmonic and subharmonic motions in the directly excited spatial mode or subharmonic motions in which both internally resonant modes appeared. Period-doubling route to chaos was shown.

Chaos and fractal theories were applied to study dynamics buckling of viscoelastic plates in reference [285]. The nonlinear integral-differential dynamic equation was reduced to an autonomic four-dimensional dynamical system. The Lyapunov spectra and fractal dimensions of strange attractors were reported.

The nonlinear mathematical theory of perforated viscoelastic thin plates using both Kármán hypotheses and Boltzmann's constitutive law of linear viscoelastic materials was derived in [73]. In particular, the nonlinear dynamic stability of a viscoelastic angular plate was studied, and a novel method of Lyapunov exponent spectrum estimation was proposed.

In reference [177] fractal dimension and the maximum Lyapunov exponent were applied to study large deflections of a simply supported rectangular plate. Fourier spectra, state-space plots, Poincaré maps, and bifurcation diagrams were computed. Various bifurcations were detected and their links to chaotic orbits occurrence were discussed.

Nonlinear equations governing dynamics of Timoshenko's viscoelastic thick plates with damage were derived and studied in [269]. The influences of load, geometry, and material parameters on the dynamical plate behavior were investigated via the Galerkin approach.

In reference [318] nonlinear equations of motion for the rectangular moderately thick plates with transverse surface penetrating crack on an elastic foundation subject to periodic load action were derived. Bifurcational and chaotic behavior of such plates were studied using the Galerkin and Runge-Kutta integration methods.

Ribeiro and Duarte [258] studied geometrically nonlinear, linear elastic dynamics of composite laminated plates. The existence of chaos was confirmed by calculating the largest Lyapunov exponent.

In reference [117] it was shown how the use of aeroelastic modes can drastically reduce the number of coupled nonlinear modal equations for the large-amplitude nonlinear panel flutter study at an arbitrary yawed supersonic flow angle and elevated temperatures. Periodic and chaotic dynamics of the panel were reported.

Ribeiro [257] studied chaotic behavior of geometrically nonlinear vibrations of linear elastic and isotropic plates under the combined effect of thermal fields and mechanical excitations. The Newmark implicit time integration method was applied to solve the governing equations in a time domain.

Finally, let us briefly review the state-of-art of irregular dynamics of shells. It is rather less investigated in comparison to other structural members, i.e., beams and plates.

In reference [259] the numerical-perturbation approach was applied to study the axisymmetric dynamic vibrations of closed spherical shells subject to external harmonic excitation having a frequency near one of the natural frequencies of a flexural mode in the presence of a two-to-one auto-parametric resonance between the excited mode and a lower flexural mode. The limit cycles born after Hopf bifurcation experienced pitchfork bifurcation and a cascade of period-doubling bifurcations leading to chaos. The subcritical Hopf bifurcation was also illustrated, and a cyclic-fold bifurcation yielding chaos was reported, among others.

Popov et al. [250] studied numerically internal auto-parametric instabilities in the free nonlinear vibrations of a cylindrical shell. Regular and chaotic behavior of two-mode interaction was analyzed, with emphasis on energy transfer between the modes.

Soliman and Gonçalves [282] investigated the axisymmetric chaotic dynamic behaviour and snap-through buckling of thin elastic shallow spherical shells under harmonic excitation. Both Galerkin and Fourier-Bessel approaches were applied to reduce the PDEs to a finite degrees-of-freedom system. Steady-state and transient stability boundaries were presented and the vertical load conditions were determined.

A shallow cylindrical shell under gravity and periodic acceleration and possessing concentrated mass was studied in reference [222]. The Galerkin method was used to reduce the problem of investigation of ODEs and the mass value influence on the system's chaotic dynamics was investigated as well.

Amabili [5] studied large vibrations of doubly curved shallow shells with rectangular base, simply supported at the four edges and subjected to harmonic excitation normal to the surface in the spectral neighborhood of the fundamental mode. Donnell's and Novozhilov's shell theories were used to compute the elastic strain energy. Shell stability under static and dynamic loads was studied, both Lyapunov exponents and dimensions were computed, and snap-through instability, subharmonic resonance, as well as the period-doubling routes to chaos were illustrated and discussed.

Pellicano and Amabili [243] studied dynamic stability of circular cylindrical shells subjected to static and dynamic axial loads. Chaotic dynamics of precompressed shell within Donnell's and Sanders-Koiter's theorems was illustrated.

Readers can supplement the knowledge devoted to the considered problems by studying our other work devoted to this branch of science such as [23, 24, 25, 26, 27, 28, 39, 46] and papers [21, 22, 29, 30, 34, 35, 36, 37, 40, 41, 42, 43, 44, 45, 160, 161, 162, 163, 164, 165]. Furthermore, the collective work, of Cz. Woźniak [328] is recommended, in which a large bibliography devoted to mechanics of elastic plates and shells with the special attention paid to the contribution of Polish researchers has been included.

Finally, let us emphasize that from the variety of methods based on MBG we will use in this monograph its variant associated with the Fourier series.

On the Bubnov-Galerkin Method

Historical Perspective

The origin of the Bubnov-Galerkin method is strictly connected with the name of the Russian ship designer Bubnov (1872–1919). At the same time, together with Krylov he was one of the creators of the Russian navy and designed 48 ships and submarines. He is known as the author of the scientific discipline called mechanics of ship construction.

In 1907 Timoshenko published a work [295] in which on the basis of an example of axially compressed rod he considered the problem of stability using the minimization principle for the rod's potential energy. Successively, Timoshenko generalized his own work and wrote the monograph titled *About Stability of Elastic Systems*, which appeared between 1910 and 1911. Furthermore, in 1910 it was published in *Works of the Kiev Polytechnic Institute* (see [294]). He was awarded for this work, receiving The Zhuravskiy Gold Medal and 2500 gold rubles. In 1913 The French Engineering Association published the aforementioned work in a French journal that was dedicated to the construction of roads and bridges [292].

It was the first and at the same time the last Zhuravskiy grant. Professors Bielelubski, Bielecki, Bubnov, and Kolosov reviewed the discussed work, and in 1913 their reviews were published in *Works of Engineers Institute* [254]. The year 1913 is considered the year of birth of the Bubnov method, which was recognized as the generalized method to the solution of differential equations.

As a matter of fact the method was formulated two years earlier (the end of May of 1911), but one can begin the exploration of its source at Timoshenko's work. The last one was awarded in 1911 and the publication considering that topic appeared in press on June 15, 1911 (in the newspaper *Rech*). Later, on June 25th of the same year a press notice appeared also in *Vestnik Putey Soobshcheniya* [125].

I.G. Bubnov proposed two variants of the method of reduction of PDEs or their systems to the algebraic and ODEs.

The Western literature indicates work with the B.G. Galerkin article [101] published in 1915, in which an analysis of equilibrium states of rods and thin shells was studied. However, the previously given overview of the real events confirms Bubnov's brilliant idea, which he had mentally formulated while writing the review of Timoshenko's work. Right at the same time, and for the first time, Bubnov observed an identity between the Rayleigh-Ritz-Timoshenko energetic method and his approach (the Bubnov-Galerkin method). Later, Bubnov rarely used that method himself.

Mainly thanks to Galerkin and the work of his co-workers the MBG was popularized in the Russian literature. In the West the method became known together with Duncan's work [84], which was devoted to the dynamics of flying objects. Bickey applied MBG to the solution of the problem of non-stationary heat transfer and with the application of an equivalent electrical circuit. He compared Duncan's results with the solutions obtained by means of the colocation and the least-squares methods.

After that eventful time a large development and propagation of MBG applications began. Various modifications of the method appeared. As mentioned, after the review of Timoshenko's work Bubnov was still aware of the relation between the Bubnov-Galerkin method and other variational ones (such as the Rayleigh-Galerkin method), which was formally proved in [133]. The relation played a crucial role in the second half of the 1960s, being strongly connected with the elaboration of the *finite elements method.*

Bubnov-Galerkin Method Against the Background of Other Projection Methods

A scheme of the application of the Bubnov-Galerkin method will be shown below (Fig. 1). Let the example be an abstractive two-dimensional problem that is governed by a linear differential equation of the form

$$L(u) = 0, \tag{1}$$

which is considered in region $D(\overrightarrow{x})$, and having the following boundary conditions:

$$S(u) = 0 \tag{2}$$

given on a curve ∂D being the boundary of region D.

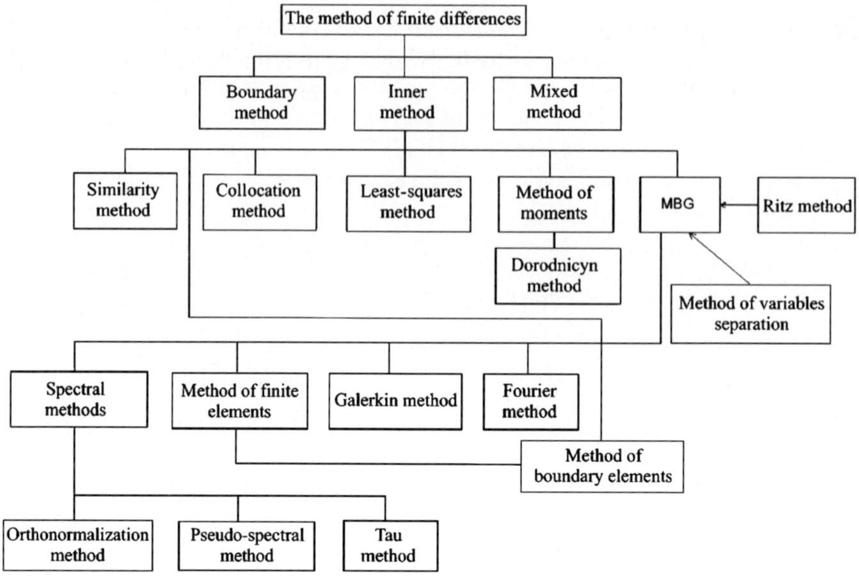

Fig. 1 MBG versus other projection methods

In agreement with the MBG method a function will be sought in the form

$$u_a = u_0(\overrightarrow{x}) + \sum_{j=1}^{N} a_j \varphi_j(\overrightarrow{x}), \tag{3}$$

where $\varphi_j(\overrightarrow{x})$ are the known analytical functions, and a_j are the coefficients to be found during the problem solution. In order to satisfy boundary conditions the form of the $u_u(x_1, x_2)$ function has been introduced. Substituting Eq. (3) in Eq. (1) yields the error $R \neq 0$

$$R(a_0, a_1, a_2, \ldots a_N, \overrightarrow{x}) = L(u_a) = L(u_0) + \sum_{j=1}^{N} a_j L(\varphi_j). \tag{4}$$

Let us introduce the following definition of *inner product*:

$$(f, g) = \iint_D fg\, dx_1\, dx_2. \tag{5}$$

In the case of the Bubnov-Galerkin method the appearance in Eq. (3) of unknown coefficients should be determined through the solution to the following system of algebraic functions:

$$(R, \varphi_k) = 0, \quad k = 1, 2, \ldots, N, \tag{6}$$

where the error R is defined by Eq. (1), and φ_k are the same analytical functions appearing in Eq. (3). Our problem has been brought to a linear one (compare with Eq. (1)); therefore Eq. (6) can be written as the following form of a matrix equation:

$$\sum_{j=1}^{N} a_j(L(\varphi_j), \varphi_k) = -L(u_0, \varphi_k). \tag{7}$$

Solving the above equation with respect to a_j, and then substituting the result in Eq. (3), one obtains the approximated solution u_a.

In the literature the notion of weighting function $w_k(\overrightarrow{x})$ is often used. The functions $\varphi_j(\overrightarrow{x})$ in Eq. (3) are sometimes referred to any test functions. Some requirements should be met after the application of MBG and the selection of weighting and test functions $\varphi_j(\overrightarrow{x})$. It is recommended to select them from the set of first N functions of the complete system of functions (see [133]). It constitutes the necessary condition of convergence of the approximate solution to the accurate one when $N \to \infty$. A theoretical background and MBG applications in theory of shells can be found in the classical Vorovich monograph [316].

The necessary conditions for the application of the classical version of the Bubnov-Galerkin method follow.

Weighting and test functions $w_k(\overrightarrow{x})$ should belong to the same family of functions $\varphi_j(\overrightarrow{x})$:

1. $w_k(\overrightarrow{x})$ and $\varphi_j(\overrightarrow{x})$ should be linearly independent;
2. $w_k(\overrightarrow{x})$ and $\varphi_j(\overrightarrow{x})$ should be the first N components of the complete system of functions;
3. $\varphi_j(\overrightarrow{x})$ should satisfy boundary conditions (or any initial conditions, if they appear).

To summarize, the first condition defines the Bubnov-Galerkin method; the second condition should be satisfied when some independent equations for the estimation of unknown a_j coefficients are going to be found; the third one has the decisive influence on the effectiveness of the MBG which can be significantly decreased owing to their violation.

Chapter 1
Theory of Non-homogeneous Shells

In this chapter a general theory of non-homogeneous shells is introduced. First, fundamental relations and assumptions are given, non-homogeneities of shells are introduced, and then the governing variational equations and equations of motion are derived. After the boundary and initial conditions are introduced, the equations are cast into non-dimensional form and the so-called variable parameters of a shell stiffness are defined. In addition, a flexural stiffness coefficient of a shell element is formally introduced. The chapter concludes with a consideration of generalized functions.

1.1 Preliminary Remarks

It is generally agreed that after creation of a shell's model, which should as closely as possible describe the real existing objects (shells), many simplifications are necessary, especially regarding the construction of a shell and the material of which it is made. Shell constructions that are observed in many technical applications can be of various types, for instance, one- or multiple-layered, reinforced by ribs, waffled, etc.

A shell material can be either isotropic or anisotropic. Analysis of the shell dynamics is usually problematic. On the other hand, proper introduction of a model simplification decreases the computational time, which leads to growth of the practical applications. The investigations that were made result in a conclusion that the selection of a shell model should be made with particular respect to the economics of computations and the sufficiently exact model of the shell.

In this book, the shells that are mainly considered are made of isotropic and elastic material. Hook's law satisfies a static and dynamic behavior of those shells. Plastic and adhesive-plastic deformations of shells are not analyzed here, because they are usually associated with the class of problems associated with other constructive non-homogeneities.

The main characteristic attribute connected with the analysis of dynamics and statics of thin shells is recognized in the theory of elasticity as a reduction of the

J. Awrejcewicz, V.A. Krysko, *Chaos in Structural Mechanics*,
DOI: 10.1007/978-3-540-77676-5_2, © Springer-Verlag Berlin Heidelberg 2008

three-dimensional problem to the two-dimensional one. If only one shell is investigated the coordinate system has to be joined with the central surface of a shell. One of the possibilities of reduction of the three-dimensional problem to a two-dimensional one is the consideration of a hypothesis devoted to the lack of straight perpendicular lines to the shell, that is, to the conservation of the Kirchhoff-Love hypothesis. According to the hypothesis, any fiber that is perpendicular to the central surface before a deformation remains perpendicular until the whole deformation process is finished. Furthermore, the length of that fiber measured along the shell's thickness remains constant. Additional assumptions are based on the observation that the stresses perpendicular to the shell can be neglected because of their negligible smallness in comparison to the basic ones. In the theory of shells, basic stresses are understood as the stresses occurring in some parallel layers that are perpendicular and tangent to the central surface of these shells.

1.2 Fundamental Relations and Assumptions

Let us consider a rectangular shell (Fig. 1.1) the central surface of which is bounded by a closed curve Γ and let us assign to it a rectangular coordinate system x, y.

Let an axis perpendicular to the central surface be denoted by z, and the positive aspect of the axis be directed to the curvature's center. Moreover, assume in our investigations a right-handed coordinate system. From the considerations assumed the shell coordinates meet the following inequalities:

$$0 \leq x \leq a, \quad 0 \leq y \leq b, \quad -h \leq z \leq h. \tag{1.1}$$

The displacement of points of the shell's center surface in directions x, y, z are denoted by u, v, w, respectively.

According to the classical theory of shells, the displacement of an arbitrarily selected point of a shell in direction z is not dependent on z and is moreover identical for all points of the considered element, which are located on the straight line that is perpendicular to the shell's central surface. Indeed, we have assumed that the shell's material is isotropic, but we will always allow the element made of a material of another elasticity modulus to be included with the analyzed isotropic shell. In addition, the shell's shape variations are possible if its thickness varies in time. In other words, the changes can be described by coordinate-dependent functions.

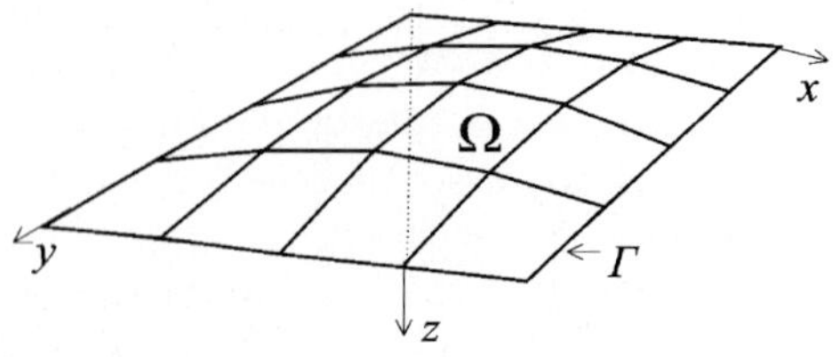

Fig. 1.1 Scheme of the analyzed shell

Deformations are mainly expected to be in agreement with Hook's law. Following the assumptions we have made, the central surface deformations for an isotropic shell material, which is characterized by elasticity modulus E, of the first type, elasticity modulus G of the second type (compression modulus), and a Poisson's coefficient μ are connected with deformations σ_{xx}, σ_{yy}, and σ_{xy} using the following relations:

$$e_{xx} = \frac{q}{E}\left(\sigma_{xx} - \mu\sigma_{yy}\right),$$
$$e_{yy} = \frac{q}{E}\left(\sigma_{yy} - \mu\sigma_{xx}\right),$$
$$e_{xy} = \frac{2(1+\mu)}{E}\sigma_{xy}. \tag{1.2}$$

Solving Eq. (1.1) with respect to the deformations that have occurred, we obtain

$$\sigma_{xx} = \frac{E}{1-\mu^2}\left(e_{xx} + \mu e_{yy}\right),$$
$$\sigma_{yy} = \frac{E}{1-\mu^2}\left(e_{yy} + \mu e_{xx}\right),$$
$$\sigma_{xy} = \frac{E}{2(1+\mu)}e_{xy}. \tag{1.3}$$

Corresponding to the Kirchhoff-Love model, the displacement of the central shell's layer points along the considered coordinates depends on x, y coordinates and on time $t:\ u = u(x,y,t),\ v = v(x,y,t)$ (displacements $w = w(x,y,t)$ are involved in a similar way).

Displacement of any point of the coordinate z before deformation is

$$u^z = u - z\frac{\partial w}{\partial x},$$
$$v^z = v - z\frac{\partial w}{\partial y},$$
$$w^z = w. \tag{1.4}$$

Fully described deformations that occurred in a layer during both extension and rotation of a shell placed at a distance from the central surface and in accordance with the nonlinear theory of shells are given below [234]:

$$\varepsilon_{xx} = \frac{\partial u^z}{\partial x} - k_x w + \frac{1}{2}\left(\frac{\partial u^z}{\partial x}\right)^2 + \frac{1}{2}\left(\frac{\partial v^z}{\partial x}\right)^2 + \frac{1}{2}\left(\frac{\partial w}{\partial x}\right)^2,$$
$$\varepsilon_{yy} = \frac{\partial v^z}{\partial y} - k_y w + \frac{1}{2}\left(\frac{\partial u^z}{\partial y}\right)^2 + \frac{1}{2}\left(\frac{\partial v^z}{\partial y}\right)^2 + \frac{1}{2}\left(\frac{\partial w}{\partial y}\right)^2,$$
$$\varepsilon_{xy} = \frac{\partial u^z}{\partial y} + \frac{\partial v^z}{\partial x} + \frac{\partial w}{\partial x}\frac{\partial w}{\partial y}. \tag{1.5}$$

In the case of thin flexible plates and shells it is assumed that the deflection angles $\partial w^z/\partial x$, $\partial w^z/\partial y$ are significantly larger than the values of derivatives $\partial u/\partial x$, $\partial u/\partial y$, etc. that are connected with the volume of material. Furthermore, squares of derivatives $(\partial w/\partial x)^2$ will be of the same order as components $\partial u^z/\partial x$, etc., which means the terms such as $(\partial u/\partial x)^2$ can be neglected.

Satisfying the above assumptions, expressions (1.5) are simplified:

$$\begin{aligned}
\varepsilon_{xx} &= \frac{\partial u^z}{\partial x} - k_x w + \frac{1}{2}\left(\frac{\partial w}{\partial x}\right)^2, \\
\varepsilon_{yy} &= \frac{\partial v^z}{\partial y} - k_y w + \frac{1}{2}\left(\frac{\partial w}{\partial y}\right)^2, \\
\varepsilon_{xy} &= \frac{\partial u^z}{\partial y} + \frac{\partial v^z}{\partial x} + \frac{\partial w}{\partial x}\frac{\partial w}{\partial y}.
\end{aligned} \tag{1.6}$$

Taking into consideration Eq. (1.4), we obtain

$$\begin{aligned}
\varepsilon_{xx} &= \frac{\partial u}{\partial x} - k_x w + \frac{1}{2}\left(\frac{\partial w}{\partial x}\right)^2 - z\frac{\partial^2 w}{\partial x^2}, \\
\varepsilon_{yy} &= \frac{\partial v}{\partial y} - k_y w + \frac{1}{2}\left(\frac{\partial w}{\partial y}\right)^2 - z\frac{\partial^2 w}{\partial y^2}, \\
\varepsilon_{xy} &= \frac{\partial u}{\partial y} + \frac{\partial v}{\partial x} + \frac{\partial w}{\partial x}\frac{\partial w}{\partial y} - 2z\frac{\partial^2 w}{\partial x \partial y},
\end{aligned} \tag{1.7}$$

or equivalently

$$\begin{aligned}
\varepsilon_{xx} &= \varepsilon_{11} - z\frac{\partial^2 w}{\partial x^2}, \\
\varepsilon_{yy} &= \varepsilon_{22} - z\frac{\partial^2 w}{\partial y^2}, \\
\varepsilon_{xy} &= \varepsilon_{12} - 2z\frac{\partial^2 w}{\partial x \partial y},
\end{aligned} \tag{1.8}$$

where the central surface deformations are governed by the following relations:

$$\begin{aligned}
\varepsilon_{11} &= \frac{\partial u}{\partial x} - k_x w + \frac{1}{2}\left(\frac{\partial w}{\partial x}\right)^2, \\
\varepsilon_{22} &= \frac{\partial v}{\partial y} - k_y w + \frac{1}{2}\left(\frac{\partial w}{\partial y}\right)^2, \\
\varepsilon_{12} &= \frac{\partial u}{\partial y} + \frac{\partial v}{\partial x} + \frac{\partial w}{\partial x}\frac{\partial w}{\partial y}.
\end{aligned} \tag{1.9}$$

By putting the variables given in Eq. (1.2) into the following expressions, which define static equivalent deformations in a section of shell

$$T_{11} = \int_{-h}^{h} \sigma_{xx}\, dz, \quad T_{22} = \int_{-h}^{h} \sigma_{yy} dz,$$

$$T_{12} = \int_{-h}^{h} \sigma_{xy}\, dz, \tag{1.10}$$

we obtain the final relations that define stresses and deformations of a mean surface of the form

$$\begin{aligned} T_{11} &= \frac{2hE}{1-\mu^2}\left(\varepsilon_{11} + \mu\varepsilon_{22}\right), \\ T_{22} &= \frac{2hE}{1-\mu^2}\left(\varepsilon_{22} + \mu\varepsilon_{11}\right), \\ T_{12} &= \frac{2hE}{2(1+\mu)}\varepsilon_{12}. \end{aligned} \tag{1.11}$$

Solving Eq. (1.11) with respect to ε_{ij}, we obtain

$$\begin{aligned} \varepsilon_{11} &= \frac{a_1}{2h}\left(T_{11} + \mu T_{22}\right), \\ \varepsilon_{22} &= \frac{a_1}{2h}\left(T_{22} + \mu T_{11}\right), \\ \varepsilon_{12} &= \frac{2(1+\mu)a_1}{2h} T_{12}. \end{aligned} \tag{1.12}$$

1.3 Non-homogeneity of a Shell

Flexural stiffness and the density of the selected part of a non-homogeneous shell can vary according to either the adjoining new material characterized by another elasticity modulus, or as a result of local change in thickness of the shell. For further considerations of all cases investigated, we will assume that toward the central shell surface the shell's shape is symmetric. Such an assumption allows for introduction of $2h_1$, a thicker additional element. As a result, the flexural stiffness D of the shell's part will change, leading as a consequence to modification of the parameter E through the contractual "rarefaction (compaction)" of its thickness between $2h_1$ and $2h$.

This procedure is used if and only if the shells of either step-variable thickness or parts of those shells made of different elasticity modulus materials are treated in the same manner (see [251]).

Such shell regions are characterized later by time-variable parameters of stiffness. Let us assume that in addition the shell can be composed of any number of rectangular parts described by different flexural stiffnesses, which are oriented in such a way that the lines bounding them are parallel to the appropriate shell edges.

Using that approach one can assume that the shell is of constant thickness, and its non-homogeneity is fully characterized by changes in its elasticity modulus E (a similar assumption is made for the density ρ of a shell material). Finally, it is considered, that $E = E(x,y)$, $\rho = \rho(x,y)$.

1.4 Variational Equations

Using energetic considerations one can obtain equations of motion for a shell, which contain some rectangular elements of the flexural being different from the basic of shells. As pointed out in Sect. 1.3, conditions $E = E(x,y)$, $\rho = \rho(x,y)$ are assumed.

A method similar to that analyzed in Volomir's monograph [304] is used for further considerations; i.e., a motion occurring in a time interval $[t_0,\ t_1]$ is examined. For this reason, some different trajectories of motion of the system characteristic points between any starting and final states will be compared. The true trajectories will be slightly different than the other ones, because the following condition, which is satisfied for the true trajectories, has to be met:

$$\int_{t_0}^{t_1} \left(\delta K - \delta V + \delta' W\right)\, dt = 0, \tag{1.13}$$

where K is the kinetic energy, V is the system potential energy, $\delta' W$ is the sum of elementary works done by all external forces.

When all forces acting on a system have a potential Π, then Eq. (1.13) takes the form

$$\delta S = \delta \int_{t_0}^{t_1} (K - V - \Pi)\, dt = 0, \tag{1.14}$$

where $S = \int_{t_0}^{t_1} (K - V - \Pi) dt$ denotes an action in Hamilton sense, and $t \in [0,\ T]$.

The last equation expresses the known Hamilton principle.

Elasticity properties of various bodies can be characterized by the energy of their deformation. Applying the most comprehensive formula of the theory of elasticity describing unity incrementation of the mechanical work of deformations and taking into consideration the lack of normal (perpendicular) deformations hypothesis, the potential energy of deformation can be estimated by means of the following rule:

$$V = \frac{1}{2} \iint_{\Omega} \int_{-h}^{h} \left(\sigma_{xx}\varepsilon_{xx} + \sigma_{yy}\varepsilon_{yy} + \sigma_{xy}\varepsilon_{xy}\right)\, dz\, ds, \tag{1.15}$$

for which it was assumed that $ds = dx\, dy$, and Ω is the integration region bounded by Γ (see Fig. 1.1).

Substituting (1.2) and (1.3) into Eq. (1.15) and carrying out an integration we obtain

$$V = V_1 + V_2, \tag{1.16}$$

where

$$V_1 = \frac{1}{2}\iint\limits_{\Omega} \frac{2hE}{1-\mu^2}\left[\varepsilon_{11}^2 + \varepsilon_{22}^2 + 2\mu\varepsilon_{11}\varepsilon_{22} + \frac{1-\mu}{2}\varepsilon_{12}^2\right] ds,$$

$$V_2 = \frac{1}{2}\iint\limits_{\Omega} \frac{2h^3E}{3(1-\mu^2)}\left[\left(\frac{\partial^2 w}{\partial x^2}\right)^2 + \left(\frac{\partial^2 w}{\partial y^2}\right)^2 \right.$$

$$\left. + 2(1-\mu)\left(\frac{\partial^2 w}{\partial x \partial y}\right)^2 + 2\mu\frac{\partial^2 w}{\partial x^2}\frac{\partial^2 w}{\partial y^2}\right] ds.$$

Let us now investigate Eq. (1.16) and rewrite its first component in the form:

$$V_1 = \frac{1}{2}\iint\limits_{\Omega} \frac{2hE}{1-\mu^2}\left(\varepsilon_{11}^2 + \varepsilon_{22}^2 + 2\mu\varepsilon_{11}\varepsilon_{22} + \frac{1-\mu}{2}\varepsilon_{12}^2\right) ds. \tag{1.17}$$

Making a step of variation of the expression (1.17) we have:

$$\delta V_1 = \iint\limits_{\Omega} \frac{2hE}{1-\mu^2}\left[\varepsilon_{11}\delta(\varepsilon_{11}) + \varepsilon_{22}\delta(\varepsilon_{11}) + \mu(\varepsilon_{11}\delta(\varepsilon_{22}) + \varepsilon_{22}\delta(\varepsilon_{11}))\right.$$

$$\left. + \frac{1-\mu}{2}\varepsilon_{12}\delta(\varepsilon_{12})\right] ds$$

$$= \iint\limits_{\Omega} \frac{2hE}{1-\mu^2}\left[(\varepsilon_{11} + \mu\varepsilon_{22})\delta(\varepsilon_{11}) + (\varepsilon_{22} + \mu\varepsilon_{11})\delta(\varepsilon_{22}) + \frac{1-\mu}{2}\varepsilon_{12}\delta(\varepsilon_{12})\right] ds. \tag{1.18}$$

Formula (1.11) yields

$$\delta V_1 = \iint\limits_{\Omega} \left[T_{11}\delta(\varepsilon_{11}) + T_{22}\delta(\varepsilon_{11}) + T_{12}\delta(\varepsilon_{12})\right] ds, \tag{1.19}$$

$$\delta V_1 = \iint\limits_{\Omega} \frac{a_1}{2h}\left[T_{11}\delta(T_{11} - \mu T_{22}) + T_{22}\delta(T_{22} - \mu T_{11})\right.$$

$$\left. + 2(1+\mu)T_{12}\delta(T_{12})\right] ds$$

$$= \iint\limits_{\Omega} \frac{a_1}{2h}\left[(T_{11} - \mu T_{22})\delta(T_{11}) + (T_{22} - \mu T_{11})\delta(T_{22})\right.$$

$$\left. + 2(1+\mu)T_{12}\delta(T_{12})\right] ds, \quad \text{where } a_1 = 1/E. \tag{1.20}$$

Let F be a function of deformations; then one obtains

$$T_{11} = \frac{\partial^2 F}{\partial y^2}, \quad T_{22} = \frac{\partial^2 F}{\partial x^2}, \quad T_{12} = -\frac{\partial^2 F}{\partial x \partial y}. \tag{1.21}$$

Applying the above to Eq. (1.20) gives

$$\delta_F V_1 = \iint\limits_{\Omega} \frac{a_1}{2h} \left[\left(\frac{\partial^2 F}{\partial y^2} - \mu \frac{\partial^2 F}{\partial x^2} \right) \frac{\partial^2}{\partial y^2} \delta(F) + \left(\frac{\partial^2 F}{\partial x^2} - \mu \frac{\partial^2 F}{\partial y^2} \right) \frac{\partial^2}{\partial x^2} \delta(F) \right.$$
$$\left. + 2(1+\mu) \frac{\partial^2 F}{\partial x \partial y} \frac{\partial^2}{\partial x \partial y} \delta(F) \right] ds. \tag{1.22}$$

Taking into consideration expressions (1.12) and (1.20), we will have found a transition from stresses to deformations, and then the following variational function of stresses is formulated:

$$\delta_F V_1 = \iint\limits_{\Omega} \left[\varepsilon_{11} \delta(T_{11}) + \varepsilon_{22} \delta(T_{22}) + \varepsilon_{12} \delta(T_{12}) \right] ds$$
$$= \iint\limits_{\Omega} \left[\varepsilon_{11} \frac{\partial^2}{\partial y^2} \delta(F) + \varepsilon_{22} \frac{\partial^2}{\partial x^2} \delta(F) - \varepsilon_{12} \frac{\partial^2}{\partial x \partial y} \delta(F) \right] ds. \tag{1.23}$$

Double integration by parts (without writing of any boundary integrals) yields

$$\delta_F V_1 = \iint\limits_{\Omega} \left[\frac{\partial^2 \varepsilon_{11}}{\partial y^2} + \frac{\partial^2 \varepsilon_{22}}{\partial x^2} + \frac{\partial^2 \varepsilon_{12}}{\partial x \partial y} \right] \delta(F)\, ds. \tag{1.24}$$

Substitution of formula (1.9) into Eq. (1.24) gives

$$\delta_F V_1 = \iint\limits_{\Omega} \left[-k_y \frac{\partial^2 w}{\partial x^2} - k_x \frac{\partial^2 w}{\partial y^2} - \frac{\partial^2 w}{\partial x^2} \frac{\partial^2 w}{\partial y^2} - \left(\frac{\partial^2 w}{\partial x \partial y} \right)^2 \right] \delta(F)\, ds$$
$$= \iint\limits_{\Omega} \left[-\nabla_k^2 w - \frac{1}{2} L(w,w) \right] \delta(F)\, ds. \tag{1.25}$$

Boundary integrals adequate to Eq. (1.24) have the following form:

$$\int_0^a \left[\varepsilon_{11} \frac{\partial}{\partial y} (\delta F)\big|_0^b - \frac{\partial \varepsilon_{11}}{\partial y} (\delta F)\big|_0^b + \frac{\partial \varepsilon_{12}}{\partial x} (\delta F)\big|_0^b \right] dx$$
$$+ \int_0^b \left[\varepsilon_{22} \frac{\partial}{\partial x} (\delta F)\big|_0^a - \frac{\partial \varepsilon_{22}}{\partial x} (\delta F)\big|_0^a - \varepsilon_{12} \frac{\partial}{\partial y} (\delta F)\big|_0^a \right] dy. \tag{1.26}$$

Since Eqs. (1.22) and (1.25) are formally equal to each other, then

$$\delta_F V_1 = \iint_\Omega \frac{a_1}{2h}\left[\left(\frac{\partial^2 F}{\partial y^2} - \mu\frac{\partial^2 F}{\partial x^2}\right)\frac{\partial^2}{\partial y^2}\delta(F) + \left(\frac{\partial^2 F}{\partial x^2} - \mu\frac{\partial^2 F}{\partial y^2}\right)\frac{\partial^2}{\partial x^2}\delta(F)\right.$$

$$\left. + 2\,(1+\mu)\frac{\partial^2 F}{\partial x\partial y}\frac{\partial^2}{\partial x\partial y}\delta(F) + \left(\nabla_k^2 w + \frac{1}{2}L(w,w)\right)\delta(F)\right] ds, \quad (1.27)$$

$$\delta_F V_1 = \iint_\Omega \frac{a_1}{2h}\left[\left(\frac{\partial^2 F}{\partial y^2} - \mu\frac{\partial^2 F}{\partial x^2}\right)\frac{\partial^2(\cdot)}{\partial y^2} + \left(\frac{\partial^2 F}{\partial x^2} - \mu\frac{\partial^2 F}{\partial y^2}\right)\frac{\partial^2(\cdot)}{\partial x^2}\right.$$

$$\left. + 2(1+\mu)\frac{\partial^2 F}{\partial x\partial y}\frac{\partial^2(\cdot)}{\partial x\partial y} + \nabla_k^2 w + \frac{1}{2}L(w,w)\right]\delta(F)\, ds. \quad (1.28)$$

The sign $(\cdot)$ means that the derivative comes from the variation of function F, and hence

$$L(w,F) = \frac{\partial^2 w}{\partial x^2}\frac{\partial^2 F}{\partial y^2} + \frac{\partial^2 w}{\partial y^2}\frac{\partial^2 F}{\partial x^2} - 2\frac{\partial^2 w}{\partial x\partial y}\frac{\partial^2 F}{\partial x\partial y},$$

$$L(w,w) = 2\left[\frac{\partial^2 w}{\partial x^2}\frac{\partial^2 w}{\partial y^2} - \left(\frac{\partial^2 w}{\partial x\partial y}\right)^2\right],$$

$$\nabla_k^2 = k_y\frac{\partial^2}{\partial x^2} + k_x\frac{\partial^2}{\partial y^2}.$$

Taking into account Eq. (1.19), and according to the central surface deformations (1.9), a conversion of parameters $\varepsilon_{11}, \varepsilon_{22}, \varepsilon_{12}$ separating all variations with respect to u, v, and w is introduced, and the following relations are obtained:

$$\delta_u V_1 = \int_0^b [T_{11}\delta(u)]|_0^a dy + \int_0^a [T_{12}\delta(u)]|_0^b dx$$

$$- \iint_\Omega \left[\frac{\partial T_{11}}{\partial x} + \frac{\partial T_{12}}{\partial y}\right]\delta(u)\, ds, \quad (1.29)$$

$$\delta_v V_1 = \int_0^a [T_{22}\delta(v)]|_0^b dx + \int_0^b [T_{12}\delta(v)]|_0^a dy$$

$$- \iint_\Omega \left[\frac{\partial T_{22}}{\partial y} + \frac{\partial T_{12}}{\partial x}\right]\delta(v)\, ds, \quad (1.30)$$

$$\delta_w V_1 = \iint_\Omega \left\{T_{11}\delta\left[-k_x w + \frac{1}{2}\left(\frac{\partial w}{\partial x}\right)^2\right] + T_{22}\delta\left[-k_y w + \frac{1}{2}\left(\frac{\partial w}{\partial y}\right)^2\right]\right.$$

$$\left. + T_{12}\delta\left[\frac{\partial w}{\partial x}\frac{\partial w}{\partial y}\right]\right\} ds = \iint_\Omega \{(-k_y T_{22} - k_x T_{11})\,\delta(w)$$

$$+ T_{11}\delta\left[\frac{1}{2}\left(\frac{\partial w}{\partial x}\right)^2\right] + T_{22}\delta\left[\frac{1}{2}\left(\frac{\partial w}{\partial y}\right)^2\right] + T_{12}\delta\left[\frac{\partial w}{\partial x}\frac{\partial w}{\partial y}\right]\Bigg\} ds$$

$$= \iint\limits_{\Omega} \Bigg\{ (-k_y T_{22} - k_x T_{11})\,\delta(w) + T_{11}\frac{\partial w}{\partial x}\frac{\partial}{\partial x}\delta(w) + T_{22}\frac{\partial w}{\partial y}\frac{\partial}{\partial y}\delta(w)$$

$$+ T_{12}\left[\frac{\partial w}{\partial y}\frac{\partial}{\partial x}\delta(w) + \frac{\partial w}{\partial x}\frac{\partial}{\partial y}\delta(w)\right]\Bigg\} ds. \tag{1.31}$$

Integration by parts yields

$$\delta_w V_1 = \iint\limits_{\Omega} \Bigg\{ -k_y\frac{\partial^2 F}{\partial x^2} - k_y\frac{\partial^2 F}{\partial x^2} - \frac{\partial^2 w}{\partial x^2}\frac{\partial^2 F}{\partial y^2} - \frac{\partial^2 w}{\partial y^2}\frac{\partial^2 F}{\partial x^2}$$

$$+ 2\frac{\partial^2 w}{\partial x \partial y}\frac{\partial^2 F}{\partial x \partial y}\Bigg\}\,\delta(w)\; ds$$

$$= \iint\limits_{\Omega} \left\{-\nabla_k^2 F - L(w,F)\right\}\delta(w)\; ds. \tag{1.32}$$

Boundary integrals counterpart to Eq. (1.31) have the form

$$\int\limits_0^a \left(T_{22}\frac{\partial w}{\partial y} + T_{12}\frac{\partial w}{\partial x}\right)\delta(w)|_0^b\, dx$$

$$+ \int\limits_0^b \left(T_{11}\frac{\partial w}{\partial x} + T_{12}\frac{\partial w}{\partial y}\right)\delta(w)|_0^a\, dy. \tag{1.33}$$

Let us examine the second component of (1.16) of the form

$$V_2 = \iint\limits_{\Omega} \Bigg\{ \frac{2h^3 E}{6(1-\mu^2)}\left[\left(\frac{\partial^2 w}{\partial x^2}\right)^2 + \left(\frac{\partial^2 w}{\partial y^2}\right)^2 + 2(1-\mu)\left(\frac{\partial^2 w}{\partial x \partial y}\right)^2\right.$$

$$\left. + 2\mu\frac{\partial^2 w}{\partial x^2}\frac{\partial^2 w}{\partial y^2}\right]\Bigg\} ds, \tag{1.34}$$

and let us separate some variations with respect to w:

$$\delta_w V_2 = \iint\limits_{\Omega} \frac{2h^3 E}{3(1-\mu^2)}\Bigg\{\frac{\partial^2 w}{\partial x^2}\frac{\partial^2(\cdot)}{\partial x^2} + \frac{\partial^2 w}{\partial y^2}\frac{\partial^2(\cdot)}{\partial y^2} + 2(1-\mu)\frac{\partial^2 w}{\partial x \partial y}\frac{\partial^2(\cdot)}{\partial x \partial y}$$

$$+ \mu\left[\frac{\partial^2 w}{\partial x^2}\frac{\partial^2(\cdot)}{\partial y^2} + \frac{\partial^2 w}{\partial y^2}\frac{\partial^2(\cdot)}{\partial x^2}\right]\Bigg\}\,\delta(w)\; ds. \tag{1.35}$$

The symbol $(\cdot)$ denotes the derivative coming from the variation of function w.

The kinetic energy of the shell with the exception of the shell element rotational inertia is

$$K = \frac{1}{2}\iint_{\Omega} \frac{2h\rho}{g}\left\{\left(\frac{\partial u}{\partial t}\right)^2 + \left(\frac{\partial v}{\partial t}\right)^2 + \left(\frac{\partial w}{\partial t}\right)^2\right\} ds. \qquad (1.36)$$

Let us now invoke the time integral of the shell kinetic energy variation. We will assume a case of variation of function u.

In the time interval $(t_0 - t_1)$ the following relation holds:

$$\int_{t_0}^{t_1} \delta_u K dt = \int_{t_0}^{t_1}\iint_{\Omega} \frac{2h\rho}{g}\frac{\partial u}{\partial t}\delta\left(\frac{\partial u}{\partial t}\right) ds\, dt. \qquad (1.37)$$

As a result of integration by parts one obtains

$$\int_{t_0}^{t_1} \delta_u K dt = \iint_{\Omega} \frac{2h\rho}{g}\left[\frac{\partial u}{\partial t}\delta(u)\right]\Bigg|_{t_0}^{t_1} ds - \int_{t_0}^{t_1}\iint_{\Omega} \frac{2h\rho}{g}\frac{\partial^2 u}{\partial t^2}\delta(u)\, ds\, dt. \qquad (1.38)$$

Any partial variations with respect to v and w are found analogously:

$$\int_{t_0}^{t_1} \delta_v K dt = \iint_{\Omega} \frac{2h\rho}{g}\left[\frac{\partial v}{\partial t}\delta(v)\right]\Bigg|_{t_0}^{t_1} ds - \int_{t_0}^{t_1}\iint_{\Omega} \frac{2h\rho}{g}\frac{\partial^2 v}{\partial t^2}\delta(v)\, ds\, dt, \qquad (1.39)$$

$$\int_{t_0}^{t_1} \delta_w K dt = \iint_{\Omega} \frac{2h\rho}{g}\left[\frac{\partial w}{\partial t}\delta(w)\right]\Bigg|_{t_0}^{t_1} ds - \int_{t_0}^{t_1}\iint_{\Omega} \frac{2h\rho}{g}\frac{\partial^2 w}{\partial t^2}\delta(w)\, ds\, dt. \qquad (1.40)$$

The elementary work of any external forces is expressed as follows:

$$\delta' W = \iint_{\Omega}\left[p_x\delta(u) + p_y\delta(v) + \left(q - \varepsilon\frac{2h\rho}{g}\frac{\partial w}{\partial t}\right)\delta(w)\right] ds, \qquad (1.41)$$

where ε denotes the damping coefficient of a surrounding medium.

Substituting all formulae previous to Eq. (1.43), which states the counterpart to the Hamilton principle, the following variational equations are obtained:

$$\int_{t_0}^{t_1}\iint_{\Omega}\left\{\left[\frac{\partial T_{11}}{\partial x}+\frac{\partial T_{12}}{\partial y}+p_x-\frac{2h\rho}{g}\frac{\partial^2 u}{\partial t^2}\right]\delta(u)\right.$$
$$+\left[\frac{\partial T_{22}}{\partial y}+\frac{\partial T_{12}}{\partial x}+p_y-\frac{2h\rho}{g}\frac{\partial^2 v}{\partial t^2}\right]\delta(v)$$
$$+\left[\frac{2h^3E}{3(1-\mu^2)}\left(\frac{\partial^2 w}{\partial x^2}\frac{\partial^2(\cdot)}{\partial x^2}+\frac{\partial^2 w}{\partial y^2}\frac{\partial^2(\cdot)}{\partial y^2}\right)+2(1-\mu)\frac{\partial^2 w}{\partial x\partial y}\frac{\partial^2(\cdot)}{\partial x\partial y}\right.$$
$$+\mu\left(\frac{\partial^2 w}{\partial x^2}\frac{\partial^2(\cdot)}{\partial y^2}+\frac{\partial^2 w}{\partial y^2}\frac{\partial^2(\cdot)}{\partial x^2}\right)-\nabla_k^2 F-L(w,F)$$
$$\left.+q-\frac{2h\rho}{g}\left(\frac{\partial^2 w}{\partial t^2}+\varepsilon\frac{\partial w}{\partial t}\right)\right]\delta(w)$$
$$+\left[\frac{a_1}{2h}\left(\left(\frac{\partial^2 F}{\partial y^2}-\mu\frac{\partial^2 F}{\partial x^2}\right)\frac{\partial^2(\cdot)}{\partial y^2}+\left(\frac{\partial^2 F}{\partial x^2}-\mu\frac{\partial^2 F}{\partial y^2}\right)\frac{\partial^2(\cdot)}{\partial x^2}\right.\right.$$
$$\left.\left.\left.+2(1+\mu)\frac{\partial^2 F}{\partial x\partial y}\frac{\partial^2(\cdot)}{\partial x\partial y}\right)+\nabla_k^2 w+\frac{1}{2}L(w,w)\right]\delta(F)\right\}ds\,dt$$
$$-\int_{t_0}^{t_1}\int_0^a\left\{\left[\varepsilon_{11}\frac{\partial}{\partial y}(\delta F)-\frac{\partial\varepsilon_{11}}{\partial y}(\delta F)+\frac{\partial\varepsilon_{12}}{\partial y}(\delta F)+T_{22}(\delta v)+T_{12}(\delta u)\right.\right.$$
$$\left.\left.+\left(T_{22}\frac{\partial w}{\partial y}+T_{12}\frac{\partial w}{\partial x}\right)\delta(w)\right]\Bigg|_0^b\right\}dx\,dt$$
$$-\int_{t_0}^{t_1}\int_0^b\left\{\left[\varepsilon_{22}\frac{\partial}{\partial x}(\delta F)-\frac{\partial\varepsilon_{22}}{\partial x}(\delta F)+\varepsilon_{12}\frac{\partial}{\partial y}(\delta F)+T_{11}(\delta u)+T_{12}(\delta v)\right.\right.$$
$$\left.\left.+\left(T_{11}\frac{\partial w}{\partial x}+T_{12}\frac{\partial w}{\partial y}\right)\delta(w)\right]\Bigg|_0^a\right\}dy\,dt$$
$$+\iint_{\Omega}\frac{2h\rho}{g}\left[\frac{\partial u}{\partial t}\delta(u)+\frac{\partial v}{\partial t}\delta(v)+\frac{\partial w}{\partial t}\delta(w)\right]\Bigg|_{t_0}^{t_1}ds=0. \tag{1.42}$$

1.5 Equations of Motion

Similarly as in Eq. (1.42), variations δu, δv, δw, δF taken as a functions of time t are arbitrarily chosen, and therefore multipliers preceding all variations of the first integral should equal zero. As a result of variational equation the following equations of motion and only one equation of deformation compatibility are presented:

$$\left[\frac{\partial T_{11}}{\partial x}+\frac{\partial T_{12}}{\partial y}+p_x-\frac{2h\rho}{g}\frac{\partial^2 u}{\partial t^2}\right]=0, \tag{1.43}$$

$$\left[\frac{\partial T_{22}}{\partial y}+\frac{\partial T_{12}}{\partial x}+p_y-\frac{2h\rho}{g}\frac{\partial^2 v}{\partial t^2}\right]=0, \tag{1.44}$$

$$\frac{2h^3 E}{3(1-\mu^2)}\left[\frac{\partial^2 w}{\partial x^2}\frac{\partial^2(\cdot)}{\partial x^2}+\frac{\partial^2 w}{\partial y^2}\frac{\partial^2(\cdot)}{\partial y^2}+2(1-\mu)\frac{\partial^2 w}{\partial x\partial y}\frac{\partial^2(\cdot)}{\partial x\partial y}\right.$$
$$\left.+\mu\left(\frac{\partial^2 w}{\partial x^2}\frac{\partial^2(\cdot)}{\partial y^2}+\frac{\partial^2 w}{\partial y^2}\frac{\partial^2(\cdot)}{\partial x^2}\right)\right]-\nabla_k^2 F-L(w,F)+q-\frac{2h\rho}{g}\left(\frac{\partial^2 w}{\partial t^2}+\varepsilon\frac{\partial w}{\partial t}\right)=0, \tag{1.45}$$

$$\frac{a_1}{2h}\left[\left(\frac{\partial^2 F}{\partial y^2}-\mu\frac{\partial^2 F}{\partial x^2}\right)\frac{\partial^2(\cdot)}{\partial y^2}+\left(\frac{\partial^2 F}{\partial x^2}-\mu\frac{\partial^2 F}{\partial y^2}\right)\frac{\partial^2(\cdot)}{\partial x^2}\right.$$
$$\left.+2(1+\mu)\frac{\partial^2 F}{\partial x\partial y}\frac{\partial^2(\cdot)}{\partial x\partial y}\right]+\nabla_k^2 w+\frac{1}{2}L(w,w)=0. \tag{1.46}$$

If there is a need to write only a variational form of the equation, then during creation of the variational Eq. (1.42) the following steps should be introduced:

(i) Only some of terms including $\delta(F)$ variations must be considered in the equation;
(ii) Equation (1.31) should be introduced instead of Eq. (1.32).

Keeping in mind the last mathematical transformations, three equations of motion of the shell at hand will be derived. The first two, i.e., (1.43) and (1.44), remain unchanged, but instead of Eq. (1.45) the following one appears:

$$\frac{2h^3 E}{3(1-\mu^2)}\left[\frac{\partial^2 w}{\partial x^2}\frac{\partial^2(\cdot)}{\partial x^2}+\frac{\partial^2 w}{\partial y^2}\frac{\partial^2(\cdot)}{\partial y^2}+2(1-\mu)\frac{\partial^2 w}{\partial x\partial y}\frac{\partial^2(\cdot)}{\partial x\partial y}\right.$$
$$\left.+\mu\left(\frac{\partial^2 w}{\partial x^2}\frac{\partial^2(\cdot)}{\partial y^2}+\frac{\partial^2 w}{\partial y^2}\frac{\partial^2(\cdot)}{\partial x^2}\right)\right]+T_{11}k_x+T_{22}k_y+\frac{\partial}{\partial x}\left(T_{11}\frac{\partial w}{\partial x}+T_{12}\frac{\partial w}{\partial y}\right)$$
$$+\frac{\partial}{\partial y}\left(T_{22}\frac{\partial w}{\partial y}+T_{12}\frac{\partial w}{\partial x}\right)+q-\frac{2h\rho}{g}\left(\frac{\partial^2 w}{\partial t^2}+\varepsilon\frac{\partial w}{\partial t}\right)=0. \tag{1.47}$$

If there is a possibility of analyzing the dynamic process omitting any elastic wave distributions, then because the inertial terms of the first two equations, are disregarded, the set of Eqs. (1.43)–(1.46) can be significantly simplified. These conditioning equations will be met if in accordance with the variables of Eq. (1.21), a function of stresses is introduced (p_x and p_y).

In this situation the equation of motion of a shell element takes the form

$$\frac{2h^3 E}{3(1-\mu^2)}\left[\frac{\partial^2 w}{\partial x^2}\frac{\partial^2(\cdot)}{\partial x^2}+\frac{\partial^2 w}{\partial y^2}\frac{\partial^2(\cdot)}{\partial y^2}+2(1-\mu)\frac{\partial^2 w}{\partial x\partial y}\frac{\partial^2(\cdot)}{\partial x\partial y}\right.$$
$$\left.+\mu\left(\frac{\partial^2 w}{\partial x^2}\frac{\partial^2(\cdot)}{\partial y^2}+\frac{\partial^2 w}{\partial y^2}\frac{\partial^2(\cdot)}{\partial x^2}\right)\right]-\nabla_k^2 F-L(w,F)+q-\frac{2h\rho}{g}\left(\frac{\partial^2 w}{\partial t^2}+\varepsilon\frac{\partial w}{\partial t}\right)=0, \tag{1.48}$$

and the equation of deformations compatibility is

$$\frac{a_1}{2h}\left[\left(\frac{\partial^2 F}{\partial y^2}-\mu\frac{\partial^2 F}{\partial x^2}\right)\frac{\partial^2(\cdot)}{\partial y^2}+\left(\frac{\partial^2 F}{\partial x^2}-\mu\frac{\partial^2 F}{\partial y^2}\right)\frac{\partial^2(\cdot)}{\partial x^2}\right.$$

$$\left.+2(1+\mu)\frac{\partial^2 F}{\partial x\partial y}\frac{\partial^2(\cdot)}{\partial x\partial y}\right]+\nabla_k^2 w+\frac{1}{2}L(w,w)=0. \qquad (1.49)$$

When a circle section cylindrical shell is analyzed, then the pair of parameters $k_x=0$ and $k_y=1/R$, where R is the curvature radius of the mean surface of a shell, should be taken into account in Eqs. (1.48)–(1.49).

Equations of a homogeneous shell, which were obtained during our analysis, are in good agreement with those contained in [304]. In the case of $E=const$, the density $\rho=const$ should be assumed and during integration by parts the computational process of variational differentiation has to be transferred to a proper function. If the functional dependence of E and ρ on the spatial coordinates relates to any perpendicular cutouts of material (snippets), then after the analogous routines are carried out, the equations will be identical, as in reference [251].

1.6 Boundary and Initial Conditions

An integration of equations of motion needs to be supported by a properly defined boundary and initial conditions. On the basis of the behavioral (dynamic) hypothesis of straight lines, which are perpendicular to a surface, each bounding point should satisfy four boundary conditions. In the case in which the displacements u, v, and w of any boundary points of a shell are known, then after deformations they suitably determine the space-located arbitrary curve. A perpendicular straight line led through a boundary point can dislocate together with that point as well as rotate about an angle in a plane that is perpendicular to the boundary line. Correspondingly, normal line placement to the surface resulting in a shell's deformation can be determined with the use of any four values. These integrals of a variational equation calculated along boundaries allow for the formulation of any appropriate additional conditions [159]. Some of them are given below for the bound of a shell at $x=const$ ($y=const$ boundary condition could be obtained following the replacement of x with y)

1. Simply supported edge:

$$w=M_{11}=T_{11}=T_{12}=0 \quad \text{if} \quad x=0,a. \qquad (1.50)$$

2. Ball-type support on the flexible but not compressed (stretched) ribs:

$$w=M_{11}=T_{11}=\varepsilon_{22}=0 \quad \text{if} \quad x=0,a. \qquad (1.51)$$

The second condition can be also given in the form

$$w = \frac{\partial^2 w}{\partial x^2} = F = \frac{\partial^2 F}{\partial x^2} = 0 \quad \text{if} \quad x = 0, a. \tag{1.52}$$

Displacement boundary conditions are called the geometric, whereas any conditions formulated for the stresses and moments of forces are called dynamic.

Some initial conditions should also be satisfied, which simultaneously with the integration of fundamental equations are related to the velocities of points belonging to the mean surface of the shell

$$\begin{aligned} w\,|_{t=t_0} &= \zeta_1(x,y), \\ \frac{\partial w}{\partial t}\,|_{t=t_0} &= \zeta_2(x,y). \end{aligned} \tag{1.53}$$

1.7 Non-dimensional Form of Equations

Let us introduce the following non-dimensional parameters:

$$\begin{aligned} &w = 2h\bar{w}, \quad x = a\bar{x}, \quad y = b\bar{y}, \quad F = E_0 (2h)^3 \bar{F}, \\ &k_x = \frac{2h}{a^2}\bar{k}_x, \quad k_y = \frac{2h}{b^2}\bar{k}_y, \quad q = \frac{E_0 (2h)^4}{a^2 b^2}\bar{q}, \\ &t = \frac{ab}{2h}\sqrt{\frac{\rho_0}{gE_0}}\bar{t}, \quad \varepsilon = \frac{2h}{ab}\sqrt{\frac{gE_0}{\rho_0}}\bar{\varepsilon}, \\ &\lambda = a/b, \quad E = E_0\bar{E}, \quad \rho = \rho_0\bar{\rho}. \end{aligned} \tag{1.54}$$

Non-dimensional Eqs. (1.48)–(1.49) at the very beginning of this section are as follows (bars above the non-dimensional quantities are omitted):

$$\begin{aligned} &\frac{E}{12(1-\mu^2)}\left[\frac{1}{\lambda^2}\frac{\partial^2 w}{\partial x^2}\frac{\partial^2 (\cdot)}{\partial x^2} + \lambda^2 \frac{\partial^2 w}{\partial y^2}\frac{\partial^2 (\cdot)}{\partial y^2} + 2(1-\mu)\frac{\partial^2 w}{\partial x \partial y}\frac{\partial^2 (\cdot)}{\partial x \partial y} \right. \\ &\left. + \mu\left(\frac{\partial^2 w}{\partial x^2}\frac{\partial^2 (\cdot)}{\partial y^2} + \frac{\partial^2 w}{\partial y^2}\frac{\partial^2 (\cdot)}{\partial x^2}\right)\right] - \nabla_k^2 F - L(w,F) + q - \rho\left(\frac{\partial^2 w}{\partial t^2} + \varepsilon\frac{\partial w}{\partial t}\right) = 0, \end{aligned} \tag{1.55}$$

$$\begin{aligned} &a_1\left[\left(\lambda^2\frac{\partial^2 F}{\partial y^2} - \mu\frac{\partial^2 F}{\partial x^2}\right)\frac{\partial^2 (\cdot)}{\partial y^2} + \left(\frac{1}{\lambda^2}\frac{\partial^2 F}{\partial x^2} - \mu\frac{\partial^2 F}{\partial y^2}\right)\frac{\partial^2 (\cdot)}{\partial x^2} \right. \\ &\left. + 2(1+\mu)\frac{\partial^2 F}{\partial x \partial y}\frac{\partial^2 (\cdot)}{\partial x \partial y}\right] + \nabla_k^2 w + \frac{1}{2}L(w,w) = 0. \end{aligned} \tag{1.56}$$

On the basis of both Kirchhoff-Love kinematic model and the non-homogeneity of a shell one can obtain from (1.55)–(1.56) some starting equations concerning static stability of a perpendicular shell but with an assumption that any time-dependent terms are neglected. It allows one to write them in the non-dimensional form:

$$\frac{E}{12(1-\mu^2)}\left[\frac{1}{\lambda^2}\frac{\partial^2 w}{\partial x^2}\frac{\partial^2(\cdot)}{\partial x^2}+\lambda^2\frac{\partial^2 w}{\partial y^2}\frac{\partial^2(\cdot)}{\partial y^2}+2(1-\mu)\frac{\partial^2 w}{\partial x\partial y}\frac{\partial^2(\cdot)}{\partial x\partial y}\right.$$
$$\left.+\mu\left(\frac{\partial^2 w}{\partial x^2}\frac{\partial^2(\cdot)}{\partial y^2}+\frac{\partial^2 w}{\partial y^2}\frac{\partial^2(\cdot)}{\partial x^2}\right)\right]-\nabla_k^2 F-L(w,F)+q=0, \tag{1.57}$$

$$a_1\left[\left(\lambda^2\frac{\partial^2 F}{\partial y^2}-\mu\frac{\partial^2 F}{\partial x^2}\right)\frac{\partial^2(\cdot)}{\partial y^2}+\left(\frac{1}{\lambda^2}\frac{\partial^2 F}{\partial x^2}-\mu\frac{\partial^2 F}{\partial y^2}\right)\frac{\partial^2(\cdot)}{\partial x^2}\right.$$
$$\left.+2(1+\mu)\frac{\partial^2 F}{\partial x\partial y}\frac{\partial^2(\cdot)}{\partial x\partial y}\right]+\nabla_k^2 w+\frac{1}{2}L(w,w)=0. \tag{1.58}$$

1.8 Variable Parameters of Stiffness

The aim of application of the method used in monograph [251] is to introduce some variable parameters of stiffness of a shell. The method that is proposed for applications is suitable for determination of equations of motion of a shell element with either any cutouts or other modifications of its thickness. There is a possibility of simplifying the problem, but with the assumption that there is only one full parallelepiped cutout, which is oriented in the investigated shell in a way securing its edges to be parallel to the appropriate external edges of the shell. Similarly as in Eq. (1.16), the integrated function depends on displacements; therefore it can be rewritten as

$$V=\frac{1}{2}\iint_{\Omega} f(u,v,w)\,ds=\frac{1}{2}\iint_{\Omega} f(x,y)\,ds. \tag{1.59}$$

Let us assume that Ω denotes the region of integration that is bounded by both internal and external edges of the deformable system under investigation

$$\Omega=S-S_1, \tag{1.60}$$

in which S is the shell's lateral surface without any cutout and S_1 is the cutout's surface. The double integral calculated on Ω surface is transformed to the two double integrals of the form

$$V=\frac{1}{2}\iint_{S} f(x,y)\,ds-\frac{1}{2}\iint_{S_1} f(x,y)\,ds. \tag{1.61}$$

Let us now take a shell element bounded by coordinates $x=a_1$, $x=a_2$, $y=b_1$, $y=b_2$. Assume that the element includes a cutout defined by the lines $x=x_1$,

$x = x_2$, $y = y_1$, $y = y_2$. In association with this, the cutout belongs to the interior of the analyzed region, so we can write

$$a_1 < x_1 < a_2, \quad a_1 < x_2 < a_2,$$
$$b_1 < y_1 < b_2, \quad b_1 < y_2 < b_2. \tag{1.62}$$

Equation (1.61) is composed of the difference of any two double integrals, which are calculated along S and S_1, but the region S_1 is included in S. Moreover, regions S and S_1 are rectangular and with mutually parallel boundaries (Fig. 1.2).

For further considerations, we will use a unity Heaviside function dependent on two variables $\Gamma_0(x - x_1,\ y - y_1)$ of the form

$$\Gamma_0\,(x - x_1; y - y_1) = \Gamma_0\,(x - x_1)\,\Gamma_0\,(y - y_1)\,, \tag{1.63}$$

where

$$\Gamma_0\,(x - x_1) = \begin{cases} 0 & \text{for } x < x_1; \\ 1 & \text{for } x > x_1. \end{cases} \tag{1.64}$$

The filtering property of the unity function mentioned is defined by means of the following expression:

$$\int_{a_1}^{a_2}\int_{b_1}^{b_2} f\,(x,y)\,\Gamma_0\,(x - x_1; y - y_1)\; dx\, dy = \int_{x_1}^{a_2}\int_{y_1}^{b_2} f\,(x,y)\; dx\, dy, \tag{1.65}$$

and occurs if $a_1 < x_1 < a_2$ and $b_1 < y_1 < b_2$.

The quoted property of the function allows for introduction in Eq. (1.61) of only one double integral instead of any two others. The reduction procedure will be examined more carefully.

The expression

$$V' = \iint_S f\,(x,y)\,ds = \int_{a_1}^{a_2}\int_{b_1}^{b_2} f\,(x,y)\; dx\, dy \tag{1.66}$$

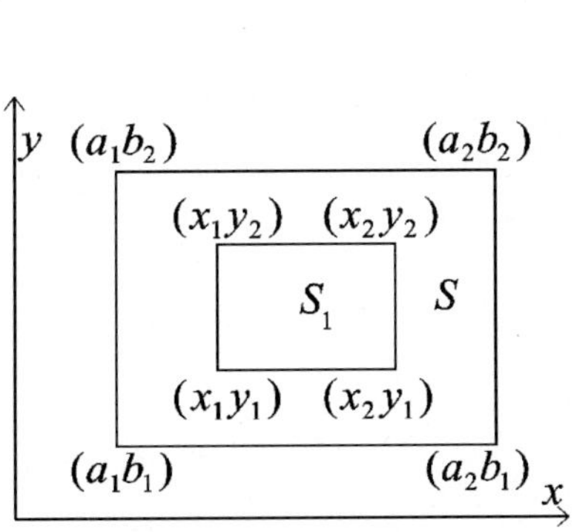

Fig. 1.2 Placement of regions S and S_1

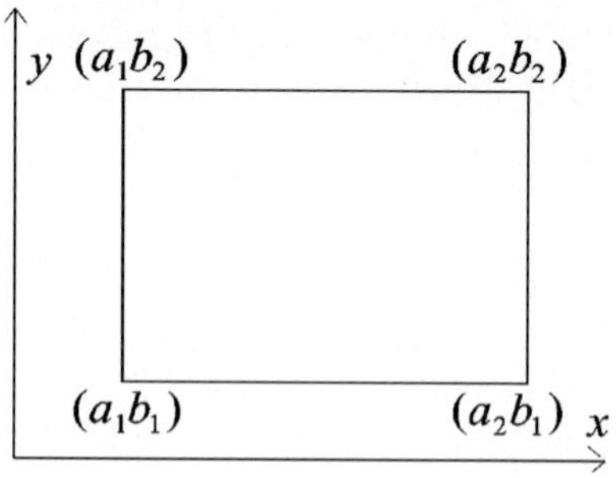

Fig. 1.3 Placement of region S

introduces an integral calculated on the S surface (Fig. 1.3). Note that

$$V' = \int_{a_1}^{a_2}\int_{b_1}^{b_2} [1-\Gamma_0(x-x_1;y-y_1)]\, f(x,y)\; dx\, dy, \tag{1.67}$$

but the above integral is calculated on the S surface excluding any separated shell element (Fig. 1.4).

The integral

$$V' = \int_{a_1}^{a_2}\int_{b_1}^{b_2} [1-\Gamma_0(x-x_1;y-y_1)+\Gamma_0(x-x_2;y-y_1)]\, f(x,y)\; dx\, dy \tag{1.68}$$

is also calculated on this surface as in the previous case, but the element excluded there has been now added (Fig. 1.5).

The expression

$$V' = \int_{a_1}^{a_2}\int_{b_1}^{b_2} [\,1-\Gamma_0(x-x_1;y-y_1)+\Gamma_0(x-x_2;y-y_1)$$
$$+\Gamma_0(x-x_1;y-y_2)\,]\, f(x,y)\; dx\, dy \tag{1.69}$$

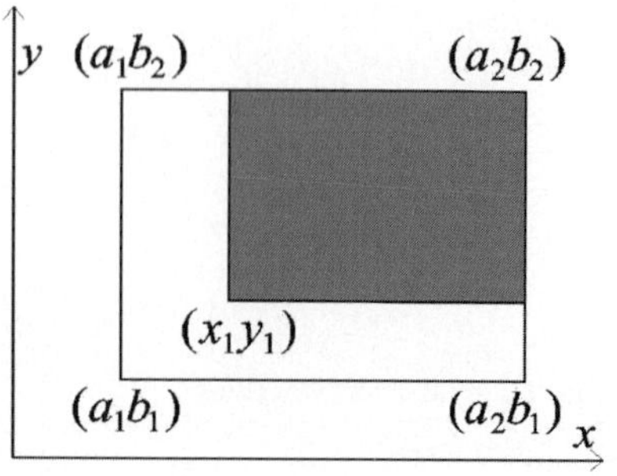

Fig. 1.4 A scheme based on calculation of integral (1.67)

Fig. 1.5 A scheme based on calculation of integral (1.68)

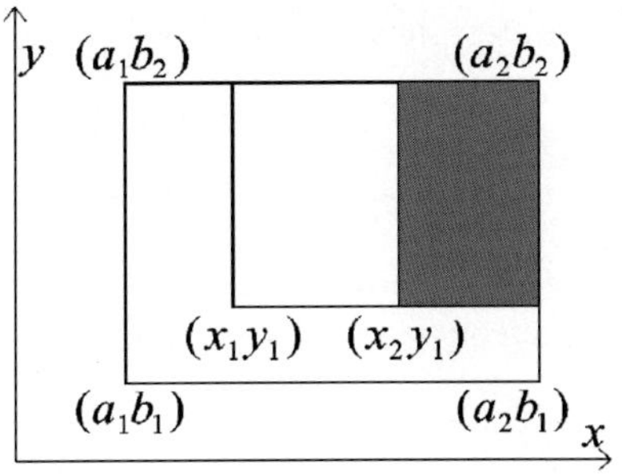

is related to the integration on this surface, as is clearly visible in Eq. (1.69) when the separated part pictured in Fig. 1.6 has been considered.

Finally, the integral

$$V' = \int_{a_1}^{a_2}\int_{b_1}^{b_2} [1 - \Gamma_0(x-x_1;y-y_1) + \Gamma_0(x-x_2;y-y_1)$$
$$+ \Gamma_0(x-x_1;y-y_2) - \Gamma_0(x-x_1;y-y_2)]f(x,y)\,dx\,dy, \quad (1.70)$$

corresponds to the integration in S with exclusion of S_1. It means that it is equal to the quantity V defined by Eq. (1.61).

Introduction of the unity function has allowed for replacement of the two double integrals by one integral calculated on the surface solely bounded by the loaded edge:

$$V_1 = \frac{1}{2}\iint_S \frac{2hE_0(1-\gamma_0)}{1-\mu^2}\left(\varepsilon_{11}^2 + \varepsilon_{22}^2 + 2\mu\varepsilon_{11}\varepsilon_{22} + \frac{1-\mu}{2}\varepsilon_{12}^2\right) ds, \quad (1.71)$$

where

$$\gamma_0 = \Gamma_0(x-x_1;y-y_1) - \Gamma_0(x-x_2;y-y_1)$$
$$- \Gamma_0(x-x_1;y-y_2) + \Gamma_0(x-x_2;y-y_2). \quad (1.72)$$

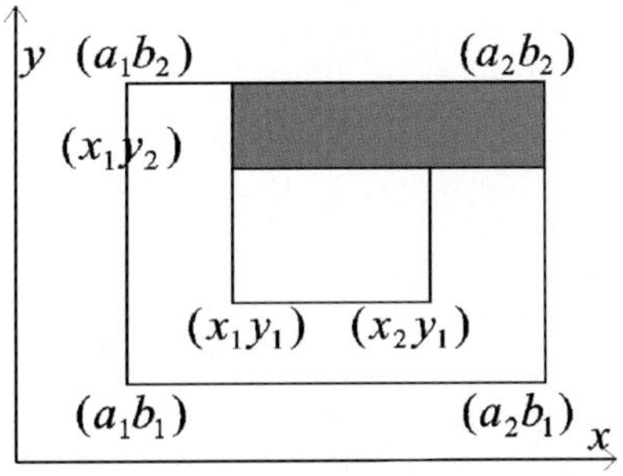

Fig. 1.6 A scheme based on calculation of integral (1.69)

Referring to Sect. 1.3, it is possible to write

$$E = E(x,y) = E_0(1-\gamma_0),$$
$$\rho = \rho(x,y) = \rho_0(1-\gamma_0). \tag{1.73}$$

This type of formula for E and ρ expresses a transition from a cutout of shell to the adequate continuous model, which has also been presented in [251].

1.9 Flexural Stiffness Coefficient of a Shell Element

Let the function $f(x)$ be as follows:

$$f(x) = \varphi(x)\lambda(x), \tag{1.74}$$

where $\varphi(x)$ is *a stepping function* changing from a to b, and $\lambda(x)$ it is an *impulse function.*

The function visible in Fig. 1.7 can be represented here by means of the unity Heaviside function as follows:

$$\begin{aligned} f(x) &= a[1-\Gamma_0(x-x_0)] + b\Gamma_0(x-x_0) \\ &= a - a\Gamma_0(x-x_0) + b\Gamma_0(x-x_0) \\ &= a - a\Gamma_0(x-x_0) + b\Gamma_0(x-x_0) \\ &= a - (a-b)\Gamma_0(x-x_0) \\ &= a\left[1 - \left(\frac{a-b}{a}\right)\Gamma_0(x-x_0)\right] \\ &= a\left[1 - \left(1-\frac{b}{a}\right)\Gamma_0(x-x_0)\right], \end{aligned} \tag{1.75}$$

and in comparison to the above we have

$$f(x) = a\left[1 - \left(1-\frac{b}{a}\right)\Gamma_0(x-x_0)\right]. \tag{1.76}$$

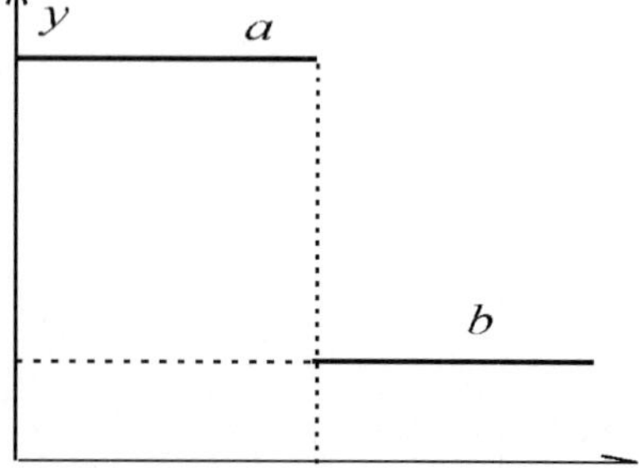

Fig. 1.7 A function $f(x)$ related to the function (1.75)

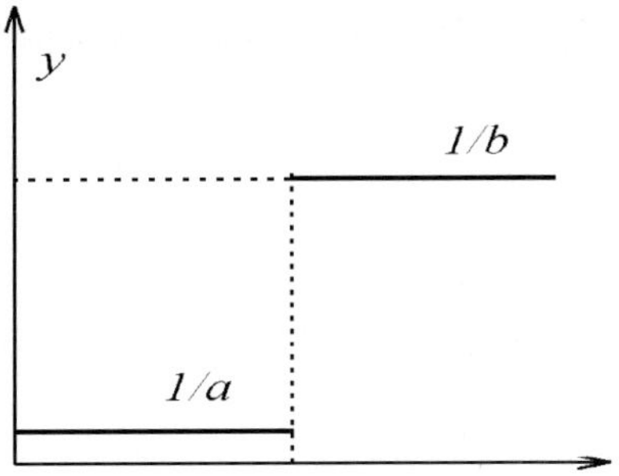

Fig. 1.8 A function $f(x)$ related to the function (1.77)

In an analogous manner, an inverse function takes the form (Fig. 1.8):

$$
\begin{aligned}
\frac{1}{f(x)} &= \frac{1}{a}\left[1-\Gamma_0(x-x_0)\right]+\frac{1}{b}\Gamma_0(x-x_0) \\
&= \frac{1}{a}+\left(\frac{1}{b}-\frac{1}{a}\right)\Gamma_0(x-x_0) \\
&= \frac{1}{a}+\left(\frac{a-b}{ab}\right)\Gamma_0(x-x_0) \\
&= \frac{1}{a}\left[1+\left(\frac{a}{b}-1\right)\Gamma_0(x-x_0)\right]. \qquad (1.77)
\end{aligned}
$$

If we assume that $a = E_0$ is the value of the modulus for a homogeneous shell, whereas $\frac{b}{a}$ is the coefficient of its changes along the shell's cutout, then the coefficient of that cutout is expressed by

$$\gamma_1 = \frac{b}{a}. \qquad (1.78)$$

The quantity $\gamma_1 = 1$ corresponds to homogeneous shells because flexural stiffnesses of both the separated cutout of the shell and the homogeneous shell are practically the same (equal), whereas $\gamma_1 < 1$ corresponds to the cutout of a smaller stiffness (or greater for $\gamma_1 > 1$) than the stiffness of the homogeneous shell. According to the previous derivations

$$
\begin{aligned}
E(x,y) &= E_0\left[1-\sum_j\left(1-\gamma_{1j}\right)\gamma_{0j}\right] \\
\frac{1}{E(x,y)} &= \frac{1}{E_0}\left[1+\sum_j\left(\frac{1}{\gamma_{1j}}-1\right)\gamma_{0j}\right], \qquad (1.79)
\end{aligned}
$$

where j denotes the number of such cutouts.

Analogously, one can obtain an expression for the density of a shell's material:

$$\rho(x,y) = \rho_0 \left[1 - \sum_j \left(1 - \gamma_{2j}\right) \gamma_{0j} \right], \tag{1.80}$$

where γ_{2j} is the density coefficient of jth cutout.

On the basis of relation (1.79) a conclusion can be drawn: if any equation contains an expression that is the inverse of the E modulus, then γ_{1j} cannot be equal to zero. It is equivalent to stating that a shell cannot possess any cutouts. Otherwise, if there is a need to consider those cutouts, the appropriate equation needs to be multiplied by E. Therefore, the general function has to be moved to the right-hand side of equation. For this case, any calculations connected with the described algorithm are rather complicated.

1.10 Generalized Functions

The term generalized function arises from the generalization of the classic function definition [172]. The term originated during analysis of some physical problems and has quickly appeared as a purely mathematical notion.

Thanks to introduction of the notion, a Fourier transform could be formally applied to a much wider class of functions, not only those integrated absolutely or in quadratures. Moreover, it allows for mathematical formulation of some idealized notions, such like density of point explosion, the material point density, a momentary impulse, etc.

We will explain it more carefully. Using a mathematical apparatus for physical phenomena investigations it is often desired to use various mathematical abstractions, in particular the term material point. For example, we can speak of a mass reduced to a space point, a force applied in a given moment of time, a point source of many various fields, etc., but after all they are the idealized formulations. Such a simplified approach to the analysis of physical phenomena turns out to be insufficient. New mathematical notions or apparatus are often required.

The first notions of generalized functions were formed in the work of Dirac and other physicist's as a type of symbolic description of the physical phenomena. To achieve a systematic method of application of the above functions it was necessary to introduce some important bases of the theory of generalized functions, which was done by Sobolev and Schwartz [267, 279], respectively.

The theory of generalized functions is a very convenient mathematical tool, which permits the solution of many problems that could not be solved in the classical way. The theory of generalized functions is currently popular in many applied sciences, as well as in pure theoretical studies.

Let us begin our present considerations from the definition of the basic term of linear space D of a function. For that reason, functions defined in $\mathbb{R}$ and that take complex values will be consequently considered.

We are interested in a space D that consists of an unlimited set of finite and differentiable functions. A property of all finite functions is observed, that their forms of combination and multiplication by numbers are still included in a linear space, but infinitely differentiable finite functions create its subspace.

The set of infinitely differentiable functions together with the formulated definition of boundary transition creates the convergent linear space, which can be directly observed.

Definition 1.1 A space of infinitely differentiable finite functions along with the invoked notion of convergence is called a space D of basic functions.

Definition 1.2 An arbitrarily chosen functional f defined in D is called the *generalized function.*

Definition 1.3 The function f that is defined along the whole real axis is called *locally integrable* if it is absolutely integrable on any finite interval.

If f is a locally integrated function and $\varphi \in D$, then the product $f\varphi$ is absolutely integrable on the whole considered axis.

Let the locally integrable function f be defined for a functional (f, φ) in D as follows:

$$(f,\varphi) = \int_{-\infty}^{+\infty} f(x)\varphi(x)\, dx. \tag{1.81}$$

The above function is both linear and continuous.

A conclusion can be made that for an arbitrary and locally integrable function $f(x)$ there exists a generalized function (f, φ). An arbitrary and locally integrable function can be considered in that sense as a generalized function (a suitable thesis can be proposed: any generalized function (f, φ) is generated by function f).

Generalized functions are in some situations denoted by $f(x)$. The symbolic notation does not denote the value of a generalized function at a point $x \in \mathbb{R}$, but emphasizes that generalized functions are the generalization of normal (locally integrable) functions and no other value of a generalized function at the point x is assigned.

For an estimation of the value of the generalized function f at the point $\varphi = \varphi(x)$ of a space D and with the exception of the notation (f, φ) the following equivalent can be used:

$$\int_{-\infty}^{+\infty} f(x)\varphi(x)\, dx. \tag{1.82}$$

It leads to the comprehensive definition:

$$\int_{-\infty}^{+\infty} f(x)\varphi(x)\, dx = (f,\varphi). \tag{1.83}$$

Equation (1.83) is a definition of the symbol (1.82), which should be formally understood in the following way: all generalized functions constitute a generalization of functionals given by definition (1.81), where f is the locally integrable function.

Let us now define *the derivative of a generalized function*. Most of all, a step-by-step explanation is necessary, bringing the meaning of the derivative of normal continuous and along the whole real axis differentiable function f treated as a functional (f', φ) in D. Let us note, moreover, that the derivative f' being continuous on the whole real axis is the locally differentiable function. Integrating by parts as well as taking into consideration the finiteness of the function $\varphi \in D$, we obtain

$$(f', \varphi) = \int_{-\infty}^{+\infty} f'(x)\varphi(x)dx = -\int_{-\infty}^{+\infty} f(x)\varphi'(x)dx = -(f, \varphi'), \qquad (1.84)$$

but $\varphi' \in D$. The conclusion is that the derivative f' is a functional in D and its values are precisely expressed by both the values of function f (considered as a functional) and the Eq. (1.84). This observation makes it possible to construct the following definition.

Definition 1.4 The derivative of a generalized function f will be called the *functional defined in D* denoted by f' and defined by the following equality:

$$(f', \varphi) = -(f, \varphi'), \quad \varphi \in D. \qquad (1.85)$$

In other words, all values of the functional f' in any point φ of the space D equal the values of the functional f with the opposite sign at point $\varphi' \in D$.

Previous studies allow drawing some conclusions, that any generalized function or any locally integrable function possesses a derivative in agreement with Definition 1.4.

On the basis of (1.84), the normal derivative of a continuous differentiable function on the whole real axis, treated as a functional in D, coincides with its derivative in the sense of generalized functions.

The procedure for calculating the derivative of a generalized function is called *differentiation* (analogously to the case of normal functions).

Higher-order derivatives of generalized functions can be calculated in an analogous manner as for normal functions:

$$f'' = (f')', \quad f''' = (f'')', \ldots,$$

in general

$$f^{(k)} = (f^{(k-1)})', \quad k = 1, 2, \ldots, \quad f^{(0)} = f. \qquad (1.86)$$

Corresponding to the above definition, a generalized function can have any derivatives of arbitrary high orders. It is often substituted with another statement claiming that they are differentiable anywhere up to infinity.

In 1933, the impulse functions of the theory of elasticity were for the first time applied in Gersevanov's work and, subsequently, in the work of Nazarov [229] and Radtsig [253]. Nevertheless, Novitskog [233], Vainberg, and Rajtfarb's [312] achievements have indicated a wide range of applicable prospects of functions in situations within the large class of mechanical problems (mainly in building engineering).

The zero-order unity function makes the fundamental concept of understanding of the zero-order impulse functions, defined as follows:

$$\Gamma_0(x-x_0)=\begin{cases}0 & \text{for } x<x_0,\\ 1 & \text{for } x\geq x_0.\end{cases} \tag{1.87}$$

Traditionally, the above function remains in contrast to the nth order impulse function of the form $\Gamma_n(x-x_0)$, called either the unity Heaviside function or the zero-order impulse function. The function at hand is locally integrable so it can be assumed officially as the generalized function.

It is worth noting that the two-variable impulse function is equal to multiplication of two other analogous variable functions of different arguments:

$$\Gamma_0(x-x_0;y-y_0)=\Gamma_0(x-x_0)\Gamma_0(y-y_0). \tag{1.88}$$

One of the more important merits of searching for solutions to many miscellaneous problems is the filtrate property of the nth derivative of delta function. It has led to an enormous popularization of that useful function. The filtrate property is characterized by the following expression:

$$\int_a^b f(x)\Gamma_1^{(n)}(x-\xi)\,dx=\begin{cases}0, & \xi<a,\\ (-1)^n f^{(n)}(\xi), & a<\xi<b,\\ 0, & \xi>b.\end{cases} \tag{1.89}$$

The often used mathematical dependencies arise from Eqs. (1.88) and (1.89):

$$\int_0^a\int_0^b \Gamma_1^{(x,y)}(x-x_1;y-y_1)f(x,y)\;dx\;dy=f(x_1,y_1),$$

$$\int_0^a\int_0^b \Gamma_1^{(y)}(x-x_1;y-y_1)f(x,y)\;dx\;dy=\int_{x_1}^a f(x,y_1)\;dx,$$

$$\int_0^a\int_0^b \Gamma_2^{(y)}(x-x_1;y-y_1)f(x,y)\;dx\;dy=-\int_{x_1}^a \left\{\frac{\partial}{\partial y}[f(x,y)]\right\}\bigg|_{y_1}\;dx,$$

$$\int_0^a\int_0^b \Gamma_2^{(x)}(x-x_1;y-y_1)f(x,y)\;dx\;dy=-\int_{y_1}^b \left\{\frac{\partial}{\partial x}[f(x,y)]\right\}\bigg|_{x_1}\;dy,$$

$$\int_0^a \int_0^b \Gamma_0(x-x_1;y-y_1)f(x,y)\,dx\,dy = \int_{x_1}^a \int_{y_1}^b f(x,y)\,dx\,dy,$$

$$\int_0^a \int_0^b \Gamma_1^{(x)}(x-x_1;y-y_1)f(x,y)\,dx\,dy = \int_{y_1}^b f(x_1,y)\,dy. \tag{1.90}$$

The terms $\Gamma_j^{()}$ denote the derivatives of impulse functions calculated in the determined coordinates.

Chapter 2
Static Instability of Rectangular Plates

This chapter deals with the static instability problems of rectangular plates. First, fundamental concepts of the theory of elastic stability are illustrated and discussed. Second, two fundamental formulas of the energy-based criterion of bifurcational stability loss of an elastic continuous mechanical system are derived. In addition, advantages and disadvantages of today's stability investigation approaches are critically revisited, with emphasis on problems not yet satisfactorily solved. In the next section various methods devoted to stability investigations are briefly addressed, exhibiting their strong and weak points regarding applications with particular attention to computational advantages of Galerkin's methods. In Sect. 2.4 the Bubnov-Galerkin method of high-order approximations and the associated numerical algorithm are presented. Rectangular form additions of other materials to a shell are described in Sect. 2.5, whereas the next section deals with static shell stability problems. Finally, the central square element, cross addition element, and perforation-type non-homogeneities are introduced.

2.1 Fundamental Concepts of the Theory of Elastic Stability

Increasing strength properties of the traditional material characteristics that are applied for various kinds of constructions as well as the use of new composite mediums characterized by extremely high-resistance features have led in the contemporary machines and building industry to a widespread application of light and economical constructions.

In the case of such constructions the role of strength estimations has significantly expanded. The main reason is the destruction of thin-walled structures closely connected with local or global loss of stability of their particular elements.

Within the last two to three decades opposing views in the theory of construction stability have emerged. Moreover, theories that were not comprehensive and trustworthy and many incorrect practically applied ideas were seen as well. For instance, we can remember the view in the scientific literature and consequently

J. Awrejcewicz, V.A. Krysko, *Chaos in Structural Mechanics*,
DOI: 10.1007/978-3-540-77676-5_3, © Springer-Verlag Berlin Heidelberg 2008

recommended by many specialists concerning the estimation of construction stability that was based on the so-called approximation of low values of critical loading. It is necessary to stress that quite recently some general theoretical basics have been worked out by the foremost mechanical engineers in the world, providing a possibility for finding suitable explanations of many problems of the theory of construction stability.

Modern construction engineers have, in spite of access to the data file of stability computations for a particular construction's elements, also quite easily accessible universal commercial numerical packages allowing for analyses of constructions that have in practice an unlimited complexity. It turns out, however, that in spite of wide range of computational techniques the present-day engineer has to understand the phenomenon of the loss of stability of thin-walled structures as well as to have a full comprehension of the series of implemented hypotheses and simplifications being made after each mathematical modeling of a problem.

The loss of stability of a deformable system is the time-dependent process expected to be investigated dynamically. Most questions of a construction's stability "allow" for application of such a static approach, in which any equilibrium conditions are formulated with the omission of inertial forces that have, after all, a significant influence on the quantity of deformations of the analyzed system.

In our further considerations the problem will be reduced to the analysis of stability of *conservative systems,* for which the static approach brings about the same results as the analysis of a complete and more complex dynamical approach [335]. The conservative system will be understood as a system composed of a deformable body being both fixed in space by means of ideal constraints and loaded by *conservative forces*. Let us also define, that to the conservative forces we will refer any other potential forces, the work done, that depends only on an initial and final system configuration but does not depend on the shape through which the system passed in the time between initial and final state. Reaction forces of *ideal constraints* do not do any work through all the possible movements of the system points in which they are attached.

To paraphrase from Lev Tolstoy's novel *Anna Karenina*, all stable mechanical systems are similar to each other, while some particular properties characterize the unstable ones. The observation refers not only to stability problems but also affects the significant question of motion stability and can also be adopted in a much narrow class of problems, for example, locus of equilibrium conception as found in [238].

The locus of equilibrium of a spring system can be expressed in many cases by two parameters, i.e., a characteristic movement v and a loading parameter P. For all existing positions of equilibrium there corresponds a curve in the v–P coordinate system. It permits prediction of the system's behavior after monotonically increasing loading or a displacement parameter, to estimate any instability regions and, in general, the critical values of parameters. Graphical presentations of these relationships are called *graphs* (or *curves*) of *equilibrium positions of a system.*

A graphical curve of equilibrium positions of a linear deformable system is represented then in the form of straight line. It turns out that the behavior of structurally nonlinear objects after deformations causes in many situations very complex relationships, but graphs of equilibrium positions of such systems can be miscellaneous and of high complexity as well.

A subclass of systems can be excluded from the main class of various spring systems in which behavior is qualitatively characterized by common features of the previously indicated P–v curve.

The most frequently encountered problems of engineering are considered below (see [238]). We will assume, when carrying out a behavioral analysis of systems, that *a loading parameter* (the case of soft loading) increases monotonically. Furthermore, there will be taken into consideration a regime of work in an environment of kinematic loading that is described by a monotonic increase of v parameter (hard loading).

1. For an arbitrary loading parameter $P < P_{cr}$, there exists one and completely stable configuration of an equilibrium locus. For any loadings exceeding P_{cr}, there exist several equilibrium positions and also with inclusion of unstable equilibrium positions (Fig. 2.1; on the drawing and in subsequent study, the unstable positions have been matched on P–v curves by crosses).
2. Up to a certain loading value P^*_{cr} there exists a single and, in addition, a stable equilibrium state. In the loading interval (P^*_{cr}, P_{cr}) we deal with three different equilibrium positions: a) stable; b) unstable; c) stable, but now different from the locus type (a). For the locus of equilibrium at $P > P_{cr}$ cases a) and b) are invalid, and then just one stable equilibrium locus remains (c) (Fig. 2.2).
3. Up to some loadings P_{cr} there exists a single locus of stable equilibrium (Fig. 2.3a), but in some situations any two equilibrium positions can coexist, concurrently, stable and unstable (Fig. 2.3b). There is no equilibrium locus for $P > P_{cr}$.
4. The starting point of the equilibrium locus is the only one for each parameter value of loading, but setting $P < P_{cr}$ the locus becomes stable, whereas for $P > P_{cr}$ it becomes the unstable locus of equilibrium (Fig. 2.4).

Let us underline the fact that when we speak of stability of any equilibrium positions, we really have in mind the stability in a "little" sense, i.e., that after a small disturbance the system gets back to the starting position.

The next study focuses on four cases.

The first one introduces a typical conditional branch (bifurcation) of equilibrium positions corresponding to the critical loading values $P = P_{cr}$. In the process of a monotonic increase of loading parameter P there will appear at $P = P_{cr}$ a transition to a new equilibrium state.

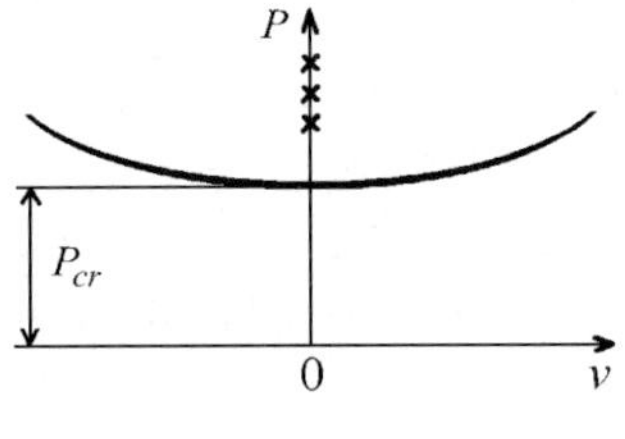

Fig. 2.1 The curve of equilibrium states (systems of first type)

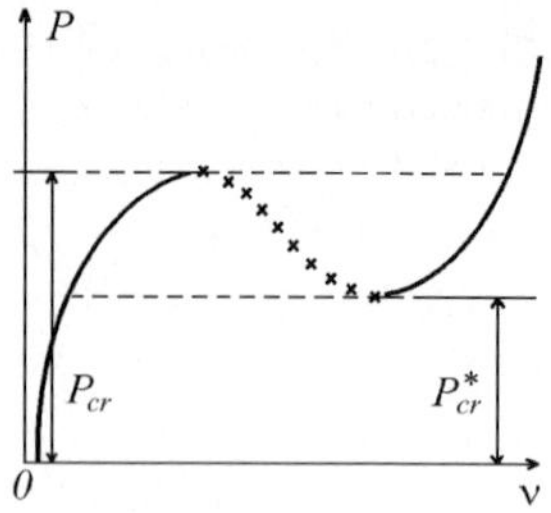

Fig. 2.2 The curve of equilibrium states (systems of the second type)

The second one introduces a monotonic increase of loading parameter the value of which achieves P_{cr}. A rapid transition (a kind of skipping) of the system to a new locus of equilibrium is observed at the time. When the skipping is done, then the v parameter is going to rise slightly. If we suppose that the parameter P decreases in time, then at $P = P^*_{cr}$ the system will skip to the initial state of equilibrium. During a successive decrease of loading the characteristic point of our curve will slide down, all the time being located in its starting section.

The equilibrium positions described by the depressing parts of the curve at hand are unstable and have no physical realization for the considered character of loading. Each locus of equilibrium belonging to the increasing parts in (P^*_{cr}, P_{cr}) are locally stable, because for sufficiently large excitations the qualitative change of the form of equilibrium state (skipping) can transform to the so-called loss of stability in the "large" sense. Of necessity, values for the realization of excitations skipping are defined by x-axis values of points of the depressing part of the P–v curve.

In a similar way we deal with two critical loadings: upper P_{cr} and lower P^*_{cr}, defining skippings after loading and unloading. In some systems the lower critical value equals zero or can take a negative sign.

In the third case the lack of locus of an equilibrium for $P > P_{cr}$ is the most characteristic feature. Therefore, as critical loading value is approached, a loss of stability cannot be observed but rather loss of the system's equilibrium state. The equilibrium state is not sustainable and the system goes to a movement state. Equilibrium positions that are possible at $P < P_{cr}$ can be either unstable without any restrictions (see Fig. 2.3b) or locally stable if a depressing part of the P–v curve can be

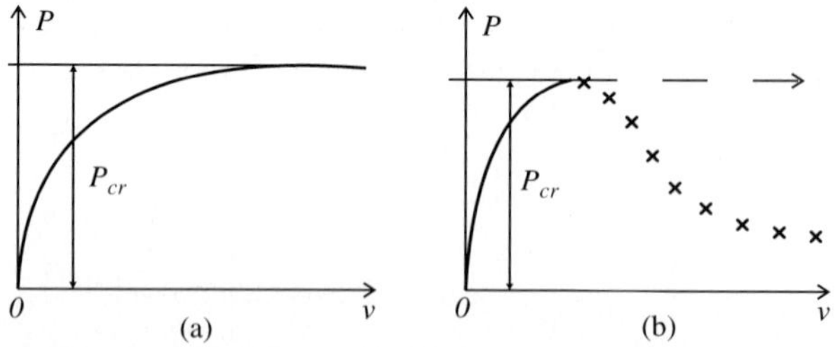

Fig. 2.3 Equilibrium position curves (systems of third type): (**a**) one locus of equilibrium at $P < P_{cr}$; (**b**) two positions of equilibrium at $P < P_{cr}$

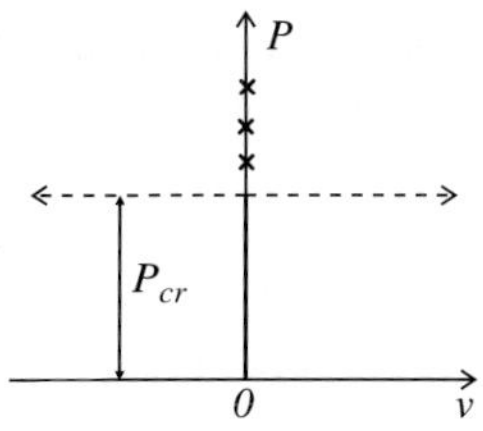

Fig. 2.4 Equilibrium position curves (systems of fourth type)

distinguished. In the second alternative case some larger excitations should force our system, which is sufficiently determined by x-axis coordinates of the depressing part of the characteristic.

Despite the existence of an initial state of equilibrium locus for $P > P_{cr}$, in the last case the position of the state of interest cannot be realized.

In some peculiar cases the curve of equilibrium locations can be represented in the form shown in Fig. 2.5, where a few values of forces can correspond to a single displacement, but no skippings are observed during a monotonically increasing parameter of loading.

Miscellaneous properties of mechanical systems and curves of the P–v type are very helpful in explanations of various existing methods, which are used to determine some particular critical states. If there is a need for long-term investigations of critical states, then the three main variants of *stability methods* can be applied quite successfully.

From a historical scientific perspective, the first variant of the analysis was proposed by Euler. In his conception, a system's instability was characterized by an equilibrium form that is closely similar to the initial one of the system. The problem of examining the initial form of the system's equilibrium is replaced by the problem of bifurcational determination of the form of initial equilibrium. A critical loading is by definition the smallest of all acting loadings that are associated with any newly created forms of equilibrium. In practical terms, an application of the stability criterion reduces to a problem that is solved by estimation of eigenvalues as well as of the corresponding eigenvectors being, in addition, connected with the analysis of any linearized differential equations. In this situation, the equations are derived from nonlinear equations, but the method of their estimation uses a variational approach or an analysis of two post-buckled equilibrium states. As opposed to a dynamic criterion, the static one is more restrictive with respect to its application possibilities. The loss of stability static criteria are used principally for conservative systems.

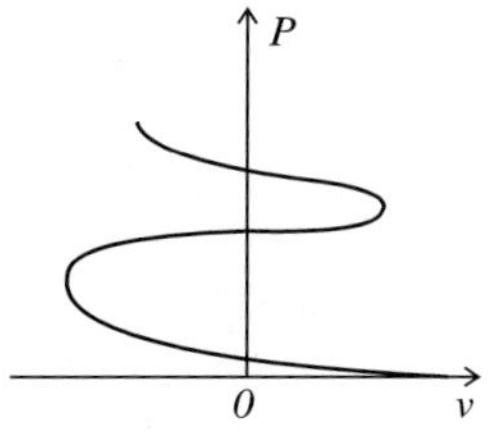

Fig. 2.5 One of the possible forms of equilibrium positions

Of course, an energetic criterion states a viable alternative of its static counterpart. The energetic criterion is based on the two fundamental principles of mechanics of a physical continuum: the principle of virtual displacements and the principle of virtual modifications (changes) of a stress state of the analyzed system. A condition of stationarity of the total potential energy of a system $\delta\Pi = 0$ can be directly found from the principle of virtual displacements. On the basis of that condition, from all the displacements satisfying boundary conditions the displacements satisfying equilibrium conditions are the only ones contributing to the quantity of stationary energy. A condition of stationarity of an extra energy can be drawn from the principle of stress state virtual changes. On the basis of that condition from all of options of virtual stresses satisfying equilibrium equations and boundary conditions, the stresses that satisfy strain continuity conditions are the only ones contributing in addition to the quantity of stationary energy.

The condition $\delta\Pi = 0$ allows for a definition of the system's equilibrium locus. The stability of the equilibrium state can be estimated by means of the Lagrange-Dirichlet theorem. If the analyzed equilibrium locus is stable then the total potential energy reaches a maximal value $(\delta\Pi = 0,\ \delta^2\Pi < 0)$; there is a constant level of potential energy that is assigned to the locus of an equilibrium $(\delta\Pi = 0,\ \delta^2\Pi = 0)$, where $\delta\Pi,\ \delta^2\Pi$ denote the first and second variation of the system's entire energy, respectively. Lagrange was the first to formulate the mentioned theorem, and its proof for systems of finite degrees of freedom was worked out by Dirichlet. The theorem can also be expressed another way [335]: for a continuous total energy of a system the locus of the system's equilibrium together with inclusion of any conservative and non-conservative forces is stable when the potential energy is positively defined. In practice, the second variation's investigations of all possible virtual displacements turn out to be very difficult.

Linear problems of stability usually consider the following necessary condition that is decisive for conservation of an equilibrium locus:

$$\delta^2\Pi = 0 \tag{2.1}$$

Let us write the system's total energy for the mentioned equilibrium locus in the form of a sum of two components, the total energy and its incremented term that has appeared during analysis of a post-buckled equilibrium state:

$$\Pi = \Pi_0 + \delta\Pi + \delta^2\Pi + \cdots, \tag{2.2}$$

where $\delta\Pi,\ \delta^2\Pi$ are the components of energy, in which first and second elements of any additional conditions that are relevant to a displacement in the post-buckled system have been included. The value $\delta\Pi = 0$ is the first variation of energy of the initial equilibrium state. Making a variation of (2.2) and assuming $\Pi_0 = 0$, we have

$$\delta(\delta^2\Pi) = 0. \tag{2.3}$$

The above condition is a kind of variational formulation of the criterion of static stability because some adequate differential equations of the criterion of static stability directly result from it. It can be noted that when an investigation of stability states the aim, conditions like (2.1) or (2.3) can be fully used. The mentioned criteria indicate that the energy increasing in a point of bifurcation is either the positively or zero defined function. A condition of existence of the second variation's minimum can be written as follows:

$$\delta^2\Pi = \min, \tag{2.4}$$

that in accordance with an energy incrementation can be simply deduced from a similar positive or zero value function. Let us observe, however, that if in a point of bifurcation the following condition $\delta^2\Pi = 0$ holds, then in the neighborhood of that point a half-positive defined function will take values greater than zero.

The described method is called the Euler method and serves for the purpose of investigation of systems for which the characteristic curve, shown in Fig. 2.1, has P–v dependence. It is possible then to determine P_{cr} value, but the curve defining the new equilibrium positions remains uninspected for $P > P_{cr}$. The essential feature of the Euler problem formulation is the introduction of an ideal system realization. For instance, after analysis of deflection a starting straight line of a rod is assumed and a compressing force should then be perfectly imposed along that rod's axis.

In the second variant of the analysis, i.e., by applying a static method, some real features (imperfections, inaccuracies, perturbations) such as initial deflections, initial non-ideal application of loads or any other additional external forces are introduced at the very beginning. The linearized forms of the equations are then used for the solution of the problem. For example, during post-buckling analysis of a rod the P–v curve can be found and drawn as well, and similar to the one in Fig. 2.3a. The curves of interest will differ from that in the figure, especially with respect to the lack of an assumed ideal interpretation of the problem, but the location of a horizontal asymptote will not be dependent on initial perturbations, which in the final stage leads to estimation of the same value of critical loading as in the previously described Euler method. Some inaccuracies of the considered method are certainly linked to a linearization process because the examined displacements are not small.

In many cases the P–v dependence has to be determined with a consideration of nonlinearities appearing during any large displacements. In particular, such an assumption could be observed during analysis of systems with skippings (Fig. 2.2), where geometrical nonlinearities play a crucial role.

We remember that in mechanical problems of a solid deformable body three types of nonlinearities can be distinguished:

1. Geometrical nonlinearity that basically appears during any large displacements that causes, as a consequence, elimination of the possibility of using any traditional simplifications like $\sin\varphi \approx \tan\varphi \approx \varphi$;
2. Physical nonlinearity that is linked with a deviation of linear material deformations;

3. Constructional nonlinearity that results from the existence of any constructively superposed restrictions (for example, detachment of a beam from an elastic base, variation of a contact surface, etc.).

The third variant of the static method is based on the *Lagrange theorem* focusing on the conservation principle of minimums of the total potential energy of a system:

> Being in an equilibrium position, the total potential energy of a conservative system has some stationary value, but the position will be stable if and only if a minimum energy can be assigned for it [335].

A well known pictorial illustration of such an approach is the behavioral analysis of a ball lying on a smooth surface (Fig. 2.6).

Such a ball's potential energy experiences some fluctuations that are proportional to that ball's vertical movement: the energy decreases as the ball moves down and is increasing as it moves up. All three of the surfaces in Fig. 2.6 include stationary points in which the ball can remain in balance. The bottom point of the deflected surface (Fig. 2.6a) relates to the ball's minimal potential energy and the position of equilibrium will be stable.

The top of the convex surface (Fig. 2.6b) refers to a stationary point having maximal potential energy and the equilibrium locus is unstable at this point. An intuitive approach as well as a simple experiment support the above assumption because the ball will roll down after any small displacement of it from that point.

A stationary point situated on the saddle surface (Fig. 2.6c) is called a transitional point (placed anywhere between extremal values). Because of the absence of a minimal potential energy at this point, the appropriate equilibrium locus will not be stable. *The equilibrium locus of a conservative system will be unstable for all cases in which the total potential energy of a system takes some constant but not minimal value.*

The total energy of the deformable conservative system is composed of both the energy U of an elastic deformation and the potential Π of external forces:

$$\mathbb{E} = U + \Pi. \tag{2.5}$$

In the equilibrium (not necessarily stable!) condition, the total potential energy does not change. The necessary condition for obtaining a minimal value yields the following:

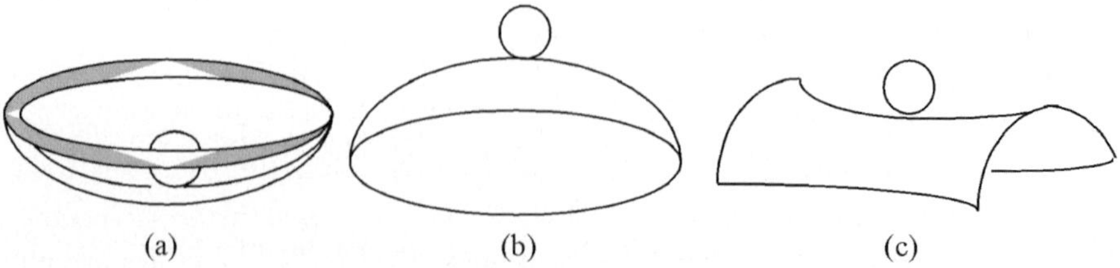

Fig. 2.6 Stable and unstable equilibrium positions

$$\delta \mathbb{E} = \delta U + \delta \Pi = 0, \tag{2.6}$$

which is called the *variational Lagrange equation.*

Let a *total energy gain* be denoted by $\Delta\mathbb{E}$ when any small deflections from the considered equilibrium position of the system occur. If the equilibrium position needs to be stable, then the following necessary and sufficient condition

$$\delta \mathbb{E} > 0 \tag{2.7}$$

must hold for any optional but sufficiently small system deflections from an equilibrium position. If we assume that for a predefined loadings and system fixing conditions there exist such system deflections from an equilibrium position, which results in satisfying (2.7), then even the equilibrium position will be unstable.

Despite both stable and unstable equilibrium position terms, another term of indifferent equilibrium position is introduced.

The increase of total energy $\Delta\mathbb{E}$ is most often estimated to an accuracy of the second power of small deflections from an equilibrium position leading as a consequence to the condition $\Delta\mathbb{E} = 0$.

Nevertheless, by finding $\Delta\mathbb{E}$ with a higher accuracy the sign of the value can be determined, explaining thereby the real features of the analyzed "indifferent equilibrium" position.

2.2 Two Fundamental Forms of the Energetic Criterion of Bifurcational Stability Loss

Let us analyze a deformation of a conservative system and assume that fixation reduces displacement of the system being investigated as a solid body. All external loadings increase proportionally to P. The total system energy is governed by (2.5), but the potential energy found in an equilibrium position reaches a stationary value. Because displacement of the whole system is not possible and the potential energy of an elastic system's deformation is always greater than zero, the equilibrium position of the non loaded system is stable. Let us assume that for $P = 0$ condition $\Delta\mathbb{E} = \Delta U > 0$ is satisfied for a system's deviations from a non-loading state. Let us investigate thereafter stability of the system's stress-deformation state, which is described in the theory of elasticity by linear equations. Such a position will be called the *initial equilibrium state*. It can be shown that in the neighborhood of point $P = 0$ the state is unstable. An initial equilibrium state can lose stability only if the loading parameter P will cross some critical value P_{cr}.

In other words, until $P < P_{cr}$ is valid, a sufficiently small deviation produces the inequality $\Delta\mathbb{E} > 0$. In addition, if $P > P_{cr}$ then deviations are also possible for which $\Delta\mathbb{E} \leq 0$. Accordingly, the following definition can derived.

Definition 2.1 The critical loading parameter P_{cr} will be called the *lower loadings value* P for which small system deviations are possible from an initial equilibrium state leading to the condition $\Delta\mathbb{E} = 0$.

The definition provides an opportunity to obtain the energetic criterion of loss of stability of the conservative system's equilibrium state. The basic steps of its derivation are given below.

We will take into consideration a known initial equilibrium position being described in two ways: via linear equations of the theory of elasticity and via first-order displacements measured from an initial equilibrium state. *The gain of the system's total potential energy* $\Delta\mathbb{E}$ *will be estimated with the accuracy of the value of square displacements and it will be written in the form* (2.8) *of two components with the first one not dependent on external loadings and the second one proportional to the* P *parameter*:

$$\Delta\mathbb{E} = W + PV. \tag{2.8}$$

In these cases, when studying a secondary (buckled) equilibrium state the condition $\Delta\mathbb{E} = 0$ is satisfied in the following form:

$$P = -\frac{W}{V}. \tag{2.9}$$

In agreement with the earlier definition, the smallest value of P of all the possible ones that are defined by the last equation equals the critical value P_{cr}. In the general case, $\Delta\mathbb{E}$, W, V are the *functionals* dependent on the first-order displacements forcing the system to achieve a new secondary position. Using the necessary condition of minimum of the ratio of two functionals, one obtains

$$\delta\left(\frac{W}{V}\right) = \frac{1}{V}\left(\delta W - \frac{W}{V}\delta V\right) = \frac{1}{V}(\delta W + P\delta V) = 0, \tag{2.10}$$

and hence

$$\delta(\Delta\mathbb{E}) = 0. \tag{2.11}$$

The obtained condition possesses a relatively simple mechanical interpretation. Observe that

$$\mathbb{E}_1 = \mathbb{E} + \Delta\mathbb{E}, \tag{2.12}$$

where $\mathbb{E}$ and $\mathbb{E}_1$ denote total potential energy in both initial and secondary equilibrium states. Making a variation to the above and then considering the initial position at $\delta\mathbb{E} = 0$, we obtain $\delta\mathbb{E}_1 = \delta(\Delta\mathbb{E})$. Therefore, condition (2.11) takes the form

$$\delta\mathbb{E}_1 = 0, \tag{2.13}$$

i.e., as the equilibrium condition for a post-buckled system in relation to an initial equilibrium position. The points of the graph representing the initial equilibrium position in the neighborhood of another equilibrium positions associated with the initial equilibrium position are called bifurcational points of the initial equilibrium state. *Condition* (2.13) *defines some bifurcations of initial location points, but the lowest bifurcational loading parameter* P *is assigned to the critical bifurcation point.*

For further analysis, first-order displacements bringing our systems from initial states to new equilibrium ones and present in the bifurcation point's neighborhood will be called *bifurcational displacements*. Condition (2.11), in which gain of the total potential energy $\Delta\mathbb{E}$ is determined with an accuracy of square bifurcational displacements, will be called *the energetic criterion of bifurcational loss of stability, or simply the energetic criterion (energetic principle) of elastic stability*.

Singularities related to calculation of $\Delta\mathbb{E}$ after the practical solution of problems will be examined on the basis of simple exemplary analysis of stability of a rectilinear equilibrium form of a thin rod [4]. Miscellaneous methods of estimation of the system's total potential energy variations $\Delta\mathbb{E}$ will be applied after calculation of the rod's stability via the energetic method. It is confirmed in mechanics of materials that a rod's longitudinal axis is not stretched and according to the assumption the close-up of its ends produced only by flexing is estimated. It gives the following variational estimation of total potential energy:

$$\Delta\mathbb{E} = \frac{1}{2}\int_0^\ell EJ\left(\frac{\partial^2 v}{\partial x^2}\right)^2 dx - P\lambda;\ \lambda = \frac{1}{2}\int_0^\ell \left(\frac{\partial v}{\partial x}\right)^2 dx, \tag{2.14}$$

where the first component is the potential energy of the deflected rod that has been expressed by twisting of its axis. The second component exhibits a variation of the external force potential.

Different expressions are also used in building engineering and in addition are designated to any variational calculations of the rod's total potential energy of the form

$$\Delta\mathbb{E} = \frac{1}{2}\int_0^\ell EJ\left(\frac{\partial^2 v}{\partial x^2}\right)^2 dx + \int_0^\ell N_0\varepsilon_2 dx;\ \varepsilon_2 = \frac{1}{2}\left(\frac{\partial v}{\partial x}\right)^2. \tag{2.15}$$

In the above, the second component describes the work of initial axial forces N_0 that is done during any additional extensions ε_2 created by the transversal displacements v. In relation to the starting position, both the secondary and buckled rod's positions will be measured from an initial state and predetermined by the displacements v_1 and u_2. The change of the total potential energy during the rod's transition to a different equilibrium position characterized by axial twisting will be represented by three successive components:

$$\Delta\mathbb{E} = \frac{1}{2}\int_0^\ell EJ\left(\frac{\partial^2 v_1}{\partial x^2}\right)^2 dx + \frac{1}{2}\int_0^\ell N_0\left(\frac{\partial v_1}{\partial x}\right)^2 dx$$

$$+\left\{\int_0^\ell N_0\frac{\partial u_2}{\partial x}dx - Pu_2(0) - \int_0^\ell qu_2 dx\right\}. \tag{2.16}$$

Additional displacements u_2 should meet the geometrical boundary conditions precisely fixed on the rod's boundaries. Therefore, the displacement u_2 can be analyzed at some variables associated with the rod's initial state. In general, the expression included in braces can be treated as the sum of shifting work done by the internal and external forces. As the rod's equilibrium position has been achieved, so with respect to the principle of possible displacements, the sum under consideration equals zero. For any arbitrary displacements restricted by constraints and with the inclusion of those additional u_2 the change of the total potential energy $\Delta\mathbb{E}$ does not depend on these displacements, so the following expression holds:

$$\Delta\mathbb{E} = \frac{1}{2}\int_0^\ell EJ\left(\frac{\partial^2 v_1}{\partial x^2}\right)^2 dx + \frac{1}{2}\int_0^\ell N_0\left(\frac{\partial v_1}{\partial x}\right)^2 dx, \qquad (2.17)$$

i.e., it is exactly defined by (2.15). The dependence found after solutions to some specific problems allows one to use any arbitrary functions u_2, which are usually selected with respect to their beneficial properties (with a simultaneously fulfilled geometrical boundary conditions).

If the geometrical equilibrium conditions are fixed only on the first border of the rod, then referring to u_0, i.e., if the problem of definition of the initial axial force is statically determinable, then the displacement u_2 can be chosen in a way such that an extra extension of the rod's axis ε_2 equals zero. The displacement is for that case estimated from the condition $\varepsilon_2 = 0$, i.e., from the following equation:

$$\frac{\partial u_2}{\partial x} = -\frac{1}{2}\left(\frac{\partial v_1}{\partial x}\right)^2. \qquad (2.18)$$

When during a longitudinal movements the second right rod's border is fixed, then on the basis of Eq. (2.18) one finds

$$u_2 = \frac{1}{2}\int_0^\ell \left(\frac{\partial v_1}{\partial x}\right)^2 dx - \frac{1}{2}\int_0^x \left(\frac{\partial v_1}{\partial x}\right)^2 dx. \qquad (2.19)$$

Expression (2.16) takes the form

$$\Delta\mathbb{E} = \frac{1}{2}\int_0^\ell EJ\left(\frac{\partial^2 v_1}{\partial x^2}\right)^2 dx - P\frac{1}{2}\int_0^\ell \left(\frac{\partial v_1}{\partial x}\right)^2 dx - \int_0^\ell q u_2\, dx. \qquad (2.20)$$

One can exclude an initial force N_0 from (2.16) even in this case if with respect to u_0 the geometrical boundary conditions have been fixed on both borders of the rod and it is no longer possible to keep the condition ε_2. Let, for example, the following boundary conditions $u_0(0) = \overline{u_0}$ i $u_0(\ell) = 0$ be analyzed, where $\overline{u_0}$ is a proposition of the initial axial displacement of the first border of the rod. The additional displacement u_2 should meet the two boundary conditions: $u_2(0) = 0$ i $u_2(\ell) = 0$.

It is worth noting that the form of mathematical representation of the total energy changes (2.15) remains valid also for the considered problem, though in the analyzed situation an initial axial force N_0 should be determined for all.

For the purpose of elimination of the initial axial force from (2.16), the following mathematical transformation is performed:

$$\int_0^\ell N_0\varepsilon_2\,dx = \int_0^\ell EF\varepsilon_0\varepsilon_2\,dx = \int_0^\ell N_2\varepsilon_0\,dx = \int_0^\ell N_2\frac{\partial u_0}{\partial x}\,dx, \tag{2.21}$$

where $EF\varepsilon_2 = N_2$.

One can treat the quantity N_2 as an extra axial force that appears in the rod during its deflections from the rectilinear form of equilibrium.

Integrating the last expression by parts, one can obtain

$$\int_0^\ell N_2\frac{\partial u_0}{\partial x}\,dx = N_2u_0\Big|_0^\ell - \int_0^\ell \frac{\partial N_2}{\partial x}u_0\,dx =$$

$$0 - \frac{\partial^2 u_0}{\partial x^2}N_2(0) - \int_0^\ell \frac{\partial N_2}{\partial x}u_0\,dx\,. \tag{2.22}$$

For the purpose of eliminating the initial displacements u_0 that appear in the last integral, the condition $\partial N_2/\partial x = 0$ has to be met. The previously given condition results in the equation used for definition of any extra displacements u_2 as the boundary conditions $u_2(0) = 0$ and $u_2(\ell) = 0$ are fulfilled.

Assuming $EJ = const$, we obtain

$$u_2 = \frac{1}{2}\left[\frac{x}{\ell}\int_0^\ell\left(\frac{\partial v_1}{\partial x}\right)^2 dx - \int_0^x\left(\frac{\partial v_1}{\partial x}\right)^2 dx\right],$$

$$N_2 = \frac{1}{2}\frac{EF}{\ell}\int_0^\ell\left(\frac{\partial v_1}{\partial x}\right)^2 dx\,. \tag{2.23}$$

Finally, Eq. (2.16) leads to an expression that does not explicitly contain the initial force N_0:

$$\Delta\mathbb{E} = \frac{1}{2}\int_0^\ell EJ\left(\frac{\partial^2 v_1}{\partial x^2}\right)^2 dx - \bar{u}_0\frac{1}{2}EF\int_0^\ell\left(\frac{\partial v_1}{\partial x}\right)^2 dx - \int_0^\ell qu_2dx, \tag{2.24}$$

where $u_2(x)$ is the function defined by Eq. (2.23).

It is assumed that *the Timoshenko form of energetic criterion* is understood as the form of the energetic criterion of the bifurcational loss of stability that is expressed by (2.14) as well as one that does not directly contain any initial stresses.

By agreement, *the Bryan form of energetic criterion* describes notation of the criterion by means of expressions like (2.15) including the initial stresses.

The aforementioned energetic method turned out to be very convenient in the rough solution of many problems focused on the stability of complex systems, but even this method cannot be treated as a universal one. It turns out that the Lagrange theorem applies solely to conservative systems whereas any true loadings are not always connected with a potential.

Although all static methods often lead to the same critical values of loading parameters, they are not completely equivalent, because they do not reflect the direct natural response to the following questions:

(i) Euler method: at which loading do any appear secondary equilibrium states?
(ii) Imperfection method (in linearized form): at which loading does a system's displacement approach infinity?
(iii) Static method: at which value of loading does the system's potential energy maintain the "minimal values" in an equilibrium position?

These questions remain unanswered.

The main flaw of the methods described is that in a sense they do not provide any general results allowing for their direct application. The *dynamic method* of stability investigation is characterized only by a full generality, which is basically reduced to the qualitative analysis of a perturbed motion resulting from changes in nature of the investigated equilibrium state. If during that motion the system still remains in the neighborhood of the equilibrium position, then such a position is called stable, and unstable for the opposite case. The dynamic method was in the past, as well as present, commonly used in the theory of stability of elastic systems, but during that long time it was considered a composite analytical object, whose results should confirm those obtained with the use of any of the static methods.

Only in the last two decades, when some abnormalities of the application of static methods have been indicated for a class of slightly different problems, has more attention been paid to the universality and applicability of the dynamic method.

2.3 Bubnov-Galerkin Methods Devoted to Shell Stability Investigations

The task of investigating shell stability naturally comprises a definition of both their forms of stability loss and their critical loadings. Integration of the stability equation in the closed form is successful only for the simpler cases of one-dimensional dynamic problems for which an initial homogeneous state is maintained when any differential equations of fixed factors come into the investigations. For the overall case any initial non-homogeneous states are reflected in the analysis of stability equations by variable coefficients, which leads to the necessity of applying various approximate methods for the solution of the discussed problems.

The partially reversible methods. The function w satisfying an equation's solution is approximated by the analytical expression that contains a finite quantity of arbitrary parameters. The series most often used has the following form:

$$w = \sum_i A_i f_i, \tag{2.25}$$

where f_i are the coordinate functions satisfying any superposed boundary as well as the internal region's conditions. The special functions are usually chosen according to the condition of the most accurate series (2.25) approximating the predicted form of the stability loss. The most advantageous ones turn out to be the complete systems of functions f_i. The coefficients A_i are determined by means of one of the approximate methods.

The most often applied methods of function approximation are the uniform approximation method, the least-squares deviation method, the collocation method, etc. However, most of the approximate methods mentioned have a significant series of drawbacks that make their application difficult. The methods of stability investigations like Rayleigh-Ritz, Bubnov, Timoshenko, Galerkin, etc. are the most popular ones.

Rayleigh-Ritz method. The method was proposed by Rayleigh to determine any vibration frequencies of elastic systems. The theoretical backgrounds of the method were worked out by Ritz but it was applied for the first time by Bryan for investigations of stability.

To approximate a function the method introduces the variational equation (2.3) and, in addition, the functions f_i should meet the geometrical boundary conditions. If, for example, the energetic criterion is used in a mixed form, then one can obtain the term $\delta^2\Pi$ of the second variation during an analysis of the two infinite closely located equilibrium positions that are specified by components (w^0, F^0, q) i $(w^0 + w, F^0 + F, q + 0)$. After a linearization the energy variation with respect to constants A_i yields the following algebraic equations:

$$\frac{\partial\left(\delta^2\Pi\right)}{\partial A_i} = 0, \quad (i = 1, 2, \ldots), \tag{2.26}$$

whose olvability predetermines the critical state of a shell. If the energetic criterion is used in a mixed form, then the function F can be defined by the solution of deformations compatibility or Eq. (2.25) with an estimated quantity of free parameters. The function F should then simultaneously satisfy the static boundary conditions. The method was used in such variant by Papkovich [240].

Bubnov method. The key to this method is utilization of the term of a function orthogonality in a predetermined region. We recall that the orthogonal functions in a predetermined region C will be called a function for which an integral of their multiplication that is fully calculated with respect to the region equals zero. A zero function, i.e., an identity function that equals zero, will be similarly orthogonal to either all f_i functions, their derivatives, or the other transformed forms connected with acting on an f_i functional operator. If the expression given in (2.25) is substituted in the equation $L(w) = 0$, then its left-hand side will not be equal to zero, but will equal some function $L(w_j)$, which is called the error function. The requirement for the function w_j to be an exact solution, i.e., to have the zero-valued error function, forces the orthogonality of $L(w_j)$ with respect to either all the functions or to any operators of those functions.

The above task is usually replaced with a much weaker one, for example, by saving in the series (2.25) only a finite number of components. In a simple variant the conditions of orthogonality are as follows:

$$\int_S L(w_j) f_i \, ds = 0. \tag{2.27}$$

Those conditions, after integration, yield the system of equations allowing estimation of A_i from (2.25).

Bubnov [62] was the first to use conditions (2.27). A short time later his method was used by Galerkin for calculation of any rod and plate deformations and for various boundary conditions [102]. Bubnov's work was unknown for a long period of time; therefore the crucial value of the cited work is the unquestionable fact that many scientists could be acquainted with his orthogonalization method.

Galerkin and Bubnov understood that the orthogonalization procedure should be conducted in a broader meaning. From the point of virtual displacement analysis the interpretation of the Bubnov method has been formulated considerably late. If $L(w_j)$ is considered as the left-hand side of the equilibrium equation and from its formal definition does exhibit the equilibrium of all internal and external forces acting on a shell element, then the components f_i successively express all of the possible displacements.

In this manner the condition (2.27) approximately defines the zero equality of the work of all internal and external forces on the virtual displacements. It is assumed that the dependence (2.27) has a complex character inclusively with a calculation of boundary integrals. In relation to the orthogonalization the Bubnov method is not concerned with any variational problems and can be with much success applied for any differential equations.

Timoshenko method. The modification of the Rayleigh-Ritz method is based on a direct application of the Dirichlet theorem [335] (see Timoshenko [293]). When the linearized equation $\delta^2\Pi = 0$ (2.1) is considered, then it implies the dependence $N = N(A_i)$ of loading on the parameters A_i. From the condition of minimal loading

$$\frac{\partial N}{\partial A_i} = 0, \quad i = 1, 2, \ldots, \tag{2.28}$$

we obtain some algebraic linear equations. In contrast, the critical loadings can be determined from the condition of existence of a nontrivial solution. In any case of conservative problems, this method leads to the same conclusions as the Rayleigh-Ritz method. The efficiency of such methods depends mainly on a proper selection of approximate functions for a predicted solution. In other words, the accuracy of any solution increases with the quantity of approximate functions. Since any upper function values are taken at the very beginning, the real critical loading can be approached. By increasing the quantity of free parameters of the unknown functions the shell is delivered with extra degrees of freedom to obtain more accurate results. It is actually possible, since a shell is a system of infinite degrees of freedom.

Galerkin methods. The description of Galerkin methods is further based on Fletcher's work [100]. Up to now, Galerkin methods were applied in solving many problems of construction mechanics, machine dynamics, fluid mechanics, etc. Applying Galerkin's projection, ordinary partial and integral differential equations have been successively analyzed. Solution of stationary and non-stationary problems as well as problems in procedures of searching for eigenvalues have also turned out to be possible with the use of Galerkin methods. In other words, *a problem that can also be described by some equations can be solved by means of one of the Galerkin method variants*.

The main characteristic features of the Galerkin method can be formulated in the following compact form. Let a two-dimensional problem be governed by the linear differential equation

$$L(u) = 0 \tag{2.29}$$

in the region $D(x,y)$ and when the conditions

$$S(u) = 0 \tag{2.30}$$

are assumed on the curve δD, which constitutes that region's boundary D. It is considered in the Galerkin method that the unknown variable u can be sufficiently well approximated by the solution

$$u_a = u_0(x,y) + \sum_{j=1}^{N} a_j \varphi_j(x,y), \tag{2.31}$$

where φ_j are the known analytical functions and the aim of introducing the u_0 function is to satisfy any boundary conditions; a_j are the coefficients that need to be determined. Substitution of (2.3.7) in (2.3.5) yields a nonzero error R of the form:

$$R(a_0,a_1,\ldots,a_N,x,y) = L(u_a) = L(u_0) + \sum_{j=1}^{N} a_j L(\varphi_j). \tag{2.32}$$

Let us define an inner product as follows:

$$(f,g) = \iint_D fg\,dx\,dy. \tag{2.33}$$

In application of the Galerkin method some unknown coefficients a_j that appeared in (2.31) are estimated in accordance with the system of equations:

$$(R,\varphi_k) = 0, \quad k = 1,\ldots,N, \tag{2.34}$$

where R defines the analyzed equation's error, and φ_k are the same as in Eq. (2.31). Because of the strong connection of the analyzed example with a linear solution's differential equation, Eq. (2.34) can be directly written in a matrix form with respect to coefficients a_j:

$$\sum_{j=1}^{N} a_j \left(L(\varphi_j), \varphi_k\right) = -\left(L(u_0), \varphi_k\right). \tag{2.35}$$

Substituting the values a_j (found by solution to Eq. (2.35)) into Eq. (2.31), the required approximate solution u_a may be determined.

The weighting errors method. Evaluation of the role of the Galerkin method against the background of other computational methods emphasizes the connections of the method with the others.

Galerkin's method belongs to a wider class of methods that are recognized as *the weighting error methods* (WEM). It can be briefly described in the following way. Let us assume that solution of the differential equation

$$L(u) = 0 \tag{2.36}$$

requires initial $I(u) = 0$ and boundary $S(u) = 0$ conditions.

Let us then introduce a rough solution u_a satisfying the following equations:

$$L(u_a) = R,\ I(u_a) = R_I,\ S(u_a) = R_b. \tag{2.37}$$

Constructing the approximate equation for u_a, selection of one of the following possibilities is allowed:

(i) The differential equation is exactly satisfied, i.e., $R = 0$. In this variant the so-called *limiting method* is taken into consideration.
(ii) Boundary conditions are exactly satisfied, i.e., $R_b = 0$. In this variant the *inner method* is taken into consideration.
(iii) Both any differential equations and the boundary conditions assumed for them are not exactly satisfied. In this variant the so-called *hybrid method* is taken into consideration.

The WEM method will be described below for the inner variant example, but the considered equations can be applied in accordance with either the inner or the hybrid variant. When the inner method is followed the approximate solution u_a can be represented in the form

$$u_a(\overrightarrow{x}, t) = u_0(\overrightarrow{x}, t) + \sum_{j=1}^{N} a_j(t)\varphi_j(\overrightarrow{x}), \tag{2.38}$$

where all of φ_j are the known analytical functions. They are often called test functions, and the expression (2.38) is called a test solution (coefficients a_j need to be found). The function $u_0(\overrightarrow{x}, t)$ is chosen possibly in an *exact* way to satisfy both initial and boundary conditions. As seen from the solution (2.38), our aim is to bring (2.36) to the form of a t-dependent differential equation. If we assume that

$\varphi_j = \varphi_j(t)$ and $a_j = a_j(\overrightarrow{x})$, then a partial differential equation will be found with elements $\overrightarrow{x}$ representing its arguments. However, if either $\varphi_j = \varphi_j(\overrightarrow{x},t)$ or a stationary problem is analyzed, then a_j are constant, causing reduction of Eq. (2.36) to a system of algebraic equations. To obtain an equation for the determination of a_j the inner product of the weighting errors is taken to equal zero:

$$(R, w_k(\overrightarrow{x})) = 0, \quad k = 1,\ldots,N, \tag{2.39}$$

so the above expression defines the introduced name of the method as the complete method. The function w_k will be called the weighting function. Estimating the unknown coefficients a_j requires having an appropriate amount of independent relations. Therefore, if the functions $w_k(\overrightarrow{x})$ are represented in an analytical form then they should be independent as well. If w_k are the parts of the absolute system of functions, then at $N \to \infty$ Eq. (2.39) allows drawing a conclusion that the R equation's error should be orthogonal to all elements of the absolute system of functions. This conclusion manages an observation showing that R converges to zero (in the interval $N \to \infty$) in the *average* sense. If such convergence holds and if (2.38) precisely defines the satisfaction of boundary conditions, then the convergence of the approximate solution u_a to the exact one (2.36) can be expected in the average sense, i.e., satisfying $\lim_{N\to\infty} \|u_a - u_e\|_2 = 0$.

The introduced convergence can be with luck compared to a uniform convergence that is defined by the condition $\lim_{N\to\infty} \|u_a - u_e\|_\infty = 0$, where

$$\|u_a - u_e\|_\infty = \max |u_a - u_\ell| .$$

Equation (2.39) is analogous to its weaker counterpart (2.36):

$$(L(u), w) = 0, \tag{2.40}$$

where w represents a general form of the weighting function.

Owing to the definition given in (2.33), the inner product used in (2.39) is continuous in the region of our interest.

Note that the inner product can be also used in a discrete form:

$$(f,g) = \sum_{i=1}^{N} f_i g_i. \tag{2.41}$$

Analogously derived considerations provide *the weighting error discrete method.* In practice, putting into the use the calculative quadratures for solving Eq. (2.39), the problem is reduced to the application of the discrete method.

Equation (2.39) is much more specific, especially with respect to a very important feature that allows joining into a single integrity many, to all appearances, not connectable methods. The difference between any separate methods is reduced to the proper choice of miscellaneous weighting functions.

2.3.1 Subdomains Method

The region in hand is now split into the n subdomains D_j that are supposed to cover each other, and one obtains

$$w_k = \begin{cases} 1 \text{ in the region } D_k, \\ 0 \text{ out of the region } D_k. \end{cases} \tag{2.42}$$

The main specific property of the method is that it reduces to an application of the classical approach using (conservation method) derivation of some partial differential equations. In fact, this method overlaps with *the method of finite volumes* commonly adopted to the problems of mechanics of gases and fluids, as well as in the theory of heat conduction. For example, during an application of the law of conservation of a mass located in some finite volume of a compressible liquid, the whole mass flow running through a surface corresponds to the diminishing rate of that flow's volume mass.

A mathematical formulation corresponding to the method of finite volume can be written as follows:

$$\int_s \rho u_n ds = -\frac{\partial}{\partial t}\iint_v \rho\, dv. \tag{2.43}$$

Using the Green formula, we have

$$\frac{\partial}{\partial t}\iint_v \left[\frac{\partial \rho}{\partial t} + \mathrm{div}(\rho \overrightarrow{u})\right] \mathrm{I}\, dv = 0, \tag{2.44}$$

which corresponds to the subdomain method applied to the finite volume v. As the representation of finite differences is analyzed a solution to the equation (2.38) is given in an implicit form.

2.3.2 Colocation Method

If the weighting function is proposed as

$$w_k(\overrightarrow{x}) = \delta(\overrightarrow{x} - \overrightarrow{x}_k), \tag{2.45}$$

where δ represents the Dirac delta function, then a solution to Eq. (2.39) is reduced to the condition $R(\overrightarrow{x}_k) = 0$. A similar property can be matched for most finite differences methods: *collocation method in an extremal point*, *methods of low order or orthogonal collocations*, and many others.

2.3.3 Least-Squares Method

In this method a weighting function is expressed in the following form:

$$w_k = \frac{\partial R}{\partial a_k}, \tag{2.46}$$

where the coefficients a_k appearing in (2.38) are quite unknown. Assuming that all of w_k are in agreement with (2.46), which is equivalent to the condition of choosing a_k, the inner product possesses a minimum. The proposed method is particularly useful in the solution to many stationary problems. Moreover, it is usually expected there that the minimal square representation of an equation's solution error implies a smallness of the expression $||u_a - u_e||_2$. An application of the least-squares method to the non-stationary problems represented by two forms visible in (2.36) and (2.37) does not have any deeper justification.

For the aim of right strengthening of the method under consideration, it is mandatory to introduce two kinds of test functions for approximation either of a time or a spatial behaviour. Coefficients a_j in (2.38) are in such an approach constant and, in addition, the inner product (2.39) extends on integration in the *time space*. If the integration takes place in an infinite time space, then a negligible meaning is assigned for the earliest times, so at this stage the solution is becoming slightly significant.

In principle, the least-squares methods with weights could be constructed in a way to introduce an extra weight for the improvement of a solution's behavior in the much earlier time interval. However, the exact criterion connected with realization of that procedure is not obvious.

2.3.4 Method of Moments

In this case, the weighting function takes the following form:

$$w_k(x) = x^k, \quad k = 0, \ldots, N. \tag{2.47}$$

This method has been (for the first time) applied to find a solution to the problem of a non-linear diffusion. It has recovered a valuable effectiveness also in an application of the method to solve equations of a laminar thin-walled layer, where it was utilized to find an analytical representation of variations of the liquid molecules' velocity in a specific diagonal direction to the thin-walled layer. Consequently, a momentum-impulse method has been elaborated that was successively exploited in various computations of laminar as well as turbulent thin-walled layers.

2.3.5 Galerkin Method

In this case, the analyzed weighting function is selected from the examined family of functions in the following way:

$$w_k(\overrightarrow{x}) = \varphi_k(\overrightarrow{x}), \quad k = 1, \ldots, N. \tag{2.48}$$

It is worth noting that in spite of the application of the Galerkin method, Michlin also mentioned in his work [209] the famous Bubnov name. On the basis of a variational analytical approach, Bubnov expanded the idea of his own solution on the eigenvalues, and even proposed a principle of the weighting functions orthogonalization. Taking into consideration the fact that the Galerkin method does not require the above principle, and besides is not connected with a variational problem formulation, then it can be satisfactorily treated completely independent from the Bubnov one.

Both weighting and test functions $\varphi_j(x)$ should be selected from a number of first N functions of the complete system of functions. In essence, it states the necessary convergence condition for an exact solution but assuming that $N \to \infty$.

We need to underline once more the conditions that must be satisfied when dealing with the traditional Galerkin method.

The weighting functions w_k are selectable from the same family of functions as the test functions φ_j.

(i) The weighting and test functions should be *linearly independent*.
(ii) The weighting and test functions should be composed of *the first N components of the complete system of functions*.
(iii) The test function should satisfy *boundary* conditions (as well as some initial conditions, if they exist) in an *exact way*.

Condition (i) specifies the Galerkin method, and to ensure the possibility of obtaining any independent equations for the determination of the unknown coefficients a_j the condition (ii) should be comprehensively satisfied. Conditions (iii) and (iv) are joined by the method's effectiveness, but lack of their satisfaction reduces the effectiveness of this method.

At the very beginning, the application of the classical Galerkin method was based on either manual calculations or an electronic calculator. Therefore, a small number of any unknown coefficients a_j was usually used to ensure an acceptable calculation's accuracy.

It is enough to assume that conditions (i)–(iv) are satisfied and a naturally derived question appears: is it possible to increase the effectiveness of the Galerkin method used?

First of all, one could predict that *a selection of some test functions* will significantly influence the accuracy of the results. As observed in practice, in comparison to the application of the higher order global functions, the use of both test and weighting local functions gives some less accurate eigenvalues.

Observe that an application of the test functions being *orthogonal* in a considered region significantly reduces the entire effort needed to find a solution for a sufficiently high N. One can predict that a selection of either weighting or test functions based on information concerning forms of exact solutions may increase the method's effectiveness.

Nevertheless, it is obvious that *there is no need to introduce any errors to the Galerkin solution below a level where the desired calculation's accuracy would be conditioned by other errors. Continuing to make an allowance for taking any nonlinear terms gives the possibility of finding the more accurate solutions as N is chosen extremely small.* Today, the Galerkin method variants, i.e., a combination of the Galerkin and spectral methods, are characterized by a high accuracy of the obtained solutions with respect, to a unity computational time of the computer-assisted calculations. Some values of the required solution's accuracy are, for instance, 0.1–1% for the modern Galerkin methods and approximately 1–10% for the traditional Galerkin methods. At present, the number of unknowns usually subjected to determination ranges from 10,000 to 20,000. At this level of computation, the concrete choice of any test functions seems to be less important especially to the application of the Galerkin method being connected with the finite differences method. The right choice of test functions is very important in the spectral method, essentially influencing the convergence rate of that method.

There also exists a modification of Galerkin method called the generalized Galerkin method. In this variant of the discussed method the weighting function appearing in Eq. (2.39) is defined by the following expression:

$$w_k(\overrightarrow{x}) = P_k(\overrightarrow{x}), \tag{2.49}$$

where $P_k(\overrightarrow{x})$ is the analytical function analogous to the weighting function φ_k. It contains some additional terms or multipliers, which are necessary for any other requirements connected with the searched estimated solution. Introduction of the generalized Galerkin method has induced a cancellation of many imperfections of the classical Galerkin method. This method has taken into account linear finite elements that were applied in investigations of the convective flows, for which any estimated algebraic equations have some disadvantageous features pertaining to the solution's instability.

2.3.6 A Comparison of the Weighting Error Methods

Let us analyze further the ordinary differential equation

$$\frac{dy}{dx} - y = 0 \tag{2.50}$$

with a boundary condition $y = 1$ at $x = 0$. We are searching for the approximate solution in the region $0 \leq x \leq 1$, but the exact solution has the form $y = e^x$.

The exact solution (often called *the test solution*) is of the form

$$y_a = 1 + \sum_{j=1}^{N} a_j x^j. \tag{2.51}$$

By using that approach the test functions x^j allow one to satisfy any uniform boundary conditions.

A conscious presentation of the solutions that meet the imposed boundary conditions can be used as an exemplary realization of the traditional Galerkin method. The described method permits one to obtain as much as possible an exact solution at the settled upon number of unknowns N. Equation (2.51) can be also written in the form

$$y_a = \sum_{j=0}^{N} a_j x^j, \tag{2.52}$$

where the coefficient a_0 has been selected in a way to satisfy the boundary condition, i.e., $a_0 = 1$.

To find a solution to the considered problem various types of the weighting errors method were applied. An answer to the question was found regarding variants of the method, which provides a high accuracy and is besides easy to use in a practical realization. In the sense of this selection criterion the test functions method is equivalent to the Galerkin method.

Applying the least-squares method (2.46) with a solution consisting of any three terms and using formulas (2.51) and (2.50) has led to the following system of equations:

$$\begin{pmatrix} \frac{1}{3} & \frac{1}{4} & \frac{1}{5} \\ \frac{1}{4} & \frac{8}{15} & \frac{2}{3} \\ \frac{1}{5} & \frac{2}{3} & \frac{33}{35} \end{pmatrix} \begin{pmatrix} a_1 \\ a_2 \\ a_3 \end{pmatrix} = \begin{pmatrix} \frac{1}{2} \\ \frac{2}{3} \\ \frac{3}{4} \end{pmatrix}. \tag{2.53}$$

An application of the Galerkin method represented by (2.51) produces the following system of equations:

$$\begin{pmatrix} \frac{1}{2} & \frac{2}{3} & \frac{3}{4} \\ \frac{1}{6} & \frac{5}{12} & \frac{11}{20} \\ \frac{1}{12} & \frac{3}{10} & \frac{13}{30} \end{pmatrix} \begin{pmatrix} a_1 \\ a_2 \\ a_3 \end{pmatrix} = \begin{pmatrix} \frac{1}{1} \\ \frac{1}{2} \\ \frac{1}{3} \end{pmatrix}. \tag{2.54}$$

An application of the method of a subregions zoning onto the parts included between 0–1/3, 1/3–2/3, and 2/3–1 gives the system of equations:

$$\begin{pmatrix} \frac{5}{18} & \frac{8}{81} & \frac{11}{324} \\ \frac{3}{18} & \frac{20}{81} & \frac{69}{324} \\ \frac{1}{18} & \frac{26}{81} & \frac{163}{324} \end{pmatrix} \begin{pmatrix} a_1 \\ a_2 \\ a_3 \end{pmatrix} = \begin{pmatrix} \frac{1}{3} \\ \frac{1}{3} \\ \frac{1}{3} \end{pmatrix}. \tag{2.55}$$

An application of the collocation method, on the assumption that the error is estimated in points 0.0, 0.5, and 1.0, leads to the following system of equations:

$$\begin{pmatrix} 1 & 0 & 0 \\ 0,5 & 0,75 & 0,625 \\ 0 & 1 & 2 \end{pmatrix} \begin{pmatrix} a_1 \\ a_2 \\ a_3 \end{pmatrix} = \begin{pmatrix} 1 \\ 1 \\ 1 \end{pmatrix}. \qquad (2.56)$$

All of the solutions that were possible to find by means of the described methods have been there supplemented with "an optimal $L_{2,d}$," which was incorporated by a selection of the parameters a_1, a_2, a_3, which in a turn guarantee a minimal error of the discrete value L_2 defining a norm between the exact and the approximate solution, i.e., the minimization of $y = e^x$ has to be kept. In this case, the solution with optimal $L_{2,d}$ describes the best one of all the possible solutions (with the use of the discrete norm L_2), when an approximate solution (2.51) contains only three fixed constants. It seems more logical to compare various weighting error methods with the analyzed solution than with its exact counterpart.

Performing an analysis of errors $\|y_a - y\|_{2,d}$ a conclusion can be drawn that both Galerkin least squares and subregions methods drive to nearly optimal results. Furthermore, the collocation method brings a solution's estimation of a bit lower accuracy, though its exactness depends on the proper choice of control points.

If one takes as the control points the Gauss type points like $x = 0$, 0.1127, 0.5, 0.8873, then an equation as well as the coherent Galerkin method's solution can be found. The result includes then a solution that was obtained with the application of Taylor series and comprising the condition of an exact solution only at point $x = 0$.

As shown in Fig. 2.7, with respect to magnitude and error distribution, the Galerkin solution is located very close to the optimal one. These solutions, which were found by means of both Taylor series and the collocation method, have in this manner all errors negative.

If a choice must be made between one of the weighting error methods and with a restriction in our considerations only to the problems of a small number of unknowns (in the case of manual computations), then the following two aspects should be respected.

First of all, it should be determined which one of the methods will be the most advantageous in accomplishing the concrete task. Table 2.1 could be very useful here.

Second, the exactness of the recovered results frequently depends on the choice of test functions. The following direction can be used here: the test functions should exactly satisfy any boundary and initial conditions and, simultaneously, should be the simplest ones to use for the adequate function selection. If a special situation occurs, then it is recommended to use the problem's symmetry. Any exact solutions of other similar problems can make easier estimation of the correct choice of the test solution.

Orthonormal functions fit very well in applications of the Galerkin method. Their most popular application of a test function is realized for polynomials.

If one assumes that even at small values of N the problem's solution is found via a computer, then adoption of the Chebyshev polynomials is recommended none the less. The right choice of the class of test functions becomes quite a creative process. Because of computer power, it presently does not have as much importance as it had earlier.

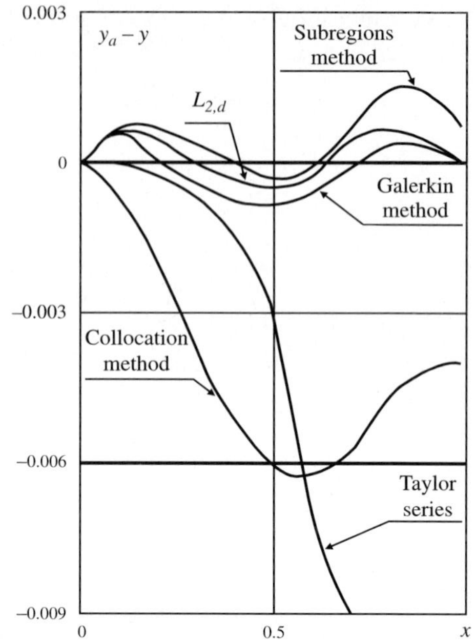

Fig. 2.7 An error distribution for the three-dimensional solution to the equation $dy/dx - y = 0$

2.3.7 Relations to Other Methods

We will now be concerned with situations when the Galerkin method relates to methods other than the weighting error methods discussed earlier. First of all, relations of the type can appear in the class of problems that are solvable with the use of the splitting variables procedure. We can meet this problem if the test functions that are used in the Galerkin method are replaced with any eigenfunctions of the analyzed problem; those obtainable by means of the splitting variables procedure are an example. Practically, such a solution is realized in the problem of N unknowns,

Table 2.1 A subjective comparison of the results received from the application of various weighting errors methods

METHODS				
Features	Galerkin	Least squares	Subregions	Collocation
Accuracy	very high	very high	high	temperate
Simplicity	temperate	low	high	very high
Remarks	Equivalent to the Ritz method	Not to apply in eigenvalues and evolutional problems	Equivalent to finite volumes methods	Orthogonal collocation leads to a high accuracy

so the application of the Galerkin method with N-termed approximations will guide the solutions coinciding with those found by means of the splitting variables procedure. In the case of the orthogonal collocation, the test functions are extracted in a way to meet the following condition of orthogonality:

$$\int_a^b w(x)P_i(x)P_j(x)\,dx = \begin{cases} 1 \text{ for } i = j, \\ 0 \text{ for } i \neq j, \end{cases} \tag{2.57}$$

at which zeros of the function $P_N(x)$ are taken as the collocation points. If the Galerkin method for weighting functions $w(x)$ and the test functions $P_j(x)$ could be applied in finding a solution of the linear differential equation $L(u) = 0$, then the inner product (2.39) takes the form

$$\int_a^b w(x)P_k(x)L(u)\,dx = 0. \tag{2.58}$$

It seems that a scheme of numerical integration could be elaborated. By means of the scheme the function's zero values $P_N(x)$ would be taken in the capacity of some node points and the integral in (2.58) then would be calculated in an exact way. It means that instead of (2.58) one can use the following finite sum:

$$\sum_i H_i w(x_i)P_k(x_i)L(u(x_i)) = 0, \tag{2.59}$$

where the coefficients H_i are weights in the quadrature scheme. It must be pointed out that the solution to equation $L(u) = 0$ with application of the orthogonal collocation method can be represented in the form

$$L(u(x_i)) = 0. \tag{2.60}$$

It goes without saying that a solution found by means of the orthogonal collocation method coincides with its Galerkin method equivalent.

The most crucial relation occurs between the Galerkin and Rayleigh-Ritz methods. It has been shown in a work [98] by Finlyson that there always exists a possibility to fit a Galerkin method correspondingly to a Rayleigh-Ritz variational method, though the inverse rule is not true. There are many examples of problems for which an appropriate variational principle does not exist even for cases when the Galerkin method can be successfully applied. For the purpose of a deep explanation of the equivalence of both the Galerkin and a variational method, more attention should be paid to the following principal observation. The approximate solution to a variational problem that has been estimated with the use of the Rayleigh-Ritz method agrees with the exact ones in a lower or upper interval, which is also obligatory for the Galerkin method. The property mentioned is useful in any error estimation procedures.

A solution of the differential equation

$$A(u) = f, \tag{2.61}$$

in which A denotes the positively defined operator, is equivalent to a task of finding of the minimum being defined by the functional

$$F(u) = (A(u), u) - 2(u, f). \tag{2.62}$$

An approximated solution to the above functional is found by means of the Rayleigh-Ritz method in the form

$$u_a(\overrightarrow{x}) = \sum_{j=1}^{N} a_j \varphi_j(\overrightarrow{x}). \tag{2.63}$$

The test functions $\varphi_j(\overrightarrow{x})$ should belong to the class of the possibly low-order complete system of functions and should be linearly independent as well. Substitution of (2.63) into (2.62) provides a relation on N of the function F. The unknown coefficients of N, means $a_1,\ a_2,\ \ldots,\ a_v$ have the form

$$F(u_a) = \left\{ \sum_{j=1}^{N} a_j A(\varphi_j), \sum_{j=1}^{N} a_k \varphi_k \right\} - 2 \left\{ \sum_{j=1}^{N} a_j \varphi_j, f \right\}$$

$$= \left\{ \sum_{j=1}^{N} \sum_{k=1}^{N} (A(\varphi_j), \varphi_k) a_j a_k - 2 \sum_{j=1}^{N} (\varphi_k, f) a_j \, . \right. \tag{2.64}$$

Minimization condition for $F(u_a)$ requires meeting

$$\frac{\partial F(u_a)}{\partial a_k} = 0, \quad k = 1, 2, \ldots, N. \tag{2.65}$$

Using (2.64), which can be rewritten as

$$F(u_a) = (A(\varphi_k), \varphi_k) a_k^2 + 2 \sum_{j \neq i} (A(\varphi_j), \varphi_k) a_j a_k - 2(\varphi_k, f) a_k,$$

we obtain

$$\frac{\partial F(u_a)}{\partial a_k} = 2 \sum_{j=1}^{N} (A(\varphi_j), \varphi_k) a_j - 2(\varphi_k, f) = 0$$

or

$$\sum_{j=1}^{N} (A(\varphi_j), \varphi_k) a_j = (\varphi_k, f), \quad k = 1, 2, \ldots, N,$$

which with respect to the coefficients a_j reduces our problem to a linear system of equations. If the Galerkin method is applied to Eq. (2.61) and the second expression (2.63) could be treated as the testing solution, then as a consequence the last equation results directly. It means that the Galerkin and Rayleigh-Ritz methods are equivalent to each other. On the other hand, the convergence of the solution obtained by means of the Rayleigh-Ritz method to the exact solution with an infinite increase of N in (2.63) is commonly known [132], if only φ_j in the expression are the elements of the complete system of functions.

It is also worth discussing the connection of the Rayleigh method with the Fourier transform. Let us examine the equation

$$A(u) = f, \tag{2.66}$$

where A is a symmetric and positively defined operator. A test solution regarded to the Galerkin method is as follows:

$$u_a = \sum_{j=1}^{N} a_j \varphi_j(x). \tag{2.67}$$

This solution is based on test functions being orthogonal in the energetic sense [200], i.e., functions

$$(A(\varphi_j), \varphi_k) = \begin{cases} 0 \text{ for } j \neq k, \\ 1 \text{ for } j = k. \end{cases} \tag{2.68}$$

Application of the Galerkin method to Eq. (2.66) produces

$$\sum_{j=1}^{N} (A(\varphi_j), \varphi_k) a_j = (f, \varphi_k), \tag{2.69}$$

or after using conditions (2.68), where $a_k = (f, \varphi_k)$, the solution (2.67) takes the form

$$u_a = \sum_{j=1}^{N} (f, \varphi_j) \varphi_j(x). \tag{2.70}$$

If with respect to energy an energetically orthonormal system of functions φ_k develops a complete system, then for an arbitrary function u (i.e., for an arbitrary solution to Eq. (2.66)), the Fourier series has the form

$$u_f = \sum_{k=1}^{\infty} (A(u), \varphi_k) \varphi_k(x), \tag{2.71}$$

and are convergent with respect to the energy to u. Convergence in an energetic sense indicates that

$$(A(u - u_f), (u - u_f)) \xrightarrow[N \to \infty]{} \varepsilon, \tag{2.72}$$

where ε is an arbitrarily chosen positive constant value. Condition (2.68) yields that the Galerkin solution (2.70) is the Fourier solution (2.71) truncated at some finite value N. As a result, the equivalence with a truncated Fourier series causes an energetic convergence of the Galerkin solution. Instead of the fully advantageous decomposition of the energetic orthonormal functions any constructional difficulties of such functions appear.

2.3.8 Theoretical Properties

After solution of problems that can be formulated in an equivalent variational form, those referred to the Rayleigh-Ritz convergence properties can be transferred to the Galerkin method. Let us suppose that our task is focused on the solution of the following operational equation:

$$A(u) = f, \tag{2.73}$$

where an operator A is symmetrical and positively defined. One can demonstrate [200] that Eq. (2.73) has only one solution. Moreover, the problem of the solution to the equation can be substituted by the problem of searching for the function that minimizes the following functional:

$$F(u) = (A(u), u) - 2(uu, f). \tag{2.74}$$

Using any energetic product's similarities, the energetic product connected to the A operator can be presented in the form

$$[u, v] \equiv (A(u), v). \tag{2.75}$$

Subsequently, if u_e represents an exact solution to Eq. (2.73), then (2.74) can be written as

$$F(u) = [u, u] - 2[u_e, u], \tag{2.76}$$

or

$$F(u) = [u - u_e, u - u_e] - [u_e, u_e], \tag{2.77}$$

or

$$F(u) = \| u - u_e \|_A^2 - \| u_e \|_A^2, \tag{2.78}$$

where the symbol $\| \|_A$ denotes the energetic norm defined by

$$\|u\|_A = (A(u), u)^{1/2}. \tag{2.79}$$

If $u = u_e$, then the functional $F(u)$ reaches the minimum, but, in fact, this minimal value is proportional to energy. The energetic norm $\|u\|_A$ is said to be finite if the operator A is positive and limited from below, as well as if the free term f has a finite norm. If A is the positive and limited from below the operator

$$(Au, u) \geq \gamma^2 \| u \|_2^2, \tag{2.80}$$

where γ is positive and constant, then the series of functions u_k, $k = 1, 2, \ldots$, represents the minimizing series if

$$\lim_{k \to \infty} F(u_k) = d, \tag{2.81}$$

where d denotes $\inf F(u)$, i.e., defines as low as possible the situated set F. Any sequence u_k satisfying condition (2.81) is with respect to energy convergent to the solution of Eq. (2.73). The term of an energetic convergence means, that $u_k(x)$ converges to u_e if

$$\| u_k - u_e \|_A \to \varepsilon \quad \text{at} \quad k \to \infty, \tag{2.82}$$

where ε is the arbitrarily chosen small positive constant. It could be shown that the Rayleigh-Ritz method allows one to obtain the series of functions $u_k(x)$ convergent in the energetic sense to u_e within the condition that u_e is the finite energy solution. Making the proof of convergence of the Rayleigh-Ritz method, the test functions in (2.63) should satisfy these two conditions:

(i) The functional series $\varphi_1, \varphi_2, \ldots, \varphi_j, \ldots, \varphi_N(x)$ is the complete system with respect to energy.
(ii) The functions φ_j are linearly independent. Condition (i) is strong in a mathematical sense but, as Michlin proves, it can be also partially weaken.

It was shown earlier that the Rayleigh-Ritz solution that minimizes functional (2.74) coincides with the Galerkin solution to Eq. (2.73). It results from the observation that a class of problems described by Eq. (2.73) is characterized by convergence properties of both Rayleigh-Ritz and Galerkin solutions.

In the case of the Rayleigh-Ritz method and resulting from the Galerkin method, Kantorowich and Krylov [132] estimated an maximal error of the N element solution to the ordinary differential equation:

$$\frac{d}{dx}\left(p\frac{dy}{dx}\right) - qy = f \tag{2.83}$$

in region $0 \le x \le 1$ with boundary and additional conditions $y(0) = y(1) = 0$; $p(x) > 0$, $q(x) \ge 0$, respectively. According to Kantorovich and Krylov, the solution to the problem is equivalent to the following functional's minimization:

$$I(y) = \int_0^1 \left[p\left(\frac{dy}{dx}\right)^2 + qy^2 + 2fy \right]. \tag{2.84}$$

Solutions obtained by the Rayleigh and Ritz method (and that of Galerkin also) in the case of Eq. (2.83) will be convergent, as shown by Kantorovich and Krylov [132] to an exact solution if and only if the test solution

$$y_a = \sum_{k=1}^{N} a_k \varphi_k(x) \tag{2.85}$$

consists of any test functions belonging to a complete system of functions. The choice of function φ_k for the considered problem should satisfy the following inequalities:

$$\left| y(x) - \sum_{k=1}^{N} a_k \varphi_k(x) \right|_{N \to \varepsilon} < \varepsilon,$$

$$\left|\frac{dy}{dx}(x)-\sum_{k=1}^{N}a_k\frac{d\varphi_k}{dx}(x)\right|_{N\to\varepsilon}<\varepsilon, \tag{2.86}$$

where ε is arbitrarily small positive constant. Kantorovich and Krylov analyzed the following test functions:

$$\varphi_k=\sin k\pi x\ \varphi_k=(1-x)x^k,\quad k=1,2,\ldots,N. \tag{2.87}$$

If we assume that sine functions are the ones we are testing, then we obtain the following estimation of the error's overfilled interval:

$$\|y_a(x)-y(x)\|_\infty\leq\frac{L}{N+1}, \tag{2.88}$$

where

$$L=\left\{\frac{\|p\|_\infty+\|q\|_\infty}{(N+1)^2\pi^2}\right\}^{1/2}\left[\left\|\frac{dp}{dx}\right\|_\infty+\frac{\|q\|_\infty}{\pi}+\pi p_{\min}\right]\left\{\frac{\int_0^1 f^2dx}{2\pi^4 p_{\min}}\right\}^{1/2}$$

and

$$\|\,\|_\infty\equiv\max|\,|.$$

A more accurate estimation of the error interval is made for the particular case $p(x)=1$, so we have

$$\|\,y_a(x)-y(x)\,\|_\infty\leq\{\,1-0,25\,\|\,q\,\|_\infty\,[6\pi(N+1)]^{-1/2}\times$$
$$\left[1+\frac{4\sqrt{2}}{N^{3/2}\pi^2}\right]\left[1-\frac{\|\,q\,\|_\infty}{\pi^2(N+1)^2}\right]^{-1}\}A^{1/2},$$

where

$$A=\frac{2\|\,q\,\|_\infty}{\pi^3(N+1)^3}\left[\int_0^1\frac{f^2}{q}dx\right]. \tag{2.89}$$

Michlin [209] considered an equation generally overlapping with a set of test functions defined by Eq. (2.83). However, with slightly smaller requirements devoted to the test functions, they are generally precised by expression (2.85). We set a condition to secure that the system φ_k is the complete one, which confirms that Eq. (2.85) should be sufficiently convenient to model any continuous functions and its derivatives to achieve the mean convergence on the interval $0\leq x\leq 1$ satisfying the condition

$$\lim_{N\to\infty}\int_0^1\left|\frac{dy_a}{dx}-\frac{dy}{dx}\right|^2dx<\varepsilon, \tag{2.90}$$

where ε is arbitrarily small positive constant. Michlin showed by an application of the Galerkin method that y is uniformly convergent to y. In the next step he considered the more general differential equation

$$(-1)^m u^{(2m)} - \lambda K(u) = f(x), \tag{2.91}$$

in which $K(u)$ is the linear differential operator of order $2m-1$, and $u^{(2m)}$ denotes a short notation for d^{2m}/du^{2m}.

The solution is searched in region $a \leq x \leq b$ at the following boundary conditions:

$$\begin{aligned} u(a) &= u'(a) = \ldots = u^{(m-1)}(a) = 0, \\ u(b) &= u'(b) = \ldots = u^{(m-1)}(b) = 0. \end{aligned} \tag{2.92}$$

The operator A_0 is defined in the following manner:

$$A_0(u) = (-1)^m u^{(2m)}. \tag{2.93}$$

A Galerkin type solution is sought because the test functions are a main part of the complete system in H_0, i.e., in the Hilbert space with a finite energetic norm. In other words, we predict that the Galerkin solution is convergent along the operator's A_0 energy. According to Michlin, the Galerkin solution is uniformly convergent to the derivatives of order $m-1$, as well as for the derivative of order m the mentioned convergence results, in the mean sense, in the positiveness and limiting from below of the operator A_0.

2.3.9 Computational Advantages of Galerkin Methods

The methods that have mainly changed the traditional Galerkin method developed in two independent branches, although the common feature of both cases was to secure maximal effectiveness of computations.

The first direction is relevant to the Galerkin and finite elements methods and has developed around an idea of local functions, i.e., weighting and test functions of low orders. In this field, the most often used mathematical types of functions were established by polynomials.

In advance, the entire effort made on the ground of construction of the combined polynomial-type solutions related from somewhere else to Fourier series were not entirely successful. When these piecewise linear polynomials were chosen as the weighting or test functions, then the obtained algebraic equations resulting from calculations of first- and second-order derivatives coincided with any adequate finite difference expressions.

If all the weighting functions could be referred to any well precised geometrical point, then a linear dependence of the required algebraical equations can be found.

If we restrict our study to the introduction of weighting functions of which region of specification is closed by a given mesh node, then we obtain a system of algebraical equations in matrix form. The main feature of the system is that the number of operations required for its matrix factorization as well as during evaluation of any nonlinear terms will be of the order $O(N)$.

By introduction of a test solution defined by finite elements and with the use of the isoparametric approach a direct solution of problems for the objects (any regions) of complex shapes is possible, particularly for other objects having boundaries that are not coincident with coordinate axes.

The second direction is connected with spectral methods, which similarly to the traditional Galerkin method belong to the *global* methods. It means that both our selected weighting and test functions comprise the whole solution region.

If we consider that the weighting and test functions are orthogonal, then with regard to a definition, the weighting functions will be linearly independent equally to the obtained algebraical equations. Except where terms of the matrix are calculated as any products of both weighting and test functions, then in the case of orthogonal functions we obtain some diagonal matrix elements.

Owing to the above, an economical aspect of the method is apparent, consisting in a reduction of dimensions of the manipulated matrices. In the case of computation of any nonlinear terms, it in addition introduces a transformation allowing in a physical surface calculation of the nonlinear factors.

2.3.10 Summary

In conclusion, some essential remarks on the Bubnov–Galerkin type method are formulated below:

1. From the point of view of the normalization procedure the Bubnov method is not connected to any variational problem and can be applied to any differential equations (see discussion on p. 56).
2. We wrote about the role of Galerkin in the application and propagation of the Bubnov's method on p. 56.
3. In 1941 when Galerkin (1871–1945) was alive Papkovich had written [239, p. 838] "It demands to point out that this often used method assigned to the Galerkin's name was for the considered problem applied a bit earlier by Bubnov and some other authors. It was long before manifesting of Galerkin's work in 1915."
4. In the *Collection of Galerkin's Scientific Works* [103, p. 12] this sentence could be found: "Two years before publication of Galerkin's work Bubnov has proposed a similar approach."
5. Grigoluk [116] noted that Bubnov in his original work indicated some differences between the two orthogonalization and energetic methods. In a sense he

proposed an alternative method. It should not be strange that on this background Bubnov did not merge his method with any variational task. The Bubnov method was originally formulated as the method of pure orthogonalization. It constitutes the great achievement that ended the existing domination of Rayleigh and Ritz's views and their influence on a wide circle of mathematicians and mechanical engineers, because the studied problems had been treated only in the range of behavioral analysis of conservative systems. Thanks to Bubnov's approach it was possible to provide a solution to the non-coupled differential equations and to analyze the non-conservative systems described by differential equations. In other words, for the first time Bubnov formulated an algorithm of solving such systems that could not been solved within the range of the concept proposed by Rayleigh.

6. It is clear on the basis of the performed considerations that the method known in the former Soviet Union literature as the Bubnov-Galerkin method and, in contrast, in the West as the Galerkin method, should be called the Bubnov method, which would be fully fair.

Today we are dealing with a huge number of particular aspects of application of the Bubnov orthogonalization method. To apply the Bubnov method correctly it is mandatory to follow some important laws and rules. Disregarding them can produce incorrect results. For instance, identical transformations of differential equations cannot be made and, in the same sense, division or multiplication of these equations by a number argument is not permitted as well. Formally, either an increase or decline of any equation's order causes a flaw in the character of an orthogonalization and with respect to any functions or their derivatives the process must be also taken into consideration.

It is assumed in the original formulation of the Bubnov method that boundary conditions of an approximated function are met. However, it is presently known that Bubnov's algorithm can be used in a slightly different modification. For example, after the post-buckled rod problem is solved the algorithm under investigation directly results from the Lagrange principle of possible displacements, where at an equilibrium point a variation of potential of the internal or external forces equals zero and after integration by parts the term regarded to the problem's boundary conditions appears. It suggests a possibility of allowance of those up-to-now not satisfied boundary conditions, which are contained in the expression describing any approximate functions, but the analyzed orthogonal differential equation being in connection with a definite boundary integral needs to be equalized to zero, and where boundary conditions are not met.

The Bubnov method is now popular and many works have been devoted to either its generalization or improvements (see [115]).

On the basis of the Bubnov method we will describe in the next part of this section an algorithm of investigation of the stability of shells.

However, respecting existing tradition, we will henceforth call the method the Bubnov-Galerkin method.

2.4 Bubnov-Galerkin Method of High-Order Approximations and the Numerical Algorithm

During solution of problems of static stability we make use of the so-called *third variant* of the static method. Using a direct solution to some nonlinear equations we will principally focus here on presentation of another alternative method leading to the determination of critical loadings. In this case, there is no need to partition the problem at hand onto the first concerning analysis of the shell's initial state and the second corresponding to the analysis of the problem's stability as it takes place after using the classical method of static analysis. Critical loadings are determined on the basis of some critical points simultaneously with the use of q–v characteristics or in some bifurcational points of a nonlinear solution. This variant of the solution often turns out to be more difficult during any realizations of numerical calculations, but allows one to obtain more information about that shell's behavior. Applying this method, one can find either lower or upper values of critical loading. First low critical loading is determined by the loading leading to a post-buckling that is directed to outside and that appears there from the first over-critical equilibrium position. The loading, in comparison to the traditionally estimated low critical loading as the smallest one from all possible lower loadings, can be assumed as a characteristic value that is useful in estimations of the stability of shells.

For the purpose of finding a solution to nonlinear equations, one can use the methods of numerical integration, various modifications of the Newton method, methods of reducing boundary problems to the Cauchy problems, methods of successive approximations, finite differences methods, Rayleigh-Ritz methods, Bubnov method, etc.

To investigate the influence of non-homogeneity of a shell on its behavior, we will assume a similar kind of problem for which boundary conditions take the form

$$w=0,\ \frac{\partial^2 w}{\partial x^2}=0,\quad F=0,\quad \frac{\partial^2 F}{\partial x^2}=0 \text{ at } x=0;1,$$

$$w=0,\ \frac{\partial^2 w}{\partial y^2}=0,\quad F=0,\quad \frac{\partial^2 F}{\partial y^2}=0 \text{ at } y=0;1. \tag{2.94}$$

For convenience, let us denote the left-hand side of Eqs. (1.57)–(1.58) by Φ_1, and Φ_2, respectively. It allows us to rewrite these equations as follows:

$$\Phi_1\left(w,F,\frac{\partial^2 w}{\partial^2 x},\frac{\partial^2 F}{\partial^2 x},q,\ldots\right)=0,$$

$$\Phi_2\left(w,F,\frac{\partial^2 w}{\partial^2 x},\frac{\partial^2 F}{\partial^2 x},\ldots\right)=0. \tag{2.95}$$

The above equations should be supplemented with the appropriate boundary conditions. An exact solution of such defined boundary problem is not possible.

Some miscellaneous approximate methods are applied to find a solution for this kind of problem formulation such as the Ritz-Timoshenko variational method, Bubnov-Galerkin method, the finite differences method, and many others.

It has been depicted in monograph [159] that for the case of investigations of the statical stability of a shell some satisfactory results are generated by the Bubnov-Galerkin variational methods and the ujeduolicic-Vlasov methods. However, with respect to the simplicity of realization, the Bubnov-Galerkin method turns out to be much more convenient. In our further investigations we will solve the initial equations by means of the higher order Bubnov-Galerkin methods. In connection with that, the functions w and F imposing any boundary conditions will be sought in the following form:

$$
\begin{gathered}
w=\sum_{i,j} A_{ij}\varphi_{ij}(x,y)\\
F=\sum_{i,j} B_{ij}\psi_{ij}(x,y)\\
i=1,2,\ldots,M_x;\quad j=1,2,\ldots,M_y.
\end{gathered}
\tag{2.96}
$$

Applying the Bubnov-Galerkin procedure to (2.95) we obtain

$$
\begin{gathered}
\int_0^1\int_0^1 \Phi_1\varphi_{vz}(x,y)\,dx\,dy=0,\\
\int_0^1\int_0^1 \Phi_2\psi_{vz}(x,y)\,dx\,dy=0,\\
v=1,2,\ldots,M_x;\quad z=1,2,\ldots,M_y,
\end{gathered}
\tag{2.97}
$$

or when respecting (2.96) we have

$$
\begin{gathered}
\sum_{vz}\left[\sum_{ij} A_{ij}I_{1,vz\,ij}-\sum_{ij} B_{ij}I_{2,vz\,ij}+\sum_{ij} qI_{3,vz\,ij}\right.\\
\left.-\sum_{ij} A_{ij}\,]\sum_{kl} B_{kl}I_{4,vz\,ij\,kl}\right]=0,\\
\sum_{vz}\left[\sum_{ij} A_{ij}I_{7,vz\,ij}+\sum_{ij} B_{ij}I_{5,vz\,ij}+\sum_{ij} A_{ij}\sum_{kl} A_{kl}I_{6,vz\,ij\,kl}\right]=0,\\
v,i,k=1,2,\ldots,M_x;\quad z,j,l=1,2,\ldots,M_y.
\end{gathered}
\tag{2.98}
$$

The sign $\sum_{vz}[*]$ at each system's Eq. (2.98) indicates that one of these equations represents a system that is composed of vz of such types of equations and the integrals of Bubnov-Galerkin method will take the form:

$$
\begin{aligned}
I_{1,vz\,ij} &= \int_0^1\int_0^1 \frac{E}{12(1-\mu^2)}\left[\frac{1}{\lambda^2}\frac{\partial^2\varphi_{ij}}{\partial x^2}\frac{\partial^2\varphi_{vz}}{\partial x^2}+\lambda^2\frac{\partial^2\varphi_{ij}}{\partial y^2}\frac{\partial^2\varphi_{vz}}{\partial y^2}\right.\\
&\quad\left.+2(1-\mu)\frac{\partial^2\varphi_{ij}}{\partial x\partial y}\frac{\partial^2\varphi_{vz}}{\partial x\partial y}+\mu\left(\left[\frac{\partial^2\varphi_{ij}}{\partial x^2}\frac{\partial^2\varphi_{vz}}{\partial y^2}+\frac{\partial^2 w}{\partial y^2}\frac{\partial^2\varphi_{vz}}{\partial x^2}\right]\right)\right] dx\,dy,\\
I_{2,vz\,ij} &= \int_0^1\int_0^1 -\left(k_y\frac{\partial^2\psi_{ij}}{\partial x^2}+k_x\frac{\partial^2\psi_{ij}}{\partial y^2}\right)\varphi_{vz}\,dx\,dy,\\
I_{3,vz\,ij} &= \int_0^1\int_0^1 \varphi_{vz}\,dx\,dy,\\
I_{4,vz\,ij\,kl} &= \int_0^1\int_0^1 L(\varphi_{ij},\psi_{kl})\,\varphi_{vz}\,dx\,dy,\\
I_{5,vz\,ij} &= \int_0^1\int_0^1 a_1\left[\left(\lambda^2\frac{\partial^2\psi_{ij}}{\partial y^2}-\mu\frac{\partial^2\psi_{ij}}{\partial x^2}\right)\frac{\partial^2\psi_{vz}}{\partial y^2}\right.\\
&\quad\left.+\left(\frac{1}{\lambda^2}\frac{\partial^2\psi_{ij}}{\partial x^2}-\mu\frac{\partial^2\psi_{ij}}{\partial y^2}\right)\frac{\partial^2\psi_{vz}}{\partial x^2}+2(1+\mu)\frac{\partial^2\psi_{ij}}{\partial x\partial y}\frac{\partial^2\psi_{vz}}{\partial x\partial y}\right] dx\,dy,\\
I_{6,vz\,ij\,kl} &= \int_0^1\int_0^1 \frac{1}{2}L(\varphi_{ij},\varphi_{kl})\,\psi_{vz}\,dx\,dy,\\
I_{7,vz\,ij} &= \int_0^1\int_0^1 -\left(k_y\frac{\partial^2\varphi_{ij}}{\partial x^2}+k_x\frac{\partial^2\varphi_{ij}}{\partial y^2}\right)\psi_{vz}\,dx\,dy. \qquad (2.99)
\end{aligned}
$$

If a transversal loading is imposed on a part of shell surface, then for some cases the integrals (2.99) with exclusion of $I_{3,vzij}$ are calculated with respect to the whole mean surface of the shell.

Finally, we find from Eq. (2.98) the system of algebraic equations of order $2(M_x * M_y)$, but the first half of those equations is nonlinear with respect to A_{ij}, while the second one is linear with respect to B_{ij}.

Imposing a load q and then solving the system of equations (2.98) using a numerical method, the A_{ij} and B_{ij} found could be substituted in (2.96), making possible the determination of the unknown functions w, F.

It is necessary to consider a crucial factor in our further considerations, which determines, in a general case, that the deflection-loading dependence becomes an equivocal function, i.e., a few values of deflection correspond to the one and the same value of loading and vice versa. There appear on the "deflection-loading" graph some characteristic loops proving the deflectional ambiguity of characteristics. Loading then must be understood in the system (2.98) as an unknown function and the deflection has to be assumed as the set value. Nevertheless, the deflection appears in the system as a composite function providing any relatively difficult analyses.

Another approach is based on the step-by-step argument presented below of taking into account an application of the approximated function of deflection to Eq. (2.96). It is assumed that in (2.98) either loading and the function of deflection, i.e., q and all of A_{ij} and B_{ij} are the unknowns, but we add to Eq. (2.98) just one equation, which defines the deflection in the earlier selected characteristic point (x_0, y_0) of the shell:

$$w_0\left(x_0, y_0\right) = \sum_{ij} A_{ij}\, \varphi_{ij}\left(x_0, y_0\right). \tag{2.100}$$

In this way, by applying $w_0(x_0, y_0)$ to (2.100) and treating loading q as well as all of A_{ij} and B_{ij} as the unknowns, we solve the system (2.98) together with (2.100). As the undefined loading q, A_{ij}, and B_{ij} are found, the substitution of these last two to (2.96) causes a final estimation of the function of deflection, which in the shell's point (x_0, y_0) coincides with the deflection in point $w_0(x_0, y_0)$.

Let us analyze the procedure in a more specific way. For that purpose, let φ_{ij}, ψ_{ij} from (2.96) be represented in the form of the two functions, where all of them depend only on a single argument and can be given in a linear combination of functions explicitly satisfying any boundary conditions of the compacted form

$$\begin{aligned} \varphi_{ij}\left(x, y\right) &= \varphi_{1ij}\left(x\right) \varphi_{2ij}\left(y\right), \\ \psi_{ij}\left(x, y\right) &= \psi_{1ij}\left(x\right) \psi_{2ij}\left(y\right). \end{aligned} \tag{2.101}$$

Putting (2.99) in (2.101) we have:

$$\begin{aligned} \varphi_{1i}\left(x\right) = \psi_{1i}\left(x\right) = \sin\left(i\pi x\right), \quad i = 1, 2, \ldots, M_x, \\ \varphi_{2j}\left(y\right) = \psi_{2j}\left(y\right) = \sin\left(j\pi y\right), \quad j = 1, 2, \ldots, M_y. \end{aligned} \tag{2.102}$$

Substituting (2.102) to (2.96) we obtain

$$\begin{aligned} w &= \sum_{i,j} A_{ij}\, \sin\left(i\pi x\right)\, \sin\left(j\pi y\right). \\ F &= \sum_{i,j} B_{ij}\, \sin\left(i\pi x\right)\, \sin\left(j\pi y\right). \end{aligned} \tag{2.103}$$

The subscripts i, j take all the possible values.

By application of the Bubnov-Galerkin method we find a system of $2(M_x * M_y)$ algebraic equations of the form:

$$\begin{aligned} &\int_0^1\!\!\int_0^1 \Phi_1\, \sin\left(v\pi x\right)\, \sin\left(z\pi y\right)\, dx\, dy = 0, \\ &\int_0^1\!\!\int_0^1 \Phi_2\, \sin\left(v\pi x\right)\, \sin\left(z\pi y\right) dx\, dy = 0, \\ &v = 1, 2, \ldots, M_x, \quad z = 1, 2, \ldots, M_y. \end{aligned} \tag{2.104}$$

The integrals of the Bubnov-Galerkin procedure can be estimated according to the equations:

$$I_{1,v} = \int\limits_{x_0-\Delta x}^{x_0+\Delta x} \sin(v\pi x)\, dx = \frac{2}{v\pi} \sin(v\pi x_0)\, \sin(v\pi\, \Delta x), \tag{2.105}$$

$$I_{2,z} = \int\limits_{y_0-\Delta y}^{y_0+\Delta y} \sin(z\pi y)\, dy = \frac{2}{z\pi} \sin(z\pi y_0)\, \sin(z\pi\, \Delta y), \tag{2.106}$$

in which x_0, y_0 denote coordinates of origin of the rectangular part of a loading imposing, and Δx, Δy are with respect to x and y of this region's half widths and lengths. In addition, one obtains

$$I_{3,vi} = \int\limits_0^1 \sin(i\pi x)\, \sin(v\pi x)\, dx = \begin{cases} \frac{1}{2}, & i = v, \\ 0, & i \neq v, \end{cases} \tag{2.107}$$

$$I_{4,zj} = \int\limits_0^1 \sin(j\pi y)\, \sin(z\pi y)\, dy = \begin{cases} \frac{1}{2}, & j = z, \\ 0, & j \neq z, \end{cases} \tag{2.108}$$

$$I_{5,vik} = \int\limits_0^1 \sin(i\pi x)\, \sin(k\pi x)\, \sin(v\pi x)\, dx$$
$$= \begin{cases} \frac{1}{4\pi} \Big[-\frac{\cos\alpha_1\pi}{\alpha_1} - \frac{\cos\alpha_2\pi}{\alpha_2} - \frac{\cos\alpha_3\pi}{\alpha_3} + \frac{\cos\alpha_4\pi}{\alpha_4} \\ \quad + \frac{1}{\alpha_1} + \frac{1}{\alpha_2} + \frac{1}{\alpha_3} - \frac{1}{\alpha_4} \Big], & \alpha_l \neq 0; \\ \left[\frac{\cos\alpha_l\pi}{\alpha_l} = 0,\ \frac{1}{\alpha_l} = 0 \right],\ l = 1,2,3; & \alpha_l = 0; \end{cases} \tag{2.109}$$

where

$$\alpha_1 = i + k - v, \quad \alpha_2 = k + v - i,$$
$$\alpha_3 = v + i - k, \quad \alpha_4 = i + k + v. \tag{2.110}$$

$$I_{6,zjl} = \int\limits_0^1 \sin(j\pi x)\, \sin(l\pi x)\, \sin(z\pi x)\, dy$$
$$= \begin{cases} \frac{1}{4\pi} \Big[-\frac{\cos\beta_1\pi}{\beta_1} - \frac{\cos\beta_2\pi}{\beta_2} - \frac{\cos\beta_3\pi}{\beta_3} + \frac{\cos\beta_4\pi}{\beta_4} \\ \quad + \frac{1}{\beta_1} + \frac{1}{\beta_2} + \frac{1}{\beta_3} - \frac{1}{\beta_4} \Big], & \beta_l \neq 0; \\ \left[\frac{\cos\beta_l\pi}{\beta} = 0,\ \frac{1}{\beta_l} = 0 \right],\ l = 1,2,3; & \beta_l = 0; \end{cases} \tag{2.111}$$

where

$$\beta_1 = j+l-z, \quad \beta_2 = l+z-j,$$
$$\beta_3 = z+j-l, \quad \beta_4 = j+l+z. \tag{2.112}$$

$$I_{7,vik} = \int_0^1 \cos(i\pi x)\,\cos(k\pi x)\,\sin(v\pi x)\,dx$$
$$= \begin{cases} \frac{1}{4\pi}\left[\frac{\cos\alpha_1\pi}{\alpha_1} - \frac{\cos\alpha_2\pi}{\alpha_2} - \frac{\cos\alpha_3\pi}{\alpha_3} - \frac{\cos\alpha_4\pi}{\alpha_4} - \right. \\ \left. -\frac{1}{\alpha_1} + \frac{1}{\alpha_2} + \frac{1}{\alpha_3} + \frac{1}{\alpha_4}\right], & \alpha_l \neq 0; \\ \left[\frac{\cos\alpha_l\pi}{\alpha_l} = 0,\ \frac{1}{\alpha_l} = 0\right],\ l = 1,2,3; & \alpha_l = 0; \end{cases} \tag{2.113}$$

$$I_{8,zjl} = \int_0^1 \cos(j\pi x)\,\cos(l\pi x)\,\sin(z\pi x)\,dy$$
$$= \begin{cases} \frac{1}{4\pi}\left[\frac{\cos\beta_1\pi}{\beta_1} - \frac{\cos\beta_2\pi}{\beta_2} - \frac{\cos\beta_3\pi}{\beta_3} - \frac{\cos\beta_4\pi}{\beta_4} - \right. \\ \left. -\frac{1}{\beta_1} + \frac{1}{\beta_2} + \frac{1}{\beta_3} + \frac{1}{\beta_4}\right], & \beta_l \neq 0; \\ \left[\frac{\cos\beta_l\pi}{\beta_l} = 0,\ \frac{1}{\beta_l} = 0\right],\ l = 1,2,3; & \beta_l = 0; \end{cases} \tag{2.114}$$

$$I_{Q,vz} = I_{1,v}I_{2,z}; \quad I_{AB,vz} = \left(z^2k_x + v^2k_y\right)\pi^2 I_{3,vi}I_{4,zj} \tag{2.115}$$

$$I_{vzijkl} = \pi^4\left[\left(i^2l^2 + j^2k^2\right)I_{5,vik}\,I_{6,zjl} - 2ijkl\,I_{7,vik}I_{8,zjl}\right] \tag{2.116}$$

Taking into account (1.86) one obtains

$$I_{11,vi} = \int_0^1 \gamma_0 \sin(i\pi x)\,\sin(v\pi x)\,dx$$
$$= \int_{x_1}^{x_2} \sin(i\pi x)\,\sin(v\pi x)\,dx$$
$$= \begin{cases} \left[\frac{\sin(i-v)\pi x_2 - \sin(i-v)\pi x_1}{2\pi(i-v)} - \frac{\sin(i+v)\pi x_2 - \sin(i+v)\pi x_1}{2\pi(i+v)}\right], & i \neq v, \\ \left[\frac{x_2 - x_1}{2} - \frac{\sin(2i\pi x_2) - \sin(2i\pi x_1)}{4i\pi}\right], & i = v, \end{cases}, \tag{2.117}$$

$$I_{12,zj} = \int_0^1 \gamma_0 \sin(j\pi y)\,\sin(z\pi y)\,dy$$
$$= \int_{y_1}^{y_2} \sin(j\pi y)\,\sin(z\pi y)\,dy$$

$$= \begin{cases} \left[\frac{\sin(j-z)\pi y_2 - \sin(j-z)\pi y_1}{2\pi(j-z)} - \frac{\sin(j+z)\pi y_2 - \sin(j+z)\pi y_1}{2\pi(j+z)}\right], & j \neq z, \\ \left[\frac{y_2-y_1}{2} - \frac{\sin(2j\pi y_2) - \sin(2j\pi y_1)}{4j\pi}\right], & j = z, \end{cases} \tag{2.118}$$

$$I_{13,vi} = \int_0^1 \gamma_0 \cos(i\pi x)\, \cos(v\pi x)\, dx$$

$$= \int_{x_1}^{x_2} \cos(i\pi x)\, \cos(v\pi x)\, dx$$

$$= \begin{cases} \left[\frac{\sin(i-v)\pi x_2 - \sin(i-v)\pi x_1}{2\pi(i-v)} + \frac{\sin(i+v)\pi x_2 - \sin(i+v)\pi x_1}{2\pi(i+v)}\right], & i \neq v, \\ \left[\frac{x_2-x_1}{2} + \frac{\sin(2i\pi x_2) - \sin(2i\pi x_1)}{4i\pi}\right], & i = v, \end{cases} \tag{2.119}$$

$$I_{14,zj} = \int_0^1 \gamma_0 \cos(j\pi y)\, \cos(z\pi y)\, dy$$

$$= \int_{y_1}^{y_2} \cos(j\pi y)\, \cos(z\pi y)\, dy$$

$$= \begin{cases} \left[\frac{\sin(j-z)\pi y_2 - \sin(j-z)\pi y_1}{2\pi(j-z)} + \frac{\sin(j+z)\pi y_2 - \sin(j+z)\pi y_1}{2\pi(j+z)}\right], & j \neq z, \\ \left[\frac{y_2-y_1}{2} + \frac{\sin(2j\pi y_2) - \sin(2j\pi y_1)}{4j\pi}\right], & j = z, \end{cases} \tag{2.120}$$

Let us introduce the following notations:

$$J_{1,vzij} = -\frac{\pi^4}{12(1-\mu^2)} \sum_{k=1}^{N} (1-\gamma_{1k}) \left\{ \left[\left(\frac{i^2}{\lambda^2} + \mu j^2 \right) v^2 + \left(j^2\lambda^2 + \mu i^2 \right) z^2 \right] I_{11,vi} I_{12,zj} + 2(1-\mu)\, ijvz I_{13,xi} I_{14,zj} \right\}. \tag{2.121}$$

$$J^{vz}_{1,vzij} = J_{1,vzij} + \frac{\pi^4}{12(1-\mu^2)} \left[\frac{v^4}{\lambda^2} + 2v^2z^2 + \lambda^2 z^4 \right] I_{3,vi}\, I_{4,zj}. \tag{2.122}$$

$$J_{2,vzij} = -\pi^4 \sum_{k=1}^{N} \left(\frac{1}{\gamma_{1k}} - 1\right) \left\{ \left[\left(\frac{i^2}{\lambda^2} - \mu j^2 \right) v^2 + \left(j^2\lambda^2 - \mu i^2 \right) z^2 \right] I_{11,vi}\, I_{12,zj} + 2\,(1+\mu)\, ijvz I13, vi I_{14,zj} \right\}. \tag{2.123}$$

$$J^{vz}_{2,vzij} = J_{2,vzij} + \left[\frac{v^4}{\lambda^2} + 2v^2z^2 + \lambda^2 z^4 \right] \pi^4 I_{3,vi}\, I_{4,zj}. \tag{2.124}$$

As the provided integrals are assumed and (2.102) holds, the system (2.97) takes the form

$$\sum_{vz}\left\{\sum_{ij,kl}\left[J^{vz}_{1,vzij}A_{ij}+I_{AB}B_{vz}-I_{Q}q-A_{ij}B_{kl}I_{vzijkl}\right]\right\}=0,$$

$$\sum_{vz}\left\{\sum_{ij,kl}\left[J^{vz}_{2,vzij}B_{ij}+I_{AB}A_{vz}+\frac{1}{2}A_{ij}A_{kl}I_{vzijkl}\right]\right\}=0. \tag{2.125}$$

The summation sign $\sum\limits_{vz}[*]$ placed at each equation of system (2.125) denotes that the assigned equation vz is actually composed of two equations of such a type. The constant N denotes the number of elements whose flexural stiffness differs significantly from the basic one. The coefficient of stiffness of a shell element is γ_{1j}.

At the end, the estimated system (2.125) is composed of $2n = 2(M_x * M_y)$ nonlinear algebraic equations and is still valid with respect to the unknown functions A_{ij}, B_{ij}.

Let us attach to the analyzed system (2.125) the following equation:

$$w_0\left(x_0,y_0\right)=\sum_{ij}A_{ij}\ \sin\left(i\pi x_0\right)\ \cos\left(jy_0\right), \tag{2.126}$$

where $w_0(x_0,y_0)$ states a deflection that is imposed in the selected point of the shell.

In the resulting system the number of $n+1$ equations A_{ij}, B_{ij}, and q are the demanding ones but the parameter $w_0(x_0,y_0)$ states an initial well defined quantity. It comprises the whole range of deflection and its changes are made with every step of length h.

Being at each iteration of the numerical scheme of solution to the investigated system of equations, there are determined those values A_{ij}, B_{ij}, and q, that are iteratively substituted into Eq. (2.103) giving in the result the unknown functions w, F having the forms of a trigonometric series. The last one mentioned allows one to estimate easily with respect to x and y all their counterpart derivatives. Step h of the numerical scheme is arbitrary, but its restriction can result only from the selected iterative scheme of solution of the nonlinear equations. The case becomes much more constrained when the selected method is sensitive on an initial approximation that was found in the preceding iteration. The procedure is elastic; therefore any scheme of corrections of the initial solution's approximation can be taken into account.

To solve a nonlinear system of equations one can apply various kinds of numerical methods. System equations describing the so-called strong nonlinearity as well as parameters λ, k_x, k_y appearing in those equations can vary in a wide range of values. The mentioned possibilities make worse convergence of the iterative process and, in addition, impose some extra restrictions on the method of solution. Taking into consideration the above remarks, our further analysis of the solution of nonlinear equations will be focused on the popular Newton (known also as the Newton-Raphson) method.

The Newton method is used in the form reported in [142]. After selection of the initial approximation $x_i^{[0]}$, we find the successive approximations $x_i^{[j+1]}$ by solution of the following system of linear equations:

$$f_i+\sum_{k=1}^{n}\frac{\partial f_i}{\partial x_k}\left(x_k^{j+1}-x_k^{j}\right)=0,\quad i=1,2,\ldots(2n+1),\quad j=0,1,2,\ldots \tag{2.127}$$

The values of functions f_i i $\partial f_i/\partial x_k$ are determined for $x_k = x_k^j$.
Let equations (2.125)–(2.127) be our representative example, where:

1. f_i are equations of the analyzed system.
2. $\partial f_i/\partial x_k$ denote derivatives with respect to the unknown A_{ij}, B_{ij}, q.

It is simplest to choose an approximation of the desired values by using their previous step values. In order to determine an initial (rough) approximation, a parabolic extrapolation can be used on the sloped segments of the "loading–deflection" characteristics. For the purpose of qualifying the ith step's initial approximation in accordance with deflection, it is mandatory to have knowledge of the successive values of each function, for which at steps $i-3$, $i-2$, $i-1$ (where $i > 3$) a rough approximation has to be imposed. In this way, we are able to estimate fluctuations of the desired characteristics of the second-order polynomial, in which values of deflection in the selected point of a shell for the three mentioned characteristic points are taken into assumption. Afterwards, values of the ith step's deflection are substituted to the determined polynomial and the result is then treated as the initial ith step's approximation of the given function. As the resulting calculations exhibit, this type of approach gives the possibility of reduction of the number of the required iterations. For example, the number of iterations for $h = 0.1$ does not exceed three.

Similar problems devoted to a systematic global numerical analysis of strongly nonlinear systems have been presented earlier in monographs such as [15, 16, 18]. On the basis of the Galerkin method an analysis of bifurcations and stability of periodic orbits was also successfully made [297, 298, 299].

2.5 Shells with Additions of Other Materials

Let us make a stability analysis of some spherical rectangular shells subjected to a uniformly distributed transversal loading.

The shell with additions in a rectangular form whose sides are parallel to a shell made from a different material will be called non-homogeneous. The added element will be called the stiffness element. The relative stiffness of such an element will be determined by its coefficient of stiffness γ_{1j}. The coefficient $\gamma_{1j} = 1$ relates to a homogeneous shell, i.e., to a shell not containing any additions. On the other hand, a shell with additions (some extra inclusive elements) of stiffness $\gamma_{1j} < 1$ will be called the soft shell, while in comparison to the homogeneous shell a shell with additions satisfying the condition $\gamma_{1j} > 1$ will be called the stiff shell.

A typical geometrical decomposition of rigid elements situated on the shell's surface (dark color) is pictured in Fig. 2.8:

(a) The origin of the square addition is relatively translated to the shell's origin and the addition at hand is completely defined by its surface.
(b) These extra elements are distributed along the axis of symmetry of the shell concurrently creating a cross shape, but that addition's parameter is expressed by the width of the cross element.

(a) (b) (c)

Fig. 2.8 Scheme of rigid elements situated on the shell's surface: (**a**) in the middle of the shell, (**b**) "cross"-type, (**c**) perforation-type

(c) The shell is regularly covered by some extra square elements, creating a perforated construction and its parameter is expressed by the quantity of additions distributed along one of the shell's sides.

2.6 Static Stability of a Shell

In order to define the static stability of a shell we will use system (2.125) and a methodology of progression described in Sect. 2.4. To begin, we will deal with the analysis of homogeneous shells.

Toward that aim we will assume in our analysis the scheme of a geometrical decomposition of the stiffness elements shown in Fig. 2.8a. Moreover, we will consider any surfaces of an extra element $S = 0$ and $\gamma_{1\hat{i}}$, or $\gamma_{1j} = 1$ for an arbitrary S. Both approaches are respected in the computational process as equivalent methods of a homogeneous shell's representation.

Figure 2.9 provides a typical $q - w$ relationship between the static loading of a square homogeneous shell and the deflection at the shell's origin. Computations have been carried out for $M_x = M_y = 5$ and for the parameter $k_x = k_y = 12$; 18; 24 (curves 1, 2, 3, respectively). It can be concluded that a homogeneous shell characterized by high values of that parameters possesses relatively high critical upper and lower loadings.

By increasing the value of parameter $k_x = k_y$ and starting from a certain value, the character of the shell's deformation undergoes some changes. During a monotonically increasing loading of a shell and starting from the deflection of order 0.2–0.3, the highest value of the responding deflection experiences translation from the middle of the shell to its quarters and, what is the most sensible, a local stability loss appears. In the next stage, a "hard" stability loss is observed after the critical values of deflections are achieved.

If we realize a normalization of the shell's deflection with respect to the deflection of its origin, i.e., by an introduction of a w/w_c parameter, and then draw a

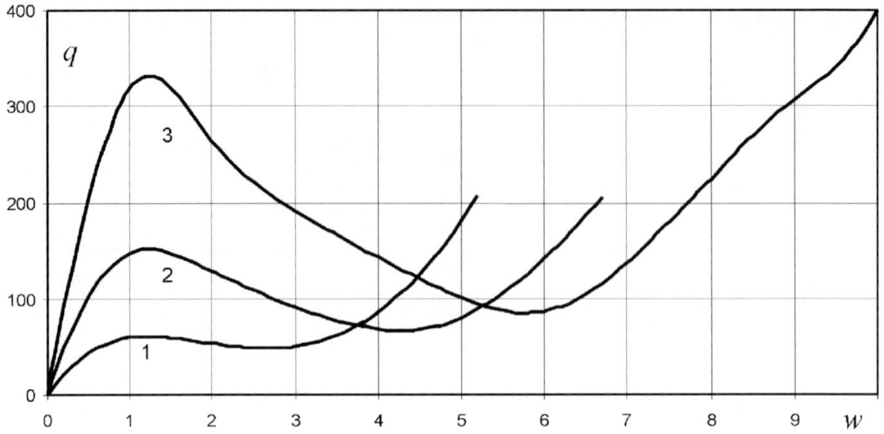

Fig. 2.9 The statical loading–deflection at origin relationship for a homogeneous shell

dependence of that parameter on the shell's origin deflection, then it could be easily seen that the obtained results confirm those of the earlier observations.

Such relationships are shown in Fig. 2.10, with curves 1, 2, 3, 4, 5 determined for the coordinates $x = y = 0.1$; 0.2; 0.25; 0.3; 0.4, which are the points belonging to the shell's surface.

It is concluded from these graphs that for a shell defined by the $k_x = k_y = 24$ parameter the biggest deflection is always achievable in the middle of the shell i.e., local loss of stability does not exist. Alternatively, for a shell defined by the $k_x = k_y = 36$ parameter the biggest deflection (see the third curve) is achieved at point $x = y = 0.25$ of the shell, i.e., the phenomenon of local loss of stability observed at the beginning.

To evaluate some qualitative properties of the shell's loss of stability it is convenient to follow fundamental changes in the shell's surface shape corresponding to variations of its loading times.

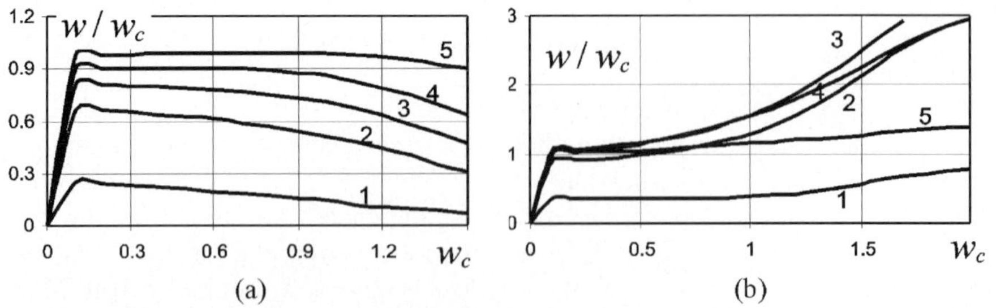

Fig. 2.10 Development of the relational deflection during imposition of a loading at the shell points: (**a**) $k_x = k_y = 24$; (**b**) $k_x = k_y = 36$

Table 2.2 Normalized loading surface of a homogeneous shell for some selected values of the loadings at the shell's origin

w	$k_x=k_y=24$	$k_x=k_y=30$	$k_x=k_y=36$
0,5	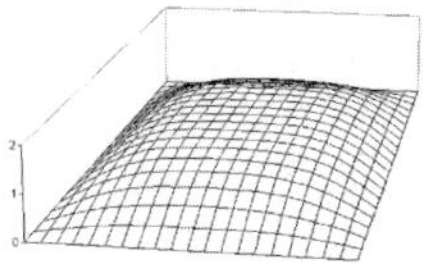	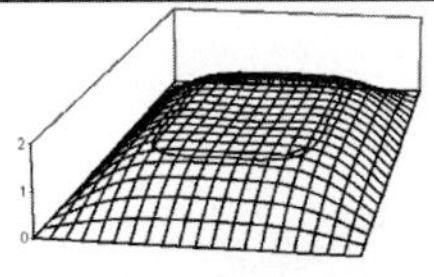	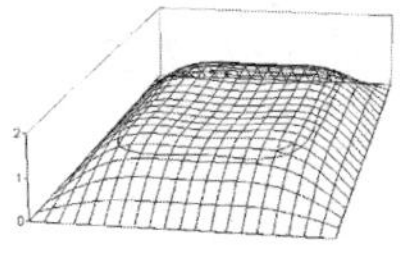
1,0	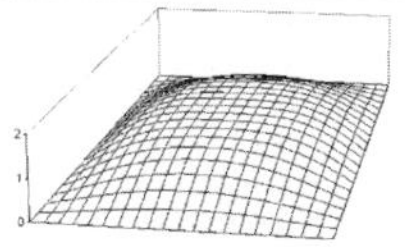	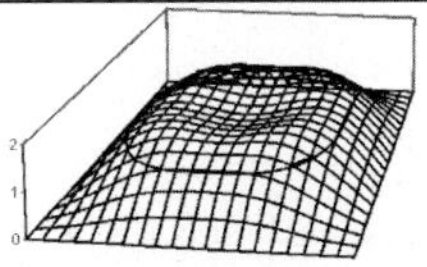	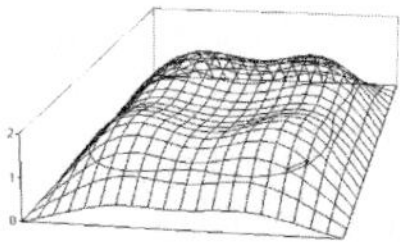
critical	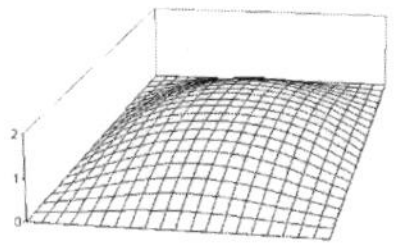	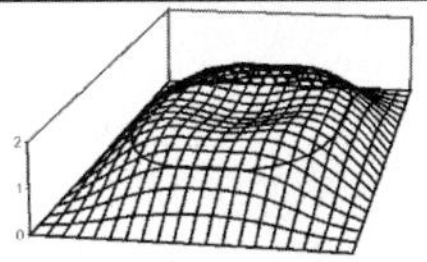	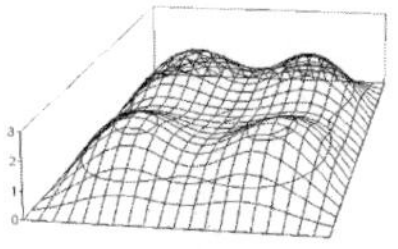
1,5	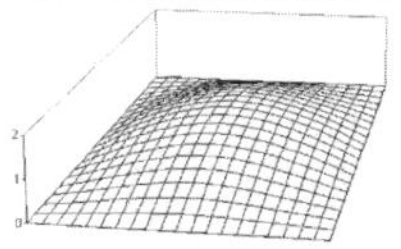	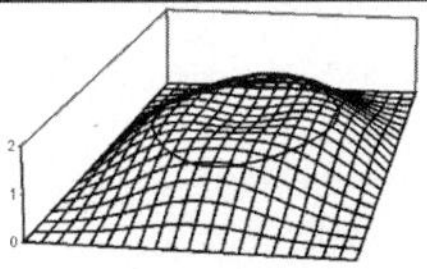	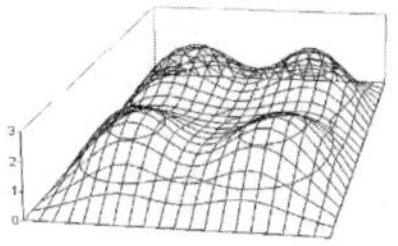
2,0	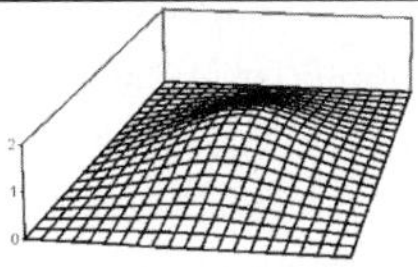	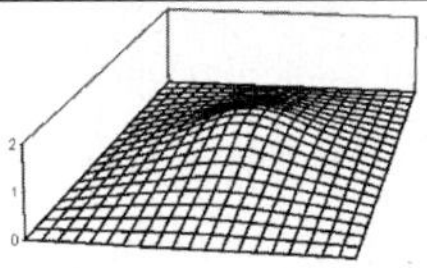	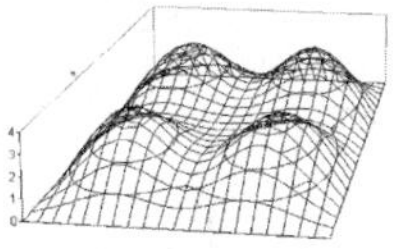

Table 2.2 refers to the middle point normalized surfaces of loading for the prescribed values of loading imposed in the middle of the shell of the parameter $k_x = k_y = 24; 30; 36$.

The given forms of the shell deformation fully and graphically confirm observations that are based on the graphs in Fig. 2.10. In the case of the shell described by $k_x = k_y = 24$, the shape of the loading surface in the entire range of loading parameters is such that the biggest loading is found at the shell's origin, and for the option $k_x = k_y = 36$ the maximum loading appears in a quarter of the shell. In the case of the shell described by $k_x = k_y = 30$ a medial value is achieved.

Figure 2.11 shows typical $q-w$ relationships for both homogeneous and non-homogeneous shells (curve 1 corresponds to curve 3 in Fig. 2.9). The shell is characterized by the non-homogeneity of the type (2.8a), and its surface $S = 0.04$. The element's stiffness coefficients $\gamma_{1j} = 0.5; 1.5$ correspond to curves 2, 3.

In Fig. 2.12 some dependencies of a static loadings (a, upper; b, lower) on the $k_x = k_y$ parameter of a non-homogeneous shell when the shell contains a central

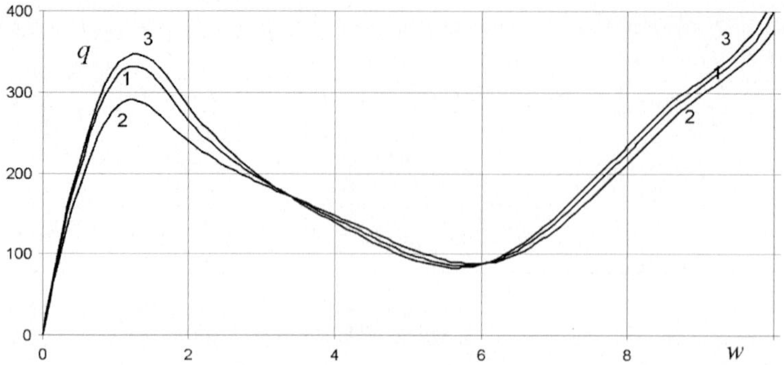

Fig. 2.11 Static loading of a non-homogeneous shell's dependence upon deflection of its center

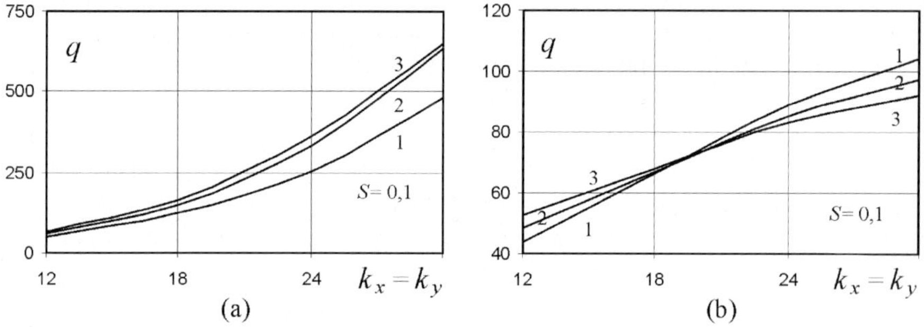

Fig. 2.12 Static critical loading of a shell's dependence on the $k_x = k_y$ parameter: (**a**) upper, (**b**) lower

element of stiffness for which surface $S = 0.1$ are shown. The element's stiffness coefficients defined by (1.79) and with $\gamma_{1j} = 0.5, 1, 1.5$ correspond to curves 1, 2, 3.

The upper critical value of loading is monotonically conservative during changes of the $k_x = k_y$ shell parameter but the soft shell is more sensitive to the stiffness parameter of the element of non-homogeneity.

The dependence of lower statical loading is represented in an analogous manner. For $k_x = k_y$ (20, for instance) such a type of non-homogeneity does not influence the critical loading. For values less than 20 the dependence is qualitatively convergent to the a case, but for the values greater than 20 it gets an inverse form, i.e., a soft shell, in contrast to its stiff equivalent characterized by higher critical loadings.

2.7 Central Square Element of Non-homogeneity

Let us structurally analyze the influence of a shell's non-homogeneity presented in Fig. 2.8a on an upper critical loading. First of all let us explain how the size of an extra element's surface at a fixed value of its stiffness can influence the significance of critical loading. It would be more convenient to consider a relative loading of a non-homogeneous shell, i.e., to split our study on these aspects related to various values q/q_0 of an homogeneous shell. Introduction of such a parameter offers the possibility of making comparisons of behavior of the non-homogeneous shells described by different $k_x = k_y$ parameters.

Figure 2.13 depicts relationships between the relative critical loading and a surface of the central element of stiffness. Curves 1, 2, 3, 4 correspond to parameters $k_x = k_y = 12, 18, 24, 30$, respectively.

Obviously, an increase of the element surface causes in the case of soft (hard) shell a reduction (increase) of the critical loading. The earlier figures confirm this observation. Therefrom, an increase of loading reaches 50%. Approaching $S = 1$, all curves inertially come closer together for various $k_x = k_y$ and after that they overlap each other. This results from the observation that if some shell's elements of non-homogeneity cover the larger part of its surface then respectively to the stiffness characteristics the shell becomes transformed into another shell. The basic role defining that shell's behavior is played by the flexible (rigid) part of the shell. At some boundary size of surface of an addition the shell again becomes homogeneous, but now is characterized by a different stiffness. As $k_x = k_y$ increases the analyzed relationship initiates a more complex structure; therefore any extremal points can be observed on the "deflection-loading" curve, indicating a higher complexity of

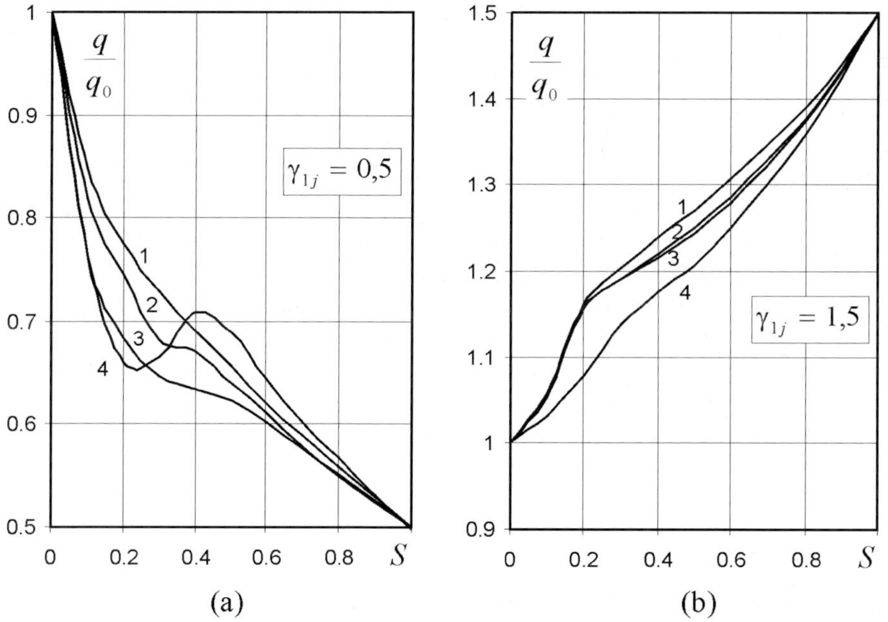

Fig. 2.13 Upper critical loading's dependence of the surface of a middle element of stiffness: (**a**) "soft" shell, (**b**) "stiff" shell

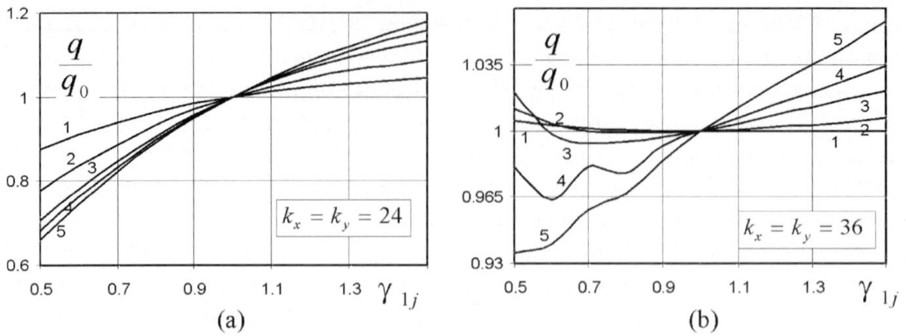

Fig. 2.14 Critical loading of an non-homogeneous shell's dependence of the coefficient of its element stiffness

interactions of this parameter as well as the shell's non-homogeneity after loss of stability.

Let us estimate now the influence of the coefficient of stiffness γ_{1j} of a non-homogeneity element on the value of the upper critical loading (Fig. 2.8a). For that purpose, at a constant surface of the addition we will only change the coefficient of stiffness.

With reference to the shell's parameter $k_x = k_y$, we will restrict our study to only two values of $k_x = k_y = 24; 36$ (see Fig. 2.14a,b, respectively).

This choice of data is guided by an observation that at assumed values the first shell loses stability in a large sense, but the second one in a small sense. Curves 1, 2, 3, 4, 5 correspond to the following extra stiffness element's surfaces $S = 0.04, 0.09, 0.16, 0.2025, 0.25$, respectively.

On the basis of the above graphs, for a "stiff" shell the "small" loss of stability does not significantly influence the "deflection-loading" characteristics, because it remains monotonic for all parameters of the surface's non-homogeneity element.

Some crucial differences appear when a "soft" nonlinearity is taken into assumption. The first type of shell is characterized by monotonic dependence of the mentioned curve, and the second one slightly increases during increases of γ_{1j} for small parameters of the surface of the addition, while it slightly decreases for large parameters of the surface. As γ_{1j} increases, i.e., after approaching the homogeneous shell, the influence of the addition's surface becomes less important. A more complex relationship between the parameter of the addition's surface S, its stiffness γ_{1j}, and the parameter $k_x = k_y$ is reflected in the appearance of a resonance character of this dependence.

2.8 Central Cross Addition of Non-homogeneity

Let us analyze the influence of the type of a shell's non-homogeneity (Fig. 2.8b) on an upper critical loading. By assumption of such a scheme of the shell's non-homogeneity one can model a shell with ribs. Observe that the assumed approach allows for the consideration of not only the variant related to increase of width of the shell along the rib's width, but also to the variants of decreasing thickness of the shell. It can be realized by assumption of an appropriate coefficient of shell's stiffness (see 2.5 – the "soft" and "stiff" shell). In application, ribs are very often arranged in a way that ensures their parallelism to the symmetry axis of the shell. This kind of non-homogeneity will be called the cross-type non-homogeneity. Here we could consider how the critical loading of a shell changes the width of such a type of cross addition, which in brief will be called the width of the cross. Later analysis will be focused on the relative loading of a non-homogeneous shell, i.e., we will divide the shell into some homogeneous parts.

Figure 2.15a,b show some relationships $q/q_0 - S$, where S denotes the width of a rib (a, "soft"; b, "stiff" shell) for non-homogeneous square shells. Curves 1, 2, 3, 4 correspond to $k_x = k_y = 18; 24; 36; 48$, respectively.

In the case of a "soft" shell as seen on the graphs, by increasing the $k_x = k_y$ shell parameter to a boundary value of local loss of stability, the dependence "loading-deflection" is monotonic, i.e., the more the control parameter reaches some high values the faster the critical loading decreases. This shell is then more sensitive to the occurrence of non-homogeneity of this type. As the parameter $k_x = k_y$ exceeds a boundary value at a local loss of stability, the considered non-homogeneity influences the shell's behavior in a quite different way.

First of all, the increase of the $k_x = k_y$ parameter causes a negligible increase of critical loading. Second, the shell of such a type is less sensitive to the non-

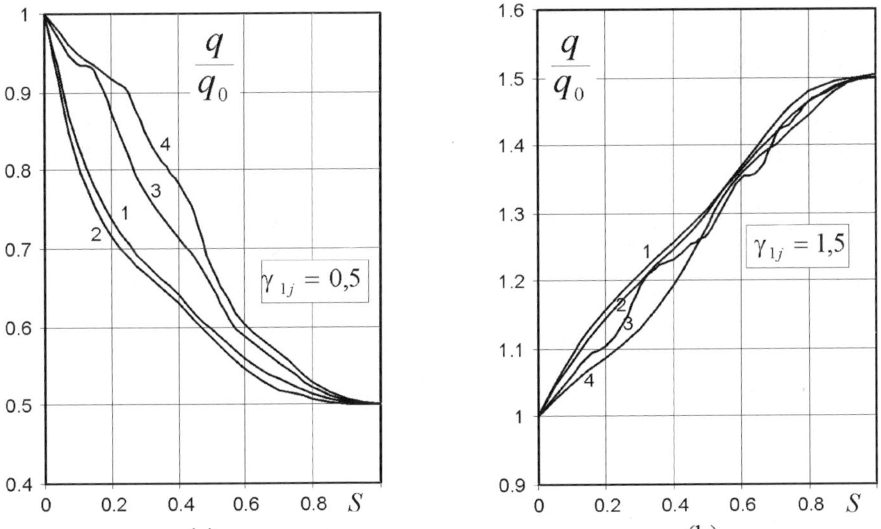

Fig. 2.15 Critical loading of a non-homogeneous shell's dependence on the width of a rib of the "cross"-type middle element (**a**) "soft" shell, (**b**) "stiff" shell

homogeneity occurrence. It can be explained by much stronger interactions between the $k_x = k_y$ parameter and the kind of non-homogeneity (in regard to the rib's width) that is described by the element coefficient of stiffness.

Similar conclusions can be drawn regarding the "soft" shell. Deviation of critical loading with respect to the shell's $k_x = k_y$ parameter is significantly lower in comparison to its "soft" counterpart. If the parameter's values are close to the appropriate boundary ones, then we obtain a resonance interaction of the parameters discussed.

In all investigated cases for a limiting width of rib at which the shell begins self-transformation to the homogeneous one (despite another stiffness parameters), all the curves begin to overlap each other. The highest deviation of critical loading is for these cases equal to about 50% of loading of the homogeneous shell.

2.9 "Perforation"-Type Non-homogeneity

Let us investigate the influence of the type of non-homogeneity of a shell (Fig. 2.8c) on the situation of an upper critical loading. For a reminder, by assumption of such a scheme of non-homogeneity the square additions regularly cover the whole shell, i.e., both homogeneous and non-homogeneous elements interleave each other.

Although application of the algorithm used allows for any choice of non-homogeneous element, all of the other elements of non-homogeneity are identical. Because the scheme of the non-homogeneity resembles a perforation process of plates and shells has been set externally, we will define this kind of non-homogeneity as "perforation," though for our case we formally do not have any holes (see 1.9).

For suitability we introduce the parameter N defining the number of elements that are arranged in a way that ensures their parallelism to any side of the shell. The relationship between the parameter and the surface defining non-homogeneity of shell has been shown in Fig. 2.16. From this figure, the basic increment of a general surface of non-homogeneity appears for $N < 8$.

Figure 2.17 shows the relationship between the upper critical loading for a non-homogeneous shell (a, "soft"; b, "stiff") and the number of elements situated along

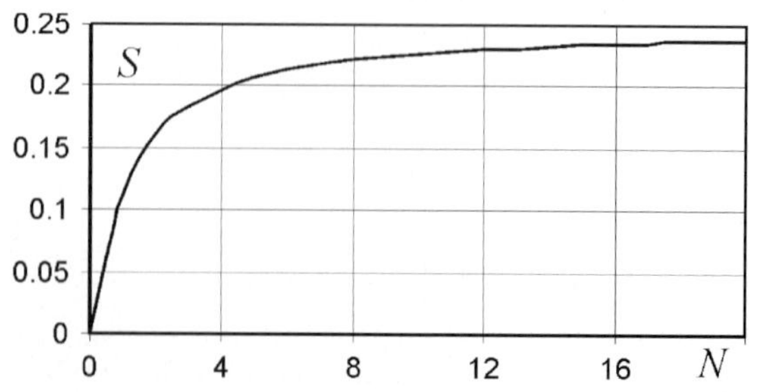

Fig. 2.16 Dependence of total surface of the stiffness elements on their amount along one of the shell's sides (the "perforation"-type scheme)

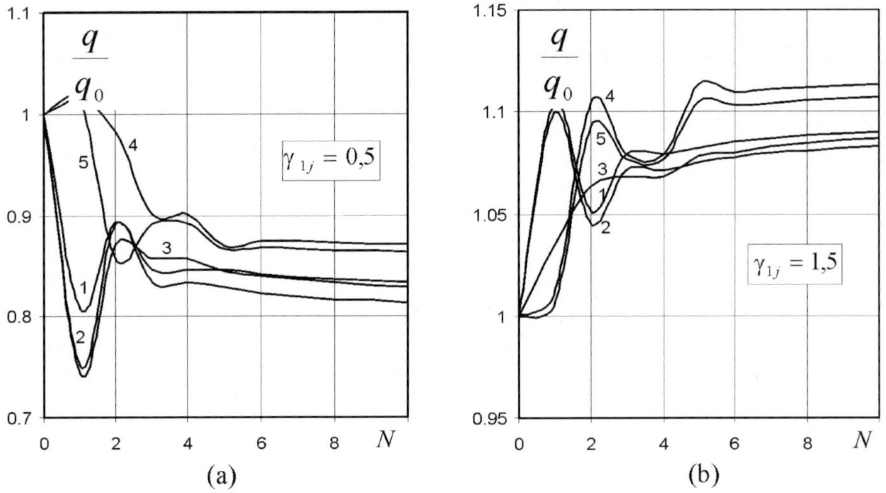

Fig. 2.17 Critical loading of a non-homogeneous shell's dependence on the number of stiffness elements along any side of a shell: (a) "soft" shell, (b) "stiff" shell

one selected side of a shell. Curves 1, 2, 3, 4, 5 on graphs correspond to the shell's parameters $k_x = k_y = 18, 24, 30, 36, 48$, respectively.

For the case of both soft and stiff shells, if the number of additions that are situated along one of the sides of the shell exceeds six, the critical loading's value is practically not dependent on the number of these additions. As a negative consequence, some differences in critical loading can be emphasized between a shell for which a local loss of stability appears and the other one for which the mentioned phenomenon does not occur. For small parameters $N = 1,2$ one can distinctly observe a resonance relation among all the values. An increment of loading becomes here most important. Extremal values are clearly distinguishable for $N = 1$ and $N = 2$, which one can explain in the following way. First, there is only one element situated in the middle of the shell. Second, four additions are symmetrically situated in relation to the middle of the shell and for this case the non-homogeneity does not extend over the shell's origin. One needs to distinguish an even as well as odd number of additions lying along any side of shell because the global shell's behavior does not result from the behavior of its part.

Chapter 3
Vibrations of Rectangular Shells

In this chapter we examine vibrations of rectangular shells. Linear and weakly nonlinear vibrations are revisited in Sect. 3.1, and then natural vibrations of nonhomogeneous shells applying the Bubnov-Galerkin method of higher order approximations are analyzed. Section 3.3 is devoted to investigation of free nonlinear vibrations of homogeneous plates and shells with respect to any choice of control parameters. The relatively extensive Sect. 3.4 addresses the spectral analysis of a stress–strain problem of any plate/shell approximated by systems of n degrees of freedom. A harmonic process convergence and spectral analysis of free nonlinear vibrations are illustrated and discussed in Sects. 3.5 and 3.6, respectively.

3.1 Linear and Weakly Nonlinear Vibrations of Mechanical Systems

Assuming for mechanical systems *a smallness* of deflections from an equilibrium position, we consider in our investigations the hypothesis of linearity of the system and then for a mathematical description theory of systems of linear differential equations is applied [238]. The theory in connection with this approach is usually called the theory of small vibrations. One can find some original premises devoted to the similar meaning of both "small vibrations" and "linear vibrations" in the "analytical mechanics" by Lagrange. From the point of view of today's mechanics these concepts become rather different.

The theory mentioned relates indeed to "linear vibrations," i.e., to the vibrations described by linear differential equations being not always the property of small vibrations. In some situations small vibrations (even "any small" ones!) have to be described via nonlinear equations to explain adequately the analyzed process. Moreover, there even exist in practical engineering applications some clear instances of a mechanical system's nonlinearities for the smaller than usual deflections of the system from an equilibrium point (see [18]).

J. Awrejcewicz, V.A. Krysko, *Chaos in Structural Mechanics*,
DOI: 10.1007/978-3-540-77676-5_4, © Springer-Verlag Berlin Heidelberg 2008

One can find in the literature notions like "frequency of free vibrations" or "frequency of natural vibrations." The latter one can be used only for linear systems in which the frequency of natural vibrations is expressed by *natural* characteristic features of a system and is not dependent on any initial conditions. Relative to the present situation, the word "amplitude" should be used exclusively in the terminology describing harmonic vibrations, because for nonlinear free vibrations the motion of a system is *not harmonic*, though periodic. A typical feature of nonlinear systems is the lack of *isohronocity*. In contrast, the frequency of free vibrations does not depend on the system's deflection.

3.2 Natural Vibrations of Non-homogeneous Shells

Our task at this point is to investigate the natural vibrations of non-homogeneous shells in comparison to any choice of control parameters.

3.2.1 The Solution Method

As an approach to the solution to the previously formulated problem we will utilize the linear variant of systems describing the dynamics of the shells expressed by Eqs. (1.55) and (1.56). We eliminate from the first equation the term of loading and assume that boundary conditions are (2.94) [38]. Hence, the linear variant of the system is

$$\begin{aligned}
&\frac{E}{12(1-\mu^2)}\left[\frac{1}{\lambda^2}\frac{\partial^2 w}{\partial x^2}\frac{\partial^2(\cdot)}{\partial x^2}+\lambda^2\frac{\partial^2 w}{\partial y^2}\frac{\partial^2(\cdot)}{\partial y^2}+2(1-\mu)\frac{\partial^2 w}{\partial x\partial y}\frac{\partial^2(\cdot)}{\partial x\partial y}\right.\\
&\quad\left.+\mu\left(\frac{\partial^2 w}{\partial x^2}\frac{\partial^2(\cdot)}{\partial y^2}+\frac{\partial^2 w}{\partial y^2}\frac{\partial^2(\cdot)}{\partial x^2}\right)\right]-\nabla_k^2 F=\rho\left(\frac{\partial^2 w}{\partial t^2}+\varepsilon\frac{\partial w}{\partial t}\right),\\
&a_1\left[\left(\lambda^2\frac{\partial^2 F}{\partial y^2}-\mu\frac{\partial^2 F}{\partial x^2}\right)\frac{\partial^2(\cdot)}{\partial y^2}+\left(\frac{1}{\lambda^2}\frac{\partial^2 F}{\partial x^2}-\mu\frac{\partial^2 F}{\partial y^2}\right)\frac{\partial^2(\cdot)}{\partial x^2}\right.\\
&\quad\left.+2(1+\mu)\frac{\partial^2 F}{\partial x\partial y}\frac{\partial^2(\cdot)}{\partial x\partial y}\right]+\nabla_k^2 w=0.
\end{aligned}\qquad(3.1)$$

The initial Eq. (3.1) will be solved with the use of the Bubnov-Galerkin method with higher order approximations. Following the above, the functions w, F satisfying boundary conditions are sought in the form

$$w=\sum_{i,j}A_{ij}(t)\sin(i\pi x)\sin(j\pi y),$$

$$F=\sum_{i,j}A_{ij}(t)\sin(i\pi x)\sin(j\pi y),$$

$$i = 1,2,\ldots M_x; \quad j = 1,2,\ldots M_y; \quad n = M_x * M_y. \tag{3.2}$$

After application of the Bubnov-Galerkin procedure to Eq. (3.1) the following systems of differential (with respect to time) and algebraic equations are obtained:

$$\sum_{vz}\left[\sum_{ij}\ddot{A}_{ij}(t)J^{vz}_{2,vzij} = \sum_{ij}A_{ij}(t)J^{vz}_{1,vzij} + B_{vz}(t)J_{3,vzij}\right],$$
$$\sum_{vz}\left[\sum_{ij}B_{ij}(t)J^{vz}_{4,vzij} = A_{vz}(t)J_{3,vzij}\right], \tag{3.3}$$

where $\sum\limits_{vz}$ specifies the number of equations in the system and the dot denotes a differentiation with respect to time.

Application of the Bubnov-Galerkin method produces the following integrals:

$$J_{1,vzij} = -\frac{\pi^4}{12(1-\mu^2)}\sum_{k=1}^{N}(1-\gamma_{1k})\left[\left(\frac{i^2}{\lambda^2}+\mu j^2\right)v^2 I_{5,vi}I_{6,zj}\right.$$
$$\left. + \left(\lambda^2 j^2 + \mu i^2\right)z^2 I_{5,vi}I_{6,zj} + 2(1-\mu)ijvzI_7,viI_{8,zj}\right],$$

$$J^{vz}_{1,vzij} = J_{1,vzij} + \frac{\pi^4}{12(1-\mu^2)}\left[\left(\frac{v^4}{\lambda^2}+\lambda^2 z^4\right)I_{1,vi}I_{2,zj} + 2v^2z^2 I_{3,vi}I_{4,zj}\right],$$

$$J_{2,vzij} = \sum_{k=1}^{N}(1-\gamma_{2k})I_{5,vi}I_{6,zj},$$

$$J^{vz}_{2,vzij} = J_{2,vzij} - I_{1,vi}I_{2,zj},$$

$$J_{3,vzij} = (k_x z^2 + k_y v^2)\pi^2 I_{1,vi}I_{2,zj},$$

$$J_{4,vzij} = \pi^4\sum_{k=1}^{N}\left(\frac{1}{\gamma_{1k}}-1\right)\left[\left(\lambda^2 j^2 - \mu i^2\right)z^2 I_{5,vi}I_{6,zj}\right.$$
$$\left. + \left(\frac{i^2}{\lambda^2}-\mu j^2\right)v^2 I_{5,vi}I_{6,zj} + 2(1+\mu)ijvzI_{7,vi}I_{8,zj}\right],$$

$$J^{vz}_{4,vzij} = J_{4,vzij} + \pi^4\left[\left(\frac{v^4}{\lambda^2}+\lambda^2 z^4\right)I_{1,vi}I_{2,zj} + 2v^2z^2 I_{3,vi}I_{4,zj}\right], \tag{3.4}$$

where

$$I_{1,vi} = \int_0^1 \sin i\pi x \sin v\pi x\, dx = \begin{cases}\frac{1}{2}, & i = v\\ 0, & i \neq v\end{cases}$$

$$I_{2,zj} = \int_0^1 \sin j\pi x \sin z\pi x\, dy = \begin{cases}\frac{1}{2}, & j = z\\ 0, & j \neq z\end{cases}$$

$$I_{3,vi} = \int_0^1 \cos i\pi x \cos v\pi x \, dx = \begin{cases} \frac{1}{2}, & i = v \\ 0, & i \neq v \end{cases}$$

$$I_{4,zj} = \int_0^1 \cos j\pi x \cos z\pi x \, dy = \begin{cases} \frac{1}{2}, & j = z \\ 0, & j \neq z \end{cases}$$

$$I_{5,vi} = \int_0^1 \gamma_{0k} \sin i\pi x \sin v\pi x \, dx$$

$$= \begin{cases} \left[\frac{\sin(i-v)\pi x_2 - \sin(i-v)\pi x_1}{2\pi(i-v)} - \frac{\sin(i+v)\pi x_2 - \sin(i+v)\pi x_1}{2\pi(i+v)} \right], & i \neq v, \\ \left[\frac{x_2 - x_1}{2} - \frac{\sin 2i\pi x_2 - \sin 2i\pi x_1}{4i\pi} \right], & i = v, \end{cases}$$

$$I_{6,zj} = \int_0^1 \gamma_{0k} \sin j\pi y \sin z\pi y \, dy$$

$$= \begin{cases} \left[\frac{\sin(j-z)\pi y_2 - \sin(j-z)\pi y_1}{2\pi(j-z)} - \frac{\sin(j+z)\pi y_2 - \sin(j+z)\pi y_1}{2\pi(j+z)} \right], & j \neq z, \\ \left[\frac{y_2 - y_1}{2} - \frac{\sin 2j\pi y_2 - \sin 2j\pi y_1}{4j\pi} \right], & j = z, \end{cases}$$

$$I_{7,vi} = \int_0^1 \gamma_{0k} \cos i\pi x \cos nv\pi x \, dx$$

$$= \begin{cases} \left[\frac{\sin(i-v)\pi x_2 - \sin(i-v)\pi x_1}{2\pi(i-v)} + \frac{\sin(i+v)\pi x_2 - \sin(i+v)\pi x_1}{2\pi(i+v)} \right], & i \neq v, \\ \left[\frac{x_2 - x_1}{2} + \frac{\sin 2i\pi x_2 - \sin 2i\pi x_1}{4i\pi} \right], & i = v, \end{cases}$$

$$I_{8,zj} = \int_0^1 \gamma_{0k} \cos j\pi y \cos z\pi y \, dy$$

$$= \begin{cases} \left[\frac{\sin(j-z)\pi y_2 - \sin(j-z)\pi y_1}{2\pi(j-z)} + \frac{\sin(j+z)\pi y_2 - \sin(j+z)\pi y_1}{2\pi(j+z)} \right], & j \neq z, \\ \left[\frac{y_2 - y_1}{2} + \frac{\sin 2j\pi y_2 - \sin 2j\pi y_1}{4j\pi} \right], & j = z. \end{cases} \quad (3.5)$$

The integrals (3.5) are calculated on the entire central surface of the shell.

The problem of vibrations frequency estimation of the shell reduces to the calculation of eigenvalues of the appropriate matrix, because after the appearance of non-homogeneous elements represented by stiff shell additions, the squares of frequency cannot be calculated directly from equations describing vibrations of the shell. The reason is that the system of equations cannot be decomposed into two separate equations describing the vibrations of appropriate harmonics. To solve this problem, let us number the single direction indices staying at summation symbols by vz and ij.

Let us introduce the following notations:

$$A = \| A_{ij} \|, \quad \ddot{A} = \| \ddot{A}_{ij} \|, \quad B = \| B_{ij} \|, \quad J_1 = \| J_{1,vzij} \|,$$
$$J_2 = \| J_{2,vzij} \|, \quad J_3 = \| J_{3,vzij} \|, \quad J_4 = \| J_{4,vzij} \|, \quad J_5 = \| J_{5,vzij} \|. \tag{3.6}$$

In the above, J_1, J_2, J_3, J_4, J_5 are the square matrices of order $2n$, and A, $\ddot{A}$, B are column matrices of type $2n \times 1$.

Putting (3.6) in Eq. (3.3) the following matrix representation is obtained:

$$\begin{aligned} J_2\ddot{A} &= J_1A + J_3B, \\ J_4B &= J_5A. \end{aligned} \tag{3.7}$$

Carrying out left-hand sided multiplication of the second equation of system (3.7) by the inverse matrix J_4^{-1}, one obtains

$$B = J_4^{-1} J_5 A. \tag{3.8}$$

Substituting condition (3.8) into the first equation of system (3.7) and then multiplying the resulting equation by the inverse matrix J_2^{-1} we obtain

$$\ddot{A} = J_2^{-1}\left(J_1A + J_3J_4^{-1}J_5A\right) = J_2^{-1}\left(J_1 + J_3J_4^{-1}J_5\right)A = DA, \tag{3.9}$$

where

$$D = J_2^{-1}\left(J_1 + J_3J_4^{-1}J_5\right). \tag{3.10}$$

Matrix D is non-symmetric because of the appearance of some additional stiffness elements described by the coefficient of density $\gamma_{2k} \neq 1$, which exists when we have to cope with a non-homogeneity of the shell density. For any stiffness coefficients of additions γ_{1k} the matrix of stiffness becomes symmetric. Its eigenvalues correspond to the squares of frequency of the natural vibrations of the shell. To determine these eigenvalues and eigenvectors of the symmetric matrix D a QR algorithm and a direct procedure $TQL2$ [314, p. 203] will be used. The $TQL2$ procedure allows one to calculate all eigenvalues and eigenvectors of a symmetric tridiagonal matrix. The tridiagonal matrix has been derived from the starting symmetrical matrix D by application of both Housholder transform and a $TRED2$ procedure [314, p. 190].

Focusing on the estimation of all eigenvalues and eigenvectors of a real non-symmetrical matrix D, which would be reduced to the Hessenberg form by means of the $ORTHES$ procedure [314, p. 298], a $HQR2$ procedure [314, p. 327] has been used.

If we want to determine only the eigenvalues of D, then [314] gives an overview and applications of some more effective procedures dedicated to the solution of the problem.

For a homogeneous shell (without any strengthening elements) the described results of the algorithm's execution coincide with an analytical solution (see [304]), because the system of equations modeling the investigated problem splits into equations for separate harmonics.

3.2.2 Description of Results

The results are valid for $M_x = M_y = 3$ and on the basis of Eq. (3.2), i.e., there were analyzed first nine forms of either homogeneous and non-homogeneous square shells in the dependency on miscellaneous model parameters, but damping of the medium has not been taken into account ($\varepsilon = 0$). All analyzed questions were classified with regard to the form, number, and a relative placement of elements of stiffness of the shell (see Fig. 2.8a,b,c):

1. *A homogeneous shell.* Figure 3.1a,b depicts a relationship between the vibration frequency of a spherical rectangular shell and the control parameters $k_x = k_y = 0 \ldots 36$. The characteristic curves $1,2,4,5,7,9$ correspond to the numbers of the modes of vibration. As the parameters $k_x = k_y$ of the shell increase, all modes of vibrations are monotonically increasing but the lower modes are more sensitive to changes in these parameters.
2. *An non-homogeneous shell.* During analysis of a non-homogeneous shell vibrations for each of its frequencies of natural vibrations we have introduced a coefficient K_d of amplification, which exhibits a relation between some appropriate frequencies of a homogeneous and non-homogeneous shells.

Figures 3.2 and 3.3 present the characteristic dependence of the parameter K_d of the central element of stiffness for a shell described by $k_x = k_y = 0$ and $k_x = k_y = 36$, respectively. The curve number covers with its counterpart number of the mode of vibration. The curves under 1 correspond to a softer plate and the shell described by $\gamma_{1k} = 0.5$, $\gamma_{2k} = 1$, whereas the curves above 1 correspond to a more stiff plate and the shell described by $\gamma_{1k} = 1.5$ $\gamma_{2k} = 1$ than the homogeneous plate adequate to them and shell identified respectively by the coefficients $\gamma_{1k} = 1$ and $\gamma_{2k} = 1$.

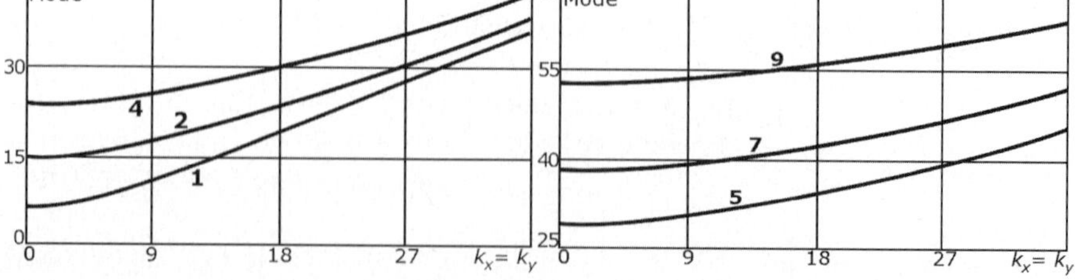

Fig. 3.1 Relationship between the frequency of shell vibrations and the parameter $k_x = k_y$

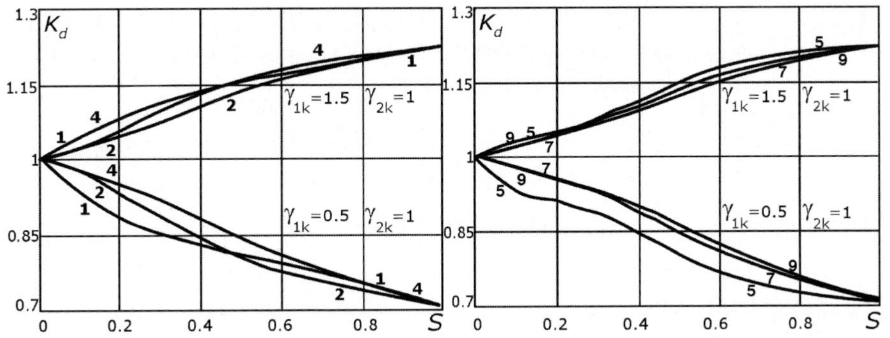

Fig. 3.2 Coefficient of dynamism of the mode of shell vibrations in relation to the surface of a center addition (see Fig. 2.8a)

On the basis of the inserted charts the coefficient of dynamism of either all forms of vibrations for the plate and the shell for a "stiff (soft)" case increases (decreases) monotonically with association to the expansion of surface S of non-homogeneity. In the critical case when a shell becomes non-homogeneous the coefficient of dynamism amounts to about 25–30%.

In Figs. 3.4 and 3.5 some dependencies are shown between the coefficient of dynamism of the mode of vibrations of the shells $k_x = k_y = 0$ and $k_x = k_y = 36$ and the coefficient of stiffness of addition γ_{1k} for the case of rib width equal to 0.1 and $\gamma_{2k} = 1$ (see Fig. 2.8b).

Curves are marked in the same way as in the preceding cases.

In the investigated domains of changes of the coefficient of stiffness the discussed relationships for a plate take practically a linear kind of shape, in contrast to the shell for which in comparison with its homogeneous adequate for any smaller coefficients of stiffness the dependence is nonlinear, thereby significantly affecting higher modes of vibration.

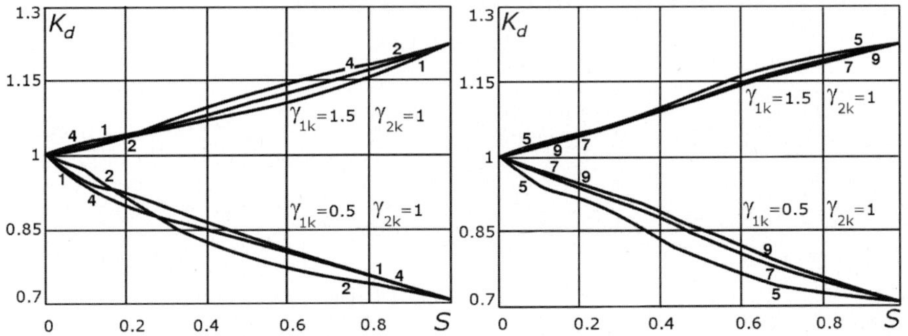

Fig. 3.3 Dependence of the coefficient of amplification of the mode of shell vibrations on the surface of the central element of stiffness (see Fig. 2.8a)

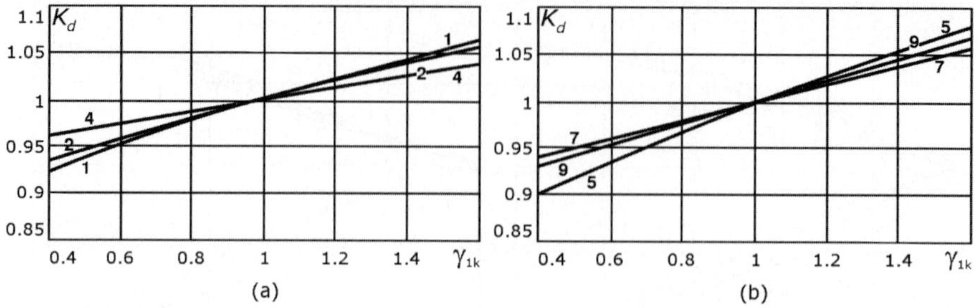

Fig. 3.4 Dependence of the coefficient of dynamic amplification of the mode of a plate vibrations on the coefficient of element stiffness (see Fig. 2.8b, the width of rib equals 0.1)

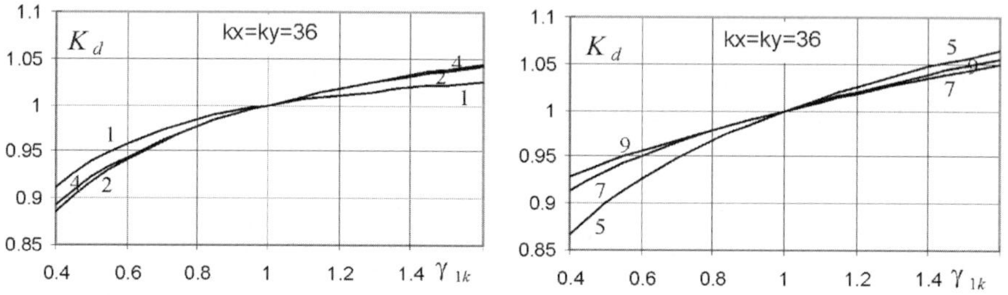

Fig. 3.5 Dependence of the coefficient of dynamic amplification of the mode of a plate vibrations on the coefficient of element stiffness (see Fig. 2.8b; the rib width equals 0.2)

The study of a shell with the "perforation"-type non-homogeneity (Fig. 2.8c) has to take into consideration the fact (see in Sect. 2.8) that all additions cover the shell in a regular manner but a side of the non-homogeneous surface of a square element equals the distance between those additional elements. Because of identical parameters of stiffness of the whole set of additions, the parameter N describing the quantity of elements that are aligned in accordance with a chosen direction of one side of a shell can be introduced. The parameter complies with the whole given non-homogeneous surface of the shell (see Fig. 2.16).

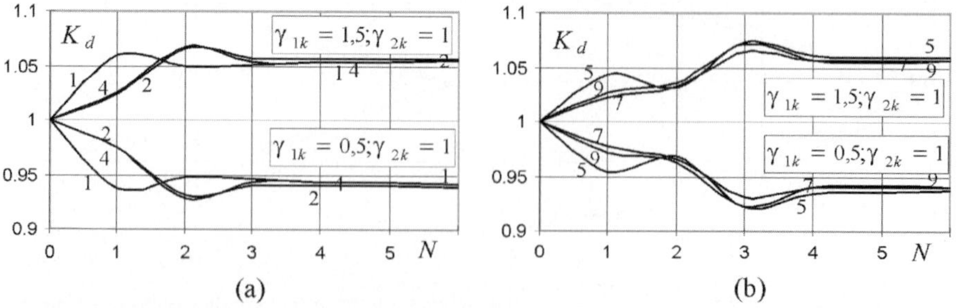

Fig. 3.6 Dependence of the coefficient of dynamic amplification of the mode of vibrations of a plate on the number N of elements of stiffness lying along one side of the plate (see Fig. 2.8c)

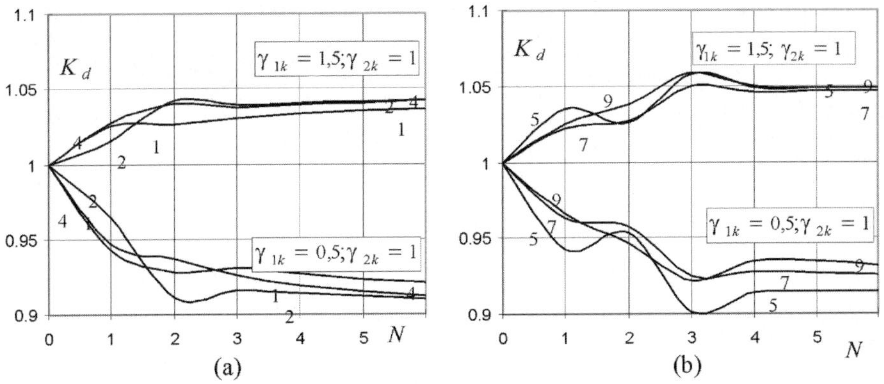

Fig. 3.7 Dependence of the coefficient of dynamic amplification of the mode of shell vibrations ($k_x = k_y = 36$) on the number N of elements of stiffness lying along one side of the shell (see Fig. 2.8c)

This results from the condition that by increasing the N parameter values above 8 we do not contribute any radical changes in the behavior of the shell. Therefore, it has to be predicted for the value a stabilization of characteristics of the analyzed non-homogeneous shell.

Figures 3.6 and 3.7 depict a dependence of the coefficient of dynamic amplification on the number N of additional elements situated along one side of the plate $k_x = k_y = 0$ and the shell $k_x = k_y = 36$.

Prior notations of curves are also actual in the case now considered. Analogously, the curves under 1 describe softer plates and shells ($\gamma_{1k} = 0.5,\ \gamma_{2k} = 1$), and the curves above 1 describe stiffer plates and shells ($\gamma_{1k} = 1.5,\ \gamma_{2k} = 1$) in comparison to homogeneous ones.

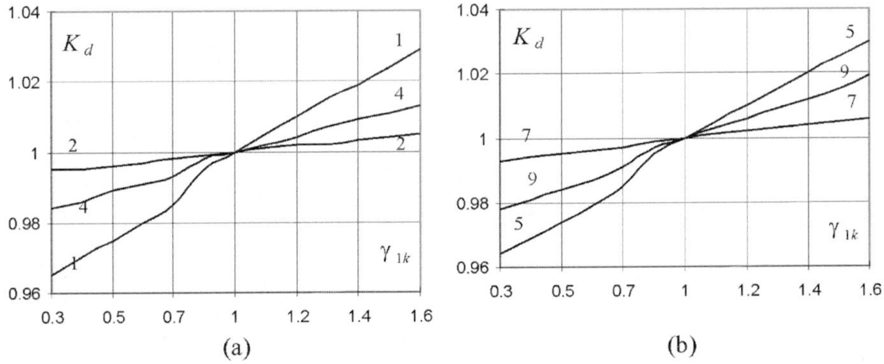

Fig. 3.8 Coefficient of dynamic amplification of the mode of vibrations of a plate versus the coefficient of stiffness of a central addition with coordinates of the element: $x_1 = 0.4,\ x_2 = 0.6,\ y_1 = 0.4,\ y_2 = 0.6$ (see Fig. 2.8a)

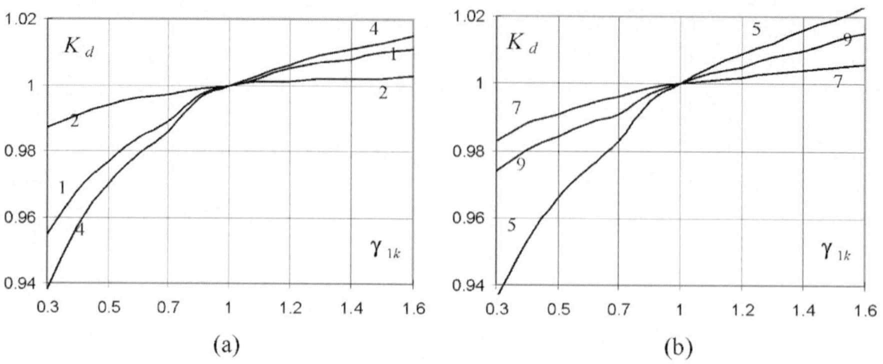

Fig. 3.9 Coefficient of dynamic amplification of the mode of vibrations of a shell versus the coefficient of stiffness of a central addition. Coordinates of the element are: $x_1 = 0.4,\ x_2 = 0.6,\ y_1 = 0.4,\ y_2 = 0.6$ (see the scheme in Fig. 2.8a)

Compared with the previous analogous dependencies, the presently analyzed case concerns some characteristics of a more intricate character that is visible during analysis of the regions of both local extremes and stabilization of the coefficient of dynamism. In the first case, for a certain value of parameter N (1, 2, 3) the total size of the non-homogeneous surface is critical, and in the second case if $N > 4$ a small expansion of the non-homogeneous surface does not practically influence the analyzed dependence.

Figures 3.8 and 3.9 show some dependencies of the coefficient of dynamism of the modes of vibrations of a non-homogeneous plate $k_x = k_y = 0$ and shell $k_x = k_y = 36$ (see Fig. 2.8a) on the coefficient of stiffness of an element γ_{1k}, where $\gamma_{2k} = 1$.

Table 3.1 Modes of vibrations of a non-homogeneous shell ($k_x = k_y = 36$) versus the coefficient of stiffness γ_{1k} ($\gamma_{2k} = 1$) of a central addition. Coordinates of the element are: $x_1 = y_1 = 0.4,\ x_2 = y_2 = 0.6$ (see Fig. 2.8a)

	Mode of vibrations								
γ_{1j}	1	2	3	4	5	6	7	8	9
0,6									
0,8									
1,0									
1,2									
1,4									

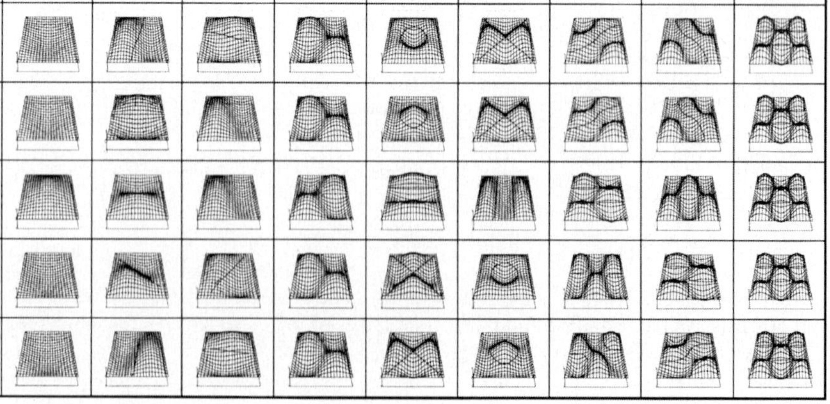

Prior notations of curves are also actual in the case now considered.

Based on the charts presented, the coefficient of element of stiffness influences more significantly the coefficient of dynamism of the mode of vibrations for the shell than for the plate, but in both cases it does not exceed 5%.

Tables 3.1–3.4 present some modes of vibrations of a homogeneous ($\gamma_{1k} = 1$, $\gamma_{2k} = 1$) and a non-homogeneous ($\gamma_{1k} \neq 1$, $\gamma_{2k} = 1$) plate $k_x = k_y = 0$ and shell $k_x = k_y = 36$ for all 9 analyzed modes in relation to the coefficient γ_{1k} of an element of stiffness.

Table 3.2 Modes of vibrations of a non-homogeneous plate versus the coefficient of stiffness γ_{1k} ($\gamma_{2k} = 1$) of a central addition. Coordinates of the element are: $x_1 = y_1 = 0.4$, $x_2 = y_2 = 0.6$ (see Fig. 2.8a)

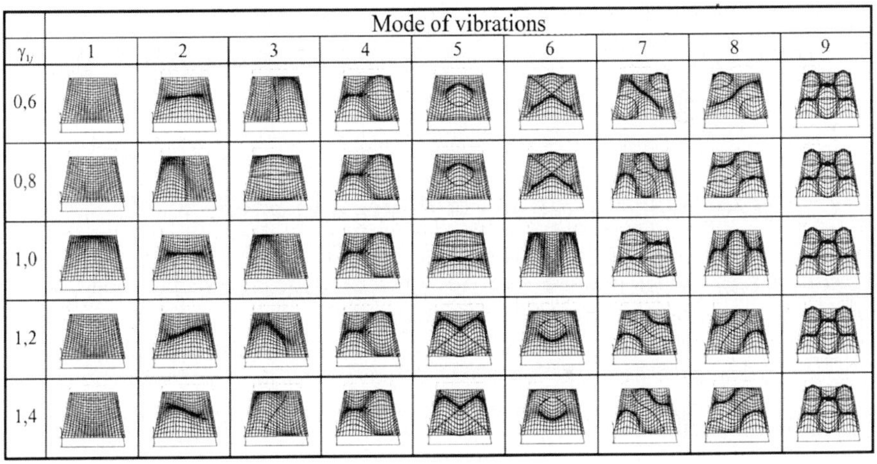

Table 3.3 Modes of vibrations of a non-homogeneous shell ($k_x = k_y = 36$) versus the coefficient of stiffness γ_{1k} ($\gamma_{2k} = 1$) for the width of rib equal to 0.1 (see Fig. 2.8b)

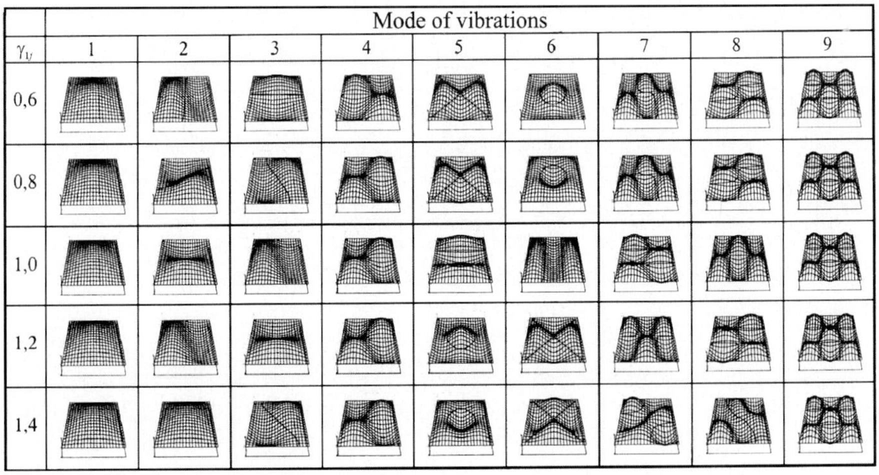

Table 3.4 Modes of vibrations of a non-homogeneous plate ($k_x = k_y = 0$) versus the coefficient of stiffness γ_{1k} ($\gamma_{2k} = 1$) for the width of rib equal to 0.1 (see Fig. 2.8b)

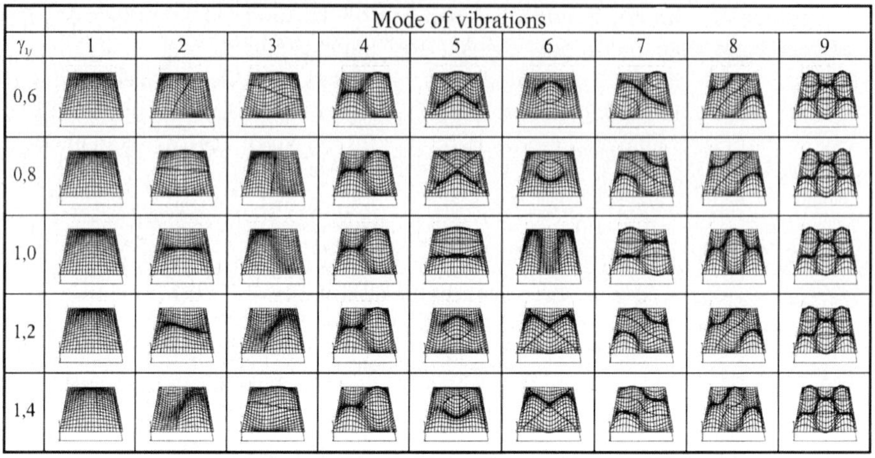

3.3 Free Nonlinear Vibrations of Plates and Shells

This section investigates of some free vibrations of homogeneous plates and shells dependent on any choice of control parameters.

3.3.1 The Solution Method

We will use for solving the problem a nonlinear variant of the equations of dynamics (1.55) and (1.56) for the parameters $q = 0,\ \varepsilon = 0,\ E = E_0 = const,\ \rho = \rho_0 = const$, and the boundary condition (2.94):

$$\frac{1}{12(1-\mu^2)}\left[\frac{1}{\lambda^2}\frac{\partial^2 w}{\partial x^2}\frac{\partial^2(\cdot)}{\partial x^2} + \lambda^2\frac{\partial^2 w}{\partial y^2}\frac{\partial^2(\cdot)}{\partial y^2} + 2(1-\mu)\frac{\partial^2 w}{\partial x\partial y}\frac{\partial^2(\cdot)}{\partial x\partial y} + \mu\left(\frac{\partial^2 w}{\partial x^2}\frac{\partial^2(\cdot)}{\partial y^2} + \frac{\partial^2 w}{\partial y^2}\frac{\partial^2(\cdot)}{\partial x^2}\right)\right] - \nabla_k^2 F - L(w,F) = \frac{\partial^2 w}{\partial t^2}, \tag{3.11}$$

$$\left(\lambda^2\frac{\partial^2 F}{\partial y^2} - \mu\frac{\partial^2 F}{\partial x^2}\right)\frac{\partial^2(\cdot)}{\partial y^2} + \left(\frac{1}{\lambda^2}\frac{\partial^2 F}{\partial x^2} - \mu\frac{\partial^2 F}{\partial y^2}\right)\frac{\partial^2(\cdot)}{\partial x^2} + 2(1+\mu)\frac{\partial^2 F}{\partial x\partial y}\frac{\partial^2(\cdot)}{\partial x\partial y} + \nabla_k^2 w + \frac{1}{2}L(w,w) = 0. \tag{3.12}$$

Initial Eqs. (3.11) and (3.12) are solved by means of the higher approximation Bubnov-Galerkin method. The desired functions w, F, satisfying our boundary conditions are assumed as follows:

$$w = \sum_{i,j} A_{ij}(t) \sin(i\pi x) \sin(j\pi y),$$
$$F = \sum_{i,j} A_{ij}(t) \sin(i\pi x) \sin(j\pi y),$$
$$i = 1, 2, \dots M_x; \; j = 1, 2, \dots M_y; \quad n = M_x * M_y. \tag{3.13}$$

By utilization of the Bubnov-Galerkin routine to Eqs. (3.11) and (3.12) regarding some spatial coordinates, we add respectively to them the two resulting descriptors: time-dependent ordinary differential equations and a system of algebraic equations.

The system of differential equations given in a normal form is then integrated by means of the fourth-order Runge-Kutta procedure. Simultaneously at each iteration, the remaining algebraic system of equations is solved with the use of the Gauss reduction method.

Initial conditions are taken in the form

$$w\Big|_{t=0} = w_0, \quad \frac{\partial w}{\partial t}\Big|_{t=0} = 0, \tag{3.14}$$

where the variable w_0 results from the solution of an associated problem of statics that is obtained by application of the method invoked in Sect. 2.4.

Figure 3.10 presents a time history of motion of a center point of the homogeneous square plate for $n = M_x * M_y = 25$ and $w_0 = 0.5,\ 1,\ 2$.

The vibrations observed are nonlinear as well as roughly harmonic for any small displacements.

Some analogous graphs are given in Fig. 3.11 but the shell is distinguished here by the following geometrical quantities: $k_x = k_y = 12$ at $w_0 = 1; 3; 5$. A nonlinear character of vibrations is here observable even for any small initial excitations.

During the search for a solution to the problem of free nonlinear vibrations with the utilization of higher order approximations one can analyze the frequency (period) of the steady-state vibrations. Though, the fundamental question about either any amplitudes and frequencies of harmonics of that process remains still opened. In other words, there is a need to create an amplitude-frequency characteristic of vibrations. For the considered plates and shells we will restrict our attempts to the construction of first-order characteristics.

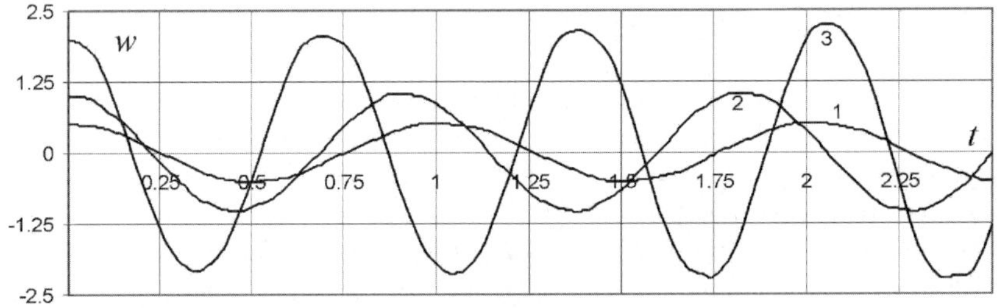

Fig. 3.10 Time history of deflection in the center of a square plate for various initial displacements

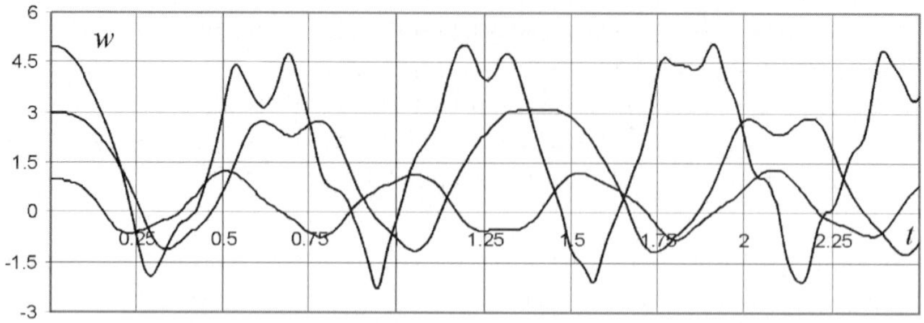

Fig. 3.11 Time history of deflection in the center of a square shell for various initial displacements

Subsequent sections of this chapter give particular attention to the method and the rationale of solving the stated problem.

3.4 Spectral Analysis of Solutions

By carrying out analysis of any plate or shell described by systems of n degrees of freedom it is possible to find a numerical solution to the stress-deflection problem of the investigated objects. One of the most primary tasks of this process is an estimation of the spectrum of frequency, of the analyzed system. We will propose an approach for solving such a problem. The approach can be successfully applied for a particular case in which the problem solution is given in the form of a graph, table, empirical formulae, or other sets of data that do not efficiently allow for the direct evaluation of frequencies and amplitudes of harmonics.

We will be concentrating on the exact solution, i.e., a solution that is represented by the double precision numerical approximation with a sufficient number of digits. In practice, this assumption is always true for any numerical integration of the problem under investigation. If the above is guaranteed then the solution obtained includes complete information about amplitudes and frequencies of harmonics.

If the period of vibrations is known then the problem is solved with the use of analysis of harmonics allowing for determination of all harmonic components of amplitudes and frequencies of the vibrating process. In general, if we do not have any prior knowledge about the period of vibrations but indeed the process possesses some periodic features, then the investigated problem can be mathematically formulated in the following manner.

One can prove that any function can be split into periodic components although they are not harmonically relevant to each other, i.e., the relation of their periods is not a rational number. The analysis aims at determination of unknown amplitudes and frequencies of each component.

Analyzing the problem, a number of function values are found for the same moment of time. One could make some proper conclusions exactly with the use of information collected from the numerical data. The problem is sometimes called *finding of hidden system frequencies*. Completing the above, the problem's solution can be found via application of Fourier transform [178].

Let us analyze a process with an odd number $2N+1$ of data. A mean value of our data series of which collection has been finished at τ will now correspond to the initial time $t=0$. As the new variable is introduced, then we can write

$$t = \frac{1}{\tau} t_1, \tag{3.15}$$

and the obtained coordinates will correspond to the following points of time:

$$t_k = 0, \pm 1, \pm 2, \ldots, \pm N.$$

The coordinates (function values) related to the moments of time are denoted by

$$f_k = f(t_k). \tag{3.16}$$

In general, the function $f(t)$ has the following shape:

$$f(t) = \sum_{\alpha=1}^{j} \left(A_\alpha \cos \theta_\alpha t + B_\alpha \sin \theta_\alpha t \right). \tag{3.17}$$

The number of terms of frequency is unknown a priori and let us denote it by j. Angular frequencies $\omega_\alpha = \theta_\alpha / \tau$ are found in a similar way. It is known that with respect to the new non-dimensional time t the angular frequencies θ_α should be bounded from above according to the inequality

$$\theta_\alpha < \pi. \tag{3.18}$$

It results from the observation that if θ_α does not satisfy the above restriction then the two different frequencies $\pi + \beta$ and $\pi - \beta$ cannot be distinguished. Let

$$\theta_\alpha = \frac{\pi}{N} p_\alpha, \tag{3.19}$$

and p_α takes values from $(0,\ N)$.

Let us separate the sine and cosine terms by introduction of summations and differences in the following form:

$$\begin{aligned} f(t) + f(-t) &= 2 \sum_{\alpha=1}^{j} A_\alpha \cos \theta_\alpha t, \\ f(t) - f(-t) &= 2 \sum_{\alpha=1}^{j} B_\alpha \sin \theta_\alpha t. \end{aligned} \tag{3.20}$$

Similarly, we split the coordinates into the two groups

$$u_k = f_k + f_{-k},$$

$$v_k = f_k - f_{-k}, \quad k = 0, 1, 2, \ldots, N. \tag{3.21}$$

For the summation, we apply a more convenient Fourier transform. The initial series of given u_k is now transformed into the $N+1$ series of amplitudes a_k at cosine terms, and the coefficients v_k into the $N-1$ series of amplitudes b_k at sine terms. The amplitudes mentioned are obtained from multiplication of the initial data by the formerly constructed matrix consisting of a proper π/N multiplicity of the sine and cosine terms

$$a_k = \sum_{\alpha=0}^{N}{}' u_\alpha \cos\frac{\pi}{N}\alpha k, \tag{3.22}$$

$$b_k = \sum_{\alpha=1}^{N-1} v_\alpha \sin\frac{\pi}{N}\alpha k. \tag{3.23}$$

The sign at $\sum'$ reflects the fact that the first and last components of the sum, means u_0 and u_N, are in the step of summation taken with weight 0.5.

Coefficients a_k and b_k, which were found according to the above, can be interpreted as a linear spectrum of the integer values p_α of the continuous parameter p. The entire further analysis will deal with an application of two series of those coefficients.

We will primarily investigate an instance in which the frequencies θ_α given from the above will satisfy the assumption that all or even some of p_α in Eq. (3.19) are any integer numbers. In this case, there is found a single peak without any collateral "tails." The isolated maximum is from both sides surrounded by zeros.

In such a situation the amplitudes of (3.22) and (3.23) determine the solution of the stated problem. In addition, most of the coefficients a_k and b_k will equal zero. Our input data will for nonzero coefficients a_k or b_k contain the following frequency:

$$\theta = \frac{\pi}{N}k. \tag{3.24}$$

The cosine function amplitude has the form

$$A = \frac{1}{N}a_k, \tag{3.25}$$

while the sine function amplitude is as follows:

$$B = \frac{1}{N}b_k. \tag{3.26}$$

If a maximum is not any integer, then the maximal amplitude borders from both sides upon the smaller amplitudes. The lowest inclination is observed in a case when the maximum is situated exactly in the half length between any two integer values p. We write at this point the following scheme:

$$\frac{1}{9}, -\frac{1}{7}, \frac{1}{5}, -\frac{1}{3}, 1, 1, -\frac{1}{3}, \frac{1}{5}, -\frac{1}{7}, \ldots. \tag{3.27}$$

The analyzed weak depreciation of amplitudes can be considerably accelerated if the input differences of amplitudes are considered. Because the signs $\pm$ appear alternately, the mentioned procedure is described as follows:

$$z_k = y_{k-1} + 2y_k + y_{k+1}. \tag{3.28}$$

The preceding scheme expressed by Eq. (3.27) takes the form

$$-\frac{8}{693}, \frac{8}{315}, -\frac{8}{105}, \frac{8}{15}, \frac{8}{3}, \frac{8}{3}, \frac{8}{15}, -\frac{8}{105}, \frac{8}{315}, \ldots. \tag{3.29}$$

The interference now yields a decrease proportional to the third power of the distance between any two peaks and finally becomes a negligible small value. The assumption must hold that the two peaks expected to be separated ones are situated effectively far from each other.

The exact position of the maximum that is calculated by means of the introduced differences (in fact, the repeating summations related to the alternate arrangement of the scheme signs) can be found through the following derivation. We identify the y_k series and verify the correctness of the alternate change of $\pm$ signs. It can be noted that in some points the alternation is disturbed by the appearance of a sequence of signs like $++$ or $--$. We estimate these irregularities and then make a conclusion that the peak should be placed between the two listed values

$$p = k \text{ and } p = k + 1. \tag{3.30}$$

In a more general case, an exact value of p related to a frequency θ_α will be placed between any two natural numbers k and $k+1$.

Subsequently, we are consider

$$p = k + \varepsilon \tag{3.31}$$

and the ε value is estimated by a scheme of interpolation. For that purpose, we will use the recurrent type of summation.

Let us construct a ratio of the two selected series z_p corresponding to $p = k$ and $p = k+1$, respectively

$$q_k = \frac{z_k}{z_{k+1}}. \tag{3.32}$$

In spite of the above, the series z_k is proportional to

$$-\frac{1}{\varepsilon - 1} + \frac{2}{\varepsilon} - \frac{1}{\varepsilon + 1} = \frac{2}{(\varepsilon + 1)\varepsilon(\varepsilon - 1)}, \tag{3.33}$$

while z_{k+1} is proportional to the following expression:

$$\frac{1}{\varepsilon} - \frac{2}{\varepsilon - 1} + \frac{1}{\varepsilon - 2} = \frac{2}{(\varepsilon - 2)(\varepsilon - 1)\varepsilon}. \tag{3.34}$$

In a result, one finds

$$q_k = \frac{z_k}{z_{k+1}} = \frac{2-\varepsilon}{\varepsilon+1}, \tag{3.35}$$

but according to the above

$$\varepsilon = -\frac{2-q_k}{1+q_k}. \tag{3.36}$$

Finally, as the values ε are determined we find the frequency θ_α in the form

$$\theta_\alpha = \frac{\pi}{N}(k+\varepsilon). \tag{3.37}$$

The coefficients A_α, B_α corresponding to the frequency depend on the amplitudes a_k and b_k of Fourier series, respectively as follows:

$$A_\alpha = \frac{a_k}{N}\frac{\pi\varepsilon}{\sin\pi\varepsilon},$$
$$B_\alpha = \frac{b_k}{N}\frac{\pi\varepsilon}{\sin\pi\varepsilon}. \tag{3.38}$$

The depicted methodology of isolation of periodic terms is notably effective if a total number of $2N+1$ observables is sufficiently large, i.e., there is a possibility to isolate any two neighboring "peak values." It is required to have the two successively appearing values p_α in Eq. (3.19) separated by at least four units. If they are situated closer than the above consideration demands, then it is not easy to separate them with a satisfactorily high accuracy.

The described method of discovering the "hidden frequencies" could be called *the spectroscope method*, because from the mathematical point of view it resembles a spectroscope mechanism. For instance, the spectroscope allows for detection of the components of radiation frequencies of a stimulated atom.

These important frequencies can be precisely determined by means of finite-width "spectral lines." The accuracy of the spectroscope measurements is dependent on a large number of optical vibrations N. It is of course possible to decrease the number N, and thereby to sustain a high accuracy by assumption of many significant digits of numbers. Although the valuable peaks are more condensed now, one can make a corrective operation by allowing the described approach to introduce an interference and analyze the influence of the adjacent peaks.

If we do not carry out sufficiently frequent observations, and our peaks lie very close to each other, then it is mandatory to apply a qualitatively different method. Understanding of the new concept demands that some crucial historical events constituting foundations of modern harmonical analysis are kept in mind.

The discovery of the acoustic meaning of vibrations as well as of their harmonics is attributed to the Greek philosopher and mathematician Pythagoras (569–475 BC). The innovative form of a series was composed of a component functions $\sin kx$ and $\cos kx$, where k is a natural number, was introduced (for the first time) in the 18th century by Euler and Lagrange. At that time, because of the still unknown definition of the limit, the concept of an infinite series could not been fully understood.

Lagrange had considered that an arbitrary superposition of any analytical, i.e., infinitely differentiable functions should lead to a complete analytical form of the resulting function. He considered on that basis that the harmonical analysis can be effectively used only for the functions that besides their continuity also have high-order derivatives.

In 1882, the brilliant Fourier's discovery abolished that restriction. A Fourier function is not given by a strict mathematical formula but, for example, it could comprise any number of snippets expressed in different analytical forms. In addition, the continuity of functions was not required. Fourier prepared and then announced some examples of harmonic analysis of a functions having a finite number of discontinuity points in the predetermined interval of time. Usually in the considered interval the function values were normalized to a new $(-\pi,+\pi)$ interval, which was sufficient for defining a function in the general form $y=f(x)$. The periodicity of function does not exist as the essential property and merely plays the principal role only outside of the basic interval of time. The theory of harmonic analysis does not require any analysis outside of the basic interval.

The second fundamental Fourier's discovery was his *Fourier integral.* With the use of the integral the method of harmonic analysis was generalized on the infinite interval $(-\infty,+\infty)$, but there was not any requirement for introduction of other periodicities of the function of decomposition. Later, for the case of a harmonic series approximation of a function, Dirichlet investigated the subject more carefully, stating some conditions to be satisfied by the *arbitrarily selected Fourier function*. These conditions, which are presently known as the *Dirichlet conditions*, are sufficient to preserve the Fourier series convergence, but they are not always the necessary ones.

Recently (in 1904), Feyer proposed a new method of summation of the Fourier series, which in a broad sense expands the domain of their applications. Using the notion of a mean arithmetic component sums, Feyer successfully realized the summation of convergent series. The only condition allowing to create this procedure was the possibility of integration of the investigated series.

The basic theorem connected with the procedure of such decomposition has the following form.

Let the function $y=f(x)$ satisfy the conditions:

1. $f(x)$ is defined for all points belonging to $-\pi \leq x \leq +\pi$;
2. $f(x)$ is unique, finite, and partially continuous. The $f(x)$ function can have a finite number of discontinuity points, and a finite interval should isolate any two successive points of the discontinuity;
3. $f(x)$ has a "finite variation," which means that in the considered interval the function cannot be characterized by an infinite number of maximums and minimums as well.

The above conditions imposed on functions like $f(x)$ are called the "Dirichlet conditions." A function that meets the Dirichlet conditions can be split into a convergent and infinite Fourier series of the form

$$f(x) = \frac{1}{2}a_0 + a_1 \cos x + a_2 \cos 2x + \cdots + \\ b_1 \sin x + b_2 \sin 2x + \cdots, \tag{3.39}$$

where the coefficients a_k, b_k are called the "Fourier coefficients" and are given by

$$a_k = \frac{1}{\pi} \int_{-\pi}^{+\pi} f(x) \cos kx dx,$$

$$b_k = \frac{1}{\pi} \int_{-\pi}^{+\pi} f(x) \sin kx dx. \tag{3.40}$$

One can define the $f(a)$ particular case in a point of discontinuity $x = a$ being the arithmetic mean of some boundary coordinates:

$$f(a) = \frac{1}{2}\left[f(a_+) + f(a_-)\right]. \tag{3.41}$$

It is worth mentioning that an infinite Fourier series that is estimated for $x = a$ converges to the above evaluated value.

If one assumes that the series in Eq. (3.39) is convergent to $f(x)$ for all points of a given interval, then one can multiply both sides of Eq. (3.39) by $\cos kx$ or $\sin kx$ and then integrate them from $-\pi$ to $+\pi$. As the result, we obtain the expressions (3.40) determining the coefficients a_k, b_k of decomposition. Excluding either $\cos kx$ or $\sin kx$, all the terms of the right-hand side of the equation at hand vanish. Formulae (3.40) are not complicated and were well known before Fourier's time. The question arises: Has the decomposition given by expressions (3.39) any sufficiently formal foundations? To find the answer, Dirichlet used the following consideration. Let us analyze the *finite series*

$$f_n(x) = \frac{1}{2}a_0 + a_1 \cos x + \cdots + a_n \cos nx \\ + b_1 \sin x + \cdots + b_n \sin nx. \tag{3.42}$$

Assuming that a_k and b_k are determined by (3.40), investigate what will happen if the value n were increased to infinity. Summing (3.42), we have

$$f_n(x) = \int_{-\pi}^{+\pi} f(t) K_n(x-t) dt, \tag{3.43}$$

where $K_n(\zeta)$ is the so-called "Dirichlet kernel"

$$K_n(\xi) = \frac{\sin(n+\frac{1}{2})\xi}{2\pi \sin(\frac{1}{2}\xi)}. \tag{3.44}$$

We are going to show that as n reaches infinity then the estimation of $f_n(x)$ approaches $f(x)$, i.e., the difference between the two final function values is any small real value. To capture the difference as small as possible a strong *focusing action* of the $K_n(\zeta)$ is expected. The function isolates a neighborhood of the $\zeta = 0$ point as well as does "the cancellation of all remainders." One can assume that the sequence of properties particularly constitute some mathematical formulations of the increasing focusing action of function $K_n(\zeta) = K_n(-\zeta)$:

$$\lim_{n\to\infty} \int_{\varepsilon}^{\pi} |K_n(\xi)|\, d\xi = 0, \tag{3.45}$$

$$\lim_{n\to\infty} \int_{-\varepsilon}^{+\varepsilon} |K_n(\xi)|\, d\xi = 1. \tag{3.46}$$

The first of the conditions mentioned that are imposed on the $K_n(\zeta)$ function preserves the *cancellation* in Eq. (3.43) of all components *excluding the most nearest neighborhood* of the $t = x$ point. The second one guarantees that the $t = x$ point takes part in integration with an appropriate weight (it is assumed here that the ε is the previously selected and arbitrarily small positive real number dependent on n).

However, the Dirichlet kernel does not satisfy the first written condition. Second, some maximal values of the function (3.44) are not sufficiently small to preserve the required property of the $\zeta = 0$ point. If the condition is met, then as a consequence, a decomposition of function $f(x)$ into an infinite and convergent Fourier series provides a restriction of class of the functions satisfying these mentioned requirements. They should be satisfactorily smooth to be under the adequate focusing action of the Dirichlet kernel. Dirichlet conditions 2 and 3 preserve the smoothness of the function under investigation.

Feyer dealt with the crucial as well as a radical change of the summation process. Consequently, it amplifies the Dirichlet kernel's focusing action and thereby expands the field of application of the Fourier series on a wider class of functions. This class of functions is restricted by only one condition, that $f(x)$ has to be "absolutely integrable" leading to preservation of existence of the integral

$$I_1 = \int_{-\pi}^{+\pi} |f(x)|\, dx. \tag{3.47}$$

The condition (3.47) is really obvious. Otherwise, the a_k and b_k Fourier coefficients determined by Eq. (3.40) would not exist.

The results obtained by Feyer come from observations and allow one to define the following kernel describing function:

$$K_n(\xi) = \frac{\sin^2(\frac{n\xi}{2})}{2\pi n \sin^2(\frac{\xi}{2})}. \tag{3.48}$$

The defined kernel is simply characterized by the strong focusing properties resulting from the study of Eqs. (3.45) and (3.46).

A short historical perspective of the analyzed problem permits drawing a such conclusion. *If we do not dispose of a sufficient information, and any peak values of investigated data series are strongly condensed lying very close to each other, then for the purpose of improvement of the obtained results by means of the stroboscope mechanism approach it is recommended in any known way to correct the focusing action of the Dirichlet kernel.*

It can be achieved most simply by application of a smoothing method, which is known in theory as the "σ method." For our case, the method relies on changing the form of input data series u_k and v_k, which are visible under the Fourier summations (3.22) and (3.23), respectively. Until formulation of these sums is made, we change the input data, simultaneously multiplying them by some properly selected weighting values and in accordance to the following expressions

$$\bar{u}_k = u_k \sigma_k,$$
$$\bar{v}_k = v_k \sigma_k, \tag{3.49}$$

in which

$$\sigma_k = \frac{\sin \frac{k\pi}{N}}{\frac{k\pi}{N}}. \tag{3.50}$$

The coefficients a_k and b_k in Eqs. (3.22) and (3.23) are now calculated with the use of these new values $\bar{u}_\alpha$, $\bar{v}_\alpha$.

The function $\frac{\sin \pi x}{\pi x}$ is via those changes transformed into the following one:

$$S(x) = \frac{1}{2\pi} \left[Si(x+\pi) - Si(x-\pi) \right]. \tag{3.51}$$

Side peaks of the function appear at the values 4.8, 2.0, 1.1%, ... measuring from the central maximum in comparison to the series of values like 21.7, 12.8, 9.1%, ... characterizing the initial function. As seen, the brand new function has a much stronger filtrate action, and its new maximums practically exhibit no dependence. At this time, peaks are more scattered in comparison to their counterpart coefficients of the initial function introduced.

The distribution of coordinates obtained has an ideal merit. In fact, one can introduce a second-order parabolic approximation going through the maximal amplitude and the two adjacent left and right maximums. As a consequence of the parabolic interpolation, a real maximum can be found. Continuing, let us examine a maximal amplitude a_k and its adjacent maximums a_{k-1} and a_{k+1}.

The variable ε in Eqs. (3.31) and (3.37) is expressed by

$$\varepsilon = \frac{1}{2} \frac{a_{k+1} - a_{k-1}}{2a_k - (a_{k+1} + a_{k-1})}, \tag{3.52}$$

and the maximal coordinate a_μ is expressed by

$$a_\mu = a_k + \frac{\varepsilon}{4}(a_{k+1} - a_{k-1}). \tag{3.53}$$

Finally, we obtain

$$A_\alpha = 1.6963\frac{a_\mu}{N} \tag{3.54}$$

(the numerical multiplier is the inverse of $S(0)$). Similarly, one can derive some calculations of the B_α amplitude of sine function by replacing a_k with b_k.

Let us estimate now the length of observation interval T for a continuous signal and with application of a fixed step of sampling. In a general case, the continuous signals $s(t)$ are taken with the use of a fixed interval Δ, so the *discrete* counterparts selected can be used in our further computations.

A discrete signal can be considered as the result of multiplication of the continuous input signal by the signal $i(t)$ that is composed of either a infinite series of singular impulses or the delta type functions [128]:

$$i(t) = \sum_{n=-\infty}^{\infty} \delta(t - n\Delta). \tag{3.55}$$

It leads to the impulse modulated signal

$$s_i(t) = s(t)i(t). \tag{3.56}$$

Consequently, utilizing the convolution theorem, we have

$$S_i(f) = \int_{-\infty}^{+\infty} S(f-g)I(g)dg, \tag{3.57}$$

where $I(g)$ is a Fourier transform of $i(t)$.

Using

$$S(f) = \frac{1}{\Delta}\sum_{n=-\infty}^{+\infty} \delta(f - \frac{n}{\Delta}), \tag{3.58}$$

and performing some transformations, one obtains

$$S_i(f) = \int_{-\infty}^{+\infty} S(f-g)\frac{1}{\Delta}\sum_{n=-\infty}^{+\infty} \delta(g - \frac{n}{\Delta})dg = \frac{1}{\Delta}\sum_{n=-\infty}^{+\infty} S(f - \frac{n}{\Delta}). \tag{3.59}$$

The last equality confirms that a discrete or an impulse-modulated signal $s_i(t)$ has a periodic $\frac{1}{\Delta}$-period Fourier transform. If $S(f)$ zeroes for $|f| \geq \frac{1}{2\Delta}$, then $S_i(f)$ is simply a periodically repeating function $S(f)$. It means that one could reproduce $S(f)$ by means of $S_i(f)$ just multiplying the $S_i(f)$ by a $H(f)$:

$$H(f) = \begin{cases} \Delta, & |f| \leq \frac{1}{2\Delta} \\ 0, & |f| > \frac{1}{2\Delta} \end{cases} \tag{3.60}$$

Concerning the region of frequency, the operation of multiplication of functions is relevant to a convolution of these functions in the time domain, therefore

$$s(t) = \int_{-\infty}^{\infty} \frac{\sin(\pi u/\Delta)}{\pi u/\Delta} s_i(t-u)du. \tag{3.61}$$

The function $\frac{\sin(\pi u/\Delta)}{\pi u/\Delta}$ can be understood as the perfect filter used for seeking the continuous signal $s(t)$ resulting from its discrete adequate $s_i(t)$. In other words, the function $\frac{\sin(\pi u/\Delta)}{\pi u/\Delta}$ is for any uniformly distributed coordinates a good example of an ideal interpolating function. The observation time T determines the degree of peak distinction for the Fourier transform, and the interval of sampling Δ is estimated by a maximal value of the corresponding frequency.

The frequency

$$f_N = \frac{1}{2\Delta} \tag{3.62}$$

is sometimes called the *Nyquist frequency*. It is identified in the theory as the most important frequency, which can be estimated on the basis of a series of numerical data that was found during application of a transient sampling period Δ.

Finally, to distinguish any two peaks with respect to the two frequencies f_1 and f_2, it is necessary for a non-rectangular window [128] to investigate the interval T of the length

$$T \geq \frac{1}{f_2 - f_1}. \tag{3.63}$$

Width of the non-rectangular windows is estimated by the inequality

$$T \geq \frac{2}{f_2 - f_1}. \tag{3.64}$$

3.5 Method Convergence

It has been theoretically proved in previous sections of this chapter that any input data like the length of observation and the interval of data handling are the decisive ones when we want to distinguish maximums (peaks) and determine the so-called Nyquist constant.

We next try to explain the convergence of the analyzed method with respect to frequency and amplitude of an isolated harmonic. Therefore, we formulate and then solve the following modelization problem.

Let a *table form periodic function of a constant step be given that has only one harmonic component. The step in the table is sensitive and varies, but in contrast the length of observation is constant.*

Using the table data as an input set of initial values, the following needs to be done:

(i) Isolation of the harmonic component, i.e., to determine numerically its frequency and amplitude;
(ii) Numerical estimation of the error of accuracy of such approximation.

As an example the following function is considered:

$$y = 5\cos\left(\frac{\pi}{7}t\right), \tag{3.65}$$

where the frequency of harmonic can be widely applied to our fundamental objects under investigation (plates and shells), i.e., it has to be the component of a low frequency as well as cosine functions in the basic interval $(-\pi, +\pi)$.

Starting from zero, the variable t takes some values with the following step of data handling:

$$\frac{\pi}{N}. \tag{3.66}$$

To solve such a problem we will use a variant of the formerly proposed method. It utilizes the feasibility of peaks division by the use of the method of side-sums. One needs to find an exact solution for such a defined problem, if N will be taken as the multiplicity of 7, i.e. $N = 7,\ 14, \ldots$, which exactly holds.

In a general case, if we do not know or cannot satisfy the required multiplicity condition, we will use a fixed step incrementation of N arbitrarily.

In our task, there is no sense in assumption for N any number that is less than 7, because the fundamental condition then will not be met to have any two neighboring peaks separated by at least four ones.

The results of this analysis are illustrated in Figs. 3.12 and 3.13. In Fig. 3.12a is shown the dependence of a harmonical frequency in the whole region of variations of the parameter N. The same graph has been enlarged in Fig. 3.12b, but with the exclusion of small values of the N parameter.

It could be concluded from the graphs that the process convergence is in respect to the frequency quite rapid. If one considers a scale of variations that is adequate to N, then it explodes on the $2N$ length intervals. In the middle of each interval the value of the harmonical frequency is extremal. The increase of number of the interval is associated with a decrease of extremes and the curves take some stationary

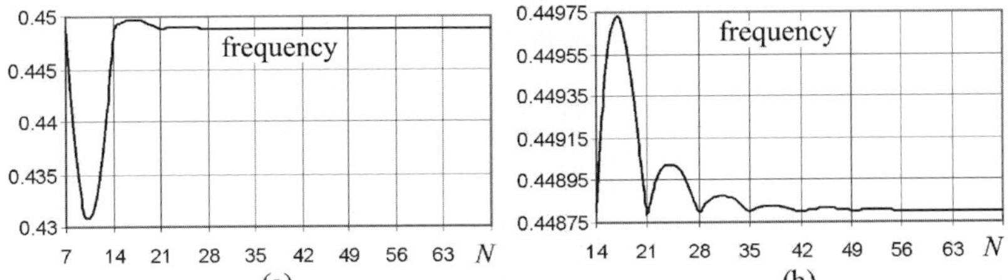

Fig. 3.12 The dependence of harmonic frequency on the number N of observations made in different intervals of time

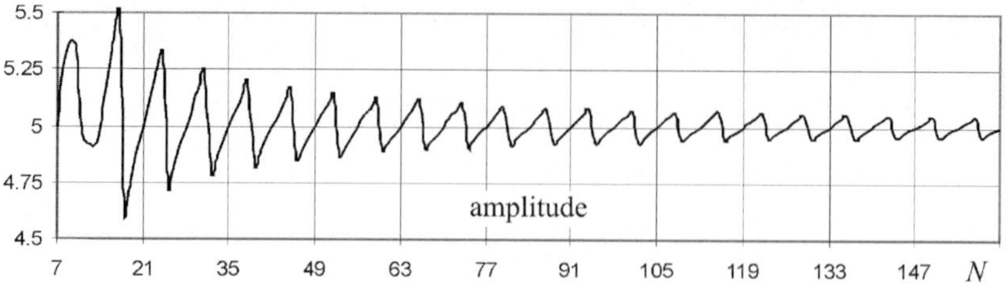

Fig. 3.13 The dependence of harmonic amplitude on the number N of observations made in different intervals of time

values, i.e., in terms of precision of numerical computations it approaches the exact value arbitrarily close.

If the analysis of harmonic process convergence with respect to an amplitude is taken into account, then it is mandatory to separate some singularities occurring here. A process of this kind is a bit slower concurrent, but in turn, from both sides, one notes a symmetrical convergence. The shape of the amplitude characteristics in each interval of length $2N$ is a transient mapping of the graph of a $\sin(x)$ function with damped amplitude. This advantage, as well as some symmetrical properties of the graph, allow for a more precise computation of the harmonic amplitude by using either average values, the whole interval, or a part of the interval.

3.6 Spectral Analysis of Free Vibrations

Nonlinear free vibrations were the subject of our study in Sect. 3.3. Let us return there and now spectrally analyze the solution obtained. Omitting the fact that any plate or shell regarded as a dynamical system has infinite degrees of freedom, we are not able to solve the problem of estimation of the entire spectrum of frequency. Instead of solving the task with higher order approximations, the concept of finite (sectioned) series for approximations of basic functions has to be realized. Possibly, any high modes of system vibrations are neglected a priori. During the Runge-Kutta scheme integration of a system of differential equations and with respect to the time variable the step of integration is chosen on the basis of the Runge principle. The interval of data handling constitutes the main parameter defining the Nyquist frequency.

The boundary problem given by (3.11)–(3.13) has been solved at $n = M_x * M_y = 25$, and the associated Cauchy problem in a time with the step $\Delta t = 0.001$. Keeping in mind that any two neighboring peaks should be isolated in (3.19) by at least four ones, then the data of our analyzed case will be separated for the frequencies included in 80–100.

To secure in Fourier transform the desired degree of peak distinction, the time of data (coordinates) monitoring should be sufficiently long, i.e., include a few periods of behavior of the investigated process of dynamical vibrations.

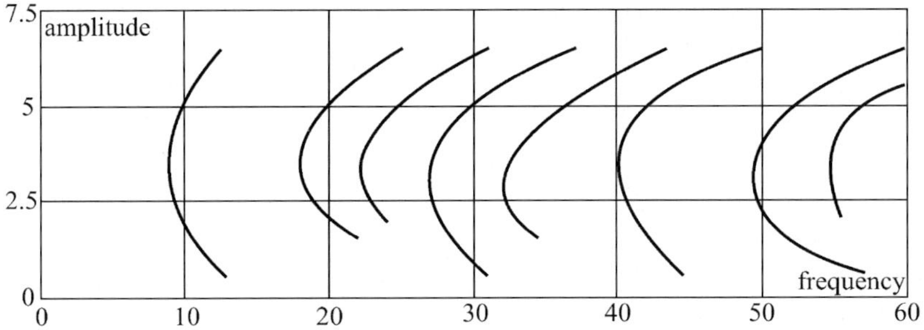

Fig. 3.14 Amplitude-frequency characteristics of free non-harmonical vibrations ($k_x = k_y = 12$)

Following the above directions, there some calculations were executed in a time for a non-dimensional form of data between 6 and 10.

Getting acquainted before with a spectrum of linear vibrations of the vibrating systems analyzed, one should predict to obtain the entire low and mean value of the demanded spectrum.

In Fig. 3.14 are shown the amplitude-frequency characteristics for a square homogeneous plate and shell ($k_x = k_y = 12$), respectively.

Summarizing, the above results provide an opportunity to draw the following conclusions:

1. The frequency of all detected harmonics for a plate and shell depends on the amplitude of free vibrations of the system.
2. At low amplitudes of free vibrations the frequency of the lower harmonical component coincides with the first frequency of linear vibrations of the system.
3. A plate can be qualitatively treated as a system described by some stiff characteristics, and any shell as a system described by some soft characteristics.
4. No harmonics of small amplitudes of free vibrations were detected, but with an increase of the amplitude some of them are easily detectable.
5. The amplitude of the first harmonic (for all of them in the analyzed problems) significantly exceeds the amplitudes of other harmonics, which at increasing number of the harmonics rapidly decrease.

Chapter 4
Dynamic Loss of Stability of Rectangular Shells

In this chapter dynamic stability loss of rectangular shells is addressed. A background containing types of dynamic buckling and prefect constructions, as well as the concept of finite-time stability, is given in Sects. 4.1–4.3. Mathematical modeling of dynamical systems, problems of synchronization, chaos, and quasi-periodicity are also briefly revisited. Sections. 4.6–4.10 refer to both static and dynamic bifurcations and their numerical estimations. Stability loss of homogeneous shells subjected to an action of transversal loads is rigorously studied in Sect. 4.11.

4.1 Types of Dynamic Buckling

Lavrientiev and Ishlinski [181] were the first to notice that the type of buckling of a construction under dynamic load can be qualitatively different from static buckling. This difference can be explained by the influence of inertial forces on the dislocations related to buckling. The construction does not "catch up" with taking the strain caused by a sudden change of environment. While solving the problem of buckling, the form that is usually determined is the one that reflects the quickest pace of displacements for a set impulse.

We conventionally distinguish the dynamic and impulse (impact) load [305]. The load will be called dynamic if during the analysis of the process of construction deformation only inertial forces representing ordinary displacements (deflections) of plates and shells can be taken into consideration. The mechanism of force transition in the central part of a construction is not the subject of our interest here. In other words, we shall assume that those changes take place in a way right in a given moment. In case of a clear impulse load, apart from the aforementioned inertial forces, the impact of inertial forces in some directions on the central surface of a shell or a plate should be taken into account.

Let us investigate the basic diagram of the relation between w — the displacement characterizing buckling process and q — a load parameter shown in Fig. 4.1. First, assume that the dynamic load increases monotonically in time. We shall begin

DOI: 10.1007/978-3-540-77676-5_5,

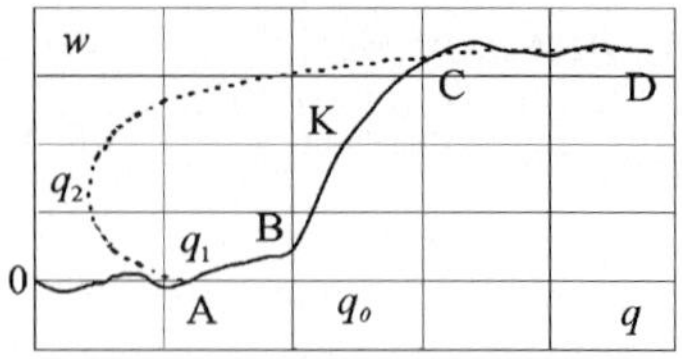

Fig. 4.1 Example of dynamic buckling

our considerations with the model of a non-perfect shell, i.e., initially buckled in relation to the set form of balance. Research in this field has shown that actual constructions or experimental stands are always characterized by certain imperfections of shape, whose size significantly depends on the technology of production. Reviewing the literature on this problem, one can reach a conclusion that for carefully prepared shells, the amplitude of initial buckling can be taken for further calculations as 0.001 of the shell's thickness.

We shall assume that the shell with initial imperfections undergoes the load quickly increasing in time, in compliance with some law, e.g., the law of linearity. The process of shell motion that is observed here can be divided into three steps.

The first step is connected with small vibrations occurring around the initial static balance point (OAB curve), and those vibrations lead mainly to the concentration of stresses in the central surface.

The following step is characterized by a relatively sudden leap of the construction to a new point of balance and is connected with significant deflections (BC curve).

Eventually, the third step (CD) involves the occurrence of nonlinear vibrations around the new point of balance or around both static points of balance.

From the practical point of view, the most important step is the second one, i.e., the transition process involving dynamic buckling. Analyzing it allows one to establish a certain conventional dynamic quantity, the so-called "buckling load" represented by q_0 in Fig. 4.1.

Unlike in the problems of statics, where the concept of upper and bottom buckling load is relatively easy to establish, in dynamics, there is no such possibility. Various researchers dealing with dynamic stability use various criteria of shell stability, emphasizing some characteristics of the analyzed problem. Usually, the quantity of deflection equal to the thickness of an analyzed shell is taken as the criterion of stability loss in a dynamic sense. The half or other part of the thickness of an analyzed shell can be assumed as a basic parameter identifying the loss of dynamic stability. Other researchers describe "buckling" load by using the position of a central point K of a curve connected with this transition step or the position of the inflection point on this curve.

If the buckling is sudden enough, then all the aforementioned points correspond to the same value of stresses on the middle surface, for instance.

The other possibility of evaluating the stability loss of a construction involves establishing certain parameters used for estimating possible danger. The moment of obtaining stress on the central surface of a shell that corresponds with the moment of flow can be taken as this parameter.

Let us emphasize here that all the examples introduce the notion of "dynamic buckling load" only conventionally. It is because in this case there is no point related to the occurrence of any bifurcation, dividing, e.g., static and non-static branches of construction or showing the transition from one static branch to the other, non-static, one. If one tried (on the basis of existing criteria) to determine the character of motion of the system, it could be observed that the whole part of the motion trajectory of a point characterizing the motion of a system on surface q–w or q–t, which is beyond the range of the upper buckling load, can be dynamically stable. However, if the buckling process proceeds quickly, it practically defines the dynamic stability loss in a general sense, as a certain unexpected process of leaving a static point of balance and motion toward another state of equilibrium.

4.2 Perfect Constructions

Let us now analyze the model of a perfect construction. The literature presents much research on the problem of analyzing such constructions [305].

One of the dominant directions in the research on stability is the analysis, in a Lyapunov sense, of the stability of a "basic" motion of a system. Some researchers conclude that the basic state turns out to be non-static if the load parameter exceeds the value of the upper buckling load.

The other approach suggests taking into account only those construction parts where, or around where, there is some influence of the load, whereas the remaining part of a construction is free from stress. The area of the dynamic buckling is determined from the condition that the critical value of a load equals the value of the upper buckling load. Both approaches mentioned above in fact conclude that the pre-critical stresses are described on the basis of dynamic relations, whereas the very process of buckling is connected with a static load.

Many researchers used parametric resonance theory in their studies of the phenomenon of buckling. It is well known that when the construction is parametrically loaded, in some conditions it will start to produce vibrations of certain amplitude and the balance point of a system will lose stability, which will consequently allow one to determine the boundaries of the system stability loss.

Parametric resonance theory can in fact be used for the description of the process of "pumping" energy from the basic motion to the energy connected with a dynamic buckling. It turns out, however, that due to the character of the very transition of a system to another state, the phenomenon of parametric resonance is different from the dynamic stability loss for a single load.

4.3 The Concept of Finite-time Stability

Let us start with the analysis conducted by Feldstein and presented in [305]. So far we have discussed possible criteria of the stability loss of an elastic system with an aperiodic load. One of the approaches to this problem involves analyzing the phenomenon of shell buckling with the transition to vibrations around the other balance point. This way of solving the problem uses certain set condition in the form of limitation to some parameter characteristic for deformation, e.g., maximal buckling as a criterion of stability loss. It must be emphasized that nowadays the concept of stability is analyzed in much deeper sense than in the case of Lyapunov theory. Limitation to the buckling size is related to fulfilling certain requirements that the construction should comply with. Stability here means rather adjusting the construction to perform certain functions under the influence of outer loads. Such a concept of stability is close to the notion of finite-time stability, according to which the system will be stable if forcing its motion does not manage to exceed a certain set quantity in an earlier established range of time.

The purposefulness of this approach will be presented using the following considerations. Let us consider an elastic system, whose motion is described by the equation

$$\Lambda \mathbf{u} + \mathbf{p} = \ddot{\mathbf{u}} + \alpha \dot{\mathbf{u}}, \tag{4.1}$$

where $\mathbf{u}$ = displacement vector, Λ = differential operator matrix, $\mathbf{p}$ = outer force vector, and a dot stands for differentiating in relation to time.

Energy dissipation was taken into account in the aforementioned model on the basis of the model of viscous linear damping. We shall consider the solutions fulfilling certain boundary conditions:

$$\Gamma(\mathbf{u}) = 0, \ \mathbf{u}(0) = \mathbf{u}_0, \ \dot{\mathbf{u}}(0) = \dot{\mathbf{u}}_0. \tag{4.2}$$

Let the system move in the way described by vector $\mathbf{U}$, different than the solution $\mathbf{u}$ of the problem of Eqs. (4.1)–(4.2) under some undefined external forces. We shall present it as

$$\mathbf{U} = \mathbf{u} + \mathbf{v}, \tag{4.3}$$

where $\mathbf{v}$ is the perturbation, whose analysis is described by the problem of analyzing the motion stability of solution $\mathbf{u}$.

Let operator Λ be presented as the following sum of a linear part Λ_1 and a nonlinear part Λ_2:

$$\Lambda = \Lambda_1 + \Lambda_2. \tag{4.4}$$

After substituting Eq. (4.3) into (4.1), one obtains

$$\begin{aligned} &(\Lambda_1 + \Lambda_2)\mathbf{u} + \mathbf{p} - \ddot{\mathbf{u}} - \alpha \dot{\mathbf{u}} \\ &= \ddot{\mathbf{v}} + \alpha \dot{\mathbf{v}} - \Lambda_1 \mathbf{v} - \Lambda_2 \mathbf{v} - [\Lambda_3(\mathbf{u})]\mathbf{v}, \end{aligned} \tag{4.5}$$

where

$$[\Lambda_3(\mathbf{u})]\mathbf{v} = \Lambda_2(\mathbf{u} + \mathbf{v}) - \Lambda_2(\mathbf{u}) - \Lambda_2(\mathbf{v}) .$$

Since $\mathbf{u}$ is the solution of Eq. (4.1) with some additional conditions of Eq. (4.2), the left-hand side of expression (4.5) equals zero. Assuming that the perturbations are minor, we shall linearize the right side of Eq. (4.5) and present the equation of perturbation motion as

$$[\Lambda_1 + \Lambda_3^*(\mathbf{u})]\mathbf{v} = \ddot{\mathbf{v}} + \alpha \dot{\mathbf{v}}. \tag{4.6}$$

Solution $\mathbf{v}$ of this equation should satisfy uniform boundary conditions and certain initial conditions. If dynamic effects are omitted in a perturbation equation, then it describes a balance point, and from the condition of the existence of some non-trivial solutions of this equation bifurcation points are determined. In dynamics, solving the problem of motion stability $\mathbf{u}$ is reduced, in a Lyapunov sense, to analyzing the behavior of solution $\mathbf{v}$ of the disturbed equation around the zero point (a trivial point of balance).

Let us present solution $\mathbf{v}$ as the sum of some complete set of function $\mathbf{v}_k$:

$$\mathbf{v} = \sum_k f_k \mathbf{v}_k. \tag{4.7}$$

Substituting Eq. (4.7) into (4.6) and requiring orthogonality of the result of substitution to all the selected functions, we obtain:

$$\sum_k \int_S \left[\ddot{f}_k \mathbf{v}_k + \alpha \dot{f}_k \mathbf{v}_k - f_k \Lambda_1 \mathbf{v}_k - f_k \Lambda_3^*(\mathbf{u})\mathbf{v}_k\right] \mathbf{v}_n \, dS = 0. \tag{4.8}$$

If $\mathbf{v}_k$ are the eigenfunctions of operator Λ_1, the last expression will be simplified:

$$\ddot{f}_n + \alpha \dot{f}_n + \omega_n^2 f_n = \sum_k f_k \varphi_{kn}, \tag{4.9}$$

where

$$\varphi_{kn} = \frac{\int_S \left[\Lambda_3^*(\mathbf{u})\mathbf{v}_k\right] \mathbf{v}_n \, dS}{\int_S \mathbf{v}_k \mathbf{v}_n \, dS}.$$

If $\mathbf{p}$ is not time-dependent, the coefficients of equations are constant. Should this quantity periodically change in time, then the Eqs. (4.9) will be Mathieu-Hill type.

Let us pay attention to a certain important quality. During the influence of a momentary impulse load, free vibration occurs in a shell and $\mathbf{u}$ will be a periodic function of time. Sometimes researchers relate the occurrence of axially asymmetric vibration of rotary shells loaded with impulse outer pressure to parametric resonance and to the vibrating ground stress state, and then they use the methods applied in analyzing periodic external load in the analysis. We should remember, however, that as a result of damping, in actual shells the axially symmetric vibration is damped, and harmonic vibration cannot significantly increase although it formally is in an unstable area. Let us also remember that the source of energy working long enough is an indispensable factor exciting parametric vibrations.

Let us come back now to Eq. (4.9) and consider finite-time loads for $t \to \infty$:

$$\lim \mathbf{p}(t) = \mathbf{p}_0. \tag{4.10}$$

In this case, the coefficients φ_{kn} also have boundaries:

$$\lim \varphi_{kn}(t) = \varphi_{kn}^0, \tag{4.11}$$

and for high enough t we can consider the following boundary system:

$$\ddot{f}_n + \alpha \dot{f}_n + \omega_n^2 f_n = \sum_k f_k \varphi_{kn}^0. \tag{4.12}$$

On the basis of the Chetayev theorem [74] the zero solution of the system of Eqs. (4.9) will be asymptotically stable if the solution of a "boundary system" is stable. Therefore, it is necessary that the roots of a characteristic equation have negative real parts.

The characteristic equation will have the following form:

$$D(\lambda) = \begin{vmatrix} \varphi_{11}^0 - \omega_1^2 & \varphi_{21}^0 & \cdots & \varphi_{m1}^0 \\ \varphi_{12}^0 & \varphi_{22}^0 - \omega_2^2 & \cdots & \varphi_{m1}^0 \\ \cdots & \cdots & \cdots & \cdots \\ \varphi_{1m}^0 & \varphi_{2m}^0 & \cdots & \varphi_{mm}^0 - \omega_m^2 \end{vmatrix} - \mathrm{Tr}(\lambda(\alpha + \lambda)) = 0. \tag{4.13}$$

In the above we have reduced our considerations to a finite system of equations. In this case, if $\mathbf{p}_0 = 0$ and $\mathbf{u}(\infty) = 0$, $\varphi_{kn} = 0$, the characteristic equation will have the following form:

$$\prod_{i=1}^{m} \left[\lambda(\alpha + \lambda) + \omega_i^2\right] = 0. \tag{4.14}$$

Real parts of the above equation are negative. We can conclude that for the loads disappearing in time, non-disturbed motion is asymptotically stable, since all perturbations vanish for $t \longrightarrow \infty$.

Nevertheless, it does not mean that those perturbations are small in any time instant. In some conditions they can be relatively large, which is proved by experimental data connected with axially asymmetric buckling of cylindrical shells under external impulse pressure. A shell may lose its capacity both under the influence of a load in finite time and because of the following loads connected with the development of the process of non-stationary perturbations. It this case it seems reasonable to determine the requirements related to the parameters characterizing stress or strain states, even though in a classical approach, the shell remains stable. In order to solve the problem of choosing a stability limit, the stability limit for solving practical problems with regard to the kind of work the construction performs and operating conditions should be taken into account. Data related to experimental research should also be considered.

4.4 Mathematical Models of Vibrating and Dynamic Systems

General laws of vibration processes in physical systems are the subject of a science called the theory of vibration. Vibration means either vibration established around a certain point of balance or the transition process between one fixed state and another. Established vibrations are usually repeatable, limited, and stable. Transition processes may be characterized by certain established (steady state) motions that they approach. The set of transitory processes related to a given established motion describes its area of attraction. A qualitative change of established vibrations, a result of the change of any parameter of a vibrating system, describes the system's bifurcation. If the change of vibrations is a sudden snap-through, we observe the "hard" formation of a new vibration regime of a system. Otherwise, the formation of a new vibration regime is called "soft" [65].

The notion of a dynamic system was introduced as the generalization of the notion of a material system, whose motion is described by Newton differential equations. Currently, the concept of a dynamic system is discussed in a wide range of fields, including physical, chemical, biological, economic, and other systems, both deterministic and stochastic. A mathematical description of dynamic systems is equally wide: it can be realized through differential equations, functions, graphs, Markov chains, etc.

At present, two approaches dominate analysis of dynamic systems. One is based on the concept of state x, characterizing system S in a certain time instant and on the notion of operator T, describing the change of this state x in time. Operator T defines the procedure that if used allows to describe $x(t)$ in a time instant t and to determine the state $x(t+\Delta t)$ of this system in a certain following time step $t+\Delta t$. If operator T does not overtly depend on time, the system S is referred to as autonomous, and otherwise as non-autonomous.

State x of system S can be regarded as a point of certain space Φ, called the phase-space of system S. A change of state x is represented in phase space Φ by the motion of a point called a representative point of a system. During this motion, a representative point describes a certain curve called a phase trajectory. The phase space Φ and operator T form a mathematical model of a dynamic system. Analyzing the behavior of a dynamic system in this approach is reduced to analyzing the division of a phase space Φ into trajectories and explaining the structure of this division in relation to physical parameters of a system.

An other approach to the analysis of dynamic systems involves analyzing the functional aspect of a studied system. Application of this approach may be the result of a lack of possibilities of understanding all inner details of the studied dynamic system. Therefore, in this case, a system is treated as a sort of "black box" for known input and output signals. A black box realizes the relation between those signals through a certain operator.

In this approach, a mathematical model is described by input and output spacer, and by the operator that clearly realizes the transition between input and output variables.

Mathematical models of dynamical systems can be classified according to the structure of their phase space Φ and the form of operator T. We distinguish between constant and discrete phase spaces, depending on the values that variables x, characterizing the state of a constant or discrete dynamical system can take. The operators T are conventionally distinguished in relation to their qualities and a set form. If the operator T has the quality of superposition, it is referred to as linear. If the operator T is nonlinear, a corresponding dynamic system is called nonlinear as well.

Among nonlinear systems, self-excited ones have a special role. The notions of "self-excited vibration" and "self-excited systems" were first introduced 50 years ago by Andronov [13]. Characteristic qualities of self-excited systems result from their nonlinearity. The right sides of differential equations that model them include nonlinear functions of phase variables x.

Studies on the behavior of a dynamic system are reduced to the analysis of its trajectory in phase space Φ. The structure of division of phase space Φ into phase trajectories is called a phase portrait of a dynamic system. From a geometric point of view, the structure of division of phase space into phase trajectories means a geometric image of phase trajectories in space Φ. It must be emphasized that a complete description of a phase portrait for an arbitrary dynamic system is a complex problem and so far has not been solved. Nevertheless, a range of basic qualities of such a structure has been studied, and for some classes of dynamic systems a full description of their phase portraits has been achieved.

One of the basic tasks in describing a phase space of a dynamical system is distinguishing between ordinary and singular phase trajectories. The flatter ones are the singular points corresponding to balance points of a system or their stationary motion, insulated closed trajectories called limit cycles and corresponding to periodic motion, separating curves, and the surfaces that are the boundaries of attraction areas to various stable singular orbits. Singular elements of phase space Φ can form integral manifolds. They can, in turn, be divided into stable, unstable, and saddle. Stable balance points and periodic solutions are simple examples of established motion.

The notion of motion stability is one of the most important ones in the theory of nonlinear vibrations. Among many existing definitions of stability, stability in a Lyapunov sense and orbital stability are most frequently applied. In the analysis of balance point stability, both concepts overlap. The point of balance $x = x^*$ is regarded as stable if for any number $\varepsilon > 0$ it is possible to find a number $\delta(\varepsilon)$ small enough for the following inequality to be satisfied for all the consecutive values of $x = x(t)$ for a different motion x^* with initial conditions different from δ, by less than t:

$$\rho(x(t), x^*) < \varepsilon, \tag{4.15}$$

where $\rho(x(t), x^*)$ is the distance between phase points with coordinates $x(t), x^*$. The state of equilibrium will be regarded as asymptotically stable if the quantity ρ additionally approaches zero during an unlimited increase of time. A type of singular point is determined by the behavior of phase trajectories in its close environment (see [17, 18]).

If autonomous dynamical systems with one degree of freedom are analyzed, then equations of their motion are generally described by two differential equations of first order:

$$\dot{x} = P(x, y), \; \dot{y} = Q(x, y), \tag{4.16}$$

whose right sides are nonlinear functions x, y. Equations (4.16) have been thoroughly analyzed with the use of the qualitative theory of differential equations. According to Eqs. (4.16), the state of a system of the second order is described by values x, y, and therefore the phase space of such system is two-dimensional, i.e., it is a certain plane. The basic problem related to the analysis of a dynamical system is presenting a complete division of phase plane into trajectories, or in other words, determining the topological structure of this division. The concept of topological structure involves the qualities that remain invariant during topological (i.e., mutually interchangeable and constant) transition of a plane into itself.

It turns out that in order to explain the qualitative image of a second-order system, one must have knowledge not of all the trajectories, but only of those regarded as singular. *These are balance points, limit cycles, and some non-closed trajectories to which at least one half-trajectory (i.e., a curve described by a representative point for $t \to +\infty$ or $t \to -\infty$ from the initial position of the point in time instant $t = t_0$) is related that is separating (a separating curve) at some balance point.* If the mutual position of these singular trajectories is known and we also know the stability of balance points and limit cycles, then we know the complete division of plane x,y into trajectories.

Physical intuition suggests that for differential Eqs. (4.16), describing the motion of a real physical system, none of the factors that we have taken into account can stay absolutely invariable in time. Therefore, the right sides of Eqs. (4.16) hang with the parameters they include. However, if these changes are small enough, as practice suggests, the physical system will not "notice" them and the qualitative characteristics of dynamics will not change. Therefore, if we want differential equations (4.16) to show this singularity, we should require their perturbation resistance (robust): for small parameter values, qualitative structure of plane division into trajectories does not change. In this way, we have distinguished a subclass of resistant dynamic systems. *The resistance of a dynamic system can be understood as the stability of a structure of the corresponding phase plane on trajectories in relation to small changes introduced to differential equations.*

Andronov and Pontriagin [12] formulated a mathematical description of the notion of resistance (robustness) for a second-order system. According to their suggestion, a dynamic system described by Eqs. (4.16) will be referred to as resistant if there is a small number $\delta > 0$ for which all the dynamic systems are described by the following differential equations:

$$\dot{x} = P(x,y) + p(x,y), \dot{y} = Q(x,y) + q(x,y) \tag{4.17}$$

in which analytical functions $p(x,y)$, $q(x,y)$ occur, satisfying the following inequality:

$$|p(x,y)| + |q(x,y)| + \left|\frac{\partial p}{\partial x}\right| + \left|\frac{\partial p}{\partial y}\right| + \left|\frac{\partial q}{\partial x}\right| + \left|\frac{\partial q}{\partial y}\right| < \delta, \tag{4.18}$$

and have the same structure of phase plane division into trajectories.

Let the right sides of the analyzed system of differential Eqs. (4.17) depend on a certain parameter λ, i.e., let them have the following form:

$$\dot{x} = P(x,y,\lambda), \ \dot{y} = Q(x,y,\lambda), \tag{4.19}$$

where $P(x,y,\lambda)$ and $Q(x,y,\lambda)$ are analytic functions of their own arguments.

If for a certain value λ a system is resistant, then, according to what was said before, for minor changes, a qualitative image on a phase plane will not change. However, this condition cannot be satisfied for all values of parameter λ. Hence, the notion of a bifurcational value of a parameter is introduced. According to definition [18], the value of parameter $\lambda = \lambda_0$ is called bifurcational if for any of its close values in relation to λ_0, i.e., for $\lambda > \lambda_0$ and $\lambda > \lambda_0$, the topological structure of a phase plane is different. The very definition of the bifurcational value of a parameter implies that for $\lambda = \lambda_0$ a dynamic system is not resistant.

As the qualitative image of a trajectory on a phase plane is determined by its singular elements (singular trajectories), only those values of parameter λ that involve non-resistant singular elements will be regarded as bifurcational.

4.5 Synchronization, Chaos, and Quasi-Periodicity

The subject of discussion of this section does not lie within the conventional field of interest of the theory of vibrations [65]. The theory of vibrations studies traditional random vibrations as a result of some stochastic disturbance. It does not take into consideration the possibility of deterministic dynamic system random (chaotic) vibrations, perhaps without stochastic waves and turbulence. One can assume that this is so because the researchers focus rather on analyzing simple instances of motion, i.e., balance (singular) points, periodic motion, and quasi-periodic motion to some extent. More complex types of motion were treated as impossible to analyze and having little importance in relation to real objects. A way of thinking resulting from the considerations connected with a phase plane did not allow discovery of such possibilities of motion, and the only premise for the existence of random motion was connected with analyzing the systems with many degrees of freedom, where everything becomes complex and confusing. The phenomenon of randomness (stochasticity) in physics and mechanics was related to many degrees of freedom and hence to many possible waves and vibrations.

It turns out, however, that this approach is not the right one. Randomness can occur in dynamic systems with only a few degrees of freedom. It is enough for the phase space to be larger than the two-dimensional one. The occurrence of such a motion is just as well-founded as the occurrence of periodic motion or the existence of balance points.

Depending on the parameter values, the same dynamic system may or may not have some qualities of randomness. For some parameter values, there is no randomness and the dynamic system is characterized by balance (equilibrium) points or periodic motion, and for other sets of parameters we have chaotic motion. For a constant transition from some parameter values to other ones, complex changes of the vibration process occur. They can be realized in a constant or sudden way. In the former case, the occurrence of chaotic vibrations can be regarded as soft; in the latter case it can be referred to as tough stiff (hard). It is in full analogy with soft and stiff excitation of self-excited vibrations for a static form of balance (equilibrium).

Let us clarify here what we mean by a general mechanism of the occurrence of chaotic vibrations. It is best to recall self-excited vibrations, with which we are already familiar. What is the mechanism and the reason for the occurrence of self-excited vibrations? In the case of soft excitement, we observe the existence of an unstable point of balance, and it is around this point that the vibrations begin, but up to a certain level only. Self-excited vibrations are the result of a compromise between instability and damping of the vibrations with higher amplitudes, which results in a periodic motion [18]. It is not the only mechanism of the occurrence of such vibrations, but it is very frequently observed. In some cases, it is possible to identify certain physical reasons and force interactions leading to instability and self-excited vibrations.

Let us proceed now to the description of one of the mechanisms of occurrence of chaotic vibration that can be presented as simultaneous lack of synchronization in the movements of particular parts of an analyzed system. It is the prolongation of changes resulting from the fact that the obtained periodic solution loses stability, and a two-dimensional integral manifold is isolated from it, consisting of two-period movements. At the same time, it is assumed that the Poincaré rotation number for such a track is rational or irrational for a high enough numerator or denominator. Next, this integral manifold loses stability and a three-dimensional integral manifold is isolated, consisting of three-periodic movements, etc., until a stable toroidal manifold consisting of m-periodic incommensurable movements is obtained.

During m-periodic motion, the law of change of each of phase variables can be presented in the following way:

$$x = \psi(\omega_1 t + \varphi_1, \ldots, \omega_m t + \varphi_m), \tag{4.20}$$

where $\psi(u_1, u_2, \ldots, u_m)$ is a periodic function with period 2π with relation to the variables $u_1, u_2, \ldots, u_m$. The process described by Eq. (4.20) is not chaotic. It is either a periodic process with a high period growing with the increase of m, or it is a quasi-periodic process. It is characterized by the quality of approximate repeatability for long enough monitor time $T(\varepsilon)$ (ε describes the accuracy of repetitions). Only for time ranges smaller than $T(\varepsilon)$ does it resemble chaotic motion. After long observation, the "chaos" of such a process can be explained by revealing its quasi-periodicity. It is yet very difficult to realize, because a kind of overlapping of small changes (fluctuations) that usually occur in a real process is very disturbing. Time $T(\varepsilon)$ is generally much higher than particular periods $2\pi/\omega_1, \ldots, 2\pi/\omega_m$ and the increase of number m does not cause the increase of such period times, and therefore small phase changes $\Delta\varphi_1, \ldots, \Delta\varphi_m$ can result in such a set of phase dislocations in time $T(\varepsilon)$ that the process repeatability is disturbed and chaos occurs.

The system with many harmonic oscillators with a weak influence on one another can be discussed as a model describing the phenomenon of chaos. The oscillators are described by the following differential equations:

$$\ddot{x}_i + \omega_i^2 x_i = \mu f_i(x_1, \ldots, x_n, \dot{x}_1, \ldots, \dot{x}_n, \mu), \tag{4.21}$$
$$i = 1, 2, \ldots, n,$$

that after the introduction of new variables ρ_i, φ_i and replacement of

$$x_i = \rho_i \sin(\varphi_i) \tag{4.22}$$

are reduced to the following system with quick-changing phases:

$$\begin{aligned} \dot{\varphi}_i &= \omega_i + \mu \Phi_i \left(\varphi_1, \ldots, \varphi_n, \rho_1, \ldots, \rho_n, \mu\right), \\ \dot{\rho}_i &= \mu R_i \left(\varphi_1, \ldots, \varphi_n, \rho_1, \ldots, \rho_n, \mu\right). \end{aligned} \tag{4.23}$$

In such a system, multi-periodic movements are possible that form stable manifolds on the torus. During periodic motion, all the component oscillators vibrate with an established frequency and established phase differences. Periodic motion may be treated as a toroidal manifolds of size one. Increasing the size of a toroidal manifolds makes the vibrations of component oscillators less and less consistent with the increase of the size of a toroidal manifolds, and eventually there is no relationship whatsoever between them for the maximal size. Reducing synchronization between the oscillators results in an increase of the randomness of vibrations (chaos). The size of a formed toroidal manifold depends on the relationship between frequencies $\omega_1, \omega_2, \ldots, \omega_n$. The occurrence of simple resonance relations between frequencies causes the reduction of the size of a toroidal manifold until synchronized vibration occurs. The concept of simple resonance relations denotes the case when for some relatively small integers k_1, $k_2, \ldots, k_n$, the following inequality is satisfied:

$$k_1 \omega_1 + k_2 \omega_2 + \ldots + k_n \omega_n = 0. \tag{4.24}$$

The more such simple resonance relations we observe, the smaller is the size of a possible toroidal manifold, and hence the higher is the degree of synchronization between oscillators. On the other hand, the lack of such simple resonance relations results in the occurrence of multifrequent vibrations, for which if the influence of fluctuations is taken into account by adding to the right sides of Eq. (4.20) minor random extortions ζ_i and η_i random phase dislocations $\varphi_1, \varphi_2, \ldots, \varphi_n$, take place that are proportional to the dispersion of disturbances ζ_i and that increase in time t with velocity $\sqrt{t}$.

If we add here that the forms of low-frequency vibrations become unstable, and the mechanism of limiting them is connected with high-frequency energy dissipation, we see how weak turbulence is formed.

It means that in the case of a model governed by Eq. (4.23), the equilibrium position at $x_1 = x_2 = \ldots = x_m = 0$ of the truncated system

$$\begin{aligned} \ddot{x}_i + \omega_i^2 x_i = \mu f_i \left(x_1, \ldots, x_m, 0, \ldots, 0, \dot{x}_1, \ldots, \dot{x}_m, 0, \ldots, 0, \mu\right), \\ i = 1, 2, \ldots, m, \end{aligned} \tag{4.25}$$

which is equivalent to a low part of spectrum of the possible frequencies of vibrations is unstable (one assumes that Eq. (4.23) take into account increasing of frequencies in a way, that $\omega_1 < \omega_2 < \ldots < \omega_n$), while the following system:

$$\ddot{x}_i + \omega_i^2 x_i = \mu f_i (0, \ldots, 0, x_{m+1}, \ldots, x_n, 0, \ldots, 0, \dot{x}_{m+1}, \ldots, \dot{x}_n, \mu), \quad i = m+1, \ldots, n, \tag{4.26}$$

equivalent to high frequencies of vibrations, is dissipative and possesses a global stable position of equilibrium.

During the process of interaction between the unstable and dissipative parts of a system, the "pumping" of energy from the low-frequency to the high-frequency forms takes place, and, as a result, a certain spectrum of vibrations is formed that is related to the distribution of amplitudes of particular oscillators with frequencies close to $\omega_1, \omega_2, \ldots, \omega_n$.

Such a distribution of amplitudes can be found if average relations describing interactions between particular oscillators (vibration forms) of a system are known.

Above, certain assumptions and premises of the "theory of quasi-periodic chaos" are given. Minor fluctuations and the unique mechanism of gathering them, a kind of "chaos enhances" are essential here. From the point of view of our considerations, the phenomena of chaos and synchronization were opposite. The occurrence of synchronization results in damping the symptoms of chaos, and the development of chaos, in turn, leads to the reduction of the importance of synchronization of particular elements (oscillators) of an entire analyzed system.

4.6 Static Bifurcations and Catastrophe Theory

In the introduction to Thompson's monograph entitled *Instabilities and Catastrophes in Science and Engineering* [290] Lighthill pointed to some basic and unique feature of nonlinear systems that is familiar to the researchers working in engineering, physics, mechanics, astronomy, and biology. What is so unique and singular is that a certain equilibrium point can become unstable during a constant parameter change, and the so far continuous process can also become discontinuous. Recently, active collaboration between mathematicians and the researchers representing the aforementioned branches of science has led to the formulation of a new bifurcation theory that has allowed us to understand many phenomena in different fields of science from a certain single point of view [173].

The mathematical theory of bifurcation and instability is historically rooted in mechanics and astronomy, and thanks to the efforts of many topology researchers, it has become an independent branch of science. Two important achievements should be emphasized here: a complete catastrophe theory by Thom [289] and Zeeman [330, 331], based on a topological concept of structural stability, and the implications of the discovery of strange attractors responsible for chaotic motion (see also [18]).

What should be also noted is the difference between static stability that is the subject of catastrophe theory, and dynamic stability occurring, e.g., in flatter-type self-excited vibrations caused by wind in planes or suspension bridges.

Rapid developments in science, especially in applied mechanics, have led to specialization and to the formulation of various versions of initially general solutions. What should be especially emphasized is the essential role of the theory of elastic stability. It is connected with the reaction of elastic bodies and constructions on mechanical loads, and it has important technological applications for estimating critical force causing the stability and carrying capacity loss of constructions in engineering. Let us recall here that there are two different kinds of mechanical systems, namely conservative systems (this category also includes systems with low-energy dissipation that are conservative without it), and nonconservative systems, usually associated with an unlimited source of energy. As far as conservative systems are concerned, it is important to pay attention to the classical research on bifurcation conducted by Koiter [137], that are based on continual formulation. After eliminating passive deformations, the system's energy is expressed by an algebraic function depending on the amplitudes of vibration forms that are responsible for the stability loss. A modern explanation of nonlinear bifurcation phenomena of continuous systems under conservative load was proposed by Budiansky [63]. In the case of nonconservative systems, some important elements of a linear classification were also introduced by Ziegler [335].

The basic relationship between modern research and the research conducted by Euler and Lagrange is connected with the contribution of such researchers as Thom [288] and Zeeman [332]. A catastrophe theory is essential in the classification of static instabilities. Some basic results obtained in the catastrophe theory are given in Table. 4.1 (see [249]).

In order to find these relationships in conservative discrete systems described by potential functions, Thom applied the topological concept of structural stability. Catastrophe theory allowed an experimental explanation of the observable forms of instability occurring with the change of control parameters. Therefore, if there is only one control parameter λ, in a general case only the catastrophe of composition type can be observed, whose local potential energy is given in a table.

If there is independent control over two parameters λ^1 and λ^2, which can represent transverse and axial load on, e.g., a rod, then "fold-type" bifurcations can

Table 4.1 Characteristics of seven elementary bifurcations

Wrinkle	$\underline{q^3} + \lambda^1 q$
Fold	$\underline{q^4} + \lambda^2 q^2 + \lambda^1 q$
Dovetail	$\underline{q^5} + \lambda^3 q^3 + \lambda^2 q^2 + \lambda^1 q$
Butterfly	$\underline{q^6} + \lambda^4 q^4 + \lambda^3 q^3 + \lambda^2 q^2 + \lambda^1 q$
Hyperbolic umbilical	$\underline{q_2^3 + q_1^3} + \lambda^1 q_2 q_1 - \lambda^2 q_2 - \lambda^3 q_1$
Elliptic umbilical	$\underline{q_2^3 - 3q_2 q_1^2} + \lambda^1 (q_2^2 + q_1^2) - \lambda^2 q_2 - \lambda^3 q_1$
Parabolic umbilical	$\underline{q_2^3 q_1 + q_1^4} + \lambda^1 q_2^2 + \lambda^2 q_1^2 - \lambda^3 q_2 - \lambda^4 q_1$

be additionally observed. In the case of independent control over three parameters λ^1, λ^2, and λ^3, one can additionally observe a "dovetail" and hyperbolic and elliptic catastrophes. For four control parameters, any catastrophe given in a table can be observed.

The examples mentioned above include all the structurally stable singularities that can be observed in the surrounding world with the use of one to four control parameters. The first four catastrophes from the list have only one active generalized coordinate q, just as for a simple stability loss, whereas the last three catastrophes are characterized by two active coordinates q_1 and q_2, just as in the case of simultaneous stability loss in relation to two forms.

A more detailed classification is possible only for bifurcation analysis with only one bifurcation parameter. Thom's prediction that pathological phenomena can be observed only for a high enough order of equation should not be forgotten.

4.7 "Wrinkle-Type" Catastrophe or a Limit Point

In the case of a "wrinkle-type" catastrophe, there is only one coordinate, so it is easy to illustrate Thom's ideas [249] in Fig. 4.2. In this figure, the letters Q and Λ denote standard variables, (V means full potential energy).

If the shape V was drawn as the shape of function Q, then the minimum and maximum values will be in the points where $\partial V/\partial Q$ equals zero. However, in this case, there will be no inflection point in which the first and second derivatives will equal zero at the same time. The occurrence of such a critical point should be treated as pathology and the probability of its occurrence would be zero. The only possibility for observing the inflection point would be drawing the family of curves parameterized by curve Λ, which is shown in the picture. In order to observe the critical state of equilibrium in which the two first derivatives equal zero, parameter Λ has to be changed, which exactly reflects the process of loading the construction until it loses stability.

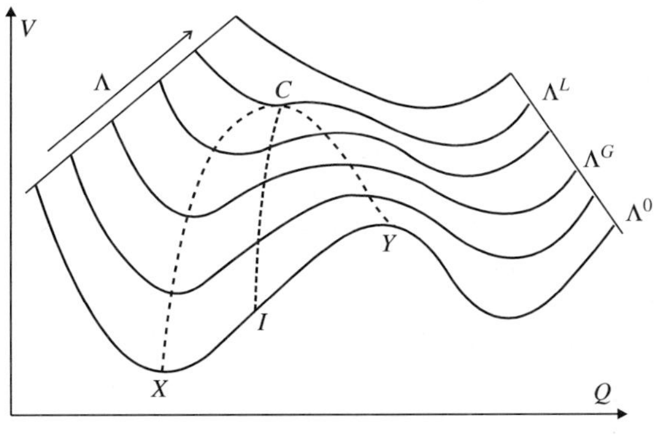

Fig. 4.2 Change of energy value for a "wrinkle-type" catastrophe

The simplest energy transformation involving overlapping and disappearing of the minimum and maximum parameter under one control is called the "wrinkle-type" catastrophe. It is represented by the balance trajectory *XCY* (see Fig. 4.2), that bends in point *C*, changing the character of stability at the same time.

4.8 A "Fold-Type" Catastrophe or Symmetric Bifurcation

If in the analysis of an unstable symmetric point of bifurcation an additional imperfection parameter ε, was introduced, the image presented in Fig. 4.3 would be obtained. The balance trajectory of a perfect (without imperfections) system is surrounded by the balance trajectories of the system with imperfections [249].

It is obvious that while loading a real, and hence always having imperfections, construction, the balance trajectory will not directly cross the branch point and it will represent one of the trajectories close to the balance trajectories of a perfect system. This means that during a one-parameter load the instability will be realized in the same way as in the case of "wrinkle-type" bifurcation. It will be possible to observe the very branch point experimentally only when load and imperfection parameters are changed simultaneously. In fact, this branch point represents the occurrence of a "fold-type" catastrophe, and the need for a two-parameter analysis confirms Thom's intuition. A stable symmetric branch point occurs, e.g., in the problem related to the analysis of stability of the Euler rod. Such a symmetric point can also be classified on the basis of terminology introduced by Thom as a "wrinkle," because of their topological similarity.

A fold of a form of symmetric bifurcations is formed in all the sciences based on mathematics. Thanks to the catastrophe theory, structural "fold-type" stability with the use of only two control parameters can be demonstrated.

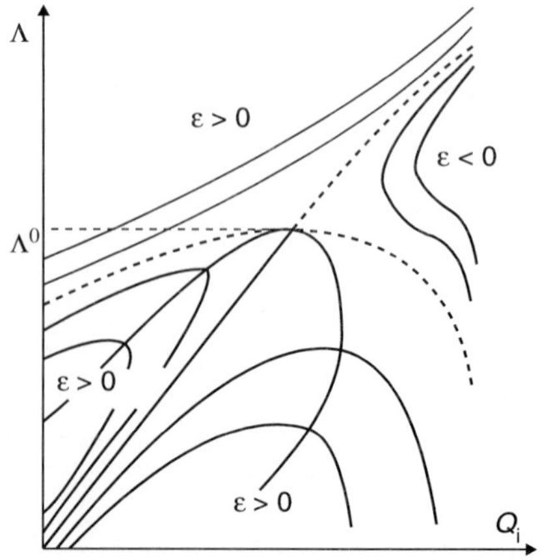

Fig. 4.3 Sensitivity for imperfection in an unstable symmetric point of bifurcation

Using Table 4.1, is easy to notice that the hyperbolic umbilici catastrophe is a symptom of one of the higher order catastrophes from Thom's list. This catastrophe involves applying three control parameters and two active coordinates, and hence the surfaces of balance lie in a five-dimensional space and it is impossible to physically present them.

As far as higher order bifurcations that are not listed in the list of seven elementary catastrophes are concerned, they often appear in the problems of optimization of constructions connected with symmetry unique to these problems. The increase in the number of forms results in generating complex secondary bifurcations.

4.9 Dynamic Bifurcations

Unlike static bifurcations related to equilibrium points, dynamic bifurcations are usually connected with trajectory changes in the form of limit cycles. If we take into account wind blowing at a uniform rate flowing around a given elastic construction, it can evoke and maintain high-amplitude vibrations, even causing the destruction of this construction. Three different mechanisms of aerodynamic instability are distinguished: one mode vibrations, resonance, and two-modes flatter.

We shall assume that all the mechanical systems analyzed below have inertial, elastic, and additionally determined dissipation forces. Taking dissipation into account allows one to construct approximate models of actual phenomena. With such an assumption concerning dissipation, the minimum of general potential energy of a discrete system is an indispensable and sufficient condition of the stability of conservative systems. The theory of dynamical systems allows one to understand the notion of structural stability and it clarifies the necessity for maintaining the nonlinearity of an accepted model (in a bifurcation point) as well as for taking damping into consideration.

Taking damping into account allows one to eliminate numerous famous paradoxes in engineering, e.g., finite destabilization for infinitely small damping. The theory of dynamical systems shows that Hopf bifurcation and saddle-node point bifurcation (or "wrinkle-type") catastrophe are possible with the change of only one control parameter.

One of the most interesting results obtained in dynamics over the last few-years is the notion of a *strange attractor* and the *occurrence of chaos* observed already in the system described by simple ordinary differential equations. Chaotic phase portraits of *strange attractors* can be observed already in case of very simple nonlinear dynamical systems in a three-dimensional space.

Already Rössler [261] noted: "If periodic vibration is typical of the behavior of a two-dimensional dynamical system, then chaos is typical of three-dimensional dynamical systems." Here, chaos will be understood as "an infinite number of unstable periodic random trajectories."

4.10 Criteria for Practical Computations

The considerations above allow one to calculate dynamic coefficients directly used in practical calculations. In the case of rods and plates, the coefficient of dynamical amplification is determined as the relation of dynamic buckling load to Euler load.

Let us recall here that in the case of a rod or a plate, the values of buckling load determined experimentally are close to those determined theoretically.

Many more difficulties appear when one tries to determine similar criteria related to thin shells. In this case, as already mentioned, buckling is caused by the process of the shell's snap. This process is related to the stability loss of a *system in a "large" sense*.

Let us focus for a while on some criteria of dynamic stability proposed by various researchers and used by them in their calculations.

Volmir [304] took the rapid bend increase with a minor change of load as a dynamic criterion.

Shian et al. [270] showed that the load reflecting the change of direction and allowing to obtain the first maximum on the load-time characteristics can be regarded as critical.

Kantor [130] analyzed axially symmetrical spherical shells with the use of the Ritz method with higher approximations and he established the condition of the bend in the middle of the shell to be larger than the shell's thickness as a dynamic criterion, which is expressed by the following formula:

$$K = \frac{2f}{2h}, \tag{4.27}$$

where f is the height of a shell's lift.

In spherical rectangular shells, the dynamic buckling loads obtained according to Kantor criterion practically overlap with those obtained according to the Volmir criterion.

In spherical shells, a non-dimensional curvature is expressed by the geometry of a shell as follows:

$$k_x = \frac{a^2}{R_x|2h|}, \; k_y = \frac{b^2}{R_y|2h|}, \tag{4.28}$$

i.e., it is a complex parameter involving all the geometrical dimensions of a shell.

In work [13] a criterion of Lyapunov stability connected with the application of the notion of the phase plane of a system is formulated. Krysko et al. [74] applied this criterion to determine the stability of a rectangular spherical shell by setting the initial imperfection on the basis of theoretical considerations or on the basis of the conditions of a technological process.

Work [76] offers the possibility to describe dynamic buckling loads in nonlinear problems of the theory of shells. The idea of this approach involves adding small perturbations to the solutions of the equations of motion. The sign before the squares of perturbation frequencies determines the system's stability.

4.11 Stability Loss of Homogeneous Shells under Transverse Loads

4.11.1 Feasibility of the Obtained Results

In order to check the feasibility of the obtained results, let us analyze the example of the problem of the stability loss of a homogeneous shell under transverse load $q = const$, whose solution will be searched for on the basis of the following tools:

1. The Bubnov-Galerkin method with higher approximation
2. The finite differences method.

In order to solve a problem formulated in this way, we shall use Eqs. (1.55) and (1.56), apply the boundary conditions of Eq. (2.94), and set the following initial conditions:

$$w = 0, \quad \frac{\partial w}{\partial t} = 0, \quad \text{dla} \quad t = 0. \tag{4.29}$$

While solving the problem with the use of Bubnov-Galerkin method, we shall search for functions *w, F* (satisfying the boundary conditions) of the following form:

$$w = \sum_{i,j} A_{ij}(t)\sin(i\pi x)\sin(j\pi y),$$
$$F = \sum_{i,j} B_{ij}(t)\sin(i\pi x)\sin(j\pi y), \tag{4.30}$$
$$i = 1,2,\ldots M_x; \quad j = 1,2,\ldots M_y; \quad n = M_x * M_y.$$

Substituting Eq. (4.30) into (1.55) and (1.56) and applying the Bubnov-Galerkin procedure with higher approximations in relation to spatial coordinates, we will obtain a system of second-order differential equations that is next reduced to an ordinary form and the system of algebraic equations in relation to $B_{ij}(t)$.

We solve the system of differential equations with the use of the fourth-order Runge-Kutta method [140] *i*-th the constant step of integration, whose value is chosen on the basis of the Runge law. During each step of integrating the ordinary system, we simultaneously solve the system of algebraic equations on the basis of the Gauss method [115].

Let us first apply the method of finite differences, and hence let us present a system of Eqs. (1.55) and (1.56) in the form of finite differences. Let us use the square mesh with step $h = 1/n$, where n denotes the number of the division of segment [0; 1].

Let us apply a thirteen-point difference diagram (Fig. 4.4). By replacing the differential operators from Eqs. (1.55) and (1.56) by approximating them as finite differences with error $O(h^2)$, we will obtain an analogical system of equations which we will solve also using Runge-Kutta and Gauss methods.

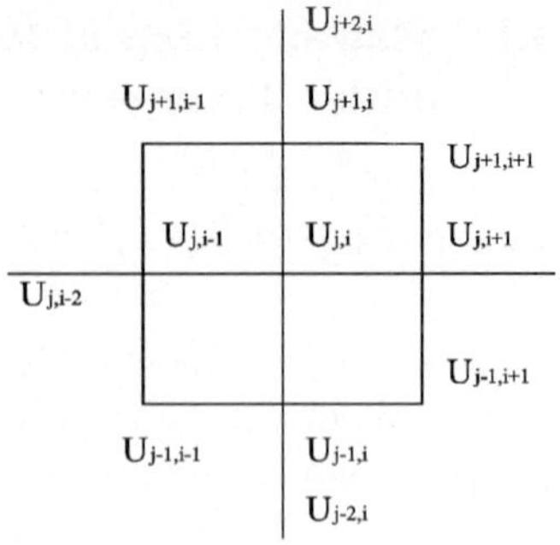

Fig. 4.4 Applied difference diagram

We will give the equations for node (j,i) as an example:

$$
\begin{aligned}
&\frac{d^2 w_{j,i}}{dt^2} + \varepsilon \frac{d w_{j,i}}{dt} = q - \frac{\lambda^2}{12h^4} w_{j-2,i} \\
&- \left[\frac{1}{6h^4} + \frac{1}{8h^4} \left(F_{j+1,i+1} - F_{j+1,i-1} - F_{j-1,i+1} - F_{j-1,i-1} \right) \right] w_{j-1,i} \\
&+ \left[\frac{1}{3h^4} + \frac{\lambda^2}{3h^4} + \frac{1}{h^4} \left(F_{j,i+1} - 2F_{j,i} + F_{j,i-1} \right) \right] w_{j-1,i} \\
&- \left[\frac{1}{6h^4} - \frac{1}{8h^4} \left(F_{j+1,i+1} - F_{j+1,i-1} - F_{j-1,i+1} - F_{j-1,i-1} \right) \right] w_{j-1,i+1} \\
&- \frac{1}{12\lambda^2 h^4} w_{j,i+2} + \left[\frac{1}{3\lambda^2 h^4} + \frac{1}{3h^4} + \frac{1}{h^4} \left(F_{j-1,i} - 2F_{j,i} + F_{j+1,i} \right) \right] w_{j,i-1} \\
&- \left[\frac{1}{2\lambda^2 h^4} + \frac{2}{3h^4} + \frac{\lambda^2}{2h^4} + \frac{2}{h^4} \left(F_{j-1,i} - 2F_{j,i} + F_{j+1,i} \right) \right. \\
&\left. + \frac{2}{h^4} \left(F_{j,i+1} - 2F_{j,i} + F_{j,i-1} \right) \right] w_{j,i} - \frac{1}{12\lambda^2 h^4} w_{j,i+2} \\
&+ \left[\frac{1}{3\lambda^2 h^4} + \frac{1}{3h^4} + \frac{1}{h^4} \left(F_{j-1,i} - 2F_{j,i} + F_{j+1,i} \right) \right] w_{j,i+1} \\
&- \left[\frac{1}{6h^2} - \frac{1}{8h^4} \left(F_{j+1,i+1} - F_{j+1,i} - F_{j-1,i+1} + F_{j-1,i-1} \right) \right] w_{j+1,i+1} \\
&- \left[\frac{1}{3h^4} + \frac{\lambda^2}{3h^4} + \frac{1}{h^4} \left(F_{j,i+1} - 2F_{j,i} + F_{j,i-1} \right) \right] w_{j+1,i} \\
&- \left[\frac{1}{6h^4} + \frac{1}{8h^4} \left(F_{j+1,i+1} - F_{j+1,i-1} - F_{j-1,i+1} + F_{j-1,i-1} \right) \right] w_{j+1,i+1} \\
&- \frac{\lambda^2}{12h^4} w_{j+2,i-1} + \frac{k_x}{h^2} \left(F_{j-1,i} - 2F_{j,i} + F_{j+1,i} \right) + \frac{k_y}{h^2} \left(F_{j,i+1} - 2F_{j,i} + F_{j,i-1} \right);
\end{aligned}
\tag{4.31}
$$

$$\lambda^2 F_{j+2,i} + 2F_{j+1,i-1} - 4\left(1+\lambda^2\right) F_{j+1,i} + 2F_{j+1,i+1} + \frac{1}{\lambda^2} F_{j,i-2}$$
$$-4\left(\frac{1}{\lambda^2}+1\right) F_{j,i-1} + \left(6\lambda^2 + \frac{6}{\lambda^2} + 8\right) F_{j,i} - 4\left(\frac{1}{\lambda^2}+1\right) F_{j,i+1}$$
$$+\frac{1}{\lambda^2} F_{j,i+2} + 2F_{j-1,i-1} - 4\left(\lambda^2+1\right) F_{j-1,i} + 2F_{j-1,i+1} + \lambda^2 F_{j-2,i}$$
$$= \left(1-\mu^2\right)\left[\frac{1}{16}\left(w_{j+1,i+1} - w_{j+1,i-1} - w_{j-1,i+1} + w_{j-1,i-1}\right)^2\right.$$
$$-\left(w_{j,i+1} - 2w_{j,i} + w_{j,i-1}\right)\left(w_{j+1,i} - 2w_{j,i} + w_{j+1,i}\right)$$
$$\left. -h^2 k_x \left(w_{j+1,i} - 2w_{j,i} + w_{j-1,i}\right) - h^2 k_y \left(w_{j,i+1} - 2w_{j,i} + w_{j,i-1}\right)\right]. \quad (4.32)$$

While solving this problem with the use of the Bubnov-Galerkin method (BG), we shall assume in Eq. (4.30) $n = 25$. While applying the method of finite differences, we divide the shell into $6*6$ and $12*12$ parts.

Figure 4.5 presents the relations between the bending in the center of a shell and time $w-t$ or the shell with parameters $k_x = k_y = 24$ and for the load intensity $q = 150$; 220; 250 (respectively, curves 1, 2, 3), and for medium damping $\varepsilon = 3$. Curve 4 represents $q = 250$ division of a shell into $6*6$ parts.

While integrating the differential equations with the use of the Runge-Kutta method, the step in time was chosen on the basis of maintaining the solution stability by complying with the Runge law. Continuous curves represent the solution with the use of the Bubnov-Galerkin method, and the point curves characterize the solutions basing on the method of finite differences. For pre-critical loads (curves 1, 2) both methods give overlapping results, and, what is more, the solutions obtained

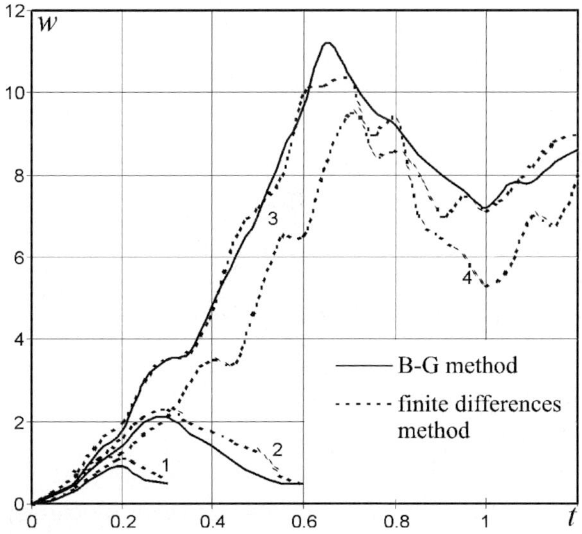

Fig. 4.5 Dynamic buckling of a homogeneous shell

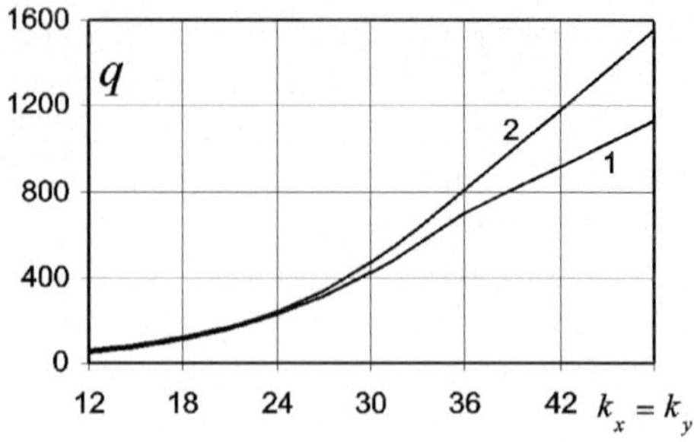

Fig. 4.6 The relation of the dynamic buckling load of a shell and parameter $k_x = k_y$

with the use of finite differences with the division into $6*6$ and $12*12$ parts also partially overlap. In the case of post-critical loads (curves 3, 4) the division into $6*6$ parts (curve 4) is not sufficient, and only the division into $12*12$ parts makes the results overlap with those obtained with the use of the BG method. The solution obtained on the basis of the finite difference method with the division of a shell into $12*12$ parts oscillates around the solution obtained with the use of the BG method. The buckling load obtained with the use of both methods is practically the same, but the time of calculations is much longer for the finite difference method. Therefore, the BG method is much more economical.

4.11.2 Buckling Load and Parameter $\mathbf{k_x = k_y}$ *of a Homogeneous Shell*

Let us analyze the relation of a dynamic buckling load of a homogeneous shell and parameter $k_x = k_y$. In order to solve the boundary problem of Eqs. (1.55), (1.56), (2.94) we shall apply the method basing on the Bubnov-Galerkin approach. We shall solve the obtained Cauchy problem with the initial conditions of Eq. (4.29) on the basis of the Runge-Kutta method. In order to determine the buckling load, we will use Volmir criterion for the stability loss of a shell. The obtained results of calculations are given in Fig. 4.6. Curve 2 was obtained by taking into account the medium's damping $\varepsilon = 3$.

The diagrams presented here show a certain monotonic relation between the buckling load and parameter $k_x = k_y$ of a shell that can be also observed in the static case. The increase of this parameter results in the increase of the buckling load. Medium damping in the analyzed case for the value $k_x = k_y < 24$ has practically no influence on the quantity of the buckling load, and in the case of large differences of this parameter value, it causes a significant increase of the buckling load.

4.12 Stability Loss of Heterogeneous Shells Under Transverse Load

Let the shell be characterized by certain heterogeneity (Fig. 2.8a), in which one element is situated in the center of the shell. Let us remember that the degree of heterogeneity is characterized by the value of the stiffness coefficient γ_{1j} and density coefficient γ_{2j} of the element, where o denotes the number of an element. In the case of a homogeneous shell, both those coefficients equal one. The value of a coefficient γ_{1j} smaller than one represents a softer shell, and the other way around. The same refers to coefficient γ_{2j}.

Let the shell undergo a load with value q that is constant in time and uniformly distributed. In such problems, the choice of the criterion of stability loss is very important. In this case, we shall comply with the concept of system stability in a finite time range for the realization of a dynamic stability loss in a "large" sense, i.e., we will be interested in this load value for which the shell undergoes buckling in an established time range. Since the shell is homogeneous, it is easier for our further considerations to introduce the notion of relative load, i.e., the load of a heterogeneous shell related to the analogous load of a homogeneous shell $\frac{q}{q_0}$ where q_0 is the load of a homogeneous shell

4.12.1 Relation Between Buckling Load and the Surface of an Extra Element

Let us explain now the way in which dynamic buckling load of a heterogeneous shell depends on the central surface of a square extra element for an established value of coefficients of stiffness γ_{1j} and density γ_{2j} of an extra element.

Figure 4.7 presents the relations between dynamic buckling load of a heterogeneous shell as related to the corresponding buckling load of a homogeneous shell

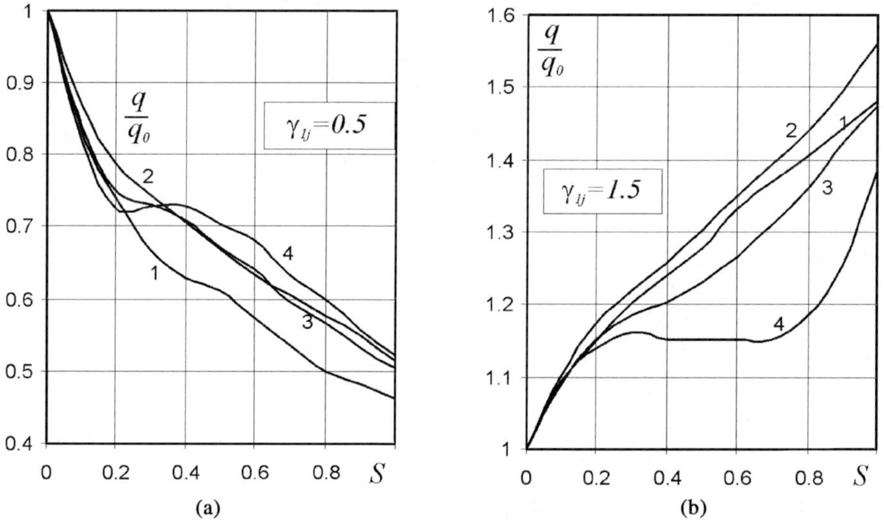

Fig. 4.7 Relation between dynamic buckling load of a heterogeneous shell and the surface of the central stiffness element: (**a**) "soft" shell, (**b**) "stiff" shell

and the surface of a central element for the cases of soft and stiff shells. $\gamma_{2j} = 1$ was assumed, and curves 1, 2, 3, 4 correspond with parameter values of a square shell with curvatures $k_x = k_y = 12$; 18; 24; 30.

The analysis of the diagrams shows that the relation of the buckling load for soft and stiff shells with parameter $k_x = k_y = 12$; 18 is practically monotonic. A further increase of this parameter results in the occurrence of bifurcations. One load value can correspond to several values of an added element surface. Vibration energy is distributed between homogeneous and heterogeneous parts of a shell, depending on the qualities of its stiffness parameters.

4.12.2 Relation Between the Buckling Load and Stiffness Coefficient of an Extra Element

In this case the value of parameter S denoting the surface of a square central element and the density coefficient is determined, and the stiffness coefficient is replaced.

Figure 4.8a,b,c shows the diagrams of the relation between the relative buckling load and the element stiffness coefficient γ_{1j}. $S = 0.4$ and $\gamma_{2j} = 1$ were assumed, and a shell parameter was assumed as $k_x = k_y = 18$; 24; 30.

One may conclude on the basis of the given diagrams that for small values of parameter $k_x = k_y$ the relation is monotonic and close to linear. For further increase of this parameter, two values of stiffness parameters $\gamma_{1j} = 0.9$; 1.1 (0.9, "soft"; 1.1, "stiff" shell) are important. Beginning with these values, the analyzed relation is significantly nonlinear, i.e., it has a significant role in energy distribution between homogeneous and heterogeneous parts of a shell. The values of buckling loads for the analyzed parameters changed by 20–25%.

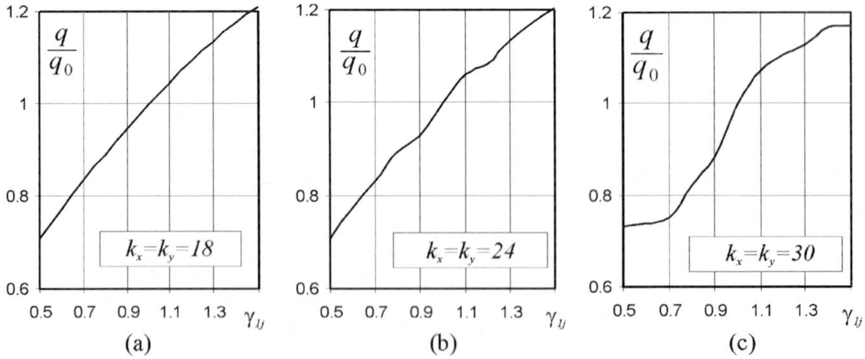

Fig. 4.8 Relation between dynamic buckling load of a heterogeneous shell and stiffness coefficient of an element

4.12.3 Relation Between Buckling Load and the Number of Reinforcement Elements Situated Along One Side of a Shell

For such a pattern of heterogeneity ("perforation-type," Fig. 2.8c) the shell is uniformly covered with square extra elements. In this case it is easier to use not the total number of extra elements covering the shell, but to use the number N of elements along only one side of the shell. The relation between the total surface of extra elements and number N is presented in Fig. 2.16.

Figure 4.9a,b shows the relation between the relative dynamic buckling load and the number of extra elements situated along one side of a "soft" shell, i.e., the stiffness coefficient of elements is $\gamma_{1j} = 0.5$; $\gamma_{2j} = 1$. Curves 1–6 correspond with the values of surface parameter $k_x = k_y = 12$; 18; 24; 30; 36; 48, respectively.

In Fig. 4.10a,b similar diagrams represent a "stiff" shell $(\gamma_{1j} = 1.5;\ \gamma_{2j} = 1)$.

All curves have a clear resonance and stationary part, where load does not practically depend on the parameter characterizing the number of stiffness elements.

If we conform the above discussion with Fig. 2.16, then it will be seen that in this range of changes of the parameter N a weak increase of the total surface of extra elements is observed. With increase of the parameter $k_x = k_y$ both for the "soft" and for the "stiff" shell, the area of the resonance part of a curve expands a little and it moves toward the increase of parameter N. The load parameter may be changed by 20%.

4.12.4 Relation Between Buckling Load and the Width of a Rib (Cross-Type Heterogeneity, Fig. 2.8b)

For such a pattern of heterogeneity, in the center of a shell there are two perpendicular "ribs" with a set width, whose stiffness coefficient can characterize a soft or stiff shell.

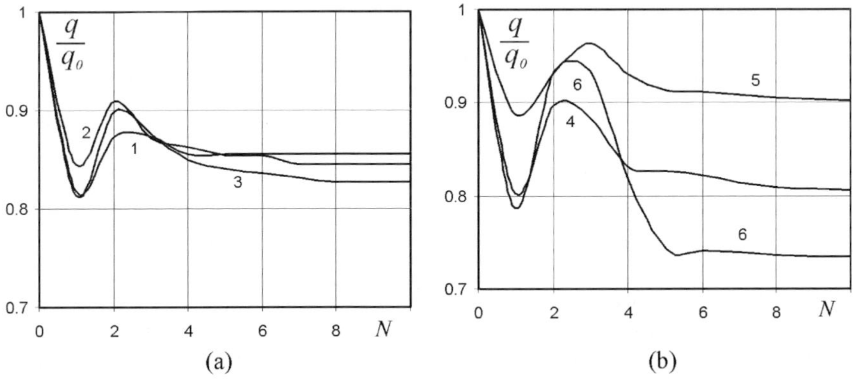

Fig. 4.9 Relation between buckling load of a "soft" heterogeneous shell and the number of stiffness elements situated along one side of a shell

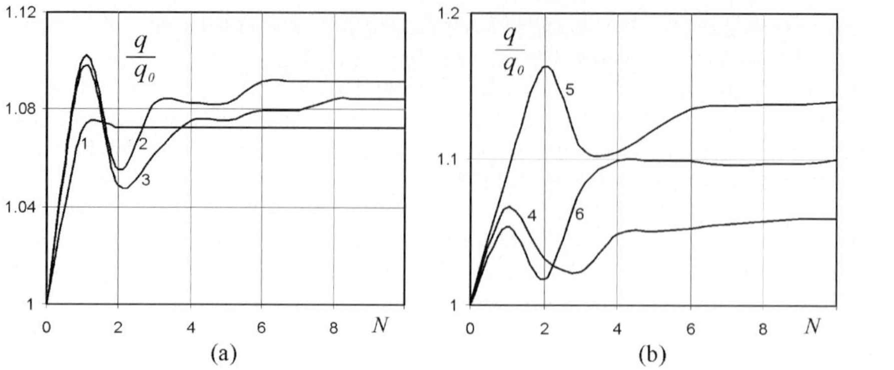

Fig. 4.10 Relation between buckling load of a "stiff" heterogeneous shell and the number of stiffness elements situated along one side of a shell

Let us establish stiffness and density coefficients ($\gamma_{1j} = 0.5$; 1.5 and $\gamma_{2j} = 1$), and we will change the width of the rib of an extra element.

Figure 4.11a,b presents the relations between the relative dynamic buckling load of a heterogeneous shell and S, the width of the rib of a stiffness element (cross-type) for the shell with parameters $k_x = k_y = 18$; 24; 30; 36 (curves 1–4, respectively).

As one can see in the diagrams, curves 1–3 practically overlap both for "soft" and "stiff" ribs, and the $k_x = k_y$ that characterizes them does not exceed 30. This leads us to the conclusion that the occurrence of heterogeneity in this kind of shell results in the influence of geometric and stiffness parameters of heterogeneity elements on each other work as the averaging influence on the shell in a qualitative sense with the influence of uniformly distributed transverse load constant in time.

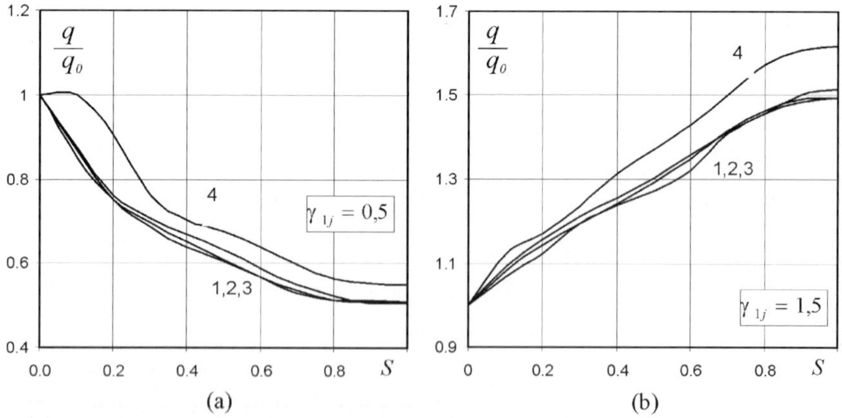

Fig. 4.11 Relation between dynamic buckling load of a heterogeneous shell and the width of the rib of a central stiffness element (cross-type): (**a**) "soft" rib, (**b**) "stiff" rib

For further increase of parameter $k_x = k_y$ geometric parameters of a shell have greater importance for the qualitative behavior of a shell.

What is important here is the occurrence of soft stability loss, i.e., local stability loss in the shell's quarters. Let us remember that for a homogeneous shell, this phenomenon occurs when the parameter $k_x = k_y$ reaches the value of 24. Then, the load can change within the range of 40–50%.

Chapter 5
Stability of a Closed Cylindrical Shell Subjected to an Axially Non-symmetrical Load

This chapter concerns stability of closed cylindrical shells subjected to an axially non-symmetrical load action. In the beginning (Sect. 5.1), equations of motion are derived, and then the influence of imperfection on the shell stability is studied. Both static and dynamic problems of buckling with the use of the Bubnov-Galerkin method of higher approximations are analyzed and many computational results are reported.

5.1 Equations of Motion

While designing the thin shell constructions, it is important to make provision for their capacity under compressing loads. From this point of view, the analysis of the stability of a cylindrical shell under axially asymmetric load is of crucial practical importance.

Based on the classical nonlinear theory of shells, we shall consider a closed cylindrical shell with a circular section and a finite length, whose stiffness and density variables undergo non-uniform outer pressure.

Equations (1.48), and (1.49) of the theory of heterogeneous shells with finite deflections of a mixed form will be the initial equations for further analysis. It should be assumed that $k_x = 0$, $k_y = 1/R$, and R is the radius of an average curvature.

The non-dimensional form of this equation is as follows

$$\frac{E}{12(1-\mu^2)}\left[\left(\frac{1}{\lambda^2}\frac{\partial^2 w}{\partial x^2}+\mu\frac{\partial^2 w}{\partial y^2}\right)\frac{\partial^2(\cdot)}{\partial x^2}+\left(\lambda^2\frac{\partial^2 w}{\partial y^2}+\mu\frac{\partial^2 w}{\partial x^2}\right)\frac{\partial^2(\cdot)}{\partial y^2}\right.$$
$$\left.+2(1-\mu)\frac{\partial^2 w}{\partial x\partial y}\frac{\partial^2(\cdot)}{\partial x\partial y}\right]-k_y\frac{\partial^2 F}{\partial x^2}-L(w,F)+k_y^2 q-\rho\frac{\partial^2 w}{\partial t^2}=0,$$

J. Awrejcewicz, V.A. Krysko, *Chaos in Structural Mechanics*,
DOI: 10.1007/978-3-540-77676-5_6, © Springer-Verlag Berlin Heidelberg 2008

$$a_1\left[\left(\lambda^2\frac{\partial^2 F}{\partial y^2}-\mu\frac{\partial^2 F}{\partial x^2}\right)\frac{\partial^2(\cdot)}{\partial y^2}+\left(\frac{1}{\lambda^2}\frac{\partial^2 F}{\partial x^2}-\mu\frac{\partial^2 F}{\partial y^2}\right)\frac{\partial^2(\cdot)}{\partial x^2}\right.$$

$$\left.+2(1+\mu)\frac{\partial^2 F}{\partial x\partial y}\frac{\partial^2(\cdot)}{\partial x\partial y}\right]+k_y\frac{\partial^2 w}{\partial x^2}+\frac{1}{2}L(w,w)=0. \tag{5.1}$$

$(\cdot)$ means that the derivative is calculated correspondingly as the variation of function w and F, $a_1=\frac{1}{E}$.

The non-dimensional values are given below:

$$w=2h\bar{w};\quad x=l\bar{x};\quad y=R\bar{y};\quad F=E_0\,(2h)^3\,\bar{F};\quad \lambda=l/R;\quad k_y=\frac{2h}{R^2}\bar{k}_y;$$

$$q=\bar{k}_y^2\frac{E_0(2h)^4}{l^2R^2}\bar{q};\quad E=E_0\bar{E};\quad t=\frac{Rl}{2h}\sqrt{\frac{\rho_0}{gE_0}}\bar{t};\quad \rho=\rho_0\bar{\rho}; \tag{5.2}$$

whereas the boundary conditions are:

$$w=0,\quad \frac{\partial^2 w}{\partial x^2}=0,\quad F=0,\quad \frac{\partial^2 F}{\partial x^2}=0,\quad \text{dla } x=0;1. \tag{5.3}$$

5.2 The Influence of Imperfection on the Stability of Shells

In order to significantly correct the solution of the problem of cylindrical shell stability, one needs to consider aberration of the actual cylindrical shell surface from the perfect shape of a model shell, i.e., the initial irregularities, or the so-called imperfection should be taken into account. The influence of imperfection on the shell behavior was first analyzed by Flügge [99], who explained the physical sense of his phenomenon and formulated the problem linearly. New terms characterizing the change of surface curvature resulting from the initial deflection were introduced to linear equations of stability.

In 1934, Donnell [77] took the initial imperfections into consideration by introducing them into nonlinear equations. Nevertheless, the solution Donnel obtained for the first time was not precise.

In 1945, Koiter [137] made a detailed analysis of the behavior of various elastic systems near the bifurcation point. For the simple and axially symmetric form of a deflection, with the wavelength equal to the length of a wave during the loss of the actual shell stability, the dependence for buckling stress was obtained, according to which even a minor axially symmetric imperfection leads to a significant buckling load.

In 1950, Donnell and Wan [78], basing on the earlier approach (of 1934), eventually proposed the method allowing one to take imperfection into account. Their method was often used by other researchers as well. It gives consideration to all the initial imperfections (geometrical, physical, etc.) by introducing certain initial deflection w_0, similar to the deflection w responsible for the stability loss.

If the shell has initial and axially symmetric imperfections, the state of stress occurs already at the load moment. The qualitative picture of the influence of initial imperfections can be best seen in the example of the analysis of cylindrical shell stability during axial compression. It is important to notice here that certain axially symmetric initial irregularities [4] occur. While solving this numeric problem, one can evaluate the buckling load q_{cr} in relation to the amplitude of initial deflection.

The results obtained by many researchers [4, 114, 138, 304] have shown that q_{cr} is very sensitive to the form of the initial axially symmetric shell imperfection.

Figure 5.1 shows the area in which the curves characterizing the buckling load depreciation as related to the non-dimensional amplitude of initial deflection f_{in}/h, with various $w_{in}(x)$ relations, are situated. In the case of a shell with the initial shape imperfection q_{cr} it can turn out to be several times smaller than the classical critical state.

In an actual shell, initial imperfections have different forms, and for thin shells the amplitudes and forms of initial imperfections are difficult to control. In the case of a shell with axially asymmetric initial imperfections, apart from the bifurcation points which can be found on the basis of linearized differential equations, there exists a range of boundary critical points, the determination of which is possible only on the basis of nonlinear theory.

If the way in which the edges of the shell are attached allows for its purely flexural deformation, then this shell presents a purely flexural form of the first kind of deflection, i.e., the one in whose surroundings there is a new equilibrium point.

While the load is approaching the critical value, additional deflections grow in time so quickly that the buckling loads determined with linearized equations for a perfectly shaped shell will be the boundary values for every actual shell (just as is the case with compressed elastic rod). The behavior of a long cylindrical shell (a pipe) under the influence of outer pressure could serve here as an illustrative example.

If the way in which the edges of the shell are attached exclude purely flexural strains, as is usually the case with real constructions, then the behavior of thin shells during stability loss is qualitatively different.

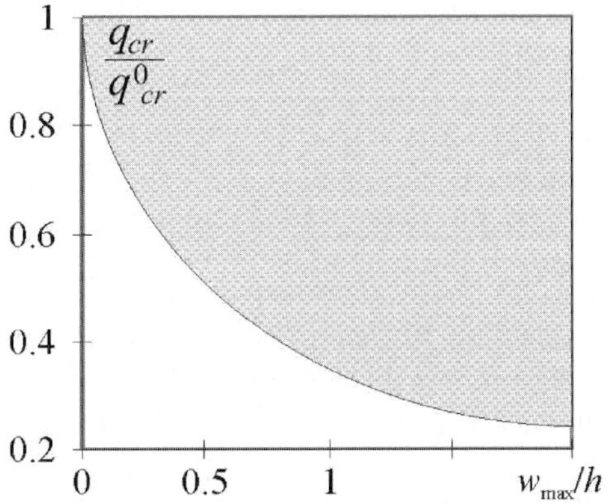

Fig. 5.1 The relation between the area of buckling load depreciation and the initial deflection amplitude

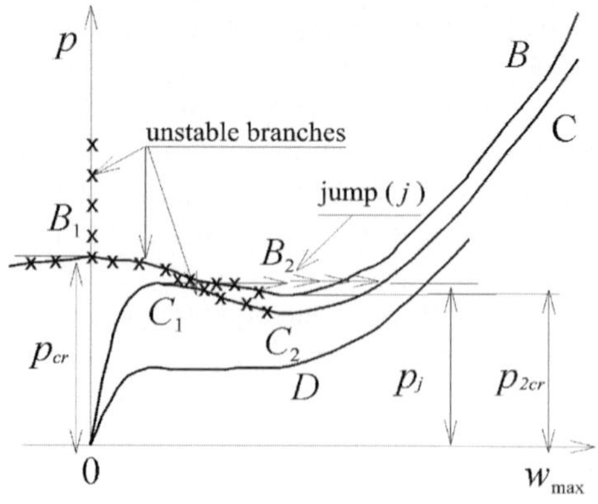

Fig. 5.2 Load-deflection characteristics

Let us analyze now the qualitative behavior of an elastic thin shell with initial imperfections, loaded with outer pressure. For the perfectly shaped shell, the load-deflection relation is characterized as presented by curve B_1B_2B shown in Fig. 5.2, where p is outer pressure (load) and w_{max} is the radial displacement of a point arranged on the wave of an additional deflection directed into the inside of a shell.

The critical point of bifurcation B_1 can be determined on the basis of linearized equations. The construction of the whole B_1B_2B curve requires a linear approach.

Again, let us pay attention to a qualitative difference between the B_1B_2B curve and the diagram for the deflections of perfect rods and plates. First, there are no new static equilibrium points around the critical point of bifurcation 1. The distance between new static equilibrium points (the segment of B_2B curve) and the initial equilibrium point (the segment of OB_1 curve) has finite values.

Therefore, the shell transition into a new state of equilibrium cannot be smooth such a transition proceeds in steps and it takes the form of shell buckling (a complete description of this phenomenon is not a subject of this study; a shell will get a new state of equilibrium after transitional vibrations die out).

Second, it is possible for the static equilibrium point to occur even before the critical pressure state is established on the basis of linearized stability equations. There is a certain energetic barrier between these new equations describing the state of equilibrium and the initial state, and this barrier becomes weaker as the load approaches critical value.

The bottom actual shell begins to deflect from its original value (OC_1 curve) as soon as a load is applied. On achieving the limit point C_1 the shell loses its stability and, by buckling, passes into a new state of equilibrium (from C_1 point to C_2C curve). If there are only elastic strains in the shell material, the further reduction of pressure makes the shell come back to the initial state of equilibrium (from the limit point C_2 to OC_1 curve), by another buckling. Let us notice that with a certain shape and size of amplitude, initial irregularities can transform the diagram of perfect shell strains OB_1B_2B into monotonic curve OD, without a limit point.

The loss of shell stability is similar in the case of other kinds of load as well. When the problem of cylindrical shell stability axially compressed is analyzed, the

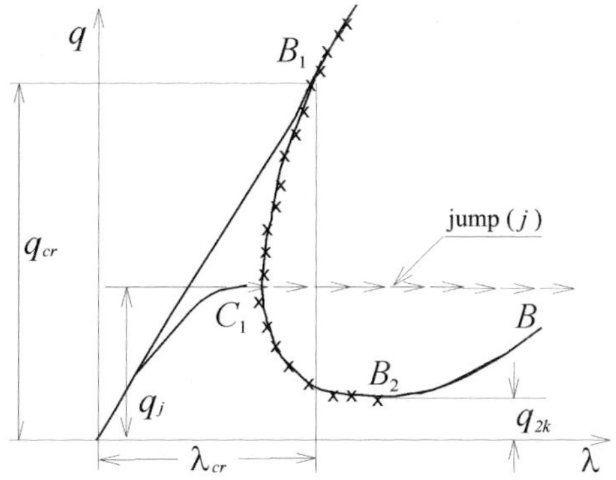

Fig. 5.3 A cylindrical shell compressed axially

diagram of strains is conventionally constructed with coordinates q, λ, where q is a compressing load; λ is the proximity of shell edges. This diagram, just as the diagram of a shell strain under outer pressure, is qualitatively different from the diagrams of rod and plate strains (Fig. 5.3) [4].

OB_1 line represents uniform compression of a perfect shell; when the value of a buckling load q is reached, this initial form is no longer static. The critical point for bifurcation B_1 turns out to be the bifurcation point around which there are no new static equilibrium points of a shell, just as is the case with the shell losing its stability under the influence of outer pressure. Initial imperfections, which often occur in actual shells, transform the diagram of a perfect shell strain, and the bifurcation point B_1 is transformed into the critical limit point C_1. When C_1 is reached, a shell loses stability with a sudden snap.

The theory of thin shell stability isolates two basic load values in a diagram.

P_{cr} is a buckling load, and when it is obtained, the initial stress–strain state of a perfect shell ceases to be static; P_j is a snap load, and when it is obtained, the snap of an actual shell with geometrical imperfections takes place.

The buckling load P_{cr}, corresponding to the bifurcation point B_1, is often called the upper buckling load P_{1cr}. The load corresponding to the limit point B_2, after which it is possible (in a perfect shell) to obtain a new state of equilibrium different from the initial one, is called the lower buckling load P_{2cr}. During unloading (unless the snap caused plastic strains), obtaining B_2 point means the occurrence of the next snap, which causes the return to the initial stress–strain state. For calculations and analysis of thin shells, the P_{2cr} value has no practical significance, since the possibility of work of actual constructions is described by the P_j value.

The value of a buckling load P_{cr} of perfect shells is evaluated on the basis of linearized equations, during axially symmetric loading of any rotary shell. Solving such an equation with or without taking the initial stress–strain state of a shell into consideration does not pose any significant difficulties. Even more complex axially asymmetric problems of shell stability can be analyzed with the use of modern possibilities of calculation.

The quantity of the load of an actual shell snap can be represented as $P_j = k_j P_{cr}$, where in many cases k_j can be significantly smaller than 1. The quantity of k_j for smooth thin shells is very sensitive for the change of a form and size of initial imperfections, which causes principal difficulties in its evaluation. k_j could be surely evaluated for ever single shell experimentally, but its quantity may change while passing from one shell to another.

In order to theoretically determine k_j the form and size of initial irregularities should be more detailed, and then k_j for a given shell could be evaluated with the use of complex and expensive calculations. The practical significance of such a solution is doubtful. First, determining initial irregularities of an actual shell is very difficult; second, they change in the process of product preparation, transport, and use.

Observe that k_j quantities are very sensitive to such random and uncontrollable factors as initial imperfections (only in the case of thin smooth shells). In rationally designed thin-walled shell construction transferring loads, their limit capacity is connected with stability loss, which requires certain k_j values. It is achieved with the use of three-layer, wafer, ribbed composite or other strengthened shells. In such constructions, it is possible to obtain stable, or even close to 1 values of k_j.

To summarize our considerations in this section, let us emphasize that the diagrams of shell equilibrium points presented in Figs. 5.2 and 5.3 are significantly simplified, since only one branch representing the equilibrium point different from the initial one is given there. In fact, a full nonlinear differential equation allows to find a certain set of such curves.

5.3 The Load Resulting from a Wind-Type Flow

Let the shell undergo a wind-type load. The distribution of pressure measured in a circular direction is described by the following formula:

$$q = q_0 (\alpha + \beta \cos y)^m; \quad \alpha + \beta = 1. \tag{5.4}$$

The increase in the value of m causes broadening of the spectrum of a load distribution of Eq. (5.4) into Fourier series $q = \sum_{i=0}^{m} A_i \cos(iy)$ and an axially symmetric component of pressure decreases (Fig. 5.4).

In the analyzed cases $q_0 = 1$ and the coordinates of an applied load are: $0 \le x \le 1$, $-\pi \le y \le \pi$. For $\beta = 0$ we shall obtain the case of uniform pressure distribution; for larger values of m the load diagrams will have the form of a narrow band, with a maximum q_0 in the point $y = 0$. A typical example of the pressure distribution diagram for $q_0 = 1$ is presented in Fig. 5.5.

Certain essential singularity should be taken into account while assessing capacity of the analyzed constructions. Unlike in the case of symmetric or similar problems, where the buckling load value was uniquely related to the loss of the construction capacity, such uniqueness is not observed here [11].

In the cases where a strong deflection dominates, the exhaustion of carrying qualities of the construction is usually the result of obtaining certain value by

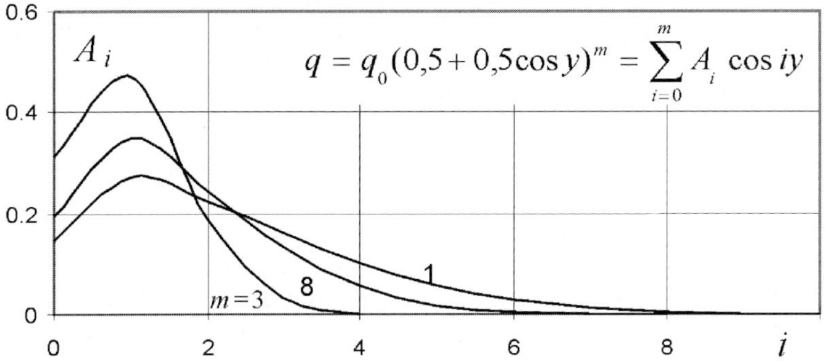

Fig. 5.4 The distribution of a load into a trigonometric series

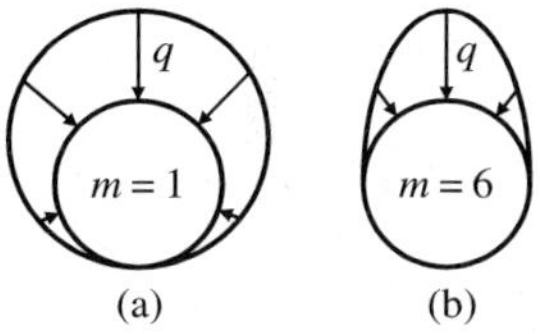

Fig. 5.5 The examples of the load distribution of a shell for $m = 1$ (a) and $m = 6$ (b)

stress or displacement of a certain determined value, although formally, there may be some other singular point of a nonlinear equation established to describe the buckling load value. Stresses corresponding to such a load significantly exceed the elastic limit and the strength limit of material. In this case, calculations should take not only geometrical, but also the physical model of construction into consideration, and the carrying quality of construction is described by limitations applied to the basic parameters of the stress–strain state. The analogical criterion of the assessment of construction capacity is also required when obtaining the critical value of a load (corresponding to a certain unique point of a solution) and transition of the construction into a post-critical state takes place without any significant changes of stress–strain state parameters. In this case, the construction can maintain its capacity under the influence of loads much larger than the critical value.

5.4 The Problem of Statics

In order to solve the static problem, one needs to eliminate the term related to time from the Eq. (5.1).

The boundary problem of Eqs. (5.1), (5.3) is solved with the use of Bubnov-Galerkin method and the aforementioned approximations. The demanded functions have the following form:

$$w = \sum_{i=1}^{M_x} \sum_{j=0}^{M_y} A_{ij} \sin(i\pi\, x)\cos(jy),$$

$$F = \sum_{i=1}^{M_x} \sum_{j=0}^{M_y} B_{ij} \sin(i\pi\, x)\cos(jy). \tag{5.5}$$

As the shell load along the x coordinate is uniform [114, 139], in Eq. (5.5) it is assumed that $M_x = 1,\ M_y = 12 - 20$. By substituting Eq. (5.5) into Eq. (5.1), we get the system of algebraic nonlinear equations with parameter q that is complemented by the following equation:

$$w(x_0, y_0) = \sum_{i=1}^{M_x} \sum_{j=0}^{M_y} A_{ij} \sin(i\pi\, x_0)\cos(jy_0). \tag{5.6}$$

After such an operation, q is treated as an unknown value, whereas $w(x_0,\ y_0)$ is a parameter. This algorithm allows one to construct a full q–w diagram. During each step of parameter changes, the algebraic system of equations was solved on the basis of the Newton method with the extrapolation of initial approximation, with the use of a parabola. The required functions were accurate to 5%.

In Fig. 5.6 the relation of a buckling load q and m-th non-uniformity of a load for a shell with parameters $k_y = 112.5,\ \lambda = 2.2$ is presented. Curves 1–5 represent the values of a load parameter 0.1–0.9 with a step 0.2.

As the non-uniformity m increases, the buckling load initially grows, obtaining the maximum, and then gets smaller, and its intensity is significantly lower than the initial one. This quality of a buckling load is much more noticeable in case of smaller parameter values α, i.e., while the role of a variable additional element β in Eq. (5.4) is increasing. For a significant value of a stable part of a load, a maximum moves into the area of a significant non-uniformity of m load.

For small values of non-uniformity, a deflection corresponding to the value of a buckling load changes within the range of tenth parts of shell thickness. The

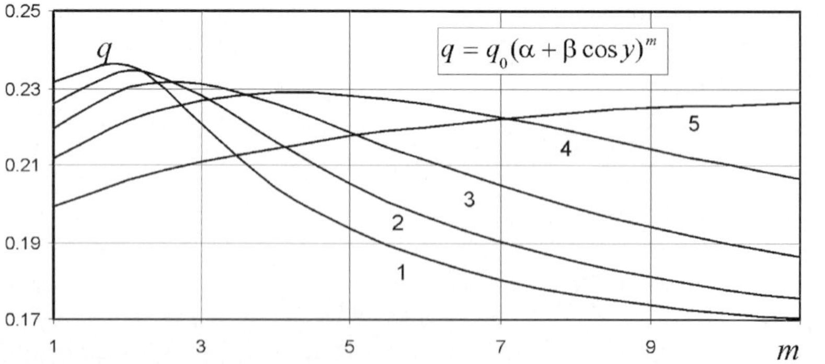

Fig. 5.6 Relation between a buckling load and m-th non-uniformity of the load of a shell with parameters $k_x = k_y = 112.5,\ \lambda = 2.2$

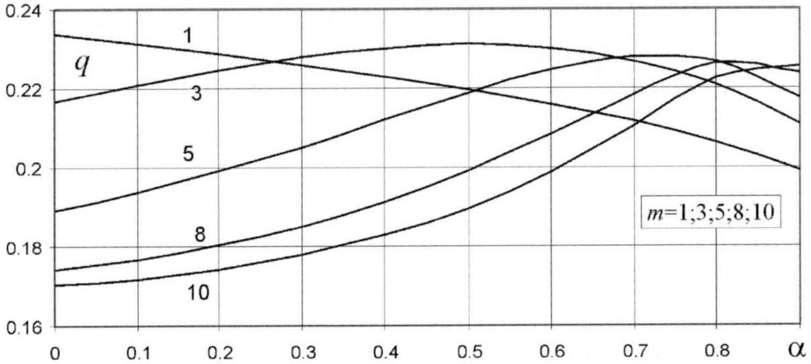

Fig. 5.7 Relation between a buckling load and the participation of outer load for various values of non-uniformity parameter α

deflection grows with the increase of m and it can reach the order of magnitude equivalent to several thicknesses of a shell. For a small participation of a variable part of a load, such a deflection does not exceed the half of shell thickness.

Figure 5.7 shows the relation between the buckling load q and the participation of a stable part of outer pressure α for various values of non-uniformity parameter m.

For a small non-uniformity m the participation of a variable part of outer pressure β is essential for the quantity of a buckling load. While his participation is being reduced, the buckling can fall monotonically.

Increasing the non-uniformity parameter of a load m changes the character of this relation. During reduction of the variable participation of outer pressure β a buckling load grows monotonically, approaching the value equivalent to the one loaded with uniform pressure.

5.5 Dynamics

In order to solve the dynamic problem, we shall use the equations of the system of Eq. (5.1).

Initial conditions are

$$w = 0, \quad \frac{\partial w}{\partial t} = 0 \quad \text{for} \quad t = 0, \tag{5.7}$$

and we search the following form of a solution

$$w = \sum_{i=1}^{M_x} \sum_{j=0}^{M_y} A_{ij}(t) \sin(i\pi\, x) \cos(jy),$$

$$F = \sum_{i=1}^{M_x} \sum_{j=0}^{M_y} B_{ij}(t) \sin(i\pi\, x) \cos(jy). \tag{5.8}$$

Due to spatial coordinates, the boundary problem of Eqs. (5.1) and (5.3) is solved with the use of the Bubnov-Galerkin method with higher approximations, and the obtained system of ordinary differential equations is solved with the use of the fourth-order Runge-Kutta method, due to the time after reducing it to a normal form.

Let a shell undergo a wind-type load described by the (5.5) type equation. Let this load have certain periodical component with frequency and have the following form:

$$q = q_0\,(\alpha + \beta \cos y)^m\,(\,1 + \sin \omega t\,);\quad \alpha + \beta = 1. \tag{5.9}$$

The form of the coordinates describing load position follows:

$$0 \le x \le 1;\quad -\pi \le y \le \pi.$$

Figure 5.8 shows the relation between the buckling load of a shell with parameters $k_y = 112.5$, $\lambda = 2.2$ and ω, i.e., the frequency of an exciting force. Curves 1, 2, 3, 4 represent the values of a load non-uniformity parameter $m = 1, 2, 3, 5$; $\alpha = \beta = 0$.

Those relations are not monotonic. In all cases, there is a maximal minimum around the value of parameter $\omega = 1$. Then, the area $\omega = 4$–12 appears, where three local maxima can be observed, and the range reflecting smooth reduction of a load and its growth follows. Together with the increase of non-uniformity parameter m certain phenomena similar to those described here occur, but the transition of the abovementioned phenomena toward larger values of the frequency of an exciting force is observed ω.

Figure 5.9 shows the relations between a dynamic buckling load and a non-uniform shell with geometric parameters: $k_y = 100$, $\lambda = 2$, depending on the value of parameter γ_{2j}, i.e., non-uniformity of an element for $\alpha = \beta = 0.5$; the coordinates

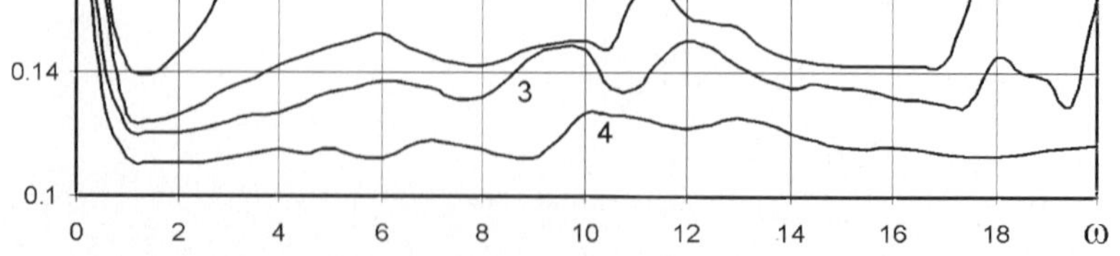

Fig. 5.8 Relation between a buckling load and the frequency of an exciting force

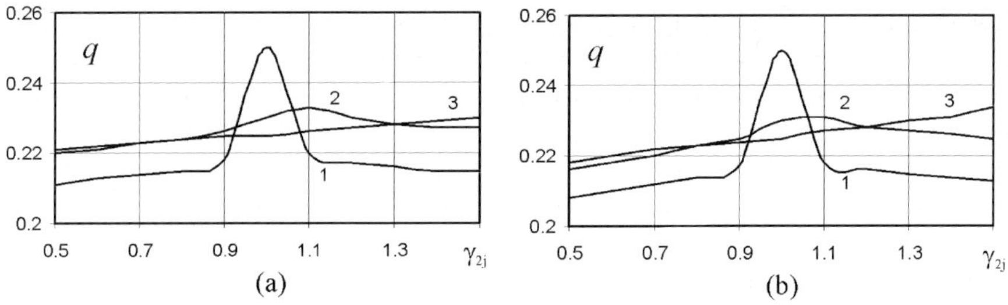

Fig. 5.9 Relation between the dynamic buckling load of a non-uniform shell and the non-uniformity parameter like element thickness; coordinates of element's position: (**a**) $x_1 = 0,\ x_2 = 1,\ y_1 = -0.1,\ y_2 = 0.1$; (**b**) $x_1 = 0,\ x_2 = 1,\ y_1 = -0.2,\ y_2 = 0.2$

of load application are: $0 \le x \le 1,\ -\pi \le y \le \pi$. Curves 1, 2, 3 represent the values of the parameter of load non-uniformity $m = 1, 2, 3$.

As the diagram shows, for small load non-uniformity values $m = 1,\ 2$, reducing or increasing the density parameter γ_{2j} results in reducing the value of a buckling load of a shell in comparison to a uniform shell. The occurrence of a maximum value of a buckling load can be observed. The maximum value is best noticeable with the minima parameter of load non-uniformity $m = 1$.

For $m \ge 3$ this relation is different: reducing (or increasing) the parameter results in reduction (increase) of the value of a buckling load.

Tables 5.1–5.3 show the forms of shell deflections in relation to a circular coordinate for various time instants t with a critical value of load for a given set of chosen parameters. Let us notice that $\gamma_{1j} = 1$ and $\gamma_{2j} = 1$ correspond to a uniform shell.

For small values of load non-uniformity m the parameter deviation γ_{2j} directed at random direction in relation to a uniform shell essentially influence the form of a shell deflection. For $\gamma_{2j} = 0.5$ concavity appears in the shell, and for $\gamma_{2j} = 0.5$ there is certain convexity in the area where the load is most intense. For $m \ge 3$ this form of a shell deflection along the circular coordinate does not depend on the

Table 5.1 Shell deflection for various time instants ($m = 1$)

γ_{2j}	$m = 1; \gamma_{1j} = 1$			
0,5	$t = 0.5$	$t = 1.0$	$t = 1.3$	$t = 1.5$
1,0	$t = 0.5$	$t = 1.0$	$t = 1.3$	$t = 1.5$
1,5	$t = 0.5$	$t = 1.0$	$t = 1.3$	$t = 1.5$

Table 5.2 Shell deflection for various time instants $(m = 2)$

γ_{2j}	$m = 2; \gamma_{1j} = 1$			
0,5	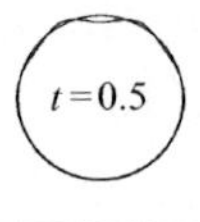	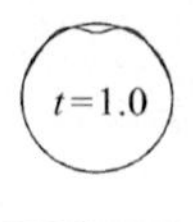	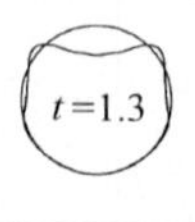	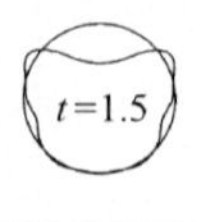
1,0	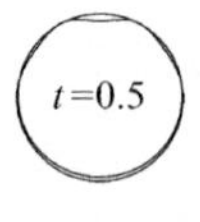	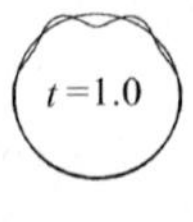	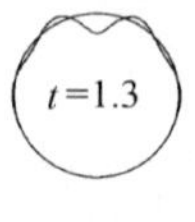	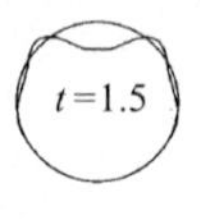
1,5	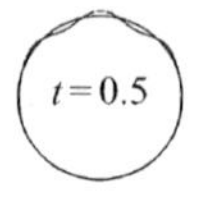	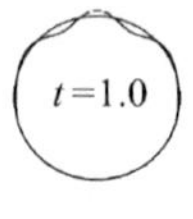	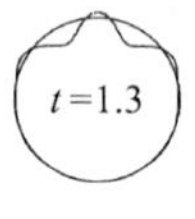	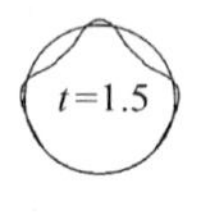

Table 5.3 Shell deflection for various time instants $(m = 3)$

γ_{2j}	$m = 3; \gamma_{1j} = 1$			
0,5	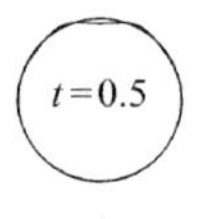	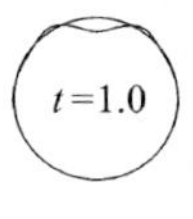	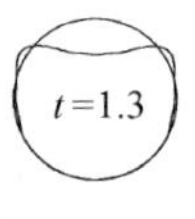	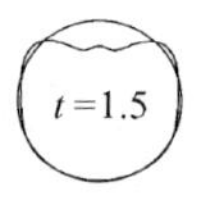
1,0	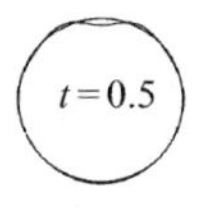	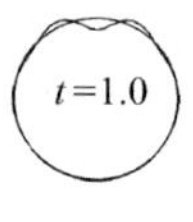	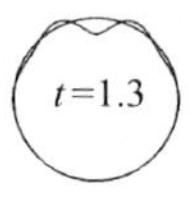	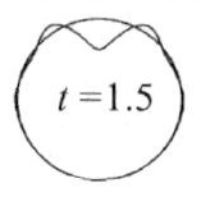
1,5	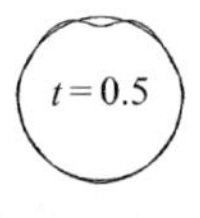	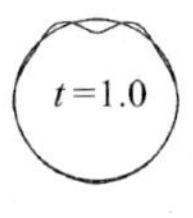	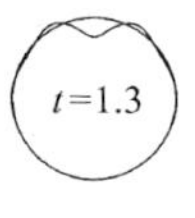	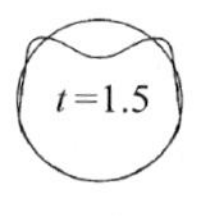

value of density parameter γ_{2j}. In this sense, a non-uniform shell behaves like a uniform one.

The results presented in Table 5.1 for $m = 1$ are qualitatively different from other results and they confirm the character of the relation described by curve 1 in Fig. 5.9.

Chapter 6
Composite Shells

Composite shells are studied in this chapter. First, equations governing the behavior of composite shells are derived, and then both static and dynamic problems of stability loss of composite shells are addressed.

6.1 Equations

The earlier used method of taking heterogeneity of shells into account in calculations can be successfully applied for introducing heterogeneity along the shell's width. In this case, heterogeneity covers the whole surface of a shell, i.e., a shell becomes homogeneous. According to the Kirchhoff-Love model, all the layers of a shell are made of isotropic material and have the possibility of random choice of parameters $E,\ \mu$, and ρ, for $E = E(z),\ \rho = \rho(z)$.

We shall assume as usual that $-h \leq z \leq h$, i.e., the thickness of the layer packet is $2h$ and this packet is symmetric in relation to the central surface of a shell. Therefore, we will set only the coordinate of the top part of the packet of each layer in relation to direction z. The Young module of an extra element is different from the module of a central layer and $z_1 < z_2$. The shell always has an asymmetrical number of layers. To make further considerations easier, we shall assume that Poisson coefficient is the same for all the layers.

The procedure of obtaining the equations governing the behavior of this type of shells is the same as before, but some corrections must be introduced:

1. In problems where integration in relation to a coordinate z must be performed, it should be assumed that $E = E(z),\ \rho = \rho(z)$.
2. In steps of solving the problem where integration in relation to the surface of shell parameters $E,\ \rho$ does not depend on x and y, they can be drawn before the integration sign.

Taking these remarks into account, we shall give only those expressions below that undergo changes.

J. Awrejcewicz, V.A. Krysko, *Chaos in Structural Mechanics*,
DOI: 10.1007/978-3-540-77676-5_7,

1. Equations of forces

$$T_{11} = \int_{-h}^{h} \frac{E(z)}{1-\mu^2}(\varepsilon_{xx} + \mu\varepsilon_{yy})\, dz$$

$$= \frac{2hE_0}{1-\mu^2}\left[1 - \sum_i (1-\gamma_{1i})\frac{z_{2i}-z_{1i}}{h}\right](\varepsilon_{11} + \mu\varepsilon_{22}),$$

$$T_{22} = \int_{-h}^{h} \frac{E(z)}{1-\mu^2}(\varepsilon_{yy} + \mu\varepsilon_{xx})\, dz$$

$$= \frac{2hE_0}{1-\mu^2}\left[1 - \sum_i (1-\gamma_{1i})\frac{z_{2i}-z_{1i}}{h}\right](\varepsilon_{22} + \mu\varepsilon_{11}),$$

$$T_{12} = \int_{-h}^{h} \frac{E(z)}{2(1+\mu)}\varepsilon_{xy} dz = \frac{2hE_0}{2(1+\mu)}\left[1 - \sum_i (1-\gamma_{1i})\frac{z_{2i}-z_{1i}}{h}\right]\varepsilon_{12}, \tag{6.1}$$

where z_{1i}, z_{2i} are the coordinates, respectively for the beginning and end of i-th layer along axis z.

2. Equations of elementary potential energy:

$$V = \frac{1}{2}\iint\limits_{\Omega}\int_{-h}^{h} (\sigma_{xx}\varepsilon_{xx} + \sigma_{yy}\varepsilon_{yy} + \sigma_{xy}\varepsilon_{xy})\; dz\, ds = V_1 + V_2,$$

where

$$V_1 = \frac{1}{2}\iint\limits_{\Omega}\int_{-h}^{h} \frac{E(z)}{1-\mu^2}\left[\varepsilon_{11}^2 + \varepsilon_{22}^2 + 2\mu\varepsilon_{11}\varepsilon_{22} + \frac{1-\mu}{2}\varepsilon_{12}^2\right] dsdz,$$

$$V_2 = \frac{1}{2}\iint\limits_{\Omega}\int_{-h}^{h} \frac{E(z)z^2}{1-\mu^2}\left[\left(\frac{\partial^2 w}{\partial x^2}\right)^2 + \left(\frac{\partial^2 w}{\partial y^2}\right)^2\right.$$

$$\left. +2(1-\mu)\left(\frac{\partial^2 w}{\partial x\partial y}\right)^2 + 2\mu\frac{\partial^2 w}{\partial x^2}\frac{\partial^2 w}{\partial y^2}\right] dsdz. \tag{6.2}$$

3. Equation of elementary kinetic energy without taking into account inertia related to the element's rotation:

$$K = \frac{1}{2}\iint\limits_{\Omega}\int_{-h}^{h} \frac{\rho(z)}{g}\left[\left(\frac{\partial u}{\partial t}\right)^2 + \left(\frac{\partial v}{\partial t}\right)^2 + \left(\frac{\partial w}{\partial t}\right)^2\right] dsdz. \tag{6.3}$$

After integration along z we obtain

$$V_1 = \frac{1}{2}\iint\limits_{\Omega} \frac{2hE_0}{1-\mu^2}\left[1-\sum_i (1-\gamma_{1i})\frac{z_{2i}-z_{1i}}{h}\right] \times \left[\varepsilon_{11}^2+\varepsilon_{22}^2+2\mu\varepsilon_{11}\varepsilon_{22}+\frac{1-\mu}{2}\varepsilon_{12}^2\right] ds, \tag{6.4}$$

$$V_2 = \frac{1}{6}\iint\limits_{\Omega} \frac{2h^3E_0}{1-\mu^2}\left[1-\sum_i (1-\gamma_{1i})\frac{z_{2i}^3-z_{1i}^3}{h^3}\right] \times \left[\left(\frac{\partial^2 w}{\partial x^2}\right)^2+\left(\frac{\partial^2 w}{\partial y^2}\right)^2+2(1-\mu)\left(\frac{\partial^2 w}{\partial x\partial y}\right)^2+2\mu\frac{\partial^2 w}{\partial x^2}\frac{\partial^2 w}{\partial y^2}\right] ds, \tag{6.5}$$

$$K = \frac{1}{2}\iint\limits_{\Omega} \frac{2h\rho_0}{g}\left[1-\sum_i (1-\gamma_{2i})\frac{z_{2i}-z_{1i}}{h}\right] \times \left[\left(\frac{\partial u}{\partial t}\right)^2+\left(\frac{\partial v}{\partial t}\right)^2+\left(\frac{\partial w}{\partial t}\right)^2\right] ds. \tag{6.6}$$

Non-dimensional equations in a hybrid form for a homogeneous shell with consideration to Eqs. (6.1)–(6.6) will have the following form:

$$\frac{1}{12(1-\mu^2)}\left[1-\sum_i (1-\gamma_{1i})\left(z_{2i}^3-z_{1i}^3\right)\right]\left[\frac{1}{\lambda^2}\frac{\partial^2 w}{\partial x^2}\frac{\partial^2 (\cdot)}{\partial x^2}+\lambda^2\frac{\partial^2 w}{\partial y^2}\frac{\partial^2 (\cdot)}{\partial y^2}\right.$$
$$\left. +\ 2(1-\mu)\frac{\partial^2 w}{\partial x\partial y}\frac{\partial^2 (\cdot)}{\partial x\partial y}+\mu\left(\frac{\partial^2 w}{\partial x^2}\frac{\partial^2 (\cdot)}{\partial y^2}+\frac{\partial^2 w}{\partial y^2}\frac{\partial^2 (\cdot)}{\partial x^2}\right)\right]$$
$$-\ \nabla_k^2 F - L(w,F)+q = \left[1-\sum_i (1-\gamma_{2i})(z_{2i}-z_{1i})\right]\left(\frac{\partial^2 w}{\partial t^2}+\varepsilon\frac{\partial w}{\partial t}\right), \tag{6.7}$$

$$\left[1-\sum_i\left(\frac{1}{\gamma_{1i}}-1\right)(z_{2i}-z_{1i})\right]\left[\left(\lambda^2\frac{\partial^2 F}{\partial y^2}-\mu\frac{\partial^2 F}{\partial x^2}\right)\frac{\partial^2 (\cdot)}{\partial y^2}\right.$$
$$\left. +\left(\frac{1}{\lambda^2}\frac{\partial^2 F}{\partial x^2}-\mu\frac{\partial^2 F}{\partial y^2}\right)\frac{\partial^2 (\cdot)}{\partial x^2}+2(1+\mu)\frac{\partial^2 F}{\partial x\partial y}\frac{\partial^2 (\cdot)}{\partial x\partial y}\right]$$
$$+\ \nabla_k^2 w+\frac{1}{2}L(w,w)=0. \tag{6.8}$$

6.2 Static Stability of Composite Shells

We add boundary conditions of Eq. (2.94) to Eqs. (6.7)–(6.8) and we solve them with the use of high-approximation Bubnov-Galerkin methods.

6.2.1 Three-Layered Shell

The packet consists of three layers of equal thickness. Each layer is made of isotropic material of the same thickness. If we assume the Young module of the inner layer as equal to 1 then for the layers above and below this layer it is easy to respectively use the relation of these parameters E_i/E_0.

Figure 6.1 shows a typical q–w relation for shells with parameter $k_x = k_y = 24$, where curves 1, 3 correspond to the following relations of Young modules 0.9; 1. Curve 1 is also presented that corresponds to a homogeneous shell. Let us remember that the thickness of the packet is the same for all analyzed cases and is $2h$. All three curves are qualitatively similar and they are dislocated along the deflection axis. For a shell whose layers have a higher (or lower) Young module, a curve is situated above (or below) the curve for a homogeneous shell.

Figure 6.2 presents the relation between a static buckling load of the top layer of a trilayered shell $k_x = k_y = 24$ and the Young modules of shell material, for E_0 being a parameter of the central (basic) layer. In the whole range of control parameter changes, this relation is close to linear.

The elementary question concerning the analysis of a trilayered shell follows: how does such a shell behave when the number of layers is increased? In our case, this question has some results. If the shell's layers have the same mechanical parameters, then this example is equivalent to the case of a homogeneous shell whose thickness is equal to the sum of thicknesses of all its layers. The situation becomes different when we analyze a multilayered shell consisting of layers that are made of different materials.

Figure 6.3 shows the relation of the static top buckling load and the number of layers of a shell with $k_x = k_y = 24$, for a relative Young module of each layer that increases by 0.1. Curve 1 corresponds with a negative value of this increase, and curve 2 corresponds with its positive value.

It is obvious that for a negative increase with step 0.1 that the number of layers cannot exceed 19. In both analyzed cases we observe a relation close to linear.

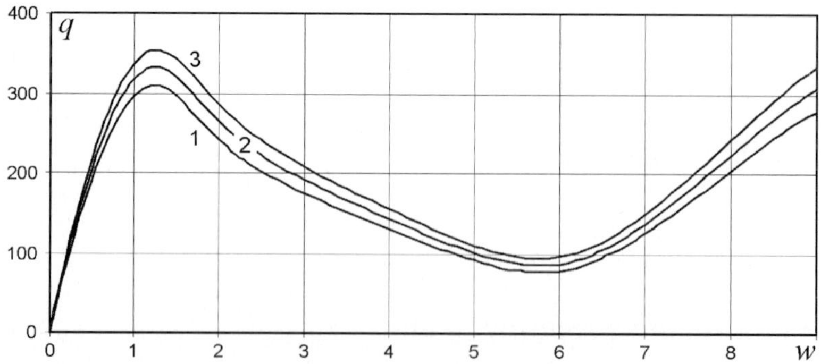

Fig. 6.1 Diagram of "load-buckling" for a trilayered shell

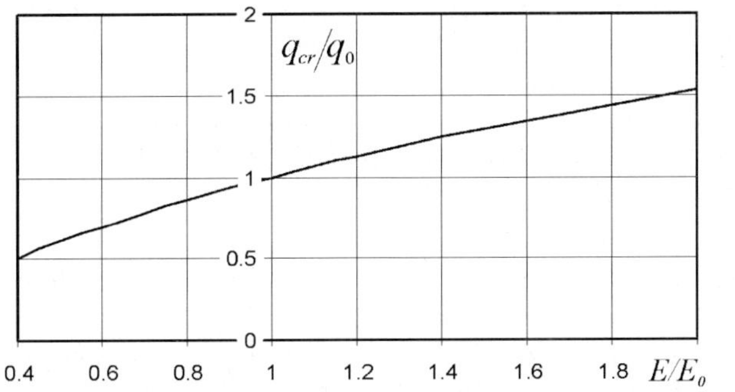

Fig. 6.2 Relation between static buckling load of a trilayered shell vs. and the Young modules of material

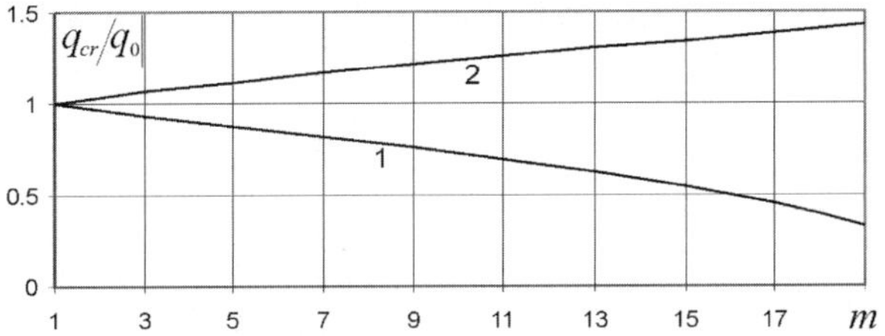

Fig. 6.3 Relation between static buckling load of a homogeneous shell vs. the number of layers

Figure 6.4 presents the same relations as shown in Fig. 6.3, but the alloys of titanium and aluminum were chosen as layer material. Curve *Ti–Al* reflects the case when the central layer is made of titanium, and then the materials alternate. The *Al–Ti* case is similar, but the central layer is made of *Al* alloy.

Taking into account the fact that the actual thickness of a packet is $2h$, and the number of layers increases, the value of the buckling load initially waves and then it approaches the stationary value characteristic of a given type of shell.

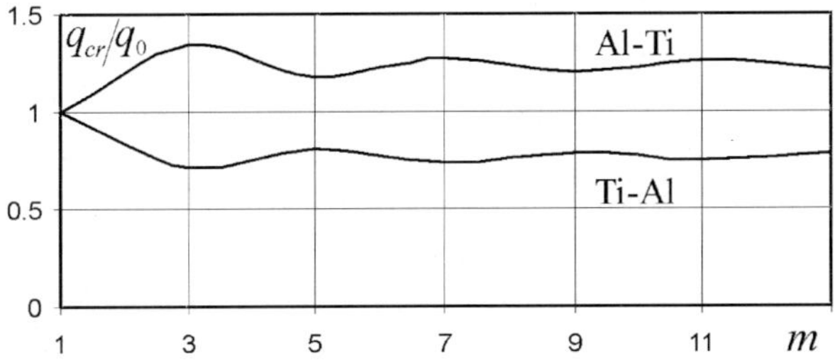

Fig. 6.4 Relation of static load of multilayered shell with parameters $k_x = k_y = 24$ vs. the number of layers

6.3 Dynamic Stability

Let us analyze the previously considered problem in terms of dynamics. In order to solve it, we shall use Eqs. (1.55,1.56), boundary conditions of (2.94), and the following initial conditions:

$$w = 0, \quad \frac{\partial w}{\partial t} = 0 \ \text{for} \quad t = 0. \tag{6.9}$$

Medium damping shall not be taken into account, i.e., we will take coefficient $\varepsilon = 0$ in Eqs. (1.55). Let a trilayered shell with parameter $k_x = k_y = 24$ undergo a transverse and uniformly distributed load independent of time. Let us remember that in this case the Young module of a central layer is assumed to equal 1. We shall solve this problem using a higher-approximation Bubnov-Galerkin method and the previous methodology. We will assume the Volmir criterion as the dynamic criterion of stability loss here.

Figure 6.5 presents diagrams of the relation "shell center deflection–time" for the following values of buckling load $q = 226, 210, 241$ (respectively, curves 1, 2, 3), where curve 1 was formed for a homogeneous shell, and curves 2, 3 for a trilayered shell. For curves 2, 3 it was assumed that $E/E_0 = 0.9, 1.1$. Since all the analyzed curves correspond with critical states, all of them are qualitatively and quantitatively close to each other. In other words, shell buckling occurs in the same way in all of

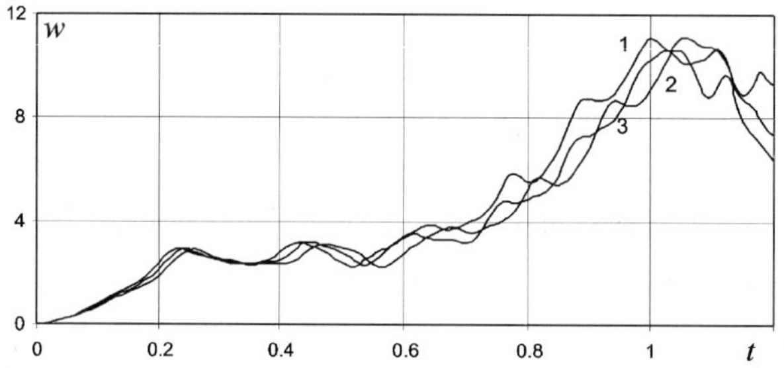

Fig. 6.5 Changes in time of the deflection of the center of a shell for $k_x = k_y = 24$

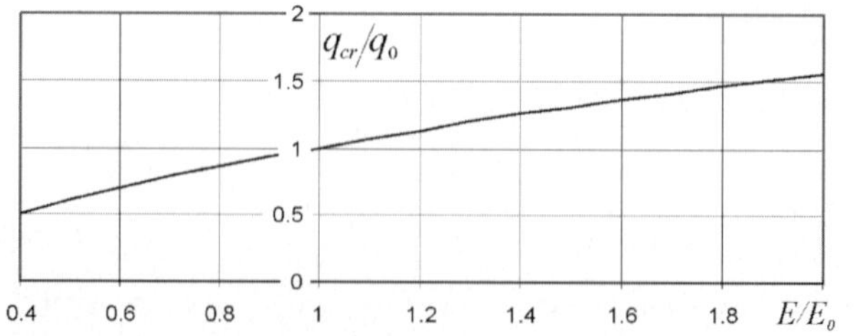

Fig. 6.6 Relation of dynamical buckling load of a trilayered shell vs. the Young modules of shell material

them, i.e., critical deflections practically overlap, and critical time is slightly shorter (longer) for a softer (stiffer) shell as a result of the dislocation of shell vibration.

Figure 6.6 presents the same relation as in Fig. 6.2, but formulated in terms of dynamics.

The diagrams practically overlap qualitatively and quantitatively, which means that the shell's three layers influences the quantity of relative buckling load in a way analogous to the one analyzed above.

Chapter 7
Interaction of Elastic Shells and a Moving Body

In this chapter the problem of interaction between flexible construction and a moving lumped body is reduced to that of essentially simpler ones, i.e., that of vibrations subject to moving force P_0 and that of displacement in the domain of the moving masses under the action of the mentioned force. Advantages of the proposed method are illustrated and discussed.

It is characteristic for the modeling of interaction between a construction and moving bodies that the impact on the construction is expressed by the weight and inertial forces of the objects moving on the studied construction. This crucial feature of the applied approach constitutes also the essential difficulty in the mathematical analysis of the problem.

Let us note that most publications in this field related to the analysis of the interaction between a shell and a moving mass are based on the application of a geometrically linear model or on the assumption that a movable mass does not tear off the leading construction. Apart from that, calculations were usually conducted with the use of approximated methods with only few first approximations taken into account. This chapter deals with this problem in a complex and detailed way and it provides the methods leading to the solution.

7.1 Vibration of Construction and Moving Lumped Body (One-Sided Constraint Case)

In general, the problem of interaction between moving objects and engineering constructions belongs to important tasks of engineering dynamics. The term "moving load" is frequently used in technology today. Movable objects can be either rigid or deformable bodies, fluids, impacting waves, heat or electromagnetic field sources, etc. However, in this contribution, a "moving load" or a "moving body" is meant to be a load acting either on a plate or a shell created through a weight or an inertial force, or/and moving medium (or bodies).

J. Awrejcewicz, V.A. Krysko, *Chaos in Structural Mechanics*,
DOI: 10.1007/978-3-540-77676-5_8, © Springer-Verlag Berlin Heidelberg 2008

Dynamic problems of plates and circled cylindrical shells subject to an action of moving objects have already a long history in mechanics. It is mainly motivated by an enormous application of cylindrical shells in the rocket and aircraft industry, as well as the industry of shipbuilding.

Without doubt, a study of the interaction between a landing (starting) airplane and an air strip plays a key role in estimating the quality of the plane and the passengers' safety.

Furthermore, in many cases the interaction between ship elements and waves is modeled through interaction of a thin-walled structure with a fast moving lumped body.

Another important field where dynamic behavior of constructions influenced by movable bodies is involved has recently been revealed due to the use of nuclear plants. In view of the safety requirements imposed on exploitation of nuclear power plants, design of constructions housing nuclear reactors must obey high standard norms. It is obvious that among other things, the probability of destruction of such objects is expected to be extremely low, and consequently, the nuclear plants situated, for instance, in the vicinity of airports must be studied with respect to the airplane impact. In fact, an isolated construction can be often modeled either as a plate or a shell, while an airplane is usually modeled as a system of coupled lumped oscillators.

A mass moving with constant velocity along a surface is subject to the load $P_D(t)$ action, which is nowhere zero in spite of the contact point between the interacting bodies, where the mass load M is [96]

$$P_{\mathrm{D}}(t) = P - M\frac{d^2 z_{\mathrm{DM}}}{dt^2}. \tag{7.1}$$

In the above, z_{DM} denotes dynamical transversal mass (lumped body) displacement as well as simultaneous vibrations of the mass and the associated surface; P is the normal component of the weighting force of the body with mass M.

With the origin of the relative coordinate system fixed on the neutral surface of the associated vibrating construction at the point under the body, the introduced coordinate axes are directed along a tangent and a normal to the neutral surface.

Assuming that the mass moves while in contact with the associated surface, the studied motion can be treated as a complex one, i.e., consisting of both the associated surface motion and the relative motion with velocity v measured with respect to the surface. In this case one may derive both the mass acceleration in the vertical direction and the vertical component of the mass trajectory using partial derivatives of the dynamical deflection of the associated surface, i.e.:

$$\frac{d^2 z_{\mathrm{D,M}}(x,t)}{dt^2} = \frac{d^2 z_{\mathrm{D}}(x,t)}{dt^2} = \frac{\partial^2 z_{\mathrm{D}}}{\partial t^2} + 2v\frac{\partial^2 z_{\mathrm{D}}}{\partial x \partial t} + v^2\frac{\partial^2 z_{\mathrm{D}}}{\partial x^2}. \tag{7.2}$$

The transversal acceleration $\frac{d^2 z_{D,M}(x,t)}{dt^2}$ of the point mass measured in the fundamental coordinates x, y (Fig. 7.1a) consists of the transitional acceleration $\partial^2 z_D/\partial t^2$ (where z_D denotes deflection of the associated surface under the mass), the Coriolis

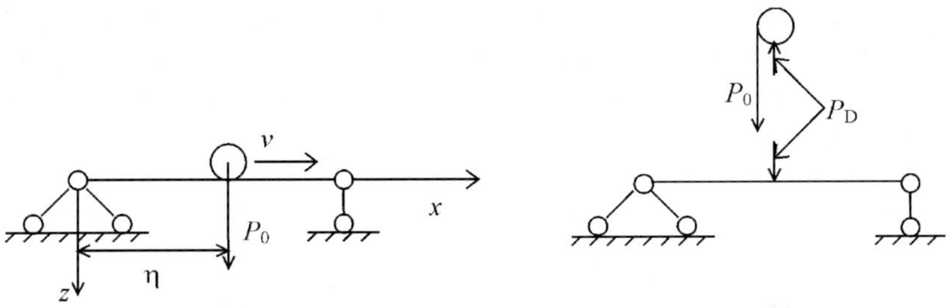

Fig. 7.1 Scheme of the investigated system: moving lumped body and the associated surface treated as one (**a**) or two separated (**b**) systems

acceleration $2v\partial^2 z_D/(\partial x \partial t)$ generated by a rotation of the relative system during vibrations of the elastic surface, and the relative (oriented into the centre) acceleration $v^2 \partial^2 z_D/\partial x^2$.

Since during a contact the body movement $z_D(x,\ t)|_{x=vt} = z_{D,M}(t)$, and since for each time instant the mass and the point of the associated surface (situated under the mass) vibrate transversally as one body (Fig. 7.1a), the mass pressure in relation (7.1) can be expressed through the function of transversal deflections of the associated surface, i.e.:

$$P_D(t) = P - M\left(\frac{\partial^2 z_D}{\partial t^2} + 2v\frac{\partial^2 z_D}{\partial x \partial t} + v^2\frac{\partial^2 z_D}{\partial x^2}\right)\Bigg|_{x=vt}. \tag{7.3}$$

Note that the beam vibrations subject to the action of the moving force (7.3) can be reduced through the Englis-Bolotin method to the system of differential equations yielding the values of transversal beam vibrations z_D [61].

On the other hand, the equation of beam vibrations with the load (7.3) can be reduced to integral-differential, integral, or algebraic equations [97]. During the mentioned reduction process the occurrence of two-sided constraints applied to the mass moving along a smooth surface guarantees the continuous contact between two objects.

However, experimental investigations indicate a possibility of the contact lack between the body and the associated beam (surface). In view of the latter observation our mechanical system consisting of the mass and the associated elastic surface is divided into two components (Fig. 7.1). In what follows two different problems will be studied.

First, the problem of transversal mass displacement subject to the forces of external and dynamical reaction P_D is examined. The second problem analyzed is that transversal vibrations of the interacting construction subject to an action of unknown moving force P_D.

The values of the pressure generated by the moving lumped body and dynamical reaction are equal, since they play the roles of action and reaction.

Comparison of the mass displacement in the direction normal to the surface and the transversal displacement of the construction under the force action allows derivation of the condition of intersection of the moving mass trajectory and the deformed surface of the associated construction.

When the mentioned interaction is realized, i.e., the mass lies on the associated surface, one gets a positive value of force P_D applying the conditions of equal displacements (force directions associated with the mass moving on the surface are shown in Fig. 7.1).

Under a one-sided constraint, the mass loses contact with the surface for $P_D \leq 0$ and exhibits its own independent motion. In this case $P_D = 0$ is taken for further consideration. Then, an impact occurs when the sign of P_D changes.

It is worth noting that the mass moving along the deformable surface can be treated as a mechanical system with non-stationary holonomic constraints of the form

$$f(x,y,z,t) = 0, \tag{7.4}$$

and hence it has two degrees of freedom. Virtual mass displacements (allowed by the constraint) will be denoted by δx, δy, δz, whereas the dynamical constraint reaction is denoted by P_D (P_{Dx}, P_{Dy}, P_{Dz}).

Note that relation (7.4) holds also for virtual displacements, i.e.:

$$f(x+\delta x, y+\delta y, z+\delta z, t) = 0. \tag{7.5}$$

Since the constraint is holonomic, formulas (7.3) and (7.4) yield

$$\delta f = \frac{\partial f}{\partial x}\delta x + \frac{\partial f}{\partial y}\delta y + \frac{\partial f}{\partial z}\delta z = 0. \tag{7.6}$$

Note that time has not been variated while deriving (7.6) because the virtual displacements are not matched with the mass motion. Since in our mechanical system the virtual work of reaction is equal to zero, one obtains

$$P_{Dx}\delta x + P_{Dy}\delta y + P_{Dz}\delta z = 0. \tag{7.7}$$

Both relations (7.7) and (7.6), are linear. Furthermore, the linear form of (7.7) is a linear combination of (7.6), and therefore

$$P_{Dx} = \lambda\frac{\partial f}{\partial x}; \quad P_{Dy} = \lambda\frac{\partial f}{\partial y}; \quad P_{Dz} = \lambda\frac{\partial f}{\partial z}, \tag{7.8}$$

where λ is the positive Lagrange multiplier (it characterizes normal reaction force) [336].

The d'Alembert principle yields

$$(X - M\ddot{x})\delta x + (Y - M\ddot{y})\delta y + (Z - M\ddot{z})\delta z = 0, \tag{7.9}$$

where X, Y, Z are components of active forces acting on mass M.

Multiplying in (7.6) δf by λ i and extracting $\lambda\delta f$ from (7.9), one obtains

$$\left(X - M\ddot{x} - \lambda\frac{\partial f}{\partial x}\right)\delta x + \left(Y - M\ddot{y} - \lambda\frac{\partial f}{\partial y}\right)\delta y + \left(Z - M\ddot{z} - \lambda\frac{\partial f}{\partial z}\right)\delta z = 0. \quad (7.10)$$

Since from the three possible displacements only two are independent, λ can be taken arbitrarily. Let us take λ satisfying the relation

$$X - M\ddot{x} - \lambda\frac{\partial f}{\partial x} = 0, \quad (7.11)$$

and assume that δy and δz are arbitrary and independent quantities.

Formula (7.10) yields

$$Y - M\ddot{y} - \lambda\frac{\partial f}{\partial y} = 0, \quad (7.12)$$

$$Z - M\ddot{z} - \lambda\frac{\partial f}{\partial z} = 0. \quad (7.13)$$

The latter two dependences and conditions (7.11) create the first-order Lagrange equations.

In a general case, when the mass moves along the deformable surface, equation (7.4) takes the form

$$z(t) - z_n(x,y,t) - z_{nier}(x,y,t) - \alpha_c(x,y,t) = 0, \quad (7.14)$$

where z is the vertical mass displacement; z_n is displacement of the surface under the mass; z_{nier} stands for the distribution of local surface irregularities of contacting bodies; and α_c denotes the bodies contact close-up.

Assuming that a rule for the mass motion on the associated surface in plane x, y is given, only Eq. (7.13) will be further studied.

For small construction deflection and for small surface irregularities, the following estimation holds $P_{Dz} \approx P_D$; $Z = P_0 = Mg$, where g is Earth gravity acceleration.

Owing to relation (7.8) vertical mass displacement (7.13) takes the following form:

$$P_0 - \frac{P_0}{g}\frac{d^2 z}{dt^2} - P_D = 0. \quad (7.15)$$

Observe that if $z_{DM} = z$, and for $P = P_0$; $M = P_0/g$, Eq. (7.15) overlaps with (7.1).

In what follows the analyzed problem of dynamic interaction of movable mass on the deformable associated surface is reduced to determination of the unknown dynamical reaction P_D from Eq. (7.14).

Note that the proposed method improves the dynamic model, allows for the contact lack between the mass and surface, and predicts the next contact associated with an impact, but allows also for the introduction of new dynamic factors. Namely, it enables the accounting for local deformations in the contact bodies as well as irregularities of the surface.

On the other hand, Eq. (7.14) generalizes the Timoshenko equation for the simple impact [96], since the mass approaching the interacting surface may have both vertical and horizontal components.

Furthermore, the case of a two-sided constraint can be also derived from Eq. (7.14). In the latter case, the change of sign of P_D does not yield lack of the contact between the lumped body and the surface.

There is one more advantage of the presented approach. A solution to the problem of dynamic impact of the moving masses on machines and construction elements is essentially simplified. It is reduced to an independent analysis of much simpler problems of construction vibrations driven by movable force P_D, and of displacement of the moving lumped body.

7.2 Moving Load Equations

Let us consider a shell with coordinate z directed to the Earth center, assuming that the body moving in the gravity field possesses the point mass M_T. In what follows both sloping $v_\Gamma \neq 0$ and transversal $v_B \neq 0$ impacts will be considered.

It is further assumed that the body may move on the shell only after a sloping impact, and its velocity is parallel either to axis x or y. In order to define the mass displacement in directions z, through Eq. (7.15) one obtains

$$M_T \frac{d^2 z}{dt^2} = G_T - P_T, \tag{7.16}$$

where G_T is the body weight, whereas P_T is the reaction of the interaction between the lumped body and the shell.

Let the body move with constant acceleration ω on the shell with initial velocity v_x, parallel to the shell's side with diameter a. The position η of the body is defined through the relation

$$\eta = \frac{\omega t^2}{2} + v_x t. \tag{7.17}$$

We are going to use the variable η instead of t in Eq. (7.16). Making use of the formula

$$\frac{dz}{dt} = \frac{dz}{d\eta}\frac{d\eta}{dt}, \tag{7.18}$$

one obtains

$$\frac{dz}{dt} = \frac{dz}{d\eta}\frac{d\eta}{dt} = \frac{dz}{d\eta}(\omega t + v_x),$$

$$\frac{d^2 z}{dt^2} = \frac{d}{dt}\left[\frac{dz}{d\eta}(\omega t + v_x)\right]$$

$$= \frac{d^2z}{d\eta^2}\frac{d\eta}{dt}(\omega t + v_x) + \frac{dz}{d\eta}\omega = \frac{d^2z}{d\eta^2}(\omega t + v_x)^2 + \frac{dz}{d\eta}. \tag{7.19}$$

Owing to (7.17) one obtains

$$2\eta = \omega t^2 + 2v_x t,$$

$$\omega t^2 + 2v_x t - 2\eta = 0,$$

or

$$t^2 + \frac{2v_x}{\omega}t - \frac{2\eta}{\omega} = 0. \tag{7.20}$$

Equation (7.17) can be rewritten in the following way:

$$\left(t + \frac{v_x}{\omega}\right)^2 - \frac{2\eta}{\omega} - \frac{v_x^2}{\omega^2} = 0,$$

or

$$(\omega t + v_x)^2 = v_x^2 + 2\omega\eta.$$

Finally, we obtain

$$\frac{d^2z}{dt^2} = \frac{d^2z}{d\eta^2}(2\omega\eta + v_x) + \frac{dz}{d\eta}\omega. \tag{7.21}$$

Equation (7.16) expressed in terms of variable η reads

$$\frac{d^2z}{d\eta^2}(2\omega\eta + v_x) + \frac{dz}{d\eta}\omega = \frac{G_T - P_T}{M_T}. \tag{7.22}$$

7.3 Non-dimensional Form of Lumped Body Equations

Since variable t and normal load parameter q occurring in the shell motion equation are transformed to the non-dimensional form (bars) owing to (1.54), and since reaction P_T of the interaction between the body and the shell is equivalent to shell load, the new non-dimensional parameters are as follows:

$$\eta = a\overline{\eta}, \quad z = (2h)\overline{z}, \quad v_x = \frac{(2h)}{b}\sqrt{\frac{Eg}{\rho}}\overline{v}_x, \quad \omega = \frac{(2h)^2}{ab^2}\frac{Eg}{\rho}\overline{\omega},$$

$$P = \frac{(2h)^4E}{a^2b^2}\overline{P}, \quad M_T = \frac{2h\rho}{g}\overline{M}_T, \quad \lambda_1 = \frac{a}{2h}, \quad \lambda_2 = \frac{b}{2h}. \tag{7.23}$$

Equation (7.16) takes the form

$$\frac{d^2\overline{z}}{d\overline{t}^2} = \frac{\rho a^2b^2}{(2h)^3E} - \frac{\overline{P}_T}{\overline{M}_T}, \tag{7.24}$$

and let us introduce a new non-dimensional quantity

$$\frac{\rho a^2 b^2}{(2h)^3 E} = \alpha. \tag{7.25}$$

Since α is a non-dimensional parameter M_T describing the ratio of the mass body and the mass shell per unit surface, the equation can be transformed to the following non-dimensional form (index "T" and bars are omitted for simplicity):

$$\frac{d^2 z}{dt^2} = \alpha - \frac{P}{M}, \tag{7.26}$$

where P is reaction force of interaction between load and shell, and M is the ratio of the lumped body and shell masses.

A solution to Eq. (7.26) in the interval $[t_0,\ t]$ has the following form:

$$z(t) = z_1(t) + z_2(t)(t - t_0) + \int_{t_0}^{t} \left[\alpha - \frac{P}{M}\right](t - \eta)d\eta, \tag{7.27}$$

where $z_1,\ z_2$ are constants defined through the initial conditions.

The continuous function P is approximated by the piecewise continuous function $P^{(n)}(t)$ with constant values within small step in time. The introduced approach enables easy computation of integral (7.27) and hence z and $\frac{dz}{dt}$ on each integration step are computed:

$$z(t) = z(t_0) + \left(\alpha - \frac{P}{M}\right)\frac{(t-t_0)^2}{2} + \frac{dz}{dt}(t_0)(t - t_0), \tag{7.28}$$

$$\frac{dz}{dt}(t) = \frac{dz}{dt}(t_0) + \left(\alpha - \frac{P}{M}\right)(t - t_0). \tag{7.29}$$

It is obvious that the values of $z(t_0)$ and $\frac{dz}{dt}(t_0)$ computed in a previous step serve as initial conditions for the next step.

7.4 Boundary and Initial Problem for a Shell

Consider a rectangular plate cylindrical panel or a spherical shell as the construction interacting with a moving load (see Fig. 7.2).

Owing to a geometrically nonlinear theory and the Kirchhoff-Love kinematical model, the associated equations of the dynamical hybrid form read [168]:

$$\frac{\partial^2 w}{\partial t^2} + \varepsilon \frac{\partial w}{\partial t} = \nabla_k^2 F + L(w, F) + q$$

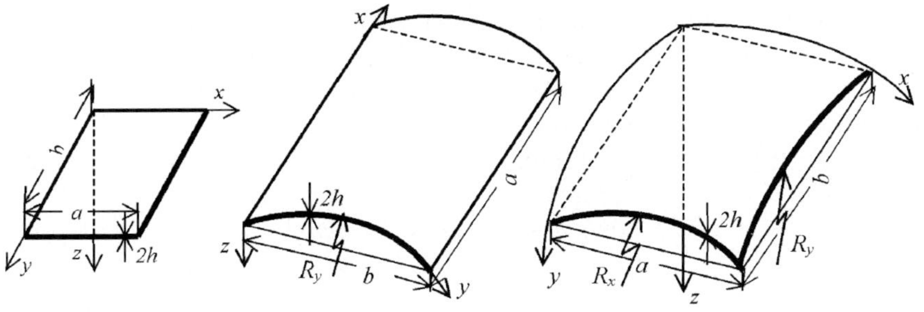

Fig. 7.2 Rectangular plate (**a**), cylindrical panel (**b**), and spherical shell (**c**) associated with a moving lumped body construction

$$-\frac{1}{12(1-\mu^2)}\left[\frac{1}{\lambda^2}\frac{\partial^4 w}{\partial x^4}+\lambda^2\frac{\partial^4 w}{\partial y^4}+2\frac{\partial^4 w}{\partial x^2\partial y^2}\right], \tag{7.30}$$

$$\frac{1}{\lambda^2}\frac{\partial^4 F}{\partial x^4}+\lambda^2\frac{\partial^4 F}{\partial y^4}+2\frac{\partial^4 F}{\partial x^2\partial y^2}=-\nabla_k^2 w-\frac{1}{2}L(w,w), \tag{7.31}$$

where the operators in (7.30) and (7.31) have the form

$$\nabla_k^2 = k_y\frac{\partial^2}{\partial x^2}+k_x\frac{\partial^2}{\partial y^2},$$

$$L(w,F)=\frac{\partial^2 w}{\partial x^2}\frac{\partial^2 F}{\partial y^2}+\frac{\partial^2 w}{\partial y^2}\frac{\partial^2 F}{\partial x^2}-2\frac{\partial^2 w}{\partial x\partial y}\frac{\partial^2 F}{\partial x\partial y}.$$

Note that for the sake of simplicity the bars are again omitted in the mentioned non-dimensional equation, and the relations between the dimensional and non-dimensional parameters are as follows:

$$w=2h\overline{w},\quad x=a\overline{x},\quad y=b\overline{y},\quad F=E\,(2h)^3\,\overline{F},$$

$$k_x=\frac{2h}{a^2}\overline{k}_x,\quad k_y=\frac{2h}{b^2}\overline{k}_y,\quad q=\frac{E(2h)^4}{a^2b^2}\overline{q},$$

$$t=\frac{ab}{2h}\sqrt{\frac{\rho_0}{gE_0}}\overline{t},\quad \varepsilon=\frac{2h}{ab}\sqrt{\frac{gE_0}{\rho_0}}\overline{\varepsilon},\quad \lambda=\frac{a}{b}. \tag{7.32}$$

In order to integrate Eqs. (7.30) and (7.31), one has to define boundary and initial conditions.

If the Kirchhoff-Love hypothesis of straight normals is applied, than each contour point should satisfy four boundary conditions. Namely, knowing displacements u, v, w of the curve of contour points, it is possible to define the position of the curve after deformation. Note that a normal associated with a contour point may be shifted together with this point and rotated by the value of a certain angle in the plane normal to the contour curve. To conclude, the normal position after shell deformation is fixed with the help of four quantities.

It is obvious that in a real shell-type construction various support types can be found, which gives a wide spectrum of their mathematical models.

Below, only some of the boundary conditions frequently met in real constructions are reported for $x = 0,\ x = a$, and $y = 0,\ y = b$:

1. Hinged support on flexible ribs non-compressed (non-stretched) in the tangential plane:

$$w = M_1 = T_1 = \varepsilon_2 = 0 \text{ for } x = 0;\ a,$$
$$w = M_2 = T_2 = \varepsilon_1 = 0 \text{ for } y = 0;\ b. \tag{7.33}$$

The above condition can be rewritten in the following form:

$$w = \frac{\partial^2 w}{\partial^2} = F = \frac{\partial^2 F}{\partial y^2} = 0 \text{ for } x = 0;\ a,$$
$$w = \frac{\partial^2 w}{\partial y^2} = F = \frac{\partial^2 F}{\partial x^2} = 0 \text{ for } y = 0;\ b. \tag{7.34}$$

2. Free edge:

$$w = M_1 = T_1 = S = 0 \text{ for } x = 0;\ a,$$
$$w = M_2 = T_2 = S = 0 \text{ for } y = 0;\ b. \tag{7.35}$$

3. Movable clamping:

a)

$$w = 0;\ \frac{\partial w}{\partial x} = 0;\ T_1 = \varepsilon_2 = 0 \text{ for } x = 0;\ a,$$
$$w = 0;\ \frac{\partial w}{\partial y} = 0;\ T_1 = \varepsilon_2 = 0 \text{ for } y = 0;\ b. \tag{7.36}$$

b)

$$w = 0;\ \frac{\partial w}{\partial x} = 0;\ T_1 = S = 0 \text{ for } x = 0;\ a,$$
$$w = 0;\ \frac{\partial w}{\partial y} = 0;\ T_2 = S = 0 \text{ for } y = 0;\ b. \tag{7.37}$$

More examples of boundary conditions within the Kirchhoff-Love model are given in monographs [143].

Integration of the fundamental equation requires satisfaction of initial conditions associated with deflections and velocities of the mean surface points, i.e.:

$$w\big|_{t=t_0} = \zeta_1(x,y),$$
$$\left.\frac{\partial w}{\partial t}\right|_{t=t_0} = \zeta_2(x,y). \tag{7.38}$$

7.5 Shell Rise

Let us begin with panel rise. Introduce rotations in Fig. 7.3, where $OD = OF = R$ is the main curvature radius of the mean panel surface; $AD = a$ is the panel dimension in the x direction. Then, $H(x_0) = KM = (OB - OC)$ is the sought quantity, which is the panel rise measured at point x_0 in direction x. From triangles OCF and OBD (see Fig. 7.3):

$$OC = \sqrt{R^2 - \frac{a^2}{4}},$$
$$OB = \sqrt{R^2 - BD^2} = \sqrt{R^2 - BK^2} = \sqrt{R^2 - \left(\frac{a}{2} - x_0\right)^2}, \tag{7.39}$$

and hence the rise at point x_0 reads

$$H(x_0) = BC = \sqrt{R^2 - \left(\frac{a}{2} - x_0\right)^2} - \sqrt{R^2 - \frac{a^2}{4}}. \tag{7.40}$$

In order to obtain the rise $H(x_0, y_0)$ at the point with coordinates (x_0, y_0), a coefficient of rise variation for $H(y_0)$ should be introduced, i.e., the ratio of the rise at a

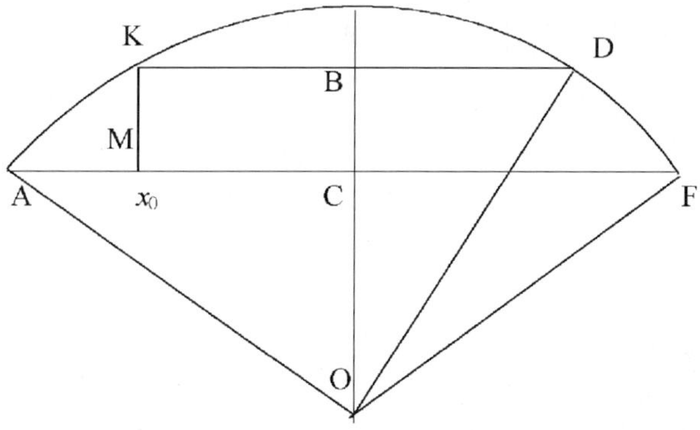

Fig. 7.3 Panel rise height

point moving along y and the largest rise associated with axis y. It reads

$$\frac{\sqrt{R_y^2 - \left(\frac{b}{2} - y_0\right)^2} - \sqrt{R_y^2 - \frac{b^2}{4}}}{R_y - \sqrt{R_y^2 - \frac{b^2}{4}}}. \tag{7.41}$$

Finally, the shell rise at point (x_0, y_0) follows:

$$H(x_0, y_0) = \left[\sqrt{R_x^2 - \left(\frac{a}{2} - x_0\right)^2} - \sqrt{R_x^2 - \frac{a^2}{4}}\right] \times \frac{\left[\sqrt{R_y^2 - \left(\frac{b}{2} - y_0\right)^2} - \sqrt{R_y^2 - \frac{b^2}{4}}\right]}{R_y - \sqrt{R_y^2 - \frac{b^2}{4}}} \tag{7.42}$$

and the same holds in the non-dimensional form:

$$H(x_0, y_0) = \lambda_1 \left[\sqrt{\frac{\lambda_1^2}{k_x^2} - \left(\frac{1}{2} - x_0\right)^2} - \sqrt{\frac{\lambda_1^2}{k_x^2} - \frac{1}{4}}\right] \times \frac{\left[\sqrt{\frac{\lambda_2^2}{k_y^2} - \left(\frac{1}{2} - y_0\right)^2} - \sqrt{\frac{\lambda_2^2}{k_y^2} - \frac{1}{4}}\right]}{\frac{\lambda_2}{k_y} - \sqrt{\frac{\lambda_2^2}{k_y^2} - \frac{1}{4}}}, \tag{7.43}$$

where again bars are omitted for the sake of simplicity.

7.6 Shell Vibrations with Two-Sided Moving Lumped Body Constraints

Let us transform the terms of Eqs. (7.30)–(7.31) into their left-hand sides and denote them by Φ_1, Φ_2, respectively. Owing to this transformation, Eqs. (7.30), (7.31) [168]:

$$\Phi_1\left(w, F, \frac{\partial^2 w}{\partial t^2}, \frac{\partial w}{\partial t}, \frac{\partial^2 w}{\partial x^2}, \frac{\partial^2 F}{\partial x^2}, q, \ldots\right) = 0,$$

$$\Phi_2\left(w, F, \frac{\partial^2 w}{\partial x^2}, \frac{\partial^2 F}{\partial x^2}, \ldots\right) = 0. \tag{7.44}$$

In general, it is impossible to attain an exact solution to these equations with the associated boundary conditions.

Recall that in order to solve the derived equations one may use the Ritz-Timoshenko variational method, the Bubnov-Galerkin approach, the finite differences method, the finite element method, etc.

Owing to the simplification, introduced earlier, of the interaction between the shell and the load moving on it, the Bubnov-Galerkin method can be further used, i.e., the governing Eqs. (7.30)–(7.31) will be solved using the Bubnov-Galerkin method with higher approximations. For this purpose, functions w, F, satisfying the boundary conditions, are sought in the form

$$w=\sum_{i,j}A_{ij}(t)\varphi_{ij}(x,y),$$

$$F=\sum_{i,j}B_{ij}(t)\psi_{ij}(x,y),\quad i=1,2,\ldots,M_x;\quad j=1,2,\ldots,M_y. \tag{7.45}$$

Applying the Bubnov-Galerkin procedure to (7.44), one arrives at

$$\int_0^1\int_0^1 \Phi_1\varphi_{vz}(x,y)\,dx\,dy=0,$$

$$\int_0^1\int_0^1 \Phi_2\psi_{vz}(x,y)\,dx\,dy=0,\quad v=1,2,\ldots,M_x;\quad z=1,2,\ldots,M_y, \tag{7.46}$$

and owing to (7.45) we obtain

$$\sum_{vz}\left[\sum_{ij}\left[\left(\frac{d^2A_{ij}}{dt^2}+\varepsilon\frac{dA_{ij}}{dt}\right)I_{8,vzij}+A_{ij}I_{1,vzij}-B_{ij}I_{2,vzij}-qI_{3,vzij}\right.\right.$$
$$\left.\left.-A_{ij}\sum_{kl}B_{kl}I_{4,vzijkl}\right]\right]=0,$$
$$\sum_{vz}\left[\sum_{ij}\left[A_{ij}I_{7,vzij}+B_{ij}I_{5,vzij}+A_{ij}\sum_{kl}A_{kl}I_{6,vzijkl}\right]\right]=0,$$
$$v,i,k=1,2,\ldots,M_x;\quad z,j,l=1,2,\ldots,M_y. \tag{7.47}$$

Note that the summation sign $\sum_{vz}[*]$ standing before each of the equations of (7.47) means that each of these equations is understood as a system of vz equations, and the associated integrals follow:

$$I_{1,vzij}=\int_0^1\int_0^1\frac{E}{12(1-\mu^2)}\left[\frac{1}{\lambda^2}\frac{\partial^2\varphi_{ij}}{\partial x^2}\frac{\partial^2\varphi_{vz}}{\partial x^2}+\lambda^2\frac{\partial^2\varphi_{ij}}{\partial y^2}\frac{\partial^2\varphi_{vz}}{\partial y^2}\right.$$

$$+2(1-\mu)\frac{\partial^2\varphi_{ij}}{\partial x\partial y}\frac{\partial^2\varphi_{vz}}{\partial x\partial y}+\mu\left(\frac{\partial^2\varphi_{ij}}{\partial x^2}\frac{\partial^2\varphi_{vz}}{\partial y^2}+\frac{\partial^2 w}{\partial y^2}\frac{\partial^2\varphi_{vz}}{\partial x^2}\right)\Bigg]\,dx\,dy,$$

$$I_{2,vzij}=\int_0^1\int_0^1-\left(k_y\frac{\partial^2\psi_{ij}}{\partial x^2}+k_x\frac{\partial^2\psi_{ij}}{\partial y^2}\right)\varphi_{vz}\,dx\,dy,$$

$$I_{3,vzij}=\int_0^1\int_0^1\varphi_{vz}\,dx\,dy,$$

$$I_{4,vzijkl}=\int_0^1\int_0^1 L(\varphi_{ij},\,\psi_{kl})\,\varphi_{vz}\,dx\,dy, \tag{7.48}$$

$$I_{5,vzij}=\int_0^1\int_0^1 a_1\left[\left(\lambda^2\frac{\partial^2\psi_{ij}}{\partial y^2}-\mu\frac{\partial^2\psi_{ij}}{\partial x^2}\right)\frac{\partial^2\psi_{vz}}{\partial y^2}\right.$$

$$\left.+\left(\frac{1}{\lambda^2}\frac{\partial^2\psi_{ij}}{\partial x^2}-\mu\frac{\partial^2\psi_{ij}}{\partial y^2}\right)\frac{\partial^2\psi_{vz}}{\partial x^2}++2(1+\mu)\frac{\partial^2\psi_{ij}}{\partial x\partial y}\frac{\partial^2\psi_{vz}}{\partial x\partial y}\right]\,dx\,dy,$$

$$I_{6,vzijkl}=\int_0^1\int_0^1\frac{1}{2}L(\varphi_{ij},\varphi_{kl})\,\psi_{vz}\,dx\,dy,$$

$$I_{7,vzij}=\int_0^1\int_0^1-\left(k_y\frac{\partial^2\varphi_{ij}}{\partial x^2}+k_x\frac{\partial^2\varphi_{ij}}{\partial y^2}\right)\psi_{vz}dx\,dy,$$

$$I_{8,vzij}=\int_0^1\int_0^1\varphi_{ij}\varphi_{vz}\,dx\,dy. \tag{7.49}$$

Integrals (7.48), perhaps apart from $I_{3,vzij}$, where the transversal load acts only on part of the shell, are computed with respect to the whole mean shell surface.

To sum up, system (7.47) consists of $M_x{*}M_y$ nonlinear second-order differential equations with respect to time, and $M_x{*}M_y$ algebraic equations which are linear with respect to B_{ij}.

Let us describe the mentioned procedure in more detail. For this purpose φ_{ij}, ψ_{ij} from (7.45) are presented in the product form of two functions, where each of them depends only on one argument and can be presented as a linear combination of functions satisfying the boundary conditions:

$$\begin{aligned}\varphi_{ij}(x,y)&=\varphi_{1ij}(x)\ \varphi_{2ij}(y),\\ \psi_{ij}(x,y)&=\psi_{1ij}(x)\ \psi_{2ij}(y).\end{aligned} \tag{7.50}$$

In order to trace the influence of the load parameters on the interaction with the shell, the following boundary conditions are further applied:

$$w = 0,\ \frac{\partial^2 w}{\partial x^2} = 0,\ F = 0,\ \frac{\partial^2 F}{\partial x^2} = 0 \text{ for } x = 0, 1,$$
$$w = 0,\ \frac{\partial^2 w}{\partial y^2} = 0,\ F = 0,\ \frac{\partial^2 F}{\partial y^2} = 0 \text{ for } y = 0, 1. \tag{7.51}$$

Owing to (7.48)–(7.50) one obtains

$$\varphi_{1i}(x) = \psi_{1i}(x) = \sin(i\pi x), \quad i = 1, 2, \ldots, M_x,$$
$$\varphi_{2j}(y) = \psi_{2j}(y) = \sin(j\pi y), \quad j = 1, 2, \ldots, M_y. \tag{7.52}$$

Putting (7.52) into (7.45), one obtains

$$w = \sum_{i,j} A_{ij}(t)\ \sin(i\pi x)\ \sin(j\pi y),$$
$$F = \sum_{i,j} B_{ij}(t)\ \sin(i\pi x)\ \sin(j\pi y), \tag{7.53}$$

where the indices $i,\ j$ may take all values.

After application of the Bubnov-Galerkin procedure, system (7.46) is recast into the following form:

$$\int_0^1\int_0^1 \Phi_1\ \sin(\nu\pi x)\ \sin(z\pi y)\ dx\,dy = 0,$$
$$\int_0^1\int_0^1 \Phi_2\ \sin(\nu\pi x)\ \sin(z\pi y)\ dx\,dy = 0, \tag{7.54}$$
$$\nu = 1, 2, \ldots, M_x, \quad z = 1, 2, \ldots, M_y.$$

The integrals of the Bubnov-Galerkin procedure read

$$I_{1,\nu} = \int_{x_0-\Delta x}^{x_0+\Delta x} \sin(\nu\pi x)\ dx = \frac{2}{\nu\pi}\sin(\nu\pi x_0)\ \sin(\nu\pi\ \Delta x), \tag{7.55}$$

$$I_{2,z} = \int_{y_0-\Delta y}^{y_0+\Delta y} \sin(z\pi y)\ dy = \frac{2}{z\pi}\sin(z\pi y_0)\ \sin(z\pi\ \Delta y), \tag{7.56}$$

where x_0, y_0 are coordinates of the center of a rectangular contact surface point, Δx, Δy denote half-width of this part with respect to x and y, respectively, and

$$I_{3,\nu i} = \int_0^1 \sin(i\pi x)\ \sin(\nu\pi x)\ dx = \begin{cases} \frac{1}{2}, & i = \nu, \\ 0, & i \neq \nu, \end{cases} \tag{7.57}$$

$$I_{4,zj} = \int_0^1 \sin(j\pi y)\ \sin(z\pi y)\ dy = \begin{cases} \frac{1}{2}, & j = z, \\ 0, & j \neq z, \end{cases} \tag{7.58}$$

$$I_{5,vik} = \int_0^1 \sin(i\pi x)\ \sin(k\pi x)\ \sin(v\pi x)\ dx$$

$$= \begin{cases} \frac{1}{4\pi}\Big[-\frac{\cos\alpha_1\pi}{\alpha_1} - \frac{\cos\alpha_2\pi}{\alpha_2} - \frac{\cos\alpha_3\pi}{\alpha_3} + \frac{\cos\alpha_4\pi}{\alpha_4} \\ \quad + \frac{1}{\alpha_1} + \frac{1}{\alpha_2} + \frac{1}{\alpha_3} - \frac{1}{\alpha_4}\Big], & \alpha_l \neq 0; \\ \left[\frac{\cos\alpha_l\pi}{\alpha_l} = 0,\ \frac{1}{\alpha_l} = 0\right],\ l = 1,2,3; & \alpha_l = 0, \end{cases} \tag{7.59}$$

where

$$\alpha_1 = i + k - v, \quad \alpha_2 = k + v - i,$$
$$\alpha_3 = v + i - k, \quad \alpha_4 = i + k + v. \tag{7.60}$$

$$I_{6,zjl} = \int_0^1 \sin(j\pi x)\ \sin(l\pi x)\ \sin(z\pi x)\ dy$$

$$= \begin{cases} \frac{1}{4\pi}\Big[-\frac{\cos\beta_1\pi}{\beta_1} - \frac{\cos\beta_2\pi}{\beta_2} - \frac{\cos\beta_3\pi}{\beta_3} + \frac{\cos\beta_4\pi}{\beta_4} \\ \quad + \frac{1}{\beta_1} + \frac{1}{\beta_2} + \frac{1}{\beta_3} - \frac{1}{\beta_4}\Big], & \beta_l \neq 0; \\ \left[\frac{\cos\beta_l\pi}{\beta} = 0,\ \frac{1}{\beta_l} = 0\right],\ l = 1,2,3; & \beta_l = 0, \end{cases} \tag{7.61}$$

where

$$\beta_1 = j + l - z, \quad \beta_2 = l + z - j,$$
$$\beta_3 = z + j - l, \quad \beta_4 = j + l + z. \tag{7.62}$$

$$I_{7,vik} = \int_0^1 \cos(i\pi x)\ \cos(k\pi x)\ \sin(v\pi x)\ dx$$

$$= \begin{cases} \frac{1}{4\pi}\Big[\frac{\cos\alpha_1\pi}{\alpha_1} - \frac{\cos\alpha_2\pi}{\alpha_2} - \frac{\cos\alpha_3\pi}{\alpha_3} - \frac{\cos\alpha_4\pi}{\alpha_4} \\ \quad - \frac{1}{\alpha_1} + \frac{1}{\alpha_2} + \frac{1}{\alpha_3} + \frac{1}{\alpha_4}\Big], & \alpha_l \neq 0; \\ \left[\frac{\cos\alpha_l\pi}{\alpha_l} = 0,\ \frac{1}{\alpha_l} = 0\right],\ l = 1,2,3; & \alpha_l = 0, \end{cases} \tag{7.63}$$

$$I_{8,zjl} = \int_0^1 \cos(j\pi x)\,\cos(l\pi x)\,\sin(z\pi x)\,dy$$

$$= \begin{cases} \frac{1}{4\pi}\left[\frac{\cos\beta_1\pi}{\beta_1} - \frac{\cos\beta_2\pi}{\beta_2} - \frac{\cos\beta_3\pi}{\beta_3} - \frac{\cos\beta_4\pi}{\beta_4}\right. \\ \left. -\frac{1}{\beta_1} + \frac{1}{\beta_2} + \frac{1}{\beta_3} + \frac{1}{\beta_4}\right], \qquad \beta_l \neq 0; \\ \left[\frac{\cos\beta_l\pi}{\beta_l} = 0,\ \frac{1}{\beta_l} = 0\right],\ l = 1,2,3; \qquad \beta_l = 0, \end{cases} \tag{7.64}$$

$$I_{q,vz} = I_{1,v}I_{2,z};\quad I_{AB,vz} = \left(z^2k_x + v^2k_y\right)\pi^2 I_{3,vi}I_{4,zj}$$

$$I_{tt,vz} = I_{3,vi}I_{4,zj};\ I_{t,vz} = \varepsilon I_{3,vi}I_{4,zj} \tag{7.65}$$

$$I_{w,vz} = \frac{\pi^4}{12(1-\mu^2)}\left(\frac{v^4}{\lambda^2} + 2v^2z^2 + \lambda^2 z^4\right)I_{3,vz}I_{4,vz} \tag{7.66}$$

$$I_{B,vz} = \left(\frac{v^4}{\lambda^2} + 2v^2z^2 + \lambda^2 z^4\right)\pi^4 I_{3,vz}I_{4,vz} \tag{7.67}$$

$$I_{vzijkl} = \pi^4\left[\left(i^2l^2 + j^2k^2\right)I_{5,vik}I_{6,zjl} - 2ijkl\ I_{7,vik}I_{8,zjl}\right]. \tag{7.68}$$

Owing to the introduced integrals, system (7.47) takes the following form:

$$\sum_{vz}\left\{\ddot{A}_{vz}I_{tt,vz} + \dot{A}_{vz}I_{t,vz} + A_{vz}I_{w,vz} + B_{vz}I_{AB,vz} - I_{q,vz}q - \sum_{ij}A_{ij}\sum_{kl}B_{kl}I_{vzijkl}\right\} = 0, \tag{7.69}$$

$$\sum_{vz}\left\{B_{ij}I_{B,vz} - A_{vz}I_{AB,vz} + \frac{1}{2}\sum_{ij}A_{ij}\sum_{kl}A_{kl}I_{vzijkl}\right\} = 0. \tag{7.70}$$

Again $\sum_{vx}[*]$ means that instead of each of the equations of system (7.69) and (7.70) the system of vz equations is taken.

The obtained system of differential equations is reduced to the normal form, and is then solved using the fourth-order Runge-Kutta method. Solving the Cauchy problem at each time step, the Gauss method is applied to solve the algebraic system (7.70).

In the case of continuous contact load movement, the pressure occurring at the place of contact between the shell and the lumped body consists of the mass weight and inertial forces generated by transversal mass vibrations together with the shell. Owing to (7.26), one obtains

$$P = M\left(\alpha - \frac{d^2z}{dt^2}\right). \tag{7.71}$$

Let the contact space between the shell and the mass be approximated by rectangle $S(x,y)$ with the sides parallel to the shell sides, where

$$x_0 - \Delta x \leq x \leq x_0 + \Delta x,$$

$$y_0 - \Delta y \leq y \leq y_0 + \Delta y. \tag{7.72}$$

Then the integrals associated with the application of the Bubnov-Galerkin method are computed through formulas (7.55)–(7.56), where x_0, y_0 describe the rectangular center, and Δx, Δy are the half-widths with respect to x, y, respectively.

Formula (7.71) may be rewritten to give $P = M\alpha - M\frac{d^2z}{dt^2}$. The inertial term $M\frac{d^2z}{dt^2}$ is then added into the inertial shell term, and $P = M\alpha$ is substituted in Eq. (7.69) for the normal load parameter q. Note that if the mass is concentrated to a point, then one may use the transition $\Delta x \to 0$, $\Delta y \to 0$, which gives

$$I_{q,\nu x} = I_{1,\nu} I_{2,z} = \Delta x \Delta y \sin(\nu\pi x_0)\sin(z\pi y_0), \tag{7.73}$$

and the product $\Delta x \Delta y$ should be included in the reaction between the mass and the shell.

Recall that in practice the so-called dynamic coefficients are often introduced. For rods and plates, a dynamic coefficient is defined by dividing the dynamic "critical" loading by the Euler-type static quantity. Furthermore, in the case of rods and plates, the critical load values estimated experimentally are close to those found theoretically.

However, in the case of thin-walled shells, derivation of a similar criterion does not belong to simple tasks, since a shell buckling is realized through a sudden jump. The latter process is associated with stability loss "in large."

Let us briefly describe some of the dynamic stability loss criterion, proposed by various authors.

Volmir defined the dynamical stability loss when a fast deflection increase corresponds to small load variations. Shian et al. proposed the first maximum of the load-time dependence as the critical one. In the Lyapunov stability criterion is applied, which is associated with the use of the phase plane of the considered system. This criterion is used for stability investigation in a rectangular spherical shell.

All the mentioned criterions are in good numerical agreement, i.e., the critical loads derived with their help are close to each other. In our further investigations the Volmir criterion is used.

In order to determine the shell stability loss of the shell-mass system, as well as to obtain some values of the critical parameters governing the interaction between the shell and the mass moving on it, a series of computation is carried out.

It has been found that the influence of the non-dimensional parameter of mass M on the shell behavior is similar to that of the normal load. There are values that can be called critical and before critical ones, i.e., the shell exhibits a stability loss. It has been observed that owing to the increase of the contact area, the change of configuration instants of shells 2, 3 in Fig. 7.4b corresponds to shell stability loss.

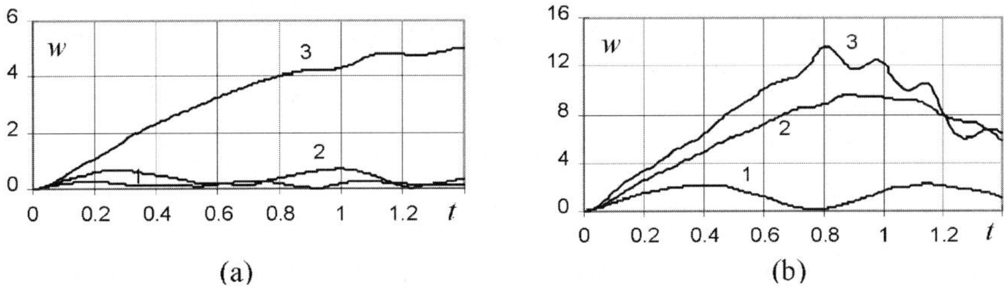

Fig. 7.4 Variation of the shell center deflection for different contacting surface areas under the lumped body (**a**) $\Delta x = \Delta y = 0.05$, (**b**) $\Delta x = \Delta y = 0.1$

In Fig. 7.4, deflection variation at the shell center with parameters $\lambda = 1$, $k_x = k_y = 24$, is reported when the center of the rectangular contact area $2\Delta x$ and $2\Delta y$ between shell and mass overlaps with the shell center. The applied approximation $M_x * M_y$ reads: $5*5 \ldots 9*9$. Curves 1,2,3 are associated with the following parameters: $\alpha = 500$, $M = 2, 4, 10$.

In Fig. 7.5 shell deflection in the cross section $y = 0.5$ for the case corresponding to Fig. 7.4b (curves 1,2) is shown. Curves 1,2,3,4 correspond to time instants $t = 0.2, 0.4, 0.6, 0.8$, respectively.

For all reported curves the time step of 0.2 has been used. For the cases when mass parameters are chosen so that the shell configuration does not change suddenly (jump), its deflection is rather small (Fig. 7.5a). It essentially grows (Fig. 7.5b), when the mentioned jumps (changes of shell configuration) occur.

On the other hand, the non-dimensional parameter α, occurring in the mass equation, may essentially influence the shell stability loss.

Figure 7.6 illustrates the dependence of critical parameter α versus the contact dimension of both bodies ($\Delta x = \Delta y$). Curves 1,2,3 correspond to lumped body mass values $M = 2, 5, 10$, respectively.

Observe that $\Delta x = \Delta y = 0.2$ is the limiting value. For $\Delta x = \Delta y > 0.2$ there is no significant influence of α, whereas for $\Delta x = \Delta y < 0.2$ this influence is important and increases with the increase of M. Owing to the mass velocity increase, the largest deflection appears almost suddenly in a zone behind the mass. The same observation

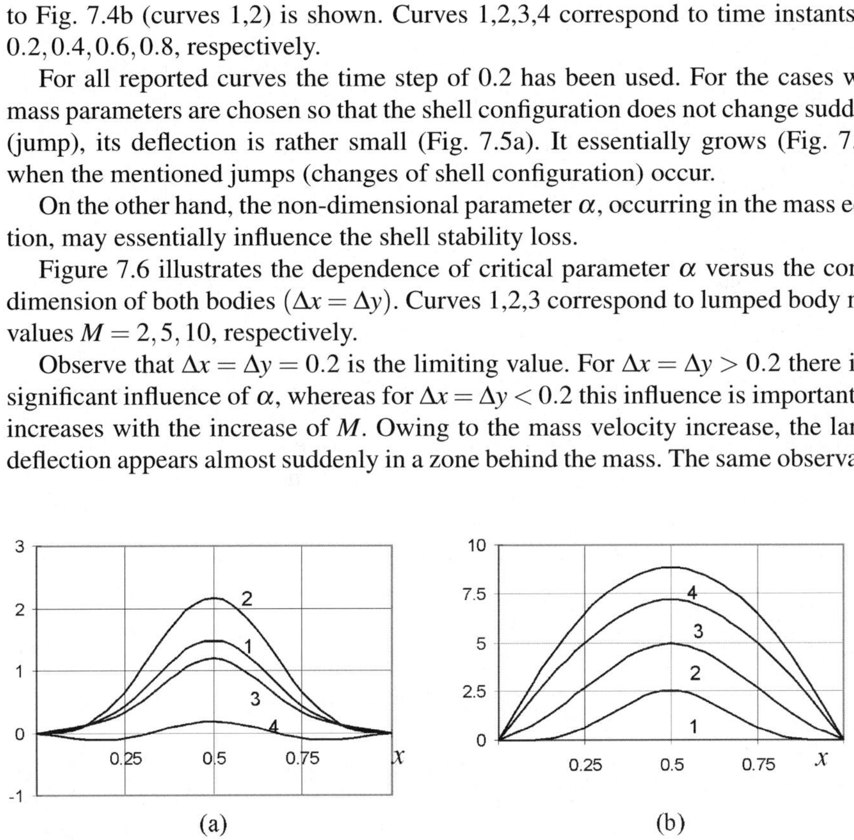

Fig. 7.5 Diagrams of deflection w in cross section $y = 0.5$: a) $M = 2$, b) $M = 4$

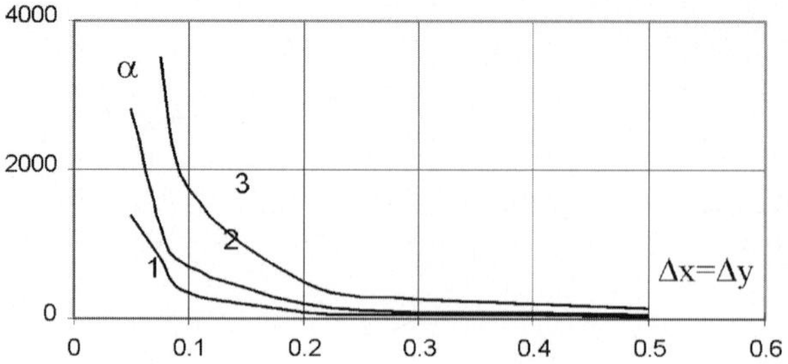

Fig. 7.6 Critical value of parameter α vs different contact areas under the lumped body

holds for the shell, although the shell-mass system is stiffer in comparison with the previously studied one.

7.7 Shell Subjected to Transversal Rigid Body Impact

In this section the dynamics of rectangular elastic shells during the impact with a rigid body are studied. Again axis z associated with a shell is directed into the Earth center. It is assumed that while interacting, a lumped body moves on the shell along a line parallel to x.

Therefore, coordinates x_0, y_0 of the first point of the impact remain constant, and only coordinate z_0 is variated. The latter is found using the motion equation for the rigid body (7.26), i.e., it is assumed that during an impact the body may have the vertical velocity component v_B.

In what follows, the process of interaction of the body and shell is considered, with the contact between them, which is either kept for a certain time, i.e.:

$$z_0 = w(x_0, y_0) + H(x_0, y_0), \tag{7.74}$$

or it is violated, i.e.:

$$z_0 < w(x_0, y_0) + H(x_0, y_0), \tag{7.75}$$

where $H(x_0, y_0)$ is the shell rise height at a chosen point.

Conditions (7.74), (7.75) governing the occurrence of one-sided constraint $z_0 \leq w(x_0,\ y_0) + H(x_0,\ y_0)$ may interleave many times. However, for a given coordinate system the interaction reaction $P \geq 0$.

To solve the defined problem it will be assumed that the shell dynamics are governed by Eqs. (7.30)–(7.31).

The Bubnov-Galerkin method of higher approximations is applied to solve the shell dynamics [168].

The system of approximating functions satisfying boundary conditions (7.33) reads

$$w = \sum_{i,j} A_{ij}(t)\ \sin(i\pi x)\ \sin(j\pi y),$$

$$F = \sum_{i,j} B_{ij}(t)\, \sin(i\pi x)\, \sin(j\pi y), \tag{7.76}$$
$$i = 1,2,3,\ldots M_x; \quad j = 1,2,3,\ldots M_y.$$

The obtained system of the second-order differential equations in time is first reduced to the normal form, and then integrated using the fourth-order Runge-Kutta method.

The following initial conditions are applied:

$$w\Big|_{t=t_0} = 0; \qquad \frac{\partial w}{\partial t}\Big|_{t=t_0} = 0. \tag{7.77}$$

The lumped body dynamics is solved exactly to yield

$$z\Big|_{t=t_0} = H(x_0, y_0); \quad \frac{dz}{dt}\Big|_{t=t_0} = v_B. \tag{7.78}$$

The unknown reaction P is estimated at each integration step solving constraint Eq. (7.74), and using the Newton method. Since the sought quantity P cannot be smaller than zero, $P = 0$ (see (7.75)) is taken when a contact loss occurs.

With known P all other quantities required for the computations are found, and then taken as the initial ones for the next computation step.

Owing to the introduced assumption that the load is either concentrated into a point or is uniformly distributed on a small surface, one has to use a large number of the series terms in (7.76) and the integration step should be taken satisfactorily small. Satisfaction of the mentioned requirements yields a good convergence of the Newton method.

To consider the problem of the dynamic shell stability loss during interaction of the shell with the lumped body, the dynamics are separated into three stages, as reported in [145].

In the first stage, the construction vibrates around an initial equilibrium state. The second stage is associated with a relatively sudden transition of the shell configuration into a new state. Finally, the third stage deals with nonlinear vibrations around the new configuration (equilibrium) position.

The second and the third stages are realized for $P > P_{cr}$, where P_{cr} is a certain critical value.

In order to investigate the influence of the mass and velocity of the impacting body on the mass-shell interaction, the series of computations are carried out. The following values are taken: $k_x = k_y = 24$, $\lambda = 1$, $x_0 = y_0 = 0.5$, $v_B = 0$, $h_t = 0.001$.

The time history of the shell center deflection for different contact surfaces (a: $\Delta x = \Delta y = 0.05$, $\alpha = 1200$; b: $\Delta x = \Delta y = 0.1$, $\alpha = 400$) is shown in Fig. 7.7.

Curves 1,2,3 correspond to mass $M = 2, 5, 10$, respectively.

A change in action between the shell and a rigid body is reported in Fig. 7.8.

From the reported figures one may conclude that the largest shell deflection depends essentially on mass M. One may also introduce a threshold (critical value) of this parameter responsible for stability loss. Increase of the shell-mass contact

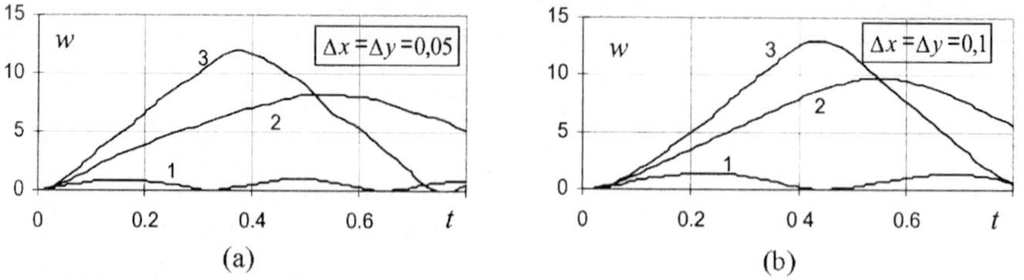

Fig. 7.7 Variation of shell center deflection for different contacting surface area under a lumped body: (**a**) $\Delta x = \Delta y = 0.05$, (**b**) $\Delta x = \Delta y = 0.1$

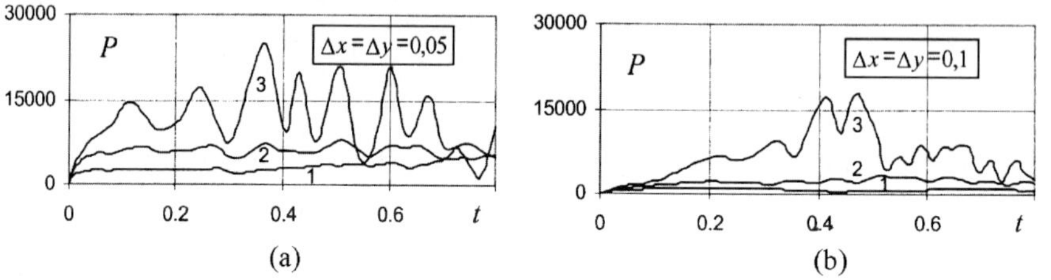

Fig. 7.8 Variation of shell center deflection for different contacting surface area under a lumped body for $k_x = k_y = 24$: (**a**) $\Delta x = \Delta y = 0.05$, (**b**) $\Delta x = \Delta y = 0.1$

decreases the values of parameter α required to achieve the same deflections. The shell-mass contact area has a negligible influence on the reaction force.

Observe that the influence of the mass value is important. For the considered parameter set, no lack of mass-shell contact is observed.

If the mass (while impacting) possesses the vertical velocity component $v_b \neq 0$, then its interaction with the shell is qualitatively different, i.e., it may lose its contact with the shell multiple times.

For $M = 2, 5, 8, 10$ (curves 1,2,3,4) the shell center deflection for $v_B = 80$ is shown in Fig. 7.9.

For the aforementioned parameters, in all cases considered, lack of contact between the mass and the shell (dotted curve) is exhibited at the beginning of the contact at the time instants at which the shell configuration is changed.

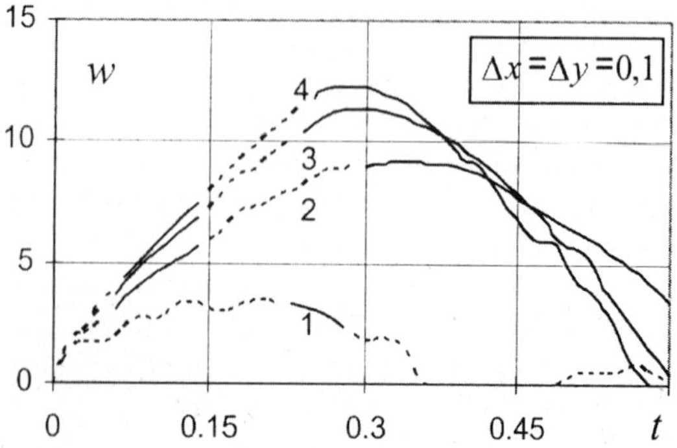

Fig. 7.9 Shell center time history

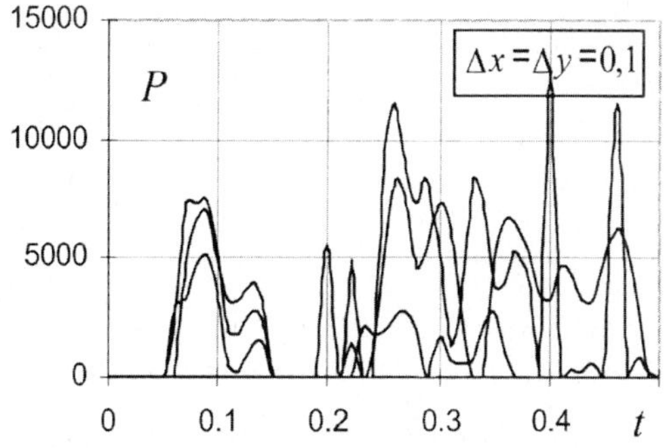

Fig. 7.10 Shell reaction time history

The time history of reaction P is shown in Fig. 7.10. Recall that for the absence of the contact zone, $P = 0$.

7.8 Shells with Constant Velocity Moving Load

The system of the Bernoulli beam with a mass moving on it belongs to those extensively studied at present. As shown in [96], inertia of the moving mass affects dynamic coefficients in an essential way. For example, the dynamic coefficient for the stress measure is introduced as the ratio of the dynamic stress and static stresses at a beam center.

Computations show that an increase of mass velocity maxima of dynamic coefficients are shifted along a mass displacement, whereas dynamic coefficients associated with deflection and stresses first increase in time, and then (after achieving their maxima) start to decrease.

Our further investigation will focus on the interaction between a load moving with constant velocity along coordinate x and with the mass uniformly distributed on a rectangular area with sides Δx, Δy, and a plate, a panel, or a spherical shell serving as the interacting structure.

Note that a moving mass trajectory is identical with the mean curve of an interacting structure, although the input point on the construction can be arbitrary, and the vertical mass velocity component is equal to zero.

Since in this case the mass moves on the shell, it is convenient to use the displacement coordinate along one of the sides of the shell, say x with $\eta = v_x t$, instead of time t (v_x is the mass velocity horizontal component). The mass dynamics is governed by Eq. (7.26), and it reads

$$v_x^2 \frac{d^2 z}{d\eta^2} = \alpha. \tag{7.79}$$

where now α is equal to $a - \frac{P}{M}$.

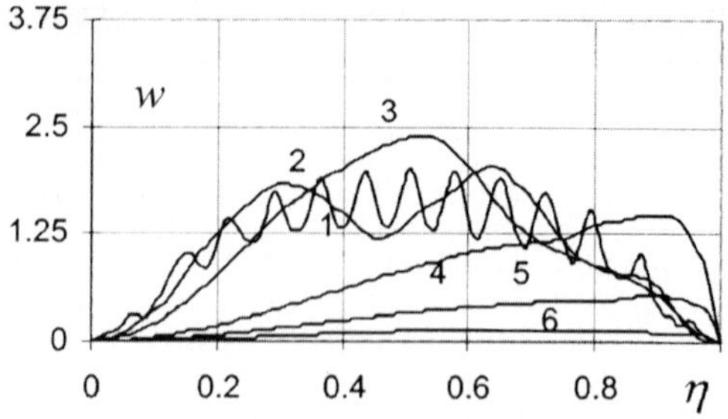

Fig. 7.11 Dynamic plate deflection caused by a moving load

In order to solve the second-order linear differential equation, the method of variation of constants will be further applied to yield the following solution (7.79)

$$z(\eta) = z(b) + \frac{\alpha}{2v_x^2}(\eta - b)^2 + \frac{dz}{d\eta}(b)(\eta - b), \tag{7.80}$$

$$\frac{dz}{d\eta}(\eta) = \frac{dz}{d\eta}(b) + \frac{\alpha}{v_x^2}(\eta - b), \tag{7.81}$$

where $h_1 = (\eta - b)$ is the integration interval (see (7.29)).

In Fig. 7.11 the squared shell deflections under the load with mass $M = 5$, $\alpha = 500$ and $v_x = 0.1, 0.5, 1, 2, 5, 8$ (for curves 1–6 respectively) are shown. For small motion velocities (curves 1,2) the interacting system vibrates with frequency strongly dependent on the mass velocity.

With the increase of mass velocity the vibrational process is damped and vanishes (curve 3), and the function that describes the process attains its maximum when the mass goes through the plate center.

Further increase of the mass velocity shifts the maximal plate deflection to its right side and simultaneously the maximal plate deflection is seriously decreased (curves 4,5,6).

In the interaction process between the mass and the plate a lack of contact can be observed. It occurs for small (large) velocities in the case of heavy (light) loading body. Zones of stable and unstable mass–shell contacts in the parameter plane M–v_x are reported in Fig. 7.12. The zone corresponding to the continuous contact between the two analyzed objects is located on the left-hand side.

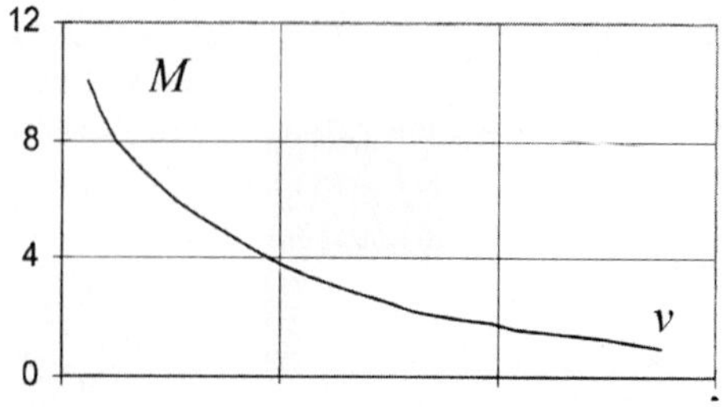

Fig. 7.12 Stable and unstable contact zones between plate and lumped body

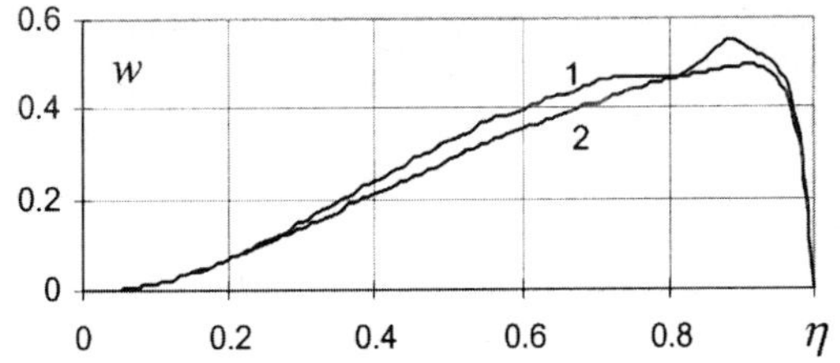

Fig. 7.13 Dynamic deflection under a lumped body (1-plate, 2-panel)

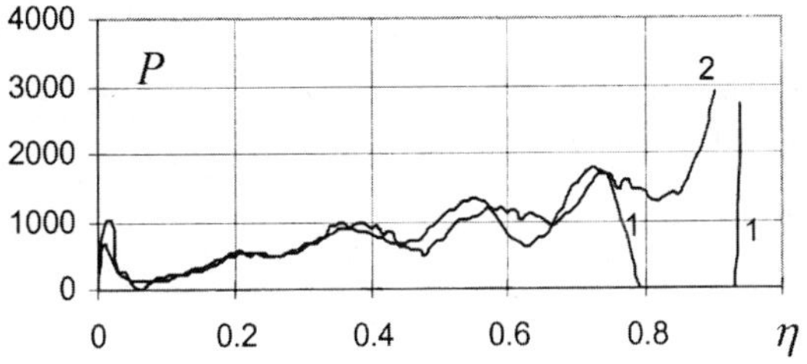

Fig. 7.14 Reaction (1-plate, 2-panel)

For parameters $M = 5$, $\alpha = 150$, $v_x = 5$ for the plate ($k_x = k_y = 0$), and the panel ($k_x = 0$, $k_y = 30$) with $\lambda = 2$ dependencies of deflection under the mass, reaction of self-interaction, forces $T_1 = \frac{\partial^2 F}{\partial y^2}$, $T_2 = \frac{\partial^2 F}{\partial x^2}$, and the moments on the interacting surface M_1, M_2 are shown in Figs. 7.13–7.16.

Although the dependencies for the panel correspond to other values, the considered systems are qualitatively similar in character.

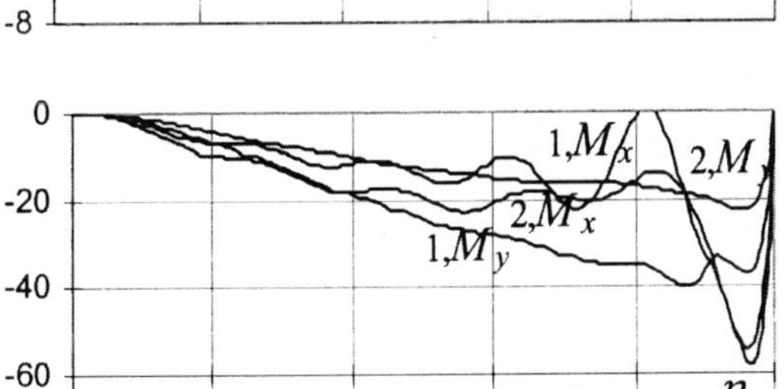

Fig. 7.15 Forces generated by moving load (1-plate, 2-panel)

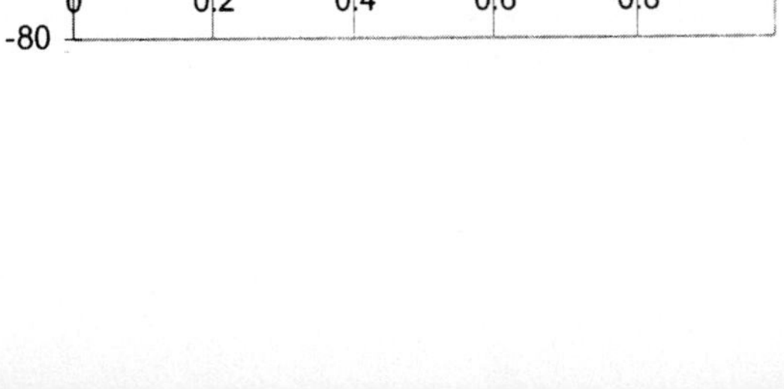

Fig. 7.16 Moments generated by moving load (1-plate, 2-panel)

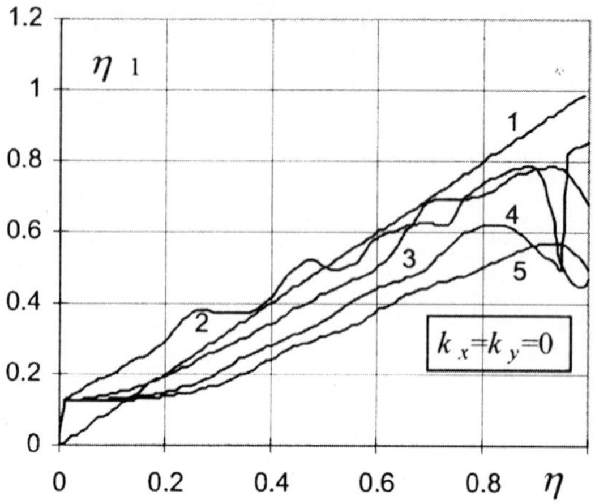

Fig. 7.17 Largest deflection plate coordinate for a moving load

In view of the computation performed, the largest plate (or panel) deflection may occur either before or behind the moving mass.

Functions of the largest deflection coordinate of the squared plate $k_x = k_y = 0$ (Fig. 7.17) and the shell (Fig. 7.18) with parameters $k_x = k_y = 24$ on a coordinate η describe the position of the uniformly moving lumped body.

The straight line 1 corresponds to the moving coordinate, whereas curves 2,3,4,5 correspond to the largest deflection coordinate in given time instants for mass velocities $v_x = 0.5, 2, 5, 10$, respectively.

An analysis of the reported figures yields the following conclusions for low-mass velocities the largest plate deflection occurs first either before or under the mass, and after crossing the plate center, it moves in a zone behind the mass.

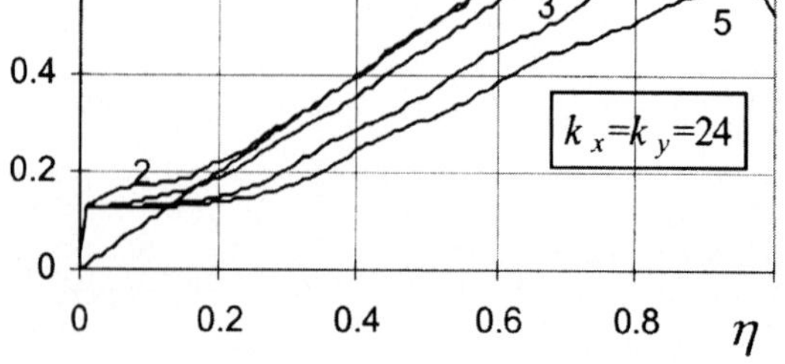

Fig. 7.18 Largest deflection shell coordinate for a moving load

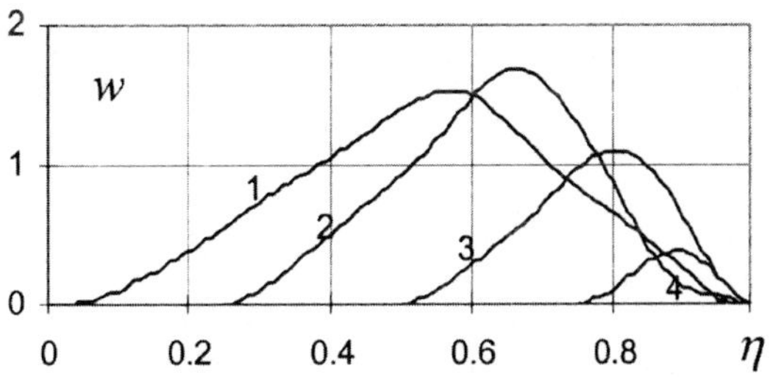

Fig. 7.19 Plate deflection for various lumped body input points

The developed algorithm allows computation of the cases for arbitrary mass point input on the interacting surface, as well as for the mass velocity with horizontal and vertical components.

Deflection variations under the mass for different input points $\eta = 0.25, 0.5; 0.75$ on the plate and panel with parameters $k_x = 0$, $k_y = 30$, $\lambda = 2$ (a: $v_x = 1$, b: $v_x = 5$) are shown in Figs. 7.19 and 7.20.

Note that the mass moves along an axial curve of the interacting surface, and its vertical velocity component is equal to 0. The graphs show that if the mass approaches the surface on the right of its center, then the dynamical deflections are essentially decreased.

In order to compare vibration regimes of the plate and the shell with a moving mass, appropriate computations have been carried out and the results are reported in the tables below.

In order to obtain the values of the investigated parameters included in the tables, which characterize the dynamical processes qualitatively, parameters α and v_x are chosen to be different for the plate and for the shell.

Table 7.1 displays w—surface deflection forms; T_1, T_2—forces, M_1, M_2—movements, of the squared plate during transition through the mass along its axial curve ($M = 5$, $\alpha = 150$) with constant velocity $v_x = 5$ for time instants corresponding to $\eta = 0.3, 0.5, 0.7, 0.9$.

In Table 7.2 the same is shown for the shell ($k_x = k_y = 24$, $\lambda = 1$, $M = 5$, $\alpha = 300$, $v_x = 2$). In both cases the mass pressure is uniformly distributed on the area $\Delta x = \Delta y = 0.1$.

Although there are peculiar similarities reported for plates and shells, they differ in stiffness which is decisive for the mass–shell (–plate) interaction.

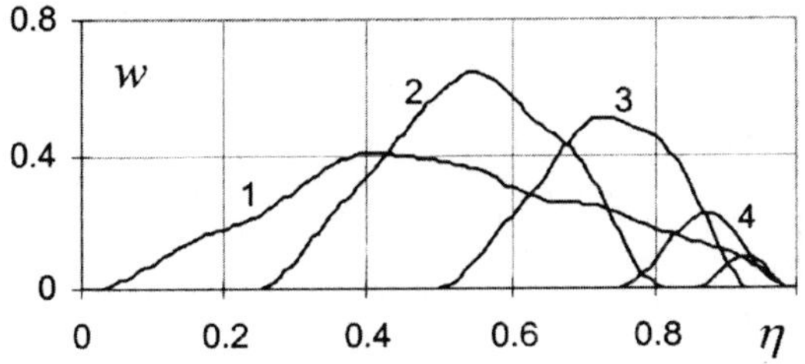

Fig. 7.20 Panel deflection for various lumped body input points

Table 7.1 Plate deflection w, forces T_1, T_2 and moments M_1, M_2 for different η

	Plate ($k_x = k_y = 0$)			
	$\eta = 0{,}3$	$\eta = 0{,}5$	$\eta = 0{,}7$	$\eta = 0{,}9$
w	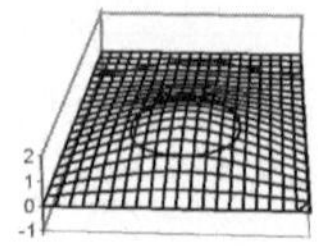	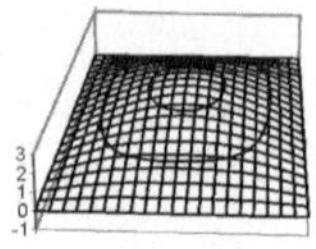	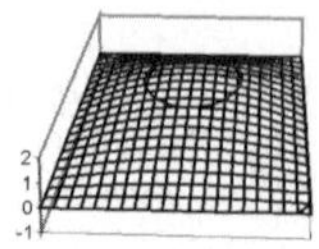	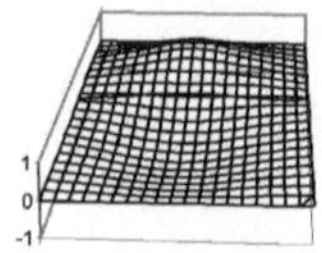
T_1	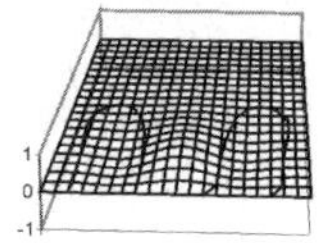	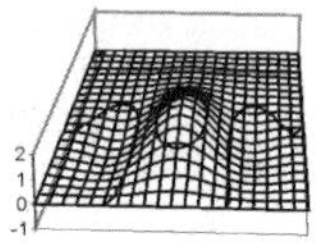	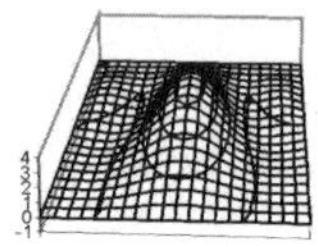	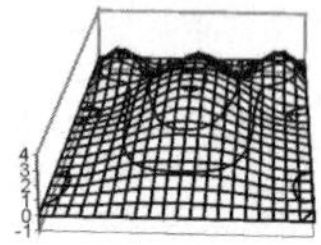
T_2	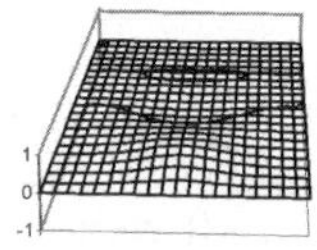	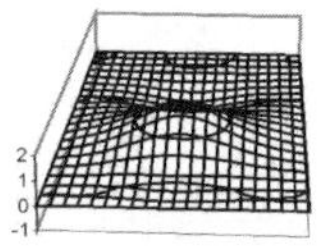	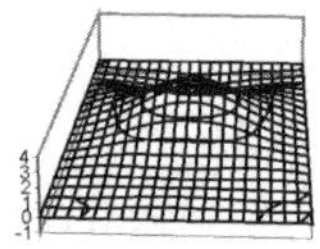	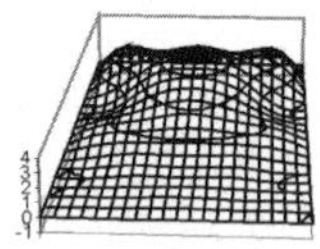
M_1	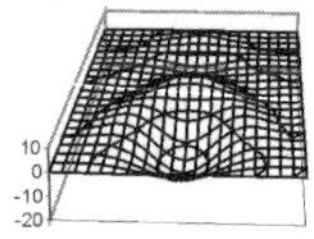	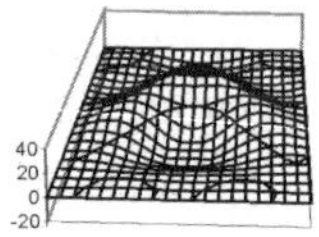	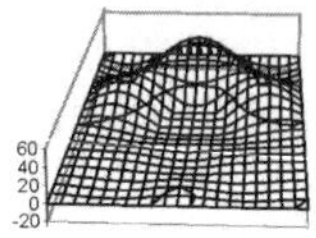	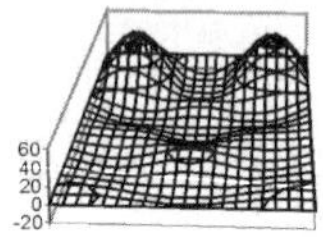
M_2	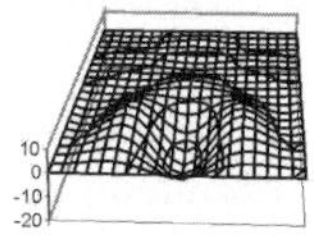	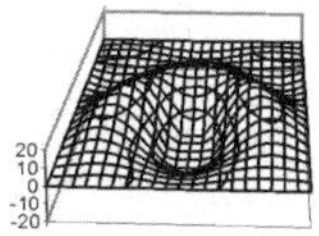	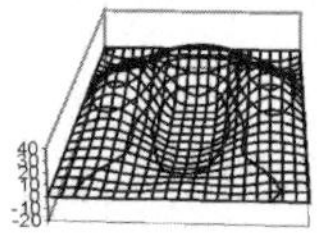	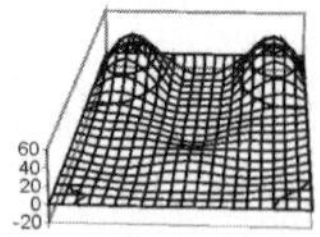

7.9 Shell and Load Moving with Constant Acceleration

When analyzing mass motion with constant acceleration (positive or negative) instead of time t, it is convenient to use a η-coordinate characterizing the load displacement along one of the shell edges (here x). Namely, let $\eta = \frac{\omega t^2}{2} + v_x t$, where v_x is the mass velocity projection on x for $t = 0$, and ω is the mass acceleration.

In this case the solution of the mass dynamics governing equation reads

Table 7.2 Shell deflection w, forces T_1, T_2 and moments M_1, M_2 for different η

Shell ($k_x = k_y = 24$)				
	$\eta = 0{,}3$	$\eta = 0{,}5$	$\eta = 0{,}7$	$\eta = 0{,}9$
w	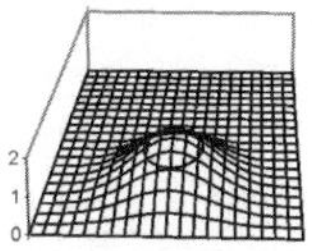	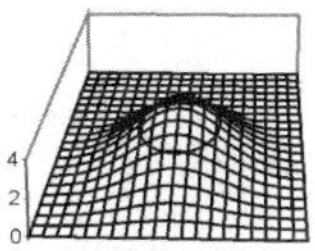	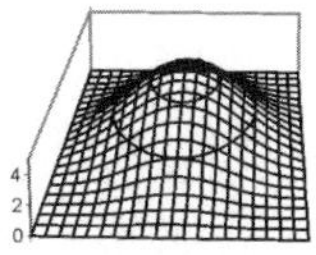	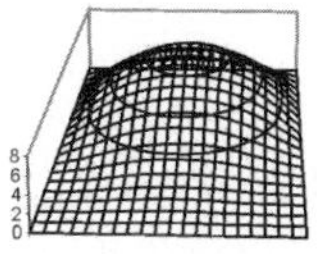
T_1	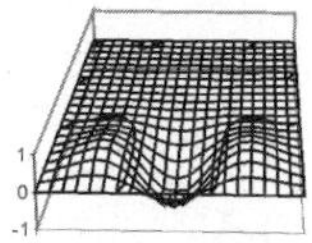	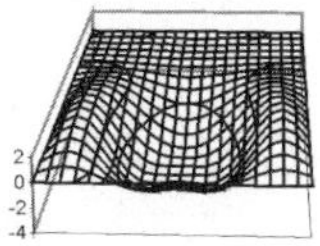	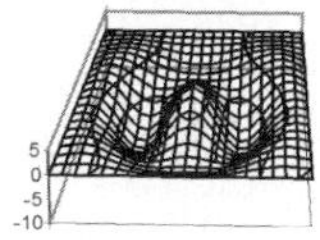	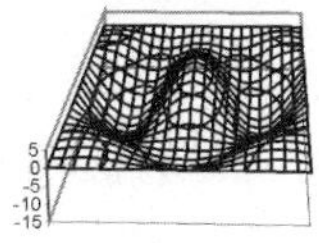
T_2	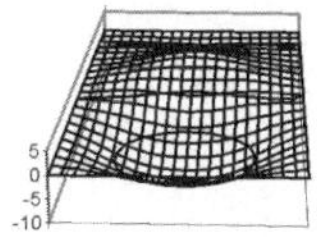	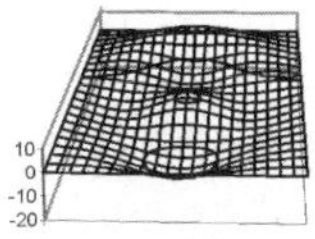	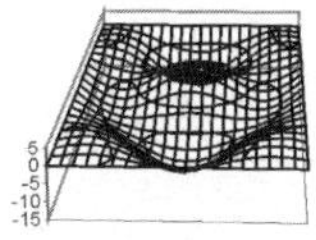	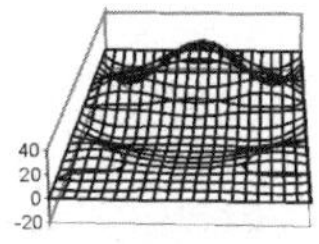
M_1	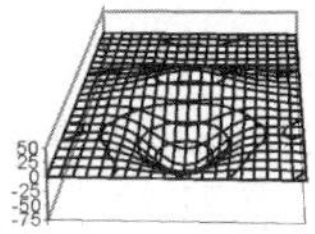	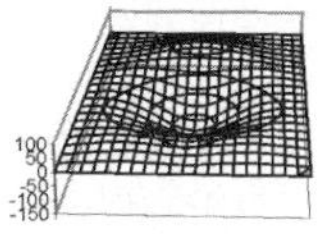	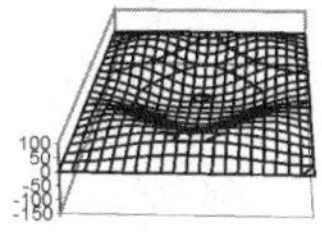	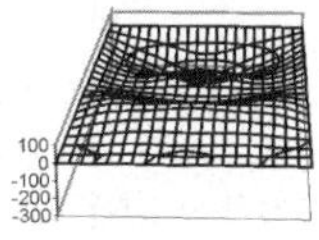
M_2	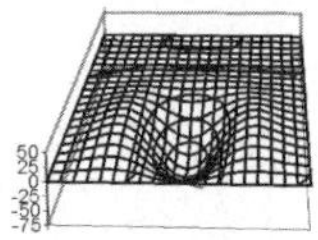	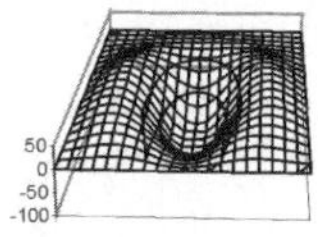	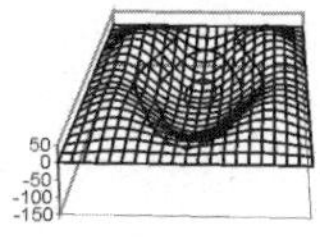	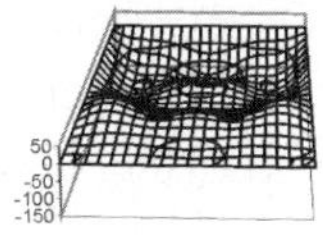

$$z(\eta) = z(b) + \frac{\alpha}{2\omega^2}\left[\sqrt{2\omega\eta + v_x^2} - \sqrt{2\omega b + v_x^2}\right]^2$$

$$+\frac{dz}{d\eta}(b)\frac{\sqrt{2\omega b + v_x^2}}{\omega}\left[\sqrt{2\omega\eta + v_x^2} - \sqrt{2\omega b + v_x^2}\right], \tag{7.82}$$

$$\frac{dz}{d\eta}(\eta) = \frac{dz}{d\eta}(b)\sqrt{\frac{2\omega b + v_x^2}{2\omega\eta + v_x^2}} + \frac{\alpha}{\omega}\left[1 - \sqrt{\frac{2\omega b + v_x^2}{2\omega\eta + v_x^2}}\right]. \tag{7.83}$$

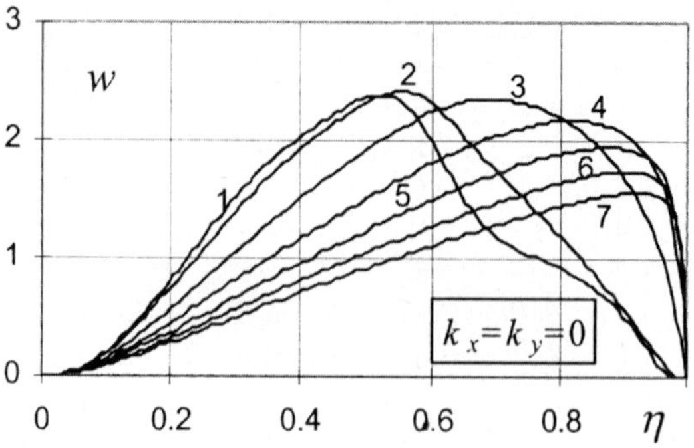

Fig. 7.21 Plate deflection under the lumped body movement with constant acceleration

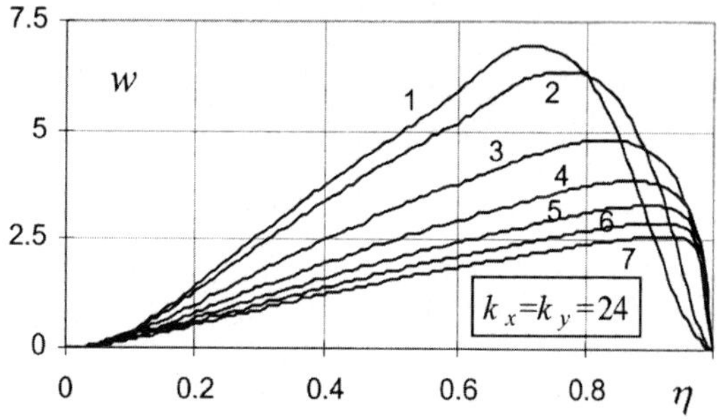

Fig. 7.22 Shell deflection under the lumped body moving with constant acceleration

The deflection development under the load for $\alpha = 150$, $M = 5$; $v_x = 1$, $\Delta x = \Delta y = 0.1$, and $\omega = 0, 1, 5, 10, 15, 20, 25$ (curves 1–7, respectively, correspond to mass movement with constant velocity) is shown for the plate in Fig. 7.21.

The same is done for the shell with the parameters $k_x = k_y = 24$, $\lambda = 1$ for $\alpha = 300$ (see Fig. 7.22).

A study of the figures allows the conclusion that for small values of positive acceleration the deflection under the mass (either for plate or shell) does not differ practically from the mass motion with constant velocity (curves 1,2).

Since acceleration increase causes mass velocity increase, deflections of either plate or shell decrease (curves 2–7). It is expected to occur since the velocity increase does not allow for sudden reaction of the interacting surface.

Furthermore, certain acceleration values may be encountered at which the mass moves with such velocity that the shell does not change its configuration at all (curves 5–7).

7.10 Shell and Load Moving with Constant Negative Acceleration

During load movement with constant negative acceleration ω, its velocity decreases to achieve the zero value for some time instant, i.e., the mass finally stops. Observe that the mass may stop at an arbitrary point of the interacting surface or out of it; the latter case means that when leaving the surface the mass has non-zero velocity.

The analysis to follow will concern the behavior of both of the interacting mass moving with negative instant acceleration and the construction, under the assumption that the mass stop is given a priori.

Development of deflection under the mass moving on a squared plate, when the mass ($\alpha = 150$, $M = 5$, $\Delta x = \Delta y = 0.1$) begins to move on the plane with the velocity and acceleration such that stops at time instants corresponding to $\eta = 0.5, 0.75, 1, 1.5$, is reported in Figs. 7.23, 7.24.

Note that for $\eta = 1.5$ the mass leaves the construction with non-zero velocity.

The same is done for the panel with parameters $k_x = 0$, $k_y = 30$, $\lambda = 2$ (see Figs. 7.25 and 7.26). Curves 1–6 correspond to velocity values $v_x = 1, 2, 3, 4, 5, 6$ of the mass motion beginning on the plate, and for various series of negative acceleration, i.e., $\omega = -1, -4, -9, -16, -25, -36$; $\omega = -0.66, -2.66, -6, -10.66, -16.66, -24$; $\omega = -0.5, -2, -4.5, -8, -12.5, -18$; $\omega = -0.33, -1.33, -3, -5.33, -8.33, -12$ (see Figs. 7.23a,b, 7.24a,b, respectively).

One may conclude from these figures that when the mass stops in the center of the interacted construction, the deflection of the construction increases monotonically until the stop of the mass is achieved.

In the case when the mass stops behind the center of the construction, an essential role is played by the velocity of the mass at its first contact with the plate. Namely,

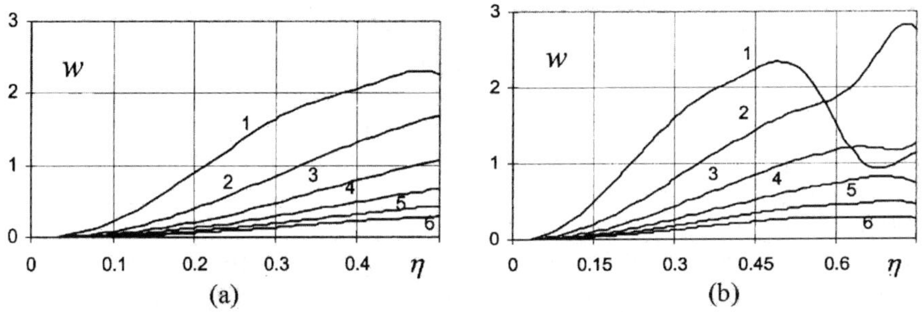

Fig. 7.23 Plate deflection under the lumped body moving with constant negative acceleration (stop of the lumped body takes place for (**a**) $\eta = 0.5$, (**b**) and $\eta = 0.75$

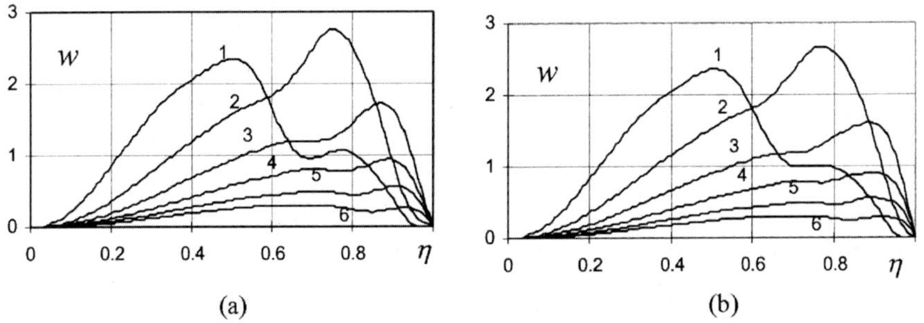

Fig. 7.24 Plate deflection under the lumped body moving with constant negative acceleration (stop of the lumped body takes place for (**a**) $\eta = 1$ and (**b**) $\eta = 1.5$)

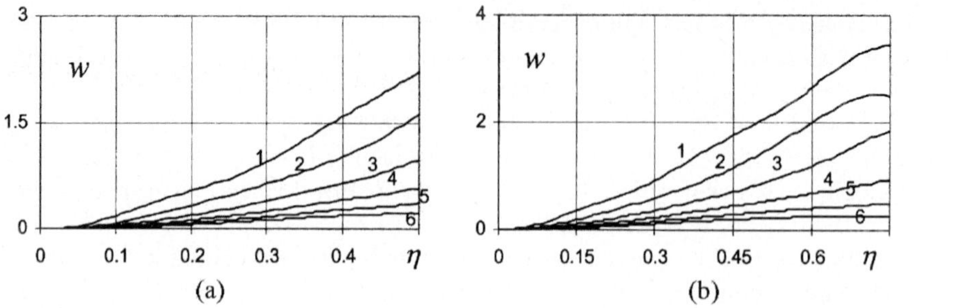

Fig. 7.25 Panel deflection under the lumped body moving with constant negative acceleration (stop of the lumped body takes place for (**a**) $\eta = 0.5$, and (**b**) $\eta = 0.75$)

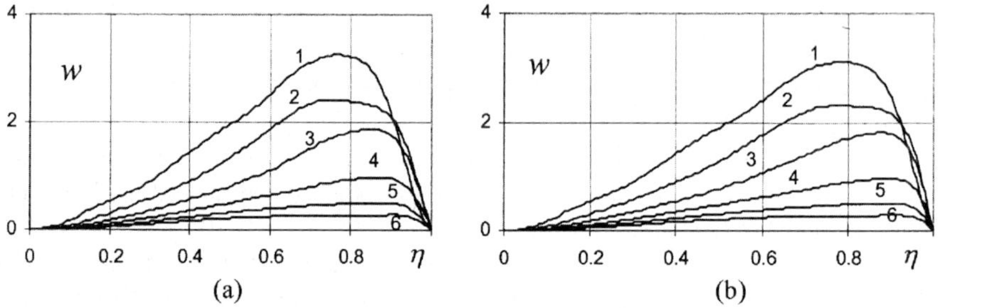

Fig. 7.26 Panel deflection under the lumped body moving with constant negative acceleration (stop of the lumped body takes place for (**a**) $\eta = 1$, and (**b**) $\eta = 1.5$)

for small mass velocities, the deflections can be relatively large, but they decrease with the increase of the velocity.

Since for the same parameter selection the panel is stiffer than the plate, the deflection does not achieve the values causing changes in the panel configurations. Such changes may occur for low mass velocities or other choices of parameters (α, M).

7.11 Conclusions

The main results reported in this chapter are briefly described below.

A solution to the problem of interaction between moving bodies and machines or construction elements is essentially simplified owing to separation of the two objects. Namely, the problem is reduced to independent solutions of considerably simpler problems of an interacting construction, i.e., the problem of vibrations subject to moving force P_D and that of displacement in the domain of the moving masses under the action of the mentioned force.

The proposed method of solution enables improvement of the dynamics modeling accuracy; account of the contact lack between interacting objects; account of successive impacts on interacting surface; and introduction of new dynamic factors useful for engineering application. The fundamental role in the method is played by Eq. (7.14) governing coupling between two bodies. It may allow for local deformations in contacting body lumped system motion on conical surface roughness with an arbitrary profile, springing support of the moving body, etc.

Since the proposed method of computation of dynamic interaction between a moving body and an associated surface assumes independent integration of the motion equations (in the case of a one-sided constraint), the choice of solution methods for the separated equations becomes simplified.

Furthermore, if the motion equations for the interacting surface are solved via the Runge-Kutta numerical method, then one may link the corresponding equation of the vertical displacement of the rigid body (7.16) to allow for the simultaneous interaction of the obtained system equation.

The integration interval of the obtained system is divided into a sufficiently large number of equal parts (usually 1000), where the reaction of interaction between objects is assumed to be unchanged within small step durations, which yield practically the exact solution to the problem.

In addition, the proposed partition of the integration interval while the dynamic reaction P_D of the interacting objects is sought from the coupling equation allows achievement of good convergence of the Newton method.

It should be emphasized that the proposed approach to the computations of dynamics of the considered objects does not require any additional restrictions. For example, in the equations governing dynamics of an interacting construction one may include new terms, involving either damping, or nonlinearity, or other kinematic models of the interacting surface.

Chapter 8
Chaotic Vibrations of Sectorial Shells

In this chapter a novel approach to study chaotic vibrations of deterministic mechanical systems represented by shallow sector-type spherical shells is proposed. Scales of vibration character of such shells being transversally and harmonically excited vs. control parameters are constructed. Scenarios to chaos are illustrated and discussed. Control of the chaotic state applying synchronous action of harmonic loading torque is proposed. Investigations are carried out using the qualitative theory of differential equations and nonlinear dynamics.

8.1 Introduction

Mathematical models of sector-type shells can be directly applied in computation of reinforced membranes which are very often sensitive elements of pressure sensors in various measuring devices. They can also be used in machine construction, etc. (see references [21, 25, 31, 32, 33, 36, 40, 46, 160, 165, 167]). However, this chapter is devoted to the analysis of chaotic dynamics of sector-type shells and its control. This problem has been investigated rather marginally, hence the work is intended to fulfill the existing gap in the research devoted to this problem.

8.2 Statement of the Problem

We consider a non-axially symmetric shallow spherical shell as a two-dimensional object in R^2 in polar coordinates bounded by the contour Γ and defined as follows: $\overline{\Omega} = \Omega + \Gamma = \{(r, \theta, z),\ r \in [0; r_n],\ \theta \in [0; \theta_k],\ z \in [-\frac{h}{2}; \frac{h}{2}]\}$. The governing equations are given in the following form:

$$\begin{aligned} w'' + \varepsilon w' &= -\nabla^2\nabla^2 w + N(w,F) + \nabla^2 F + 4q, \\ \nabla^2\nabla^2 F &= -\nabla^2 w - N(w,w), \end{aligned} \tag{8.1}$$

J. Awrejcewicz, V.A. Krysko, *Chaos in Structural Mechanics*,
DOI: 10.1007/978-3-540-77676-5_9, © Springer-Verlag Berlin Heidelberg 2008

where

$$\nabla^2(\cdot) = \frac{\partial^2(\cdot)}{\partial r^2} + \frac{1}{r}\frac{\partial(\cdot)}{\partial r} + \frac{1}{r^2}\frac{\partial^2(\cdot)}{\partial\theta^2},$$

$$\nabla^2\nabla^2(\cdot) = \frac{\partial^4(\cdot)}{\partial r^4} + \frac{2}{r}\frac{\partial^3(\cdot)}{\partial r^3} - \frac{1}{r^2}\frac{\partial(\cdot)}{\partial r^2} + \frac{1}{r^3}\frac{\partial(\cdot)}{\partial r}$$

$$+ \frac{2}{r^2}\frac{\partial^4(\cdot)}{\partial\theta^2\partial r^2} - \frac{2}{r^3}\frac{\partial^3(\cdot)}{\partial\theta^2\partial r} + \frac{4}{r^4}\frac{\partial^2(\cdot)}{\partial\theta^2} + \frac{1}{r^4}\frac{\partial^4(\cdot)}{\partial\theta^4},$$

$$N(w,F) = \frac{\partial^2 w}{\partial r^2}\left(\frac{1}{r}\frac{\partial F}{\partial r} + \frac{1}{r^2}\frac{\partial^2 F}{\partial\theta^2}\right) + \frac{\partial^2 F}{\partial r^2}\left(\frac{1}{r}\frac{\partial w}{\partial r} + \frac{1}{r^2}\frac{\partial^2 w}{\partial\theta^2}\right)$$

$$-2\frac{\partial}{\partial r}\left(\frac{1}{r}\frac{\partial w}{\partial\theta}\right)\frac{\partial}{\partial r}\left(\frac{1}{r}\frac{\partial F}{\partial\theta}\right),$$

$$N(w,w) = 2\frac{\partial^2 w}{\partial r^2}\left(\frac{1}{r}\frac{\partial w}{\partial r} + \frac{1}{r^2}\frac{\partial^2 w}{\partial\theta^2}\right) - 2\left[\frac{\partial}{\partial r}\left(\frac{1}{r}\frac{\partial w}{\partial\theta}\right)\right]^2.$$

Next, the non-dimensional parameters are introduced: $\bar{t} = \omega_0 t$, $\omega_0 = \sqrt{\frac{Eg}{\gamma R^2}}$, $\bar{\varepsilon} = \sqrt{\frac{g}{\gamma E}}\frac{R}{h}\varepsilon$, $\bar{F} = \eta\frac{F}{Eh^3}$, $\bar{w} = \sqrt{\eta}\frac{w}{h}$, $\bar{r} = b\frac{r}{c}$, $\bar{q} = \bar{q}_3 = \frac{\sqrt{\eta}}{4}\frac{q_3}{E}\left(\frac{R}{h}\right)^2$, $\eta = 12(1-\nu^2)$, $b = \sqrt[4]{\eta}\frac{c}{\sqrt{Rh}}$, where t denotes time; ε is the damping coefficient of surrounding medium, F is the stress function, w denotes the displacement function, R and c are the main curvature radius of shell resistance contour and the radius of resistance contour in circled direction, respectively; h is shell thickness, b is sloping parameter, ν is the Poisson coefficient, r is the distance between a rotation axis and a point on the shell middle surface, q is the external load parameter, and ω_0 is the frequency of linear vibrations. In Eq. (8.1) bars are omitted. Differentiation with respect to time t is denoted by a dash. The following boundary and initial conditions are associated with system (8.1).

1. Ball-type clamping of an arc slice:

$$w = 0, \quad \frac{\partial^2 w}{\partial r^2} + \frac{\nu}{r}\frac{\partial w}{\partial r} = 0, \quad F = 0, \quad \frac{\partial F}{\partial r} = 0. \tag{8.2}$$

2. Ball-type clamping of radial slices:

$$w = 0, \quad \frac{\partial^2 w}{\partial\theta^2} = 0, \quad F = 0, \quad \frac{\partial^2 F}{\partial\theta^2} = 0. \tag{8.3}$$

3. Sliding clamping of an arc slice:

$$w = 0, \quad \frac{\partial w}{\partial r} = 0, \quad F = 0, \quad \frac{\partial F}{\partial r} = 0. \tag{8.4}$$

4. Sliding clamping of radial slices:

$$w = 0, \quad \frac{\partial w}{\partial \theta} = 0, \quad F = 0, \quad \frac{\partial^2 F}{\partial \theta^2} = 0. \tag{8.5}$$

The initial conditions are as follows:

$$w = f_1(r, \theta) = 0, \quad w' = f_2(r, \theta) = 0, \tag{8.6}$$

for time instant $t = 0$.

In order to reduce our continual system of Eqs. (8.1)–(8.6) to that of lumped parameters, a finite difference method of approximation $O(\Delta^2)$ with respect to spatial variables r and θ (see Fig. 8.1) is used. The difference form of Eqs. (8.1)–(8.6) is:

$$-\Lambda(\Lambda w) + \Lambda_{rr} w(\Lambda F + \Lambda_{rr} F) + \Lambda_{rr} F(\Lambda w + \Lambda_{rr} w)$$

$$-2\Lambda_{r\theta} w \Lambda_{r\theta} F + \Lambda F + 4q_i = (w_{tt} + \varepsilon w_t)_{i,j},$$

$$\Lambda(\Lambda F) = -\Lambda_{rr} w(\Lambda w + \Lambda_{rr} w) + (\Lambda_{r\theta} w)^2 - \Lambda w,$$

where:

$$\Lambda(\cdot) = \Lambda_{rr}(\cdot) + \Lambda_r(\cdot), \quad \Lambda_r(\cdot) = \frac{1}{r_i^2}(\cdot)_r, \quad \Lambda_{rr}(\cdot) = (\cdot)_{rr},$$

$$\Lambda_{r\theta}(\cdot) = -\frac{1}{r_i^2}(\cdot)_\theta + \frac{1}{r_i}(\cdot)_{r\theta}, \quad \Lambda_{rr}(\cdot) = \frac{1}{\Delta_r^2}[(\cdot)_{i+1,j} - 2(\cdot)_{i,j} + (\cdot)_{i-1,j}],$$

$$\Lambda_r(\cdot) = \frac{1}{2\Delta_r r_i^2}[(\cdot)_{i+1,j} - (\cdot)_{i-1,j}], \quad \Delta_r = \frac{b}{n}, \quad \Delta_\theta = \frac{\theta_k}{m}. \tag{8.7}$$

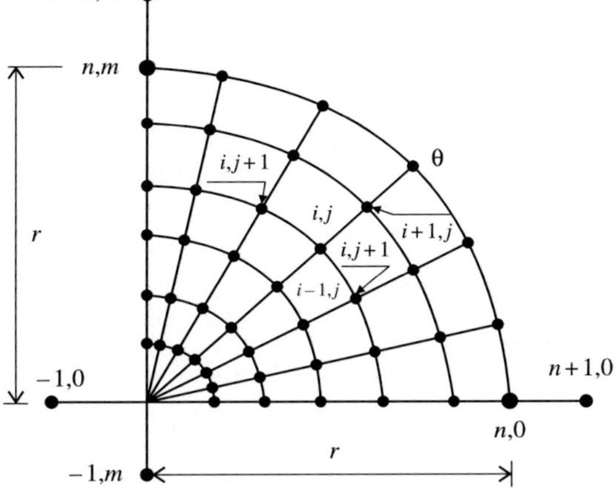

Fig. 8.1 Finite difference approximation of the investigated shell

The boundary conditions are as follows:

1. Ball-type clamping of an arc slice:

$$w_{n,j} = 0, \quad \Lambda_{rr}w - \frac{\nu}{b}\Lambda_r w = 0, \quad F_{n,j} = 0, \quad \Lambda_r w = 0, \quad j = \overline{1, m-1}. \tag{8.8}$$

2. Ball-type clamping of radial slices:

$$w_{n,j} = 0, \quad \Lambda_{\theta\theta}w = 0, \quad F_{i,j} = 0, \quad \Lambda_{\theta\theta}F = 0, \quad j = 0, m, \quad i = \overline{0, n}. \tag{8.9}$$

3. Sliding clamping of an arc slice:

$$w_{n,j} = 0, \quad \Lambda_r w = 0, \quad F_{n,j} = 0, \quad \Lambda_r F = 0, \quad j = \overline{1, m-1}. \tag{8.10}$$

4. Sliding clamping of radial slices:

$$w_{i,j} = 0, \quad \Lambda_{\theta\theta}w = 0, \quad F_{i,j} = 0, \quad \Lambda_{\theta\theta}F = 0, \quad j = 0, m, \quad i = \overline{0, n}. \tag{8.11}$$

Two supplementary conditions are required for system of Eqs. (8.7)–(8.11), i.e., in the shell top and the so-called compatibility conditions. In majority of cases solved by numerical methods, it is assumed that a shell has a central hole of small diameter, which has a minor influence on the obtained results at a sufficient distance from the shell vertex. In our approach, while solving non-axially symmetric problems $\theta = 2\pi$, the sought functions in point $r = 0$ are defined by an interpolation of second-order Lagrange formula. As a result one obtains

$$f_{0,j} = 3f_{1,j} - 3f_{2,j} + f_{3,j}, \tag{8.12}$$

where $f_{i,j} = f(r_i)_j$, $r_i = ih$ for $i = (0,\ 1,\ 2,\ 3)$, $0 \leq j \leq m-1$, and h denotes the distance between interpolating nodes.

For an out-contour point the following symmetry condition holds:

$$f_{-1,j} = f_{1,j} \quad \text{for} \quad 0 \leq j \leq m-1. \tag{8.13}$$

Compatibility conditions for non-axially symmetric problems of $\theta = 2\pi$ read:

$$w_{i,j} = w_{i,m+j}, \quad F_{i,j} = F_{i,m+j} \quad \text{for} \quad j = 0, 1, \quad 0 \leq i \leq n-1. \tag{8.14}$$

Then, the Cauchy problem (8.7)–(8.14) is solved applying the fourth-order Runge-Kutta method. The computational step is yielded by the Runge rule.

Although in the developed algorithm the applied load can be taken in an arbitrary manner, further harmonic excitation of the form $q = q_0 \sin(\omega_p t)$ is used, where ω_p denotes the excitation frequency.

8.3 Static Problems and Reliability of Results

The developed numerical algorithm provides a solution to various static and dynamic problems. In order to solve static problems, the so-called "set-up" method is applied [25, 166].

In relations $q_0(w_{\max})$ for a shell characterized by the angle sector of $\theta_k = \frac{3}{2}\pi$ (Fig. 8.2), that of $\theta_k = \pi$ (Fig. 8.3), and that of $\theta_k = \frac{\pi}{2}$ (Fig. 8.4) for various sloping parameters: for $\theta_k = \frac{3}{2}\pi$ and $\theta_k = \pi$ we take $b = 5$–10, whereas for $\theta_k = \frac{\pi}{2}$ sloping $b = 7$–12 (for smaller sloping values the sector shell behaves like a plate) are in Figs. 8.2–8.4.

As shown in the graphs, beginning from a certain parameter b, the limiting points occur on the curves. Owing to the computations carried out for the shell characterized by $\theta_k = \frac{\pi}{2}, \pi, \frac{3\pi}{2}$, the sloping values of $b = 8, 9, 11$ refer to critical ones, i.e., they are associated with a "shell jump occurrence."

Table 8.1 gives curves of equal deflections (isoclines) for all considered angles $\theta = \frac{\pi}{2}, \pi, \frac{3\pi}{2}$. Let us compare the curves for critical and post-critical loads q_0 for different θ and b. For $\theta = \frac{\pi}{2}$ and $\theta = \pi$ and for an arbitrary sloping parameter

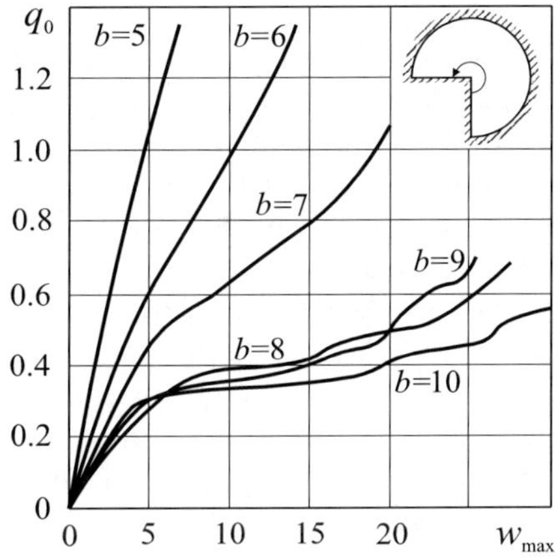

Fig. 8.2 Dependence $q_0(w_{\max})$ for $\theta_k = \frac{3}{2}\pi$

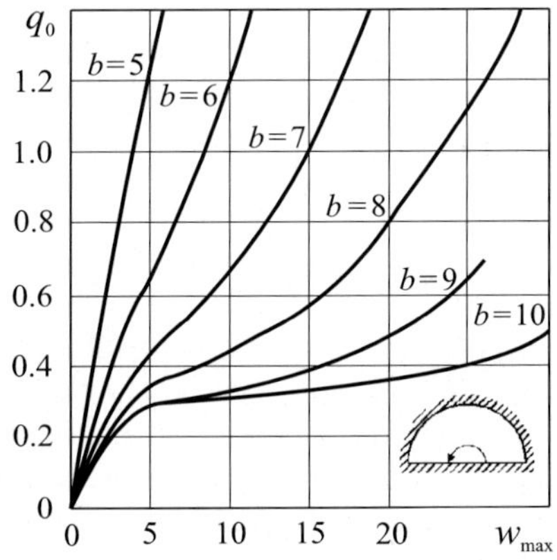

Fig. 8.3 Dependence $q_0(w_{\max})$ for $\theta_k = \pi$

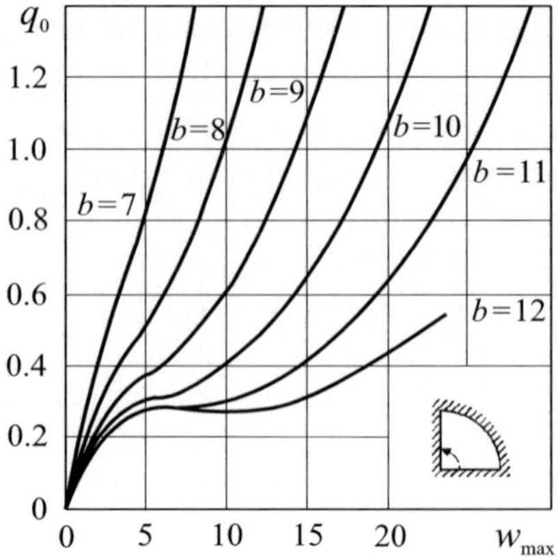

Fig. 8.4 Dependence $q_0(w_{\max})$ for $\theta_k = \frac{\pi}{2}$

the analyzed pictures coincide. A maximum deflection is achieved in the point of intersection of the angle θ bisectrix and central shell radius. For $\theta = \frac{3\pi}{2}$ and for $b = 7$ the occurrence of a "jump" is not observed and an increasing load does not change qualitatively the analyzed pictures. For the latter sector angle and for sloping $b = 9,\ 10$ the "jump" phenomenon is shown in Fig. 8.2, and the curves of equal deflections for critical and post-critical loads are clearly different. In the case of a critical load, the maximum shell deflection occurs in the middle of the bisectrix. In the case of the post-critical load, two zones of maximum deflections appear. They are symmetric with respect to the angle bisectrix.

Since the developed algorithm allows a wide class of problems without an exact solution, checking its reliability is highly required. The reliability of the obtained results depends on their comparison with numerical solutions obtained in [166]. Fig. 8.5 shows the relation $q(w_{\max})$ for a sector shell with movable resistance contour clamped along radial and arc directions for $\theta_k = \frac{3\pi}{2}$ (sloping parameter $b = 5$–8, numbers of partition with respect to a radius and angle are $n = m = 10$, respectively and $\nu = 0.3$). Curves denoted by points correspond to results given in reference [166], whereas solid curves correspond to our results. Good coincidence of the results is clearly indicated.

8.4 Convergence of a Finite Difference Method Along Spatial Coordinates for Non-stationary Problems

In order to trace the behavior of spatial systems from a common point of view a concept of phase space is applied. Partial differential equations governing dynamics of the analyzed objects, Eqs. (8.1)–(8.6), are substituted by equations governing dynamics of lumped systems applying the finite difference method (see Eqs. (8.7)–(8.14)). Below, we outline one of the most dangerous points during a transition from PDEs to ODEs. In other words, instead of infinite dimensional systems we are going to consider finite dimensional ones. It is assumed that beginning from a certain approximation a further increase of the number of equations does not introduce anything new to the obtained results and the system behaves similarly.

Table 8.1 Shell deflection isoclines for various θ, b, and q_0

$\theta = \frac{\pi}{2}$, $b=12$		$\theta = \frac{\pi}{2}$, $b=11$	
$q_0 = 0.245$	$q_0 = 0.246$	$q_0 = 0.245$	$q_0 = 0.246$
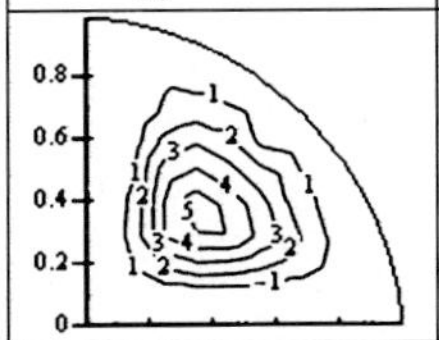	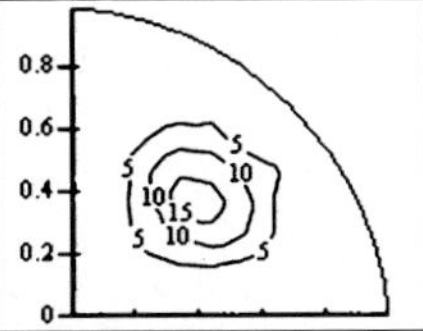	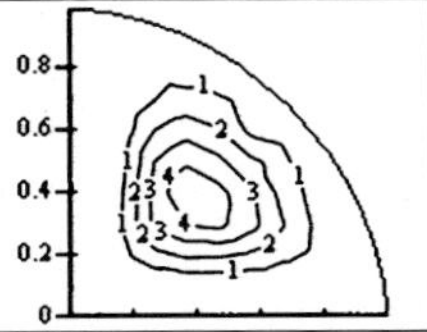	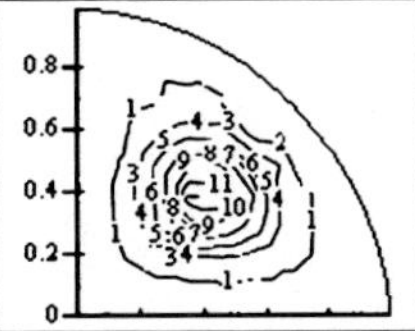
$\theta = \frac{\pi}{2}$, $b=10$		$\theta = \pi$, $b=10$	
$q_0 = 0.27$	$q_0 = 0.6$	$q_0 = 0.245$	$q_0 = 0.246$
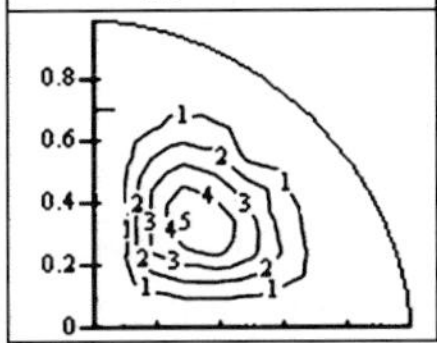	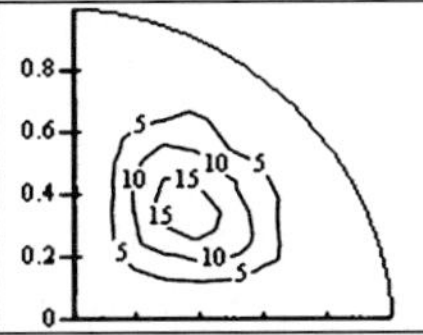	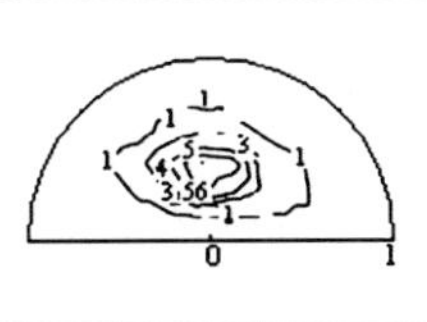	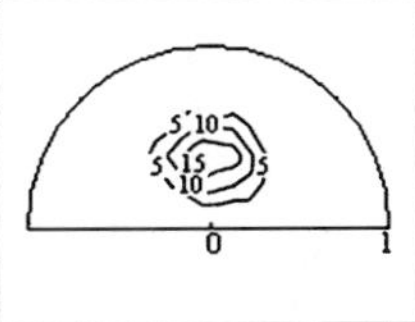
$\theta = \pi$, $b=9$		$\theta = \frac{3\cdot\pi}{2}$, $b=10$	
$q_0 = 0.245$	$q_0 = 0.246$	$q_0 = 0.27$	$q_0 = 0.28$
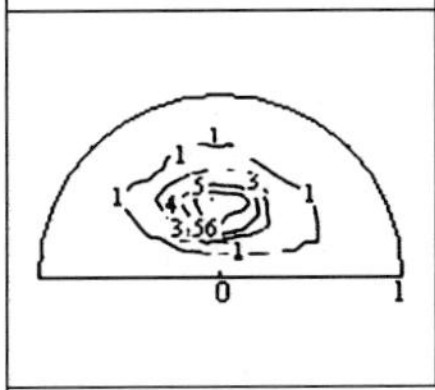	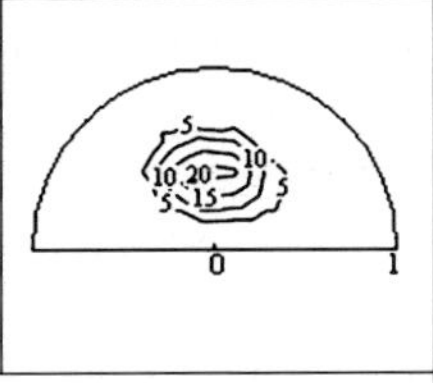	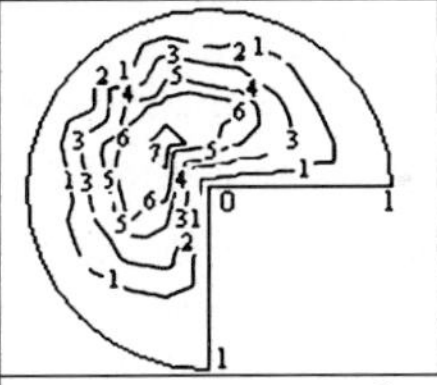	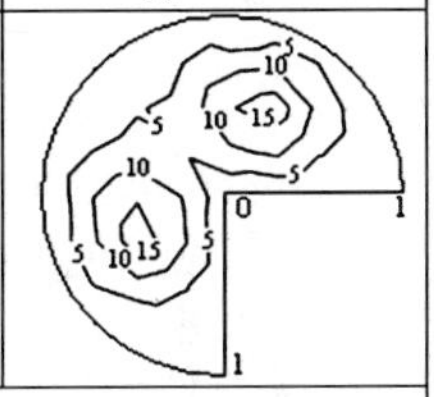
$\theta = \frac{3\cdot\pi}{2}$, $b=9$		$\theta = \frac{3\cdot\pi}{2}$, $b=7$	
$q_0 = 0.3$	$q_0 = 0.31$	$q_0 = 0.3$	$q_0 = 0.6$
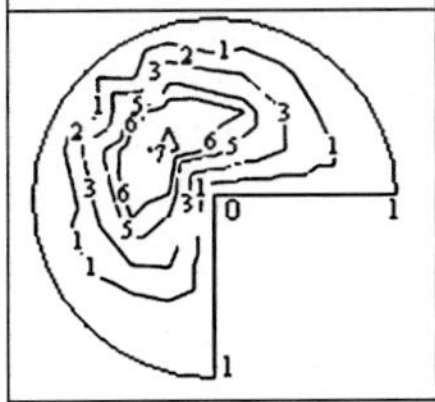	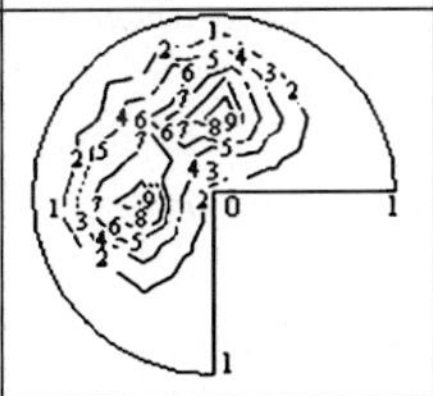	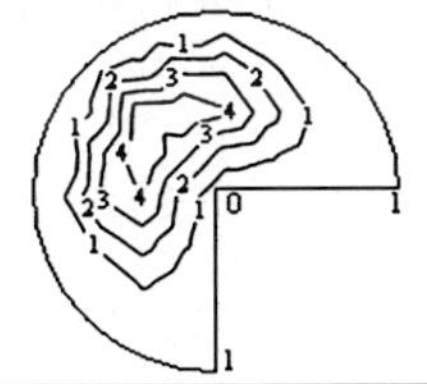	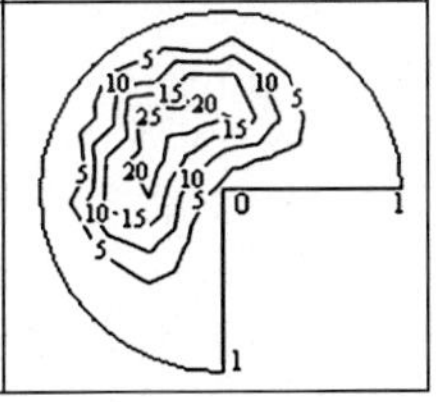

A crucial role plays here the dimension of an analyzed attractor. However, if the dimension of the studied attractors is bounded, truncation procedures applied to Eqs. (8.7)–(8.14) may play an important role. For instance, in the case of an unsuitable choice of radius and arc partition into m and n parts, while applying the finite difference scheme, the truncated system obtained may exhibit attractors that have properties qualitatively different than the real system attractors.

In order to trace the behavior of shells harmonically excited by $q = q_0 \sin(\omega_p t)$, a package of routines has been developed allowing for construction of scales of vibration characters depending on the control parameters $\{q_0, \omega_p\}$, where ω_p is

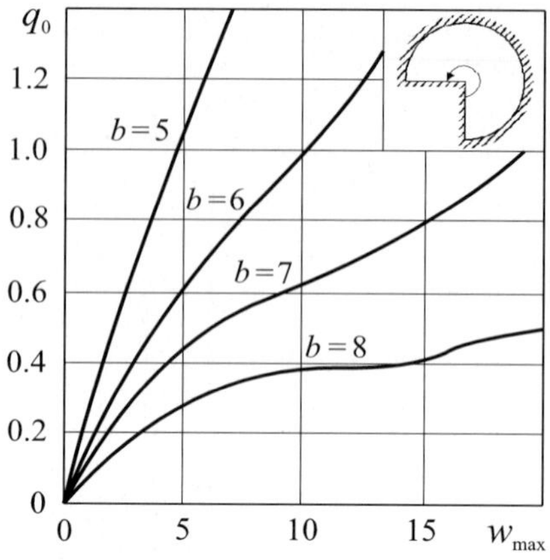

Fig. 8.5 Dependence $q_0(w_{\max})$ for $\theta_k = \frac{3}{2}\pi$

fixed. We have $\omega_p = \omega_0$, which stands for the frequency of linear vibrations. In order to construct a scale with respect to amplitude values of more than 200 points, i.e., $2 \cdot 10^2$ problems of dynamics must be considered, and power spectra should be studied for each choice of the control parameters $\{q_0, \omega_0\}$. The algorithm allows us to distinguish the zones of harmonic vibrations, the Feigenbaum type zones, chaotic zones, as well as the zones of modified Ruelle-Takens-Newhouse scenarios. The latter one is detected and illustrated further in this chapter. Below, we discuss the mentioned modified Ruelle-Takens-Newhouse scenario. After harmonic vibrations with external forcing frequency and moving along the parameter q_0, a new independent frequency appears, and a transition to chaos is realized owing to the series of combinations of two frequencies. This scenario will be analyzed in more detail.

Consider vibrations of a sector-type shell movable clamped along radial and arc slice that depend on a number of partitions of integration interval with respect to coordinates $r \in [0; r_n]$ and $\theta \in [0; \theta_k]$ using the finite difference method with approximation $O(h^2)$. Intervals $[0; r_n]$ and $[0; \theta_k]$ have been divided into 5; 10; 15; 20; 25 parts, i.e., the number of degrees of freedom is increased. The load is uniformly distributed along the shell surface and has the following form: $q = q_0 \sin(\omega_p t)$. Both relations $w_{\max}(q_0)$ and scales of vibration character $\{q_0, \omega_0\}$ are constructed owing to monitoring of frequency spectra and Lyapunov exponents. Power spectra $S(\omega_p)$ are depicted for bisectrix of the sector-type shell. Since vibrations in all points of intervals $[0; r_n]$ and $[0; \theta_k]$ have been synchronized, a further analysis is carried out for one point only. Relations $w_{\max}(q_0)$ are constructed for two sector-type shells: $\theta_k = \pi$, $b = 9$ (Fig. 8.6) and $\theta_k = \frac{\pi}{2}$, $b = 10$ (Fig. 8.7). Scales of vibration character versus a number of partition $n = m$ are included under the graphs. The notation applied now and used further is given in Fig. 8.6. All drawings refer to free vibration frequency of the system. Already for $n = m = 15$ the character of vibrations does not change with the variation of shell radius and angle partition. Only a small shift of the bifurcation zone is observed on vibration type scales. In Fig. 8.6, relations $w_{\max}(q_0)$ practically coincide.

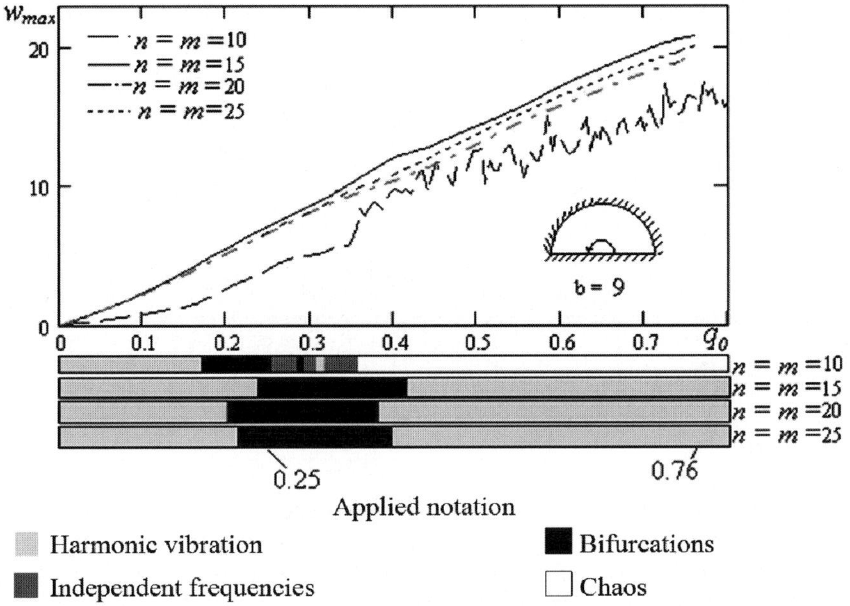

Fig. 8.6 Dependence $w_{\max}(q_0)$ and scales of vibration character for $b = 9$ and $\theta_k = \pi$

On the vibration scales (Fig. 8.6) some points taken for further study are depicted, and the associated graphs $S(\omega_p)$ versus the number of partition $n = m$ are given in Table. 8.2. Let us consider a point on the vibration scale, which for $n = m = 20$ is in the bifurcation zone for the corresponding load parameters:

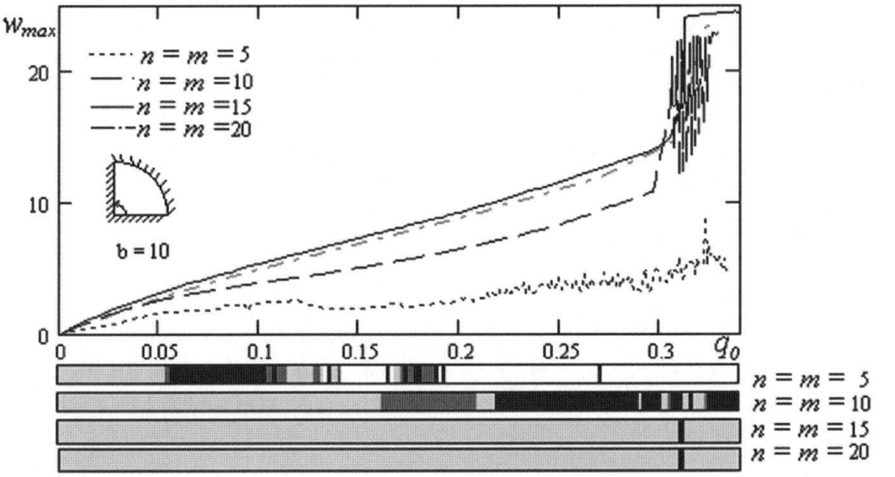

Fig. 8.7 Dependence $w_{\max}(q_0)$ and scales of vibration character for $b = 9$ and $\theta_k = \frac{\pi}{2}$

Table 8.2 Shell deflection isoclines for various θ, b, and q_0

$\theta=\pi,\ b=9,\ q_0=0.25$			
$n=5$	$n=10$	$n=15$	$n=20$
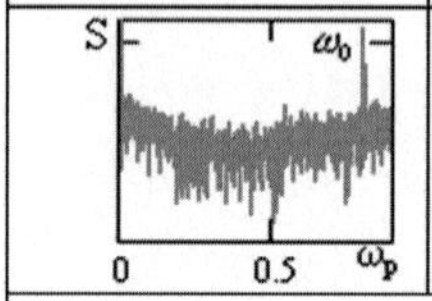	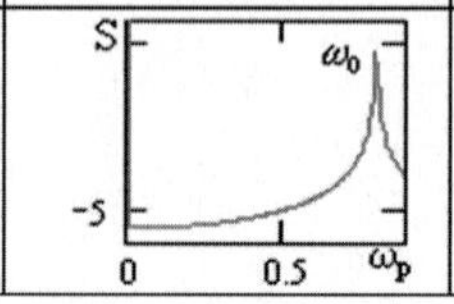	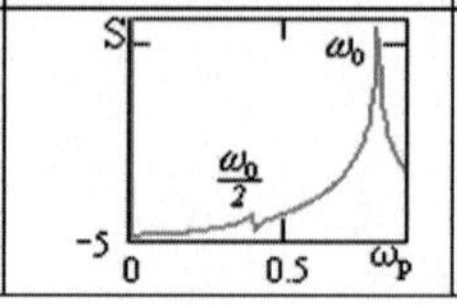	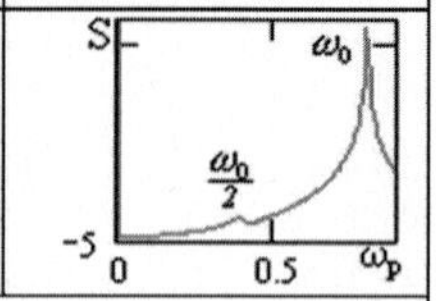
$\theta=\pi,\ b=9,\ q_0=0.76$			
$n=5$	$n=10$	$n=15$	$n=20$
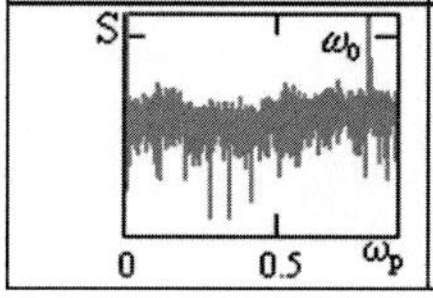	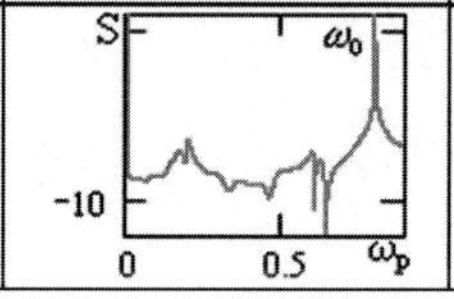	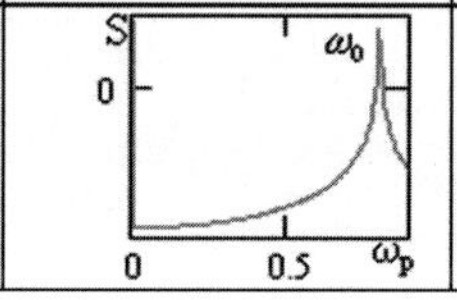	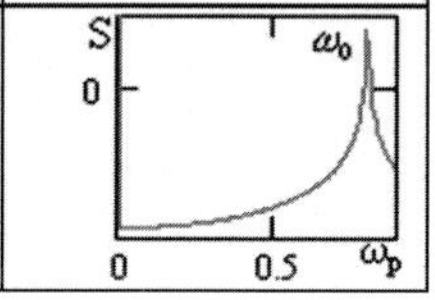

$\omega_0=0.8$; $\theta_k=\pi$; $b=9$; $q_0=0.25$; (Fig. 8.8) and $q_0=0.76$ (Fig. 8.9). In Fig. 8.8 signals $w\left(\frac{r_n}{2};\ \frac{\theta_k}{2};\ t\right)$, $320\le t\le 360$ are reported, whereas Table 8.2 gives power spectra $S(\omega_0)$ for a chosen point. For $n=m=5$ chaotic vibrations are observed: for $n=m=10$ harmonic vibrations appear; for $n=m=15$; 20 first period doubling bifurcation occurs. Relations $w\left(\frac{r_n}{2};\ \frac{\theta_k}{2};\ t\right)$ shown in Fig. 8.8 coincide well. Owing to the reported results one may conclude that beginning from $n=m\le 15$ the real process is approximated adequately and hence in further consideration we take $n=m=15$, i.e., on each computational step 450 first-order ordinary differential and 225 linear algebraic equations are solved. Table 8.2 gives analogous results for a sector-type shell with the same parameters and amplitude of excitation $q_0=0.76$. The corresponding time history versus partition number $n=m$ is shown in Fig. 8.9.

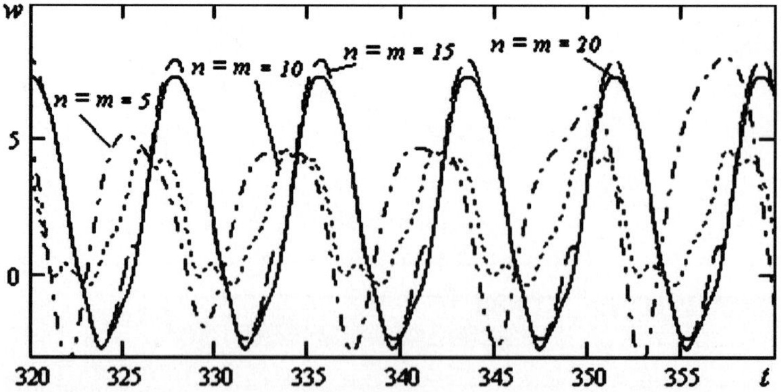

Fig. 8.8 Time histories for different partition numbers $n=m$ and $q_0=0.25$

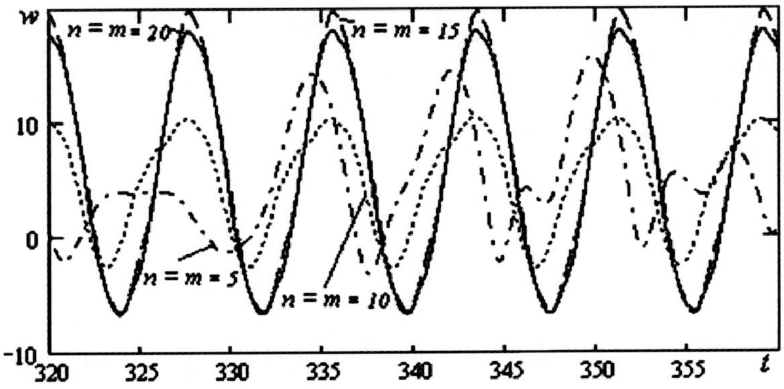

Fig. 8.9 Time histories for different partition numbers $n = m$ and $q_0 = 0.76$

In the above, the convergence of the proposed approach depending on n (number of partitions of r and θ_k) has been studied, but a problem related to N (partition number of the set $\{q_0; \omega_0\}$) remains open. The mentioned problem becomes extremely important, since chaotic dynamics depends significantly on the amplitude q_0 and frequency ω_p of excitation. A correct choice of N yields reliable construction of scales of vibration character of the investigated system. Initially, the following problem has been analyzed. How does the character of vibrations and the function $\{q_0; \omega_0\}$ change with respect to the number N of interval $(0; q_0)$ partition? Fig. 8.10 shows relations $w_{\max}(q_0)$ for scale "1": $N = 100$; scale "2": $N = 200$; scale "3": $N = 400$.

Comparing first and second scales, one may observe that an increasing number of partitions of the amplitude of exciting force causes small bifurcation zones. At increasing number of partitions up to $N = 400$, the character of vibrations does not change; hence further investigations are carried out for $N = 200$.

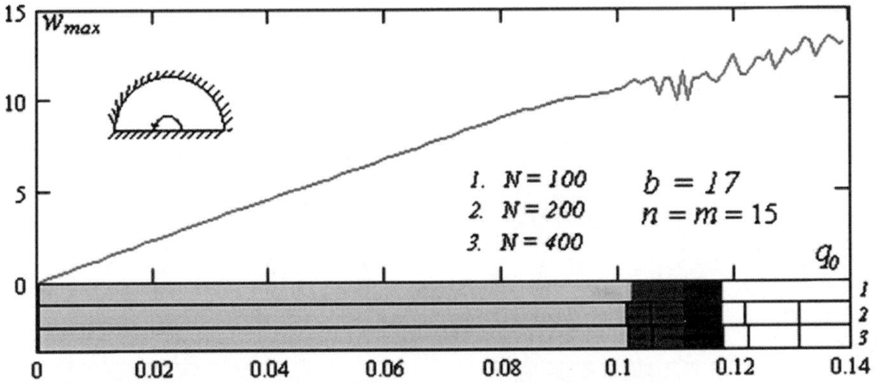

Fig. 8.10 Dependence $w_{\max}(q_0)$ and vibration scales for different partition numbers N

8.5 Investigation of Chaotic Vibrations of Spherical Sector-Type Shells

8.5.1 Boundary Conditions

A sector-type shell characterized by both the angle $\theta = \frac{3\pi}{2}$ and the space Ω is divided into $n = m = 15$ points and the sloping parameter $b = 8$. Next, we investigate how the shell's vibrations change depending on the boundary conditions movable free contour (8.8, 8.9) and movable clamping (8.10, 8.11). Relations $w_{\max}(q_0)$ are illustrated in Fig. 8.11. Scales of vibration character are shown in the graphs. Consider the graph $w_{\max}(q_0)$, constructed for sector-type shell harmonically loaded and with ball-type resistance contour (8.8, 8.9). For $q - 0 \leq 0.1$ the relation $w_{\max}(q_0)$ (see Fig. 8.11a) is linear, whereas for $q_0 = 0.1$ a first-order discontinuity associated with the Andronov-Hopf bifurcation occurs.

Graph $w_{\max}(q_0)$ (see Fig. 8.11b) for a movable clamped shell (8.10, 8.11) increases uniformly, i.e., shell vibrations are harmonic ones. Analysis of the results given in Figs. 8.11a and 8.11b yields the following conclusion boundary conditions essentially modify the character of vibrations.

A movable clamped contour is associated with complex vibrations, i.e., bifurcations and chaos interlace and also jump phenomena are observed. In the case of the clamped contour, vibrations are harmonic, and therefore a modification of boundary conditions may cancel the zones of bifurcation and chaos (compare Figs. 8.11a and 8.11b).

8.5.2 The Influence of Sector Angle

Let us investigate the influence of the opening sector angle of the sector-type shell on vibrations. For this purpose one type of boundary conditions will be analyzed, i.e., movable ball-type supported contour (8.8, 8.9). Let us fix the sloping parameter $b = 15$, and for $n = m = 15$ we increase the angle $\theta = \frac{\pi}{4}; \frac{\pi}{2}; \pi; \frac{3\pi}{2}$ (Fig. 8.12a–d,

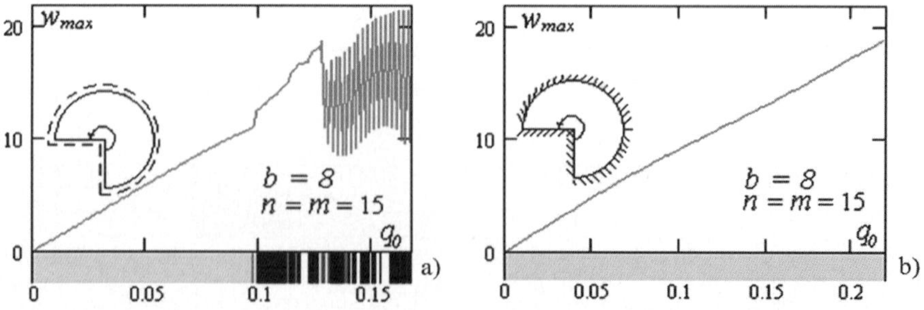

Fig. 8.11 Dependence $w_{\max}(q_0)$ for movable free **(a)** and clamped **(b)** shell contour

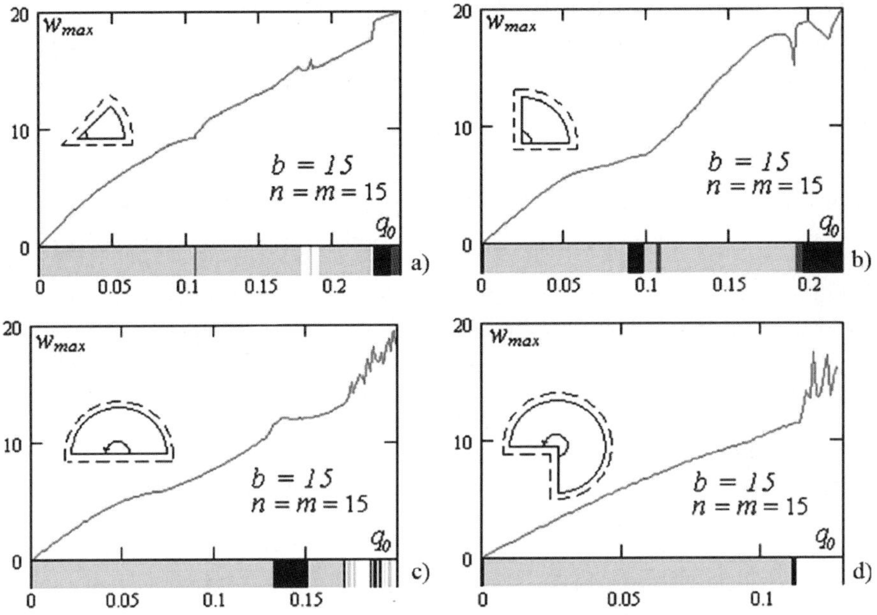

Fig. 8.12 Dependence $w_{\max}(q_0)$ for various $\theta_k = \frac{\pi}{4}$ **(a)**; $\frac{\pi}{2}$ **(b)**; π **(c)** and $\frac{3\pi}{2}$ **(d)**

respectively). In the graph $w_{\max}(q_0)$ describing the system behavior for $\theta = \frac{\pi}{4}$, large zones of independent frequencies and their linear combinations may be observed. Already for a small amplitude of excitation the investigated system exhibits vibrations of two frequencies. In the graph $w_{\max}(q_0)$, there are a few first-order discontinuities, where vibration type changes. With an increase of the angle $\theta = \frac{\pi}{2}$, the zone of harmonic vibrations increases, and chaos vanishes. Let us study the graph $w_{\max}(q_0)$ for angle $\theta = \pi$. Zones of chaotic vibrations reappear, the number of first-order discontinuities increases, but also the zone of harmonic vibrations increases owing to a decrease of the bifurcation zone. Due to an increase of the sector angle $\theta = \frac{3\pi}{2}$, again the zone of harmonic vibrations increases, and then in the chaotic zone a series of first-order discontinuities appears and vibration character changes.

An increase of the sector angle causes an increase of the first-order discontinuities, and a difference between pre-critical and post-critical states also increases.

8.5.3 Vibrations of Sector-Type Shells Versus Sloping Parameter

Further on, we consider vibrations of slice of the shell movable clamped on a radial and arc sector with $\theta_k = \pi$ versus sloping b. Both relations of $w_{\max}(q_0)$ and scales of vibration character were monitored (Fig. 8.13a–d). The shell associated with $b = 10$ works in the regime of harmonic vibrations. An increase of sloping up

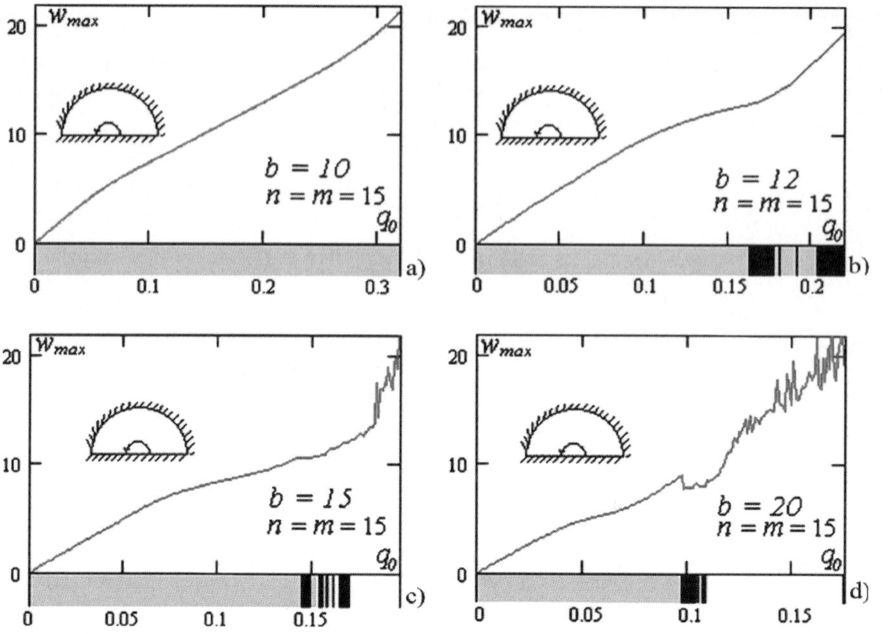

Fig. 8.13 Dependencies $w_{\max}(q_0)$ for different values of b: 10 **(a)**; 12 **(b)**; 15 **(c)**; 20 **(d)** and vibration scales

to $b = 12$ produces a zone of independent frequencies and bifurcations. For $b = 15$; 20 a zone of chaotic vibrations on the scales of vibration types and first-order discontinuities are observed in the relation $w_{\max}(q_0)$, and a zone of bifurcations and harmonic vibrations decreases.

It is rather expected that an increase of the sloping parameter yields the system to a less stable state, which is manifested by a large number of stiff stability losses and an increase of the chaotic zone.

8.6 Transitions from Harmonic to Chaotic Vibrations

Four models of transition from harmonic to chaotic vibrations are known: the Feigenbaum scenario [93], the Ruelle-Takens-Newhouse scenarios [262], and the Pomeau-Manneville [248] and the Landau scenarios [180]. However, none of the mentioned scenarios has been detected during analysis of our sector-type shell subjected to uniformly distributed sign changing load for either an arbitrary angle or sloping parameter. As already illustrated, the magnitude of sloping and sector-angle has an important influence on the system evolution. For movable clamped sector-type shell with sloping parameter $b = 12$; 15; 20, the novel scenario of transition from harmonic to chaotic vibrations has been discovered, which was named the modified Ruelle-Takens-Newhouse scenario [167]. The latter one is characterized in the following manner. First harmonic vibrations with the excitation frequency appear, and then increasing the control parameter q_0 yields a new independent frequency, and transition to chaos is realized by a series of linear combinations of two frequencies.

Investigating the sector-type shell subject to the action of sign changeable distributed load with a movable resistance contour, and characterized by the sector angle $\theta_k = \pi$ and the sloping parameter $b = 17$, the Feigenbaum scenario is manifested (see Tables 8.3 and 8.4). An Andronov-Hopf bifurcation cascade (five period doubling bifurcations) is observed after a periodic window occurred, beginning from $q_0 = 0.11785$ (see five points in the Poincaré map). At increasing q_0, a cascade of period doubling bifurcations is observed. In the Poincaré map, five groups of points are created, and every new bifurcation that causes one more point occurs in each group.

Note that for all initial inputs the modified Ruelle-Takens-Newhouse scenario has been found.

A more detailed analysis follows (see Table 8.5):

1. $q_0 = 0.1$. The system exhibits one-frequency harmonic vibrations with exciting frequency equal to a free system frequency. The phase portrait consists of a periodic orbit and the Poincaré map is represented by a point.
2. At an increasing amplitude of excitation $q_0 = 0.105$, the first independent frequency occurs, and the ratio of exciting to independent frequency is equal to $\frac{\omega_p}{\omega_1} = 0.9231\ldots$. An increase of the part of phase portrait exhibits seven times rotating cycle, and the Poincaré section is composed of seven arbitrarily located points.
3. At $q_0 = 0.108$ a new linearly independent frequency equal to $\omega_p - \omega_1$ is born.
4. An increase of the load $q_0 = 0.113$ yields first a Hopf bifurcation. The Poincaré map is composed of three groups of points and four points arbitrarily located, and in the power spectrum the local maximum associated with frequency $\omega_p = \frac{\omega_0}{2}$ is observed.
5. For the amplitude of excitation equal to $q_0 = 0.118$, the second frequency ω_2 occurs, where $\frac{\omega_p}{\omega_1} = \frac{\omega_1}{\omega_2} = 0.9231\ldots$.
6. While increasing the control parameter further up to $q_0 = 0.119$, the third frequency ω_3 appears, and the ratio of second to third frequencies is again equal to that of two earlier cases, i.e., $\frac{\omega_p}{\omega_1} = \frac{\omega_1}{\omega_2} = \frac{\omega_2}{\omega_3} = 0.9231\ldots$.

Table 8.3 Bifurcation sequence

No. of. bifurcation	1st	2nd	3rd	4th	5th
$q_{0,n}$	0.1179	0.11794	0.117949	0.11795	0.1179513
d_n	–	4.66977...	4.669175...	4.669165...	–

Table 8.4 Power spectra, time histories and phase portraits for different q_0

q_0	Power spectra $S(\omega_p)$	Time history $w(t)$	Phase portrait $w(w')$
0.11785	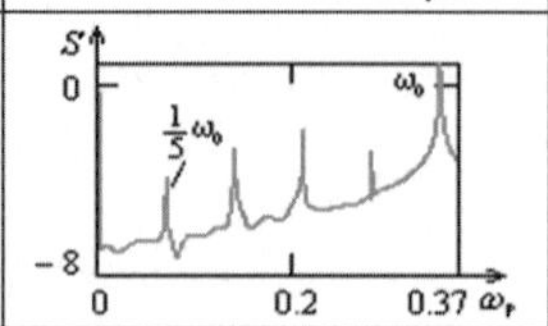	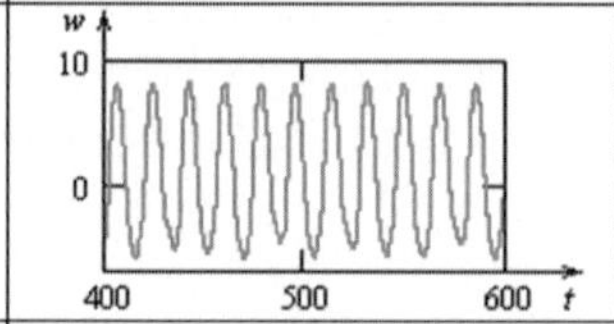	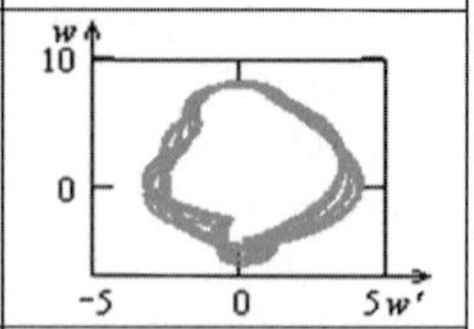
0.117898	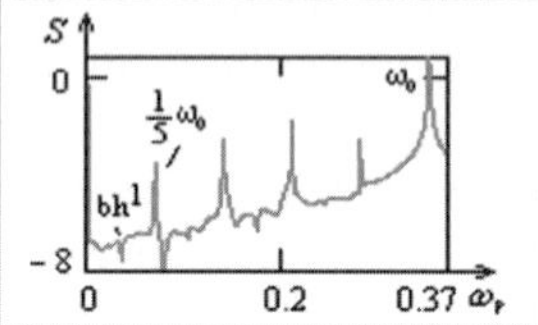	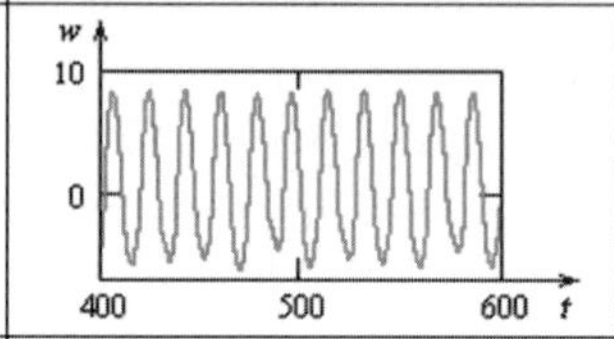	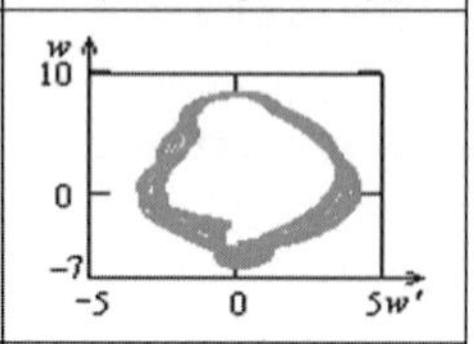
0.11794	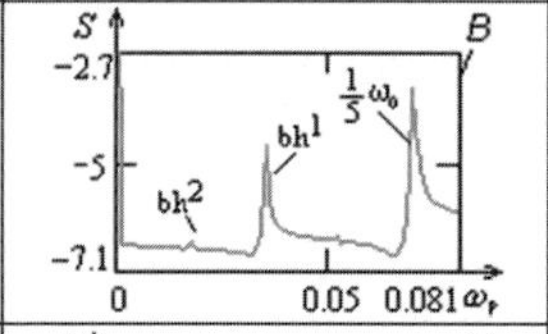	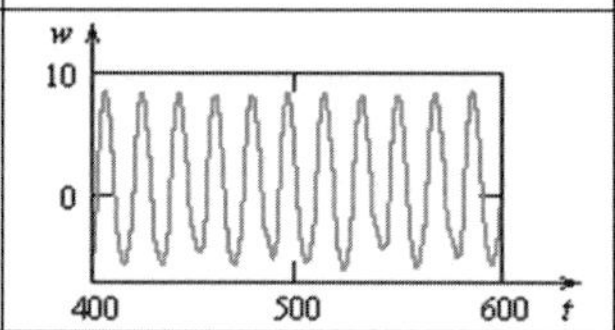	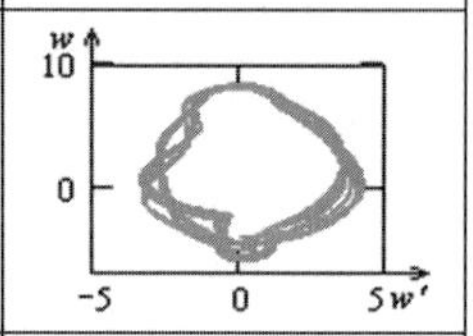
0.117894899	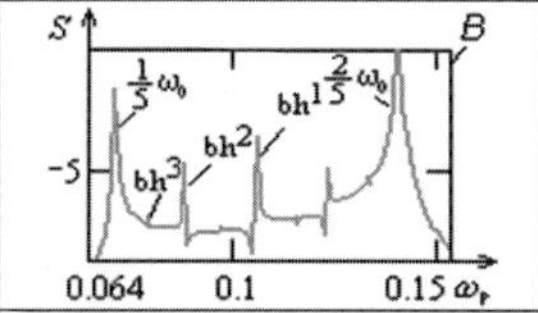	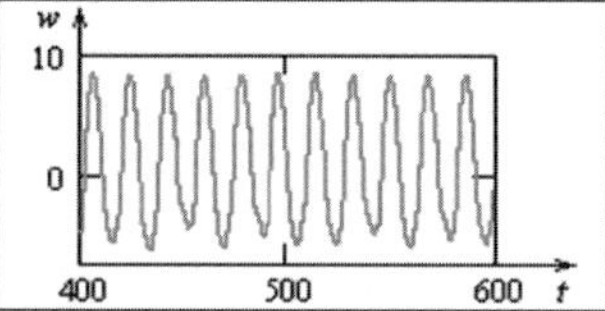	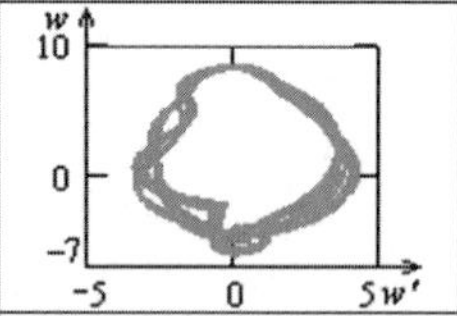
0.117795092	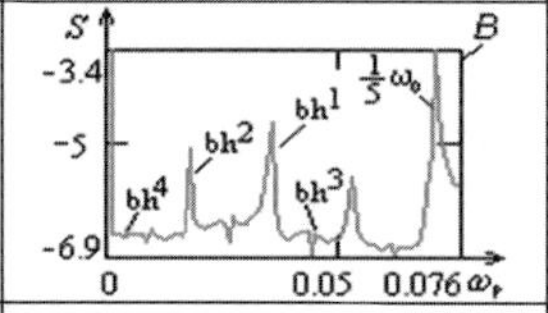	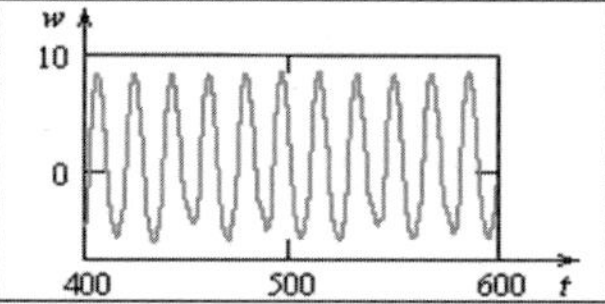	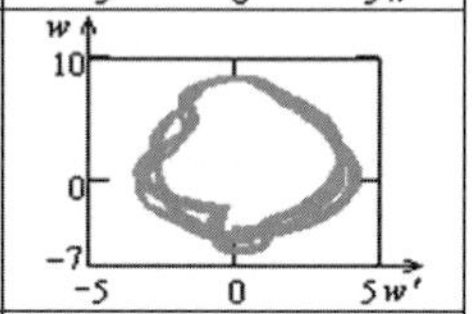
0.117795133	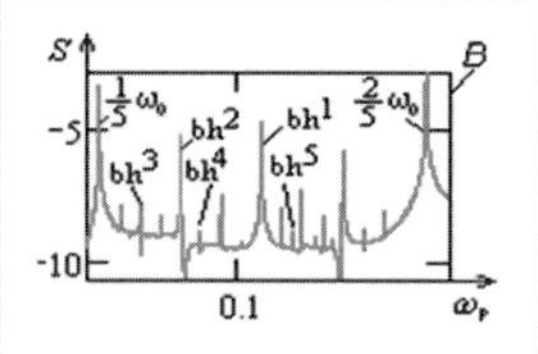	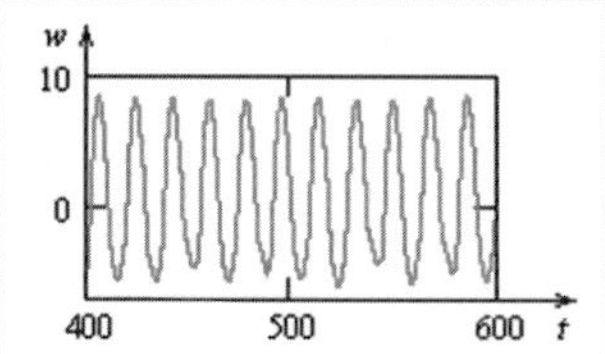	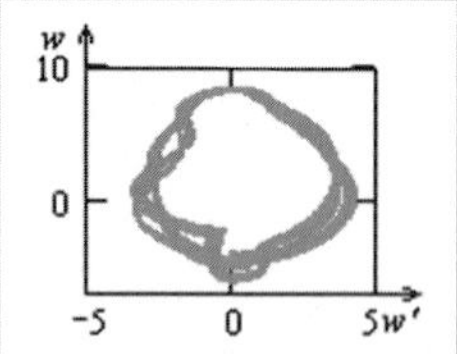

8.7 Control of Chaotic Vibrations of Flexible Spherical Sector-Type Shells

It is well known that a chaotic attractor is composed of a countable set of saddle-type cycles with different periods, and that during time evolution a phase point stays in vicinity of each of them. When a saddle-type cycle begins to stabilize, the trajectory remains in its neighborhood and the system starts to move in a periodic manner. In this perspective a control is understood as a stabilization of orbits embedded into a chaotic attractor. On the other hand, the system interaction is associated with problems of synchronization control. Applying target-oriented excitations on chaotic

Table 8.5 Bifurcation sequence

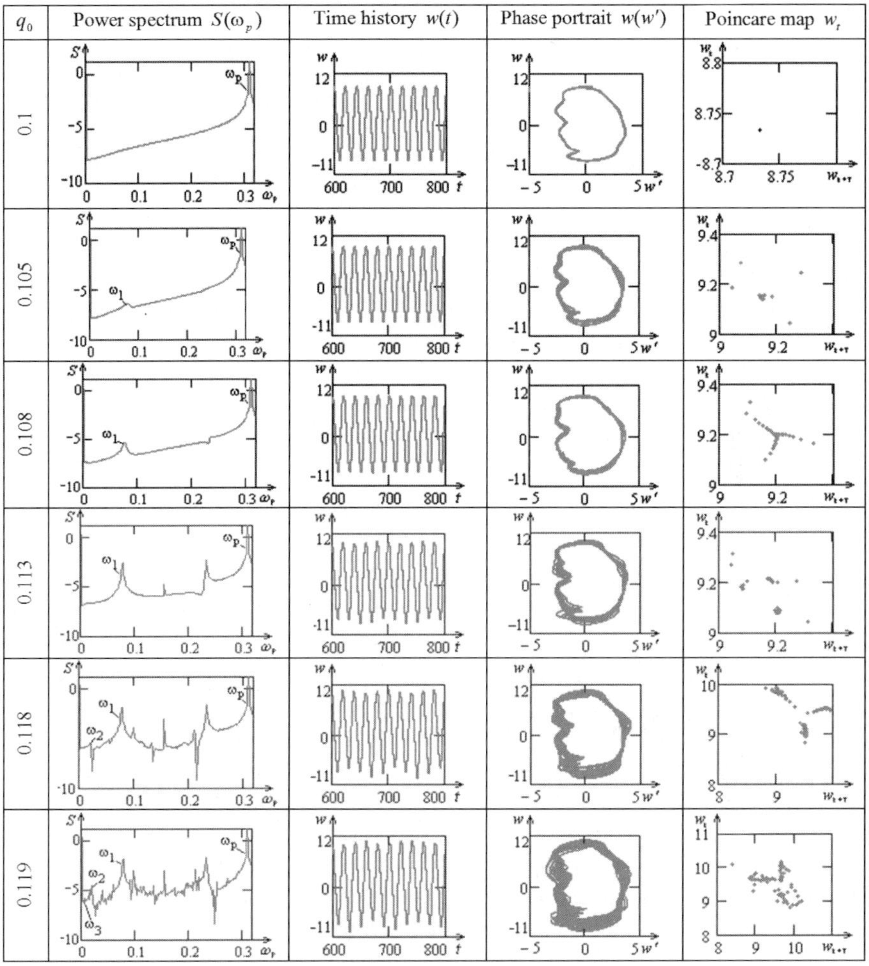

subsets associated with synchronized motions of identical systems, some of them may be transformed into stable ones, keeping the rest as unstable and vice versa. As a result, a controlled transition from non-synchronized chaotic vibrations into the regime of full synchronization of chaos is achieved.

In this work, the control is realized with a help of a target-oriented excitation of the sector-type shell uniformly loaded by both harmonic excitation $q = q_0 \sin \omega_p t$ and resistance time-dependent torque. This type of excitation is realized during the synchronization of frequencies.

Analysis of the system behavior is carried out on the basis of numerical experiments presented in a graphical form, i.e., maximum shell deflection versus the amplitude of excitation and scales of vibration type using colors. The working regime is identified using both power spectra and Lyapunov exponents.

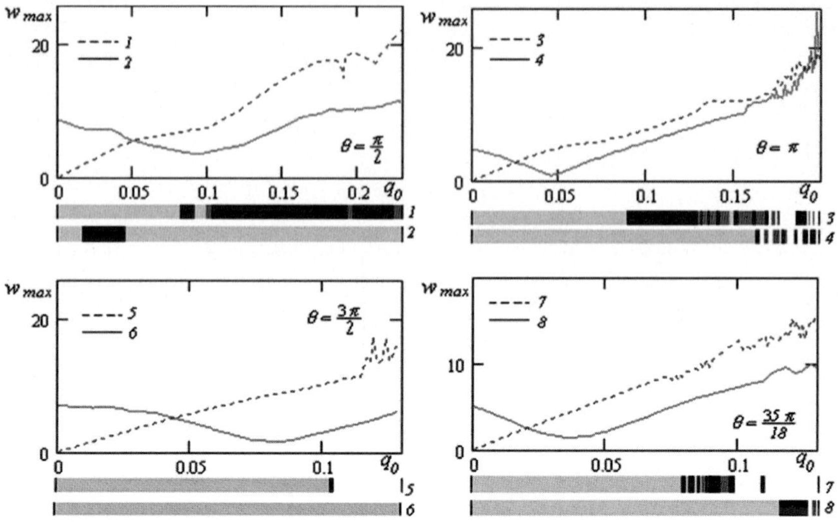

Fig. 8.14 Dependencies $w_{\max}(q_0)$ for vibration scales (see text for more details)

Consider a sector-type spherical shell with ball-type movable resistance contour, the sloping parameter $b = 15$ and shell angle $\theta = \frac{\pi}{2}$; π; $\frac{3\pi}{2}$; $\frac{35\pi}{18}$. Fig. 8.14 shows relations $w_{\max}(q_0)$ and scales of vibration character of the shell subjected to uniformly distributed load $q = q_0 \sin \omega_0 t$, where ω_0 is the frequency of free vibrations (Fig. 8.14, scales 1, 3, 5, 7). It illustrates also the graphs presenting dynamics under the action of distributed load and time-dependent moment with a free system vibration frequency excitation (scales 2, 4, 6, 8). The dashed curves denote graphs associated with the action of distributed time-changeable load (the excitation frequency is equal to the free vibration frequency), whereas solid curves correspond to the mentioned cases but with an additional action of the resistance torque $M = M_0 \sin \omega_0 t$, where ω_0 is the frequency of free vibrations. All the loads are given in Table 8.6, and the corresponding number of curves in Fig. 8.14.

Table 8.6 Shell parameters related to results given in Fig. 8.14

No. of. problem	Shell angle	Sloping parameter	Distributed load	Resistance torque
1	$\theta = \frac{\pi}{2}$	$b = 15$	$q = q_0 \sin(0.43t)$	–
2	$\theta = \frac{\pi}{2}$	$b = 15$	$q = q_0 \sin(0.43t)$	$M = 0.3 \sin(0.43t)$
3	$\theta = \pi$	$b = 15$	$q = q_0 \sin(0.33t)$	–
4	$\theta = \pi$	$b = 15$	$q = q_0 \sin(0.33t)$	$M = 0.5 \sin(0.33t)$
5	$\theta = \frac{3\pi}{2}$	$b = 15$	$q = q_0 \sin(0.3t)$	–
6	$\theta = \frac{3\pi}{2}$	$b = 15$	$q = q_0 \sin(0.3t)$	$M = 4 \sin(0.3t)$
7	$\theta = \frac{35\pi}{18}$	$b = 15$	$q = q_0 \sin(0.29t)$	–
8	$\theta = \frac{35\pi}{18}$	$b = 15$	$q = q_0 \sin(0.29t)$	$M = 5 \sin(0.29t)$

Analyzing graphs 1, 3, 5, 7, first-order discontinuities are visible in the zone of chaotic vibrations, which is also indicated by the vibration scale. Additional excitation of sign-changeable moment (Fig. 8.14, scales 2, 4, 6, 8) causes smoothing of drawings, vibrations are harmonic with the excitation frequency, and the amplitude of shell vibrations decreases. Note that graphs 2, 4, 6, 8 indicate the most important control of chaos results through the use of harmonic torque in the case, when the minimum deflection in the relation $w_{\max}(q_0)$ is achieved in the vicinity of $q_0 = \frac{q_{0max}}{2}$.

Recall that by the control of chaos we mean a transformation of the chaotic behavior of our analyzed system into another harmonic or into chaotic motions but with other properties using synchronization of frequencies.

Chapter 9
Scenarios of Transition from Harmonic to Chaotic Motion

In this chapter scenarios of transition from harmonic to chaotic motions are illustrated and discussed. First, a historical background of the problem is described. The Landau-Hopf scenario; the Ruelle, Takens, and Newhouse scenario; the Feigenbaum scenario; and the Pomeau-Manneville scenario are addressed, among others.

9.1 Historical Background

The problem of turbulence has been of a great interest for researchers for several centuries. A famous hydromechanic, Sir G. Lamb, in 1932 expressed his pessimistic opinion about the thorough solution to this problem: "I am old now, and I hope to shed more light on two problems as soon as I reach Heaven. The first one involves quantum mechanics, whereas the second one is about the phenomena of turbulence in liquids. I feel optimistic about the first one only."

Although these words were uttered during the meeting of the British Association a long time ago, they had been valid for nonlinear mechanics until 2005. Chaos is the result of a complex self-dynamics of a system (and it is not caused by random noise or other random disturbance), which in some sense can be treated as the process of formation of turbulence. It is widely known that there are enough "scenarios" of transformation into turbulence that are usually related to Navier-Stokes equations describing the motion of incompressible fluids and whose form is

$$\frac{\partial u}{\partial t}+(u\cdot\nabla)u-\nu\nabla^2 u=-\frac{1}{\rho}\nabla p+f, \tag{9.1}$$

$$div\, u=0, \tag{9.2}$$

$$u=0 \text{ in } D, \tag{9.3}$$

where D is the boundary of the space in which the fluid is, $u=u(x_i,t)$, $(i=1,2,3)$, p denotes pressure, ρ the density of fluid, f denotes external forces, if there are such, and ν is kinematic viscosity. The possibility of energy dissipation is expressed in the existence of the element $\nu\nabla^2 u$.

J. Awrejcewicz, V.A. Krysko, *Chaos in Structural Mechanics*,
DOI: 10.1007/978-3-540-77676-5_10, © Springer-Verlag Berlin Heidelberg 2008

Equation (9.1) describes a three-dimensional differential equation with partial derivatives for velocity u as related to a fixed system of coordinates (Euler approach), Eq. (9.2) describes the condition for the lack of compressibility, and Eq. (9.3) define the boundary condition.

It must be emphasized here that in case of Eqs. (9.1)–(9.3) not only do we not know any turbulent solution, but there is no theorem stating the existence of such a solution either. Such proof exists for a two-dimensional case. The physical possibility of existence of such solutions of Navier-Stokes equations is well analyzed. The first important work in this field was conducted in 1880 by Reynolds, who introduced the following parameters: Reynolds number $R = \frac{UL}{\nu}$, u/u_x nondimensional velocities, x/L nondimensional coordinates, $t = L/u$ nondimensional time, $\overrightarrow{p} = p/\rho U^2$ nondimensional pressure, and he also introduced the following differential equation:

$$\frac{\partial u}{\partial t} + (u \cdot \nabla)u - \frac{1}{R}\nabla^2 u = -\nabla p. \tag{9.4}$$

Reynolds showed that together with the increase of R the character of a flow can change from a regular (laminar) flow to a chaotic (turbulent) flow.

In hydrodynamics, the notions of turbulence were included in the description of the state of time-spatial chaos. This means that chaos in fluid can occur in all possible scales, both in space and in time. A mathematical description of this process is one of the most complex tasks of modern mathematics.

What a chaotic attractor of a turbulent flow should look like has not been made clear so far. It is in the analysis of simple dynamic systems described by a limited number of ordinary differential equations of low order and differential mappings (cascades) that has been the achievement in chaos analysis over the last few years. However, it should be emphasized that in such simple systems only time chaos is observed, which allows for the analysis of the occurrence of turbulence in some approximation when the velocity field begins to change in time in an irregular way, leaving an ordered space behind.

Our aim here will be to formulate the criteria (scenarios) of transformation of spatial constructions such as beams, plates, and shells into time-spatial chaos.

The following scenarios of transformation of vibration of dynamic systems from harmonic to chaotic has been obtained for simple systems. We recommend the following references to complete the material described in this chapter: [15, 16, 17, 18, 19, 47, 48].

9.2 Landau-Hopf Scenario (LH)

This scenario was first presented by Landau [179] in 1944 and independently of that by Hopf in 1948. The idea of this scenario is that during the crossing of Reynolds number (or some other parameter characterizing the flow) through the critical value of R laminar motion loses its stability (Fig. 9.1). For $R \to \infty$ the space of occurrence

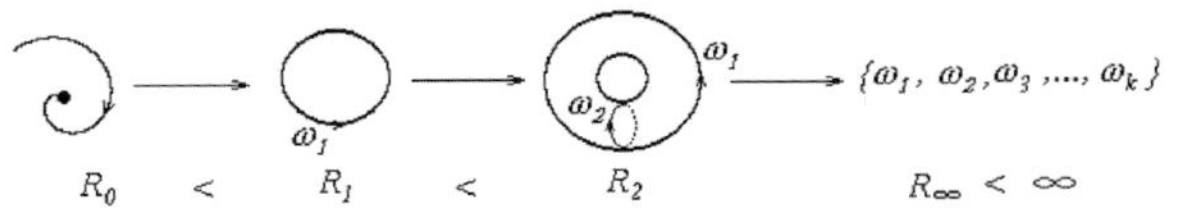

Fig. 9.1 The scheme of the Landau-Hopf (LH) scenario

of new frequencies increases ($P_n = 0, 1$), i.e., the solution can be presented as:

$$u(x,t) = \sum_{n=1}^{\infty} A_m(x)\ell^{im(\omega t+\delta)},$$

where

$$\omega = \{\omega_1, \omega_2, ..., \omega_n\}; n \to \infty; R \to \infty.$$

The frequency relationship is irrational, but the spectrum becomes practically continuous and similar to chaotic, i.e., the process of infinite quasi-period turbulence takes place that unfortunately has never been proved in experimental research (Couette flow [235] and Rayleigh-Bénard convection [59]).

The Landau-Hopf scenario is the basis for Hopf bifurcation theory. This idea is based on the following reasoning. Let us analyze the system of differential equations

$$\frac{dx}{dt} = F_p(x),\ x\,\varepsilon\,\mathrm{R}^{\mathrm{k}}, \tag{9.5}$$

where p denotes a certain parameter of a dynamic system (e.g., p the input amplitude). Points $x = x^x$ are critical points of Eq. (9.1) and in these points

$$\frac{d\overrightarrow{x}^x}{dt} = 0,\ \text{tj.}\ F_p(\overrightarrow{x}^x) = 0, \tag{9.6}$$

are linearized equations near the beginning of the coordinate system.

The stability of (9.6) is described by the analysis of characteristic values $\lambda = \lambda(p)$ (see also [308]).

For the condition that all λ are situated on the left side of a complex half-plane, i.e., all of them have a negative value, the critical point is the simple point of equilibrium (Fig. 9.2a).

A Hopf bifurcation occurs for the condition that the couple of conjugated characteristic values moves with non-zero velocity from the left complex half-plane into the right one, i.e., real parts of characteristic values become positive. For the critical value $p = p^+$ (value p, for which λ moves into the right half-plane) and when additional conditions assigned to the second and third derivatives of $F_p(x^x)$ are satisfied, as a result of bifurcation, a limit point transforms into a stable periodic orbit (limit cycle, Fig. 9.2b).

The transition from stable limit point (unstable limit cycle) to stable limit cycle (unstable limit point) is presented in Fig. 9.3 (Fig. 9.4) with the use of visualization of molecule motion within potential field where together with the increase of p the second minimum occurs.

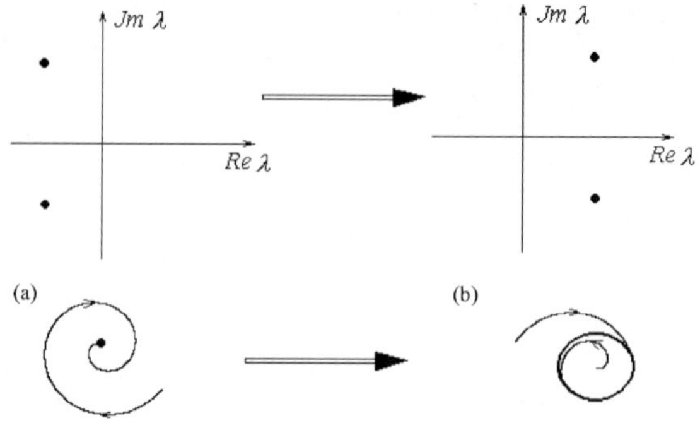

Fig. 9.2 Stable singular point (**a**) vs. stable periodic orbit (**b**)

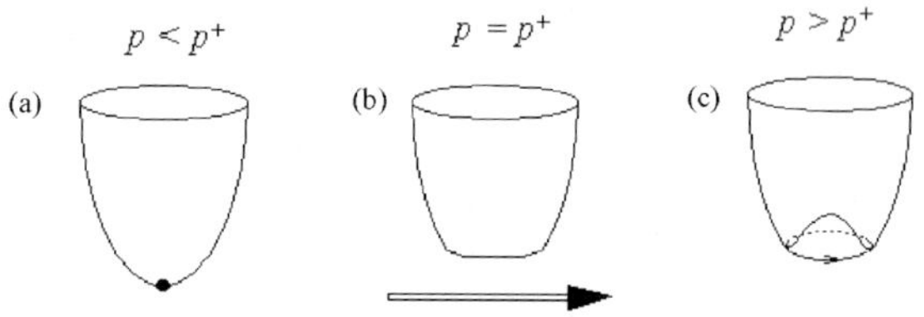

Fig. 9.3 The scheme of ordinary Hopf bifurcation during the motion of a molecule within a potential field from stable limit point to stable limit cycle

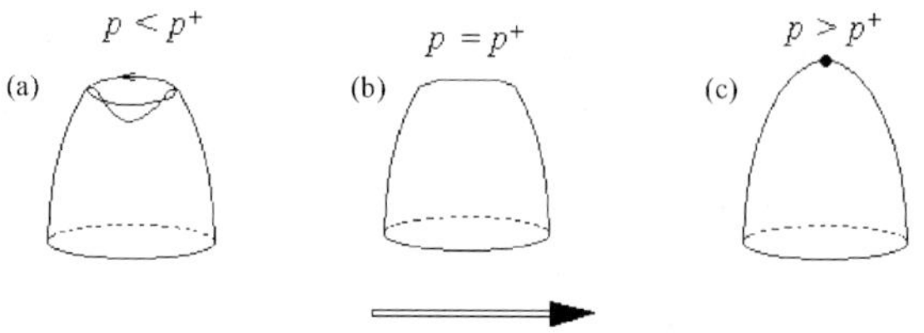

Fig. 9.4 The scheme of reversed Hopf bifurcation during the motion of a molecule within a potential field from unstable limit cycle to unstable limit point

Figures 9.2, 9.3, and 9.4 complement each another.

9.3 Scenario by Ruelle, Takens, and Newhouse

The scenario by Ruelle, Takens and Newhouse (RTN) is scientifically vital because it was the first one to view and criticize the LH scenario.

In 1971 Ruelle and Takens [262] showed that unlike the LH scenario shows, in order to obtain chaotic motion, an infinite number of Hopf bifurcations is not necessary because a few are enough. Initially, they presented the way in which after three Hopf bifurcations track T^3 can become unstable and transform into a strange chaotic attractor.

Then (in 1978) Ruelle, Takens, and Newhouse [232] proved the theorem according to which a strange attractor can occur after two bifurcations (Fig. 9.5).

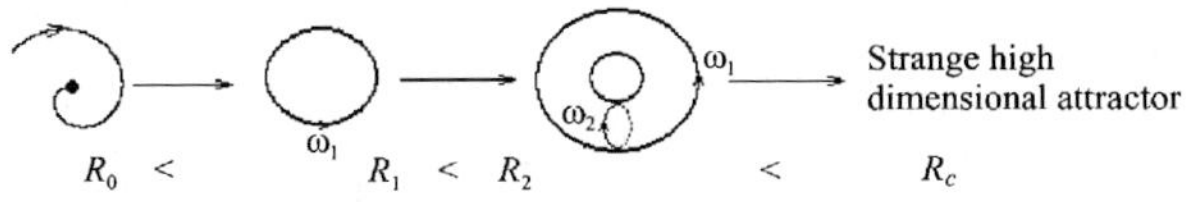

Fig. 9.5 Ruell-Takens-Newhouse (RTN) scenario

In fact, they assumed that after two Hopf bifurcations the motion is limited by manifolds having complex topology. Such manifolds have been called strange attractors. They present manifolds that do not have an integer dimension, i.e., manifolds located between plane and space. The notion of fractional dimensions was analyzed by Mandelbrot [195] within the analysis of fractals. The attractor obtained according to the RTN scenario should satisfy certain conditions, i.e., it should belong to the category of "axiom A" attractors (practice has shown that this hardly ever happens), and then the motion is chaotic. For such a motion, a strong sensitivity to the changes of initial conditions is observed.

Two group of researchers—Feigenbaum, Kadanoff, Shenker in 1982 [94] and Rand et al. in 1982 [256]—independently analyzed the problem of how quasi-periodic motion with two independent frequencies ω_1 and ω_2 on the torus becomes periodic after adding a perturbation (Fig. 9.6).

For rational values of $\omega_1/\omega_2 = p/q$ the trajectory closes after q cycles (state of synchronization); see also [20]. For the irrational relation ω_1/ω_2 motion is quasi-periodic, the trajectory does not close anywhere, and it covers the whole torus. The RTN scenario was confirmed in experiments (Rayleigh-Bénard instability and Rayleigh-Taylor instability) [325]. For hydrodynamic turbulence, RTN scenario becomes very "soft," i.e., it is related to high harmonic resonance.

In this case, two basic processes are realized:

(i) Stretching that guarantees sensitivity to initial conditions;
(ii) Composing, thanks to which attraction is possible. The example of such a complex system is the mapping of the Smale horseshoe.

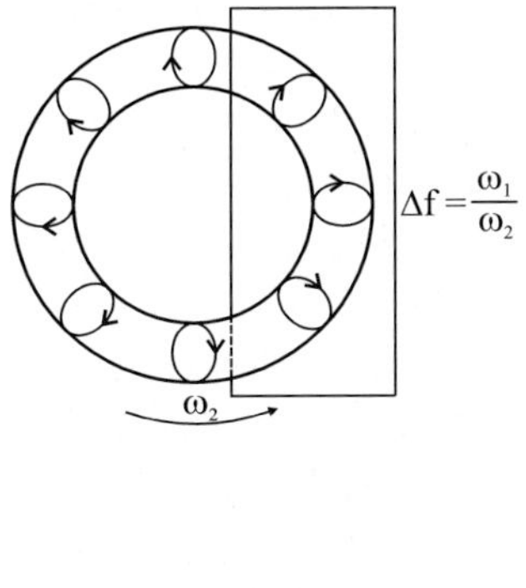

Fig. 9.6 Motion on a two-dimensional torus

9.4 Scenario by Feigenbaum

In 1978 Feigenbaum [92] presented a universal mechanism of transformation into chaos during an infinite series of doubling the period of initial motion. Such a complex behavior is displayed even by one-dimensional mappings such as

$$x_{n+1} = f(x_n) \ . \tag{9.7}$$

It should be emphasized that $f(x)$ should satisfy certain conditions. An example of such a mapping follows:

$$x_{n+1} = 4\lambda x_n(1-x_n) \quad 0 < x < 1, \tag{9.8}$$

where λ is a certain control parameter.

The above equation was at first used mainly in biology as a simple model describing the dynamics of population growth.

Feigenbaum showed geometrical convergence of the order of period doubling defining the quantity of

$$\lim_{n\to\infty} \delta_n = \lim_{n\to\infty} \frac{\lambda_{n+1}-\lambda_n}{\lambda_{n+2}-\lambda_{n+1}} = 4.6692016... \ , \tag{9.9}$$

i.e., he obtained a certain "universal number" that was confirmed in models by Lorenz and Hénon and in many real experiments.

We shall illustrate the dynamics of mapping of (9.8) for the order of value of λ. For every value of λ the integration of the mapping is conducted until all the "transformation processes" disappear and the trajectory changes its "asymptotic location" (i.e., 2-periodic cycle; 4-periodic cycle, 2^n-periodic cycle, ..., 3-periodic cycle or aperiodic attractor). Let us emphasize the following characteristics of this mapping:

(i) 2^n periodic cycles $(n = 1, 2, \ldots)$ in space λ undergo the growing process of compression;
(ii) For $\lambda > \lambda_\infty$ chaotic areas appear;
(iii) 3-periodic cycles, or odd-order cycles appear in chaotic regime, (In the study by Li-Yorke [188] there is an example showing that an orbit with period three can also lead to chaos);
(iv) For $\lambda = 1$ chaotic dynamics is not fully recognized.

This problem is analyzed more thoroughly later, for ordinary differential equations.

The value λ, in which new limit cycles appear, is given in Table 9.1. Beyond the area of doubling of period λ_∞ the structure of manifold is very rich. Grebogi and his colleagous [113] introduced new terminology, namely, they called the narrow areas of the occurrence of chaos subduction, and the corresponding widened areas an internal crisis, and the final widening of the area for $\lambda = 1$ was called a crisis.

Table 9.1 Limit cycles

	Type	λ_n (occurrence of limit cycle)
1	2. limit cycle	0.75
2	4. limit cycle	0.86237
3	8. limit cycle	0.88602
4	16. limit cycle	0.89218
5	32. limit cycle	0.8924728
...	...	...
∞	aperiodic attractor	0.892486418

9.5 Scenario by Pomeau-Manneville

The fourth scenario of transition into chaos was presented by Pomeau and Manneville in 1979 [197]. There were many materials concerning chaos at that time and it was shown for many dynamic systems that the transition from periodic motion to chaos can take place by a stroke as a result of one bifurcation. Such a transition is called hard and it is connected with the phenomenon of intermittency. Intermittency is the process in which regular vibrations and relatively short irregular extras interlace in a random way. The increase of a control parameter leads to the increase in the number of chaotic extras until the signal is totally chaotic.

This phenomenon was discovered by Pomeau and Manneville while solving differential equations describing the Lorenz model [196] and it was explained in the following way. For the value of a control parameter smaller than the critical one we observe a stable critical point. After a control parameter crosses its critical value, this point becomes unstable. The transition to an unstable state can take place according to one of three scenarios, the so-called intermittency of 1st, 2nd, and 3rd type. For all types of intermittencies, the modules of characteristics of a linearized Poincaré transformation are higher than one.

Table 9.2 presents the following characteristics for the three types of intermittencies: the character of motion and type of mapping, characteristic (eigenvalue), the form of a signal. In Table 9.2 parameter ε represents the parameter characterizing the overcritical state of a system.

In the case of the first type of intermittency, for $\varepsilon = 0$ we can observe the occurrence of contact bifurcation. Vertical and horizontal lines in the Poincaré diagram show the construction on a Lamerey diagram of double asymptotic trajectory of saddle-node point. For $\varepsilon > 0$ around the previous fixed point there is a so-called canal along which the point of trajectory moves relatively long, which corresponds to the laminar phase of intermittency. Leaving the canal by the point evokes turbulence.

In case of the second type of intermittency, for $\varepsilon = 0$ a subcritical Andronov-Hopf bifurcation occurs in the mapping, and in the case of the third type of bifurcation, for $\varepsilon = 0$ a one-dimensional mapping shows a subcritical bifurcation of period doubling of the first limit cycle.

Table 9.2 Characteristics of selected one-dimensional mappings

Type	Characteristic feature and Poincaré mapping	System behavior after bifurcation $\varepsilon < 0$ $\quad -\varepsilon > 0$	Laminar signal	Eigenvalue
1.	Real eigenvalue intersects unit circle at +1 $x_{n+1} = \varepsilon + x_n + ux_n^2$	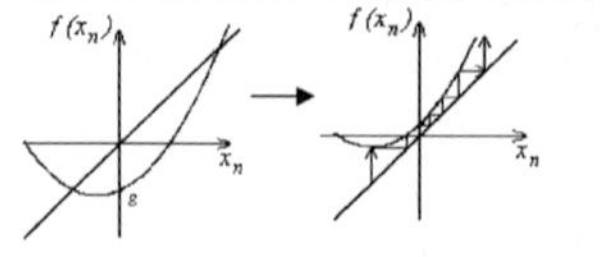	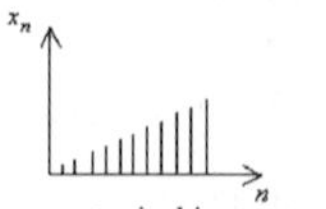 monotonical increase	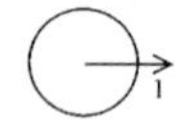
2.	Two complex conjugate values intersect unit circle $r_{n+1} = (1+\varepsilon)r_n + ur_n$ $\theta_{n+1} = \theta_n + \Omega$	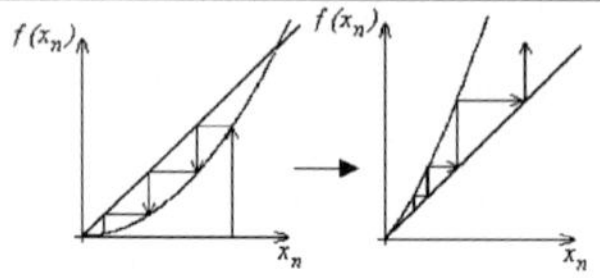	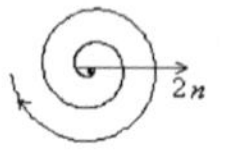 spiral increase	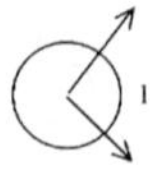
3.	Real eigenvalue intersects unit circle at -1 $x_{n+1} = -(1+\varepsilon)x_n - ux_n^3$	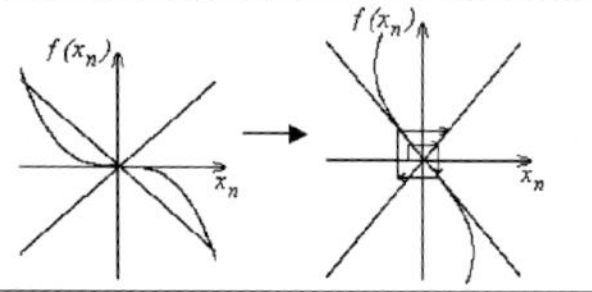	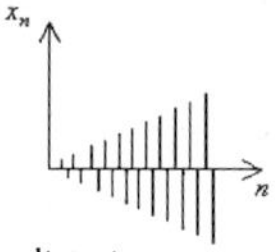 alternate	

The third type of intermittency was initially observed in experiments with Bérnárd convection in a small cubical cell. In 1984, the measurement of horizontal gradient of temperature was taken in relation to the modulation of the intensity of a light beam crossing this cell [198]. The Belousov-Zhabotinsky reaction is an example of the second-type of intermittency [2].

9.6 Synchronization of Frequencies

It is known from Rayleigh-Bénard's experiments that the transition from quasi-periodicity to chaos can take place also in the case of the loss of synchronization of frequencies ω_1, ω_2 (Fig. 9.7).

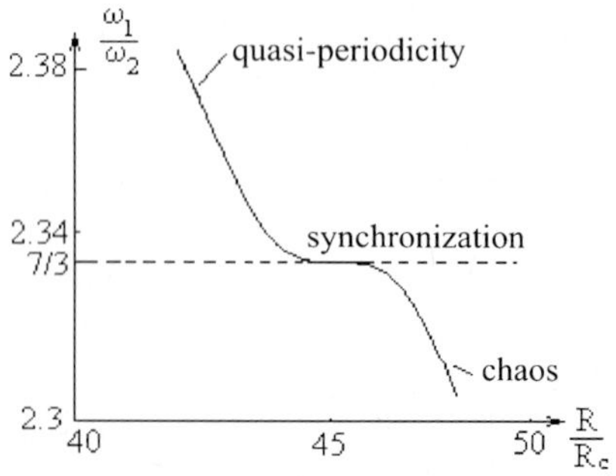

Fig. 9.7 Synchronization of frequencies observed during Bénard's experiment [112]

The transition from the state of synchronization of frequencies is a very important problem of chaos dynamics. There are several possible types of synchronization-chaos or synchronization-intermittency-chaos transition.

We shall present a few definitions of chaos for dynamic systems described by integrable mappings.

Basic characteristics of chaos are:

(i) High sensitivity to changes of initial conditions;
(ii) Intermittency (transitivity);
(iii) Regularity condition, measured by the density of periodic points.

Mapping $f: x \to x$ will be referred to as chaotic if the following conditions are satisfied:

(i) f is sensitive to initial conditions;
(ii) f is transitive;
(iii) Periodic points are densely distributed in X.

Below, we shall provide the definition for these characteristics.

1. Let $x \in X$, and let U be an open set including x. Mapping f is sensitive to initial conditions if $(\forall \delta > 0)$, $(\exists n > 0): d(f^{(n)}(x), f^{(n)}(y)) > \delta$.
2. If f is transitive, then $\forall (u, v)$ of open sets $(\exists n > 0)\ f^{(n)}(u) \Pi v \neq 0$.
3. The characteristic of the density of periodic points means that in any surroundings of any point in X there is at least one periodic point.

In practice, we deal with chaos when there is an easily observable sensitivity to initial conditions or there is at least one positive Lyapunov exponent.

Chapter 10
Dynamics of Closed Flexible Cylindrical Shells

Complex vibrations of closed cylindrical shells of infinite length and circular cross section subjected to transversal local load in the frame of the classical non-linear theory are studied. A transition from partial differential equations (PDEs) to ordinary differential equations (ODEs) is carried out using a higher order Bubnov-Galerkin approach and Fourier representation. On the other hand, the Cauchy problem is solved using the fourth-order Runge-Kutta method.

In the first part of this work static problems of the theory of closed cylindrical shells are studied. The reliability of the obtained results is verified by comparing them with the results taken from the literature. The second part is devoted to the analysis of stability, bifurcation, and chaos of closed cylindrical shells. In particular, an influence of sign-changeable external pressure and the control parameters such as magnitude of pressure measured by φ_0, relative linear shell dimension $\lambda = \frac{L}{R}$, frequency ω_p, and amplitude q_0 of external transversal load, on the shell's nonlinear dynamics is studied.

10.1 Introduction

Problems related to vibration of shell-type structures are encountered in many branches of industry, including aeronautical engineering, ocean engineering, and civil engineering [192, 260]. Nonlinear vibrations of thin circular cylindrical shells are of special interest in aerospace (design of rocket and launch vehicle structures) [212, 284], in which the structures must have a weight as low as possible and a strength as high as possible, and hence may exhibit large amplitudes of vibrations. According to the linear theory of vibrations, the natural frequencies and mode shapes are independent of the amplitude of vibration. However, in many cases, if the amplitude of vibration is large, such an assumption will not be justified due to one or another nonlinear effect. In general, the interest in vibration of nonlinear systems is focused on geometrical nonlinearities occurring at large displacement amplitudes, which yields nonlinear strain-displacement relationships.

J. Awrejcewicz, V.A. Krysko, *Chaos in Structural Mechanics*,
DOI: 10.1007/978-3-540-77676-5_11,

One of the most fascinating features encountered in the study of nonlinear vibrations in general is the occurrence of new and totally unexpected phenomena in the sense that they are not predicted or even hinted by a linear theory. On the other hand, an explanation of many experimental observations cannot be understood without taking into account nonlinear behavior. This is why nonlinear dynamics of plates and shells is still largely unresolved by researchers working in the fields of mathematics, mechanics and physics [6, 7, 8, 52, 70, 123, 228, 291, 312]. Many important specific topics of vibrations of plates and shells have been already addressed and some of them are listed below:

(i) Problems related to variation of the resonant frequencies depending on the amplitude of vibration [53, 284, 312];
(ii) Amplitude dependence on the mode shapes [53, 312];
(iii) Jump phenomenon and its corresponding multi-values region in the nonlinear frequency response curve [70, 228, 291];
(iv) Harmonic distortion of the nonlinear response to harmonic excitation, and its spatial distribution [6, 7, 52];
(v) Shift to the right of the nonlinear random frequency response curves [9, 313];
(vi) Study of internal resonances [8, 9, 14, 123];
(vii) Occurrence of sub- or super-harmonic response phenomena [228];
(viii) Occurrence of chaotic vibration [9, 228, 291];
(ix) Existence of bifurcation points [123, 228, 291];
(x) Coupling, due to the nonlinearity, between transverse and in-plane displacements (see reference [118] for a plate case, and reference [220] for a shell case);
(xi) Participation of the companion mode, in addition to the driven and axisymmetric modes, in the nonlinear forced response of shells [6, 8, 70].

Modeling and simulation of the behavior of complex aerospace structures are perhaps the more challenging shell analysis tasks to date. Following the accident of the space shuttle *Challenger*, the definition of large-scale nonlinear analysis changed as a result the classical approaches performed on the solid rocket boosters. A new design for the space shuttle external tank and other cryogenic fuel tanks for hypersonic vehicles has also challenged shell analysts [231].

The mode shapes are of particular interest in the dynamic behavior of a structure since the axial and bending strains are dependent upon the first and second derivatives of the mode shapes. Therefore, accurate prediction methods are needed to determine, at large vibration amplitudes, the nonlinear mode shapes and the corresponding resonance frequencies of shell-type structures. Moreover, the investigation of the geometrically nonlinear vibrations of shells is intended to give not only useful information about the nonlinear frequencies and mode shapes, but also to lead to important indications on the dangerous zones where the stresses (axial and bending) are concentrated. This is due to the fact that the distribution of these stresses at large vibration amplitudes may be completely different quantitatively as well as qualitatively from that obtained within a frame of the linear theory. On the other hand, in view of the increasing recourse in engineering to modal testing techniques, it can be noticed that qualitative description of the nonlinear behavior can be very useful in understanding data provided by modal testing, and can open the way to the development of more appropriate modal testing models, taking into account the nonlinear effects.

In the first investigations dedicated to shell stability under the influence of a non-symmetric external time-independent load, small inhomogeneities have been taken into account, and (in order to describe a pre-critical state) a so-called momentousless model has been analyzed. The mentioned investigations led to a conclusion that an amplitude of critical uniform pressure exceeds its level in a way proportional to that of non-uniformity increase. It is clear that a further improvement of the model is associated with inclusion of the momentous-type pre-critical state. It has been shown that one of the most important factors characterizing the level of critical static load is that associated with an order of distance between pre-critical and critical state variations and a buckling form. This effect is extremely well exhibited by a dependence of critical static loads on non-homogeneities under the influence of discontinuous loads.

It occurs also that the most dangerous cases of loading stability loss of a construction are manifested by a strong pre-critical bending state. In addition, this behavior is associated with an occurrence of nonlinear deformation. Therefore, the problem related to the estimation of errors introduced via the application of linear modeling of a pre-critical state plays an important role. The linearization of pre-critical bending causes a decrease of the level of maximum displacements and excludes the non-linear effects from consideration. Analysis of the examples of full nonlinear computations of the static problems indicates an importance of divergence of the obtained results from the corresponding data obtained using the linear model. The application of nonlinear computation enables an essential improvement of monitoring of a construction behavior subject to external loading. A remarkable contribution to the development of shell stability investigation in the case of non-symmetric deformation was introduced by Andreev et al. [11], among others.

In reference [105] an investigation of the nonlinear free flexural vibrations of a circular cylindrical shell was carried out including the effects of anisotropy, thickness, and various shell geometries using the Wilson-Θ numerical integration method coupled with a modified Newton-Raphson technique.

Chaotic vibrations of flexible shells are analyzed in references [21, 31]. On the other hand, problems related to dynamic stability investigations with inclusion of geometric nonlinearities under the influence of non-axially symmetric deformations of cylindrical shells have been discussed in a few studies. The aim of this chapter is to fill the existing gap, particularly in the class of problems devoted to the analysis of complex nonlinear vibrations of cylindrical shells subject to a sign-changeable transversal pressure.

10.2 Fundamental Equations

Below, a closed cylindrical shell of circular cross section and finite length subjected to non-uniform sign-changeable external pressure in the frame of the classical nonlinear theory is studied. The following coordinates are introduced: axis x is directed

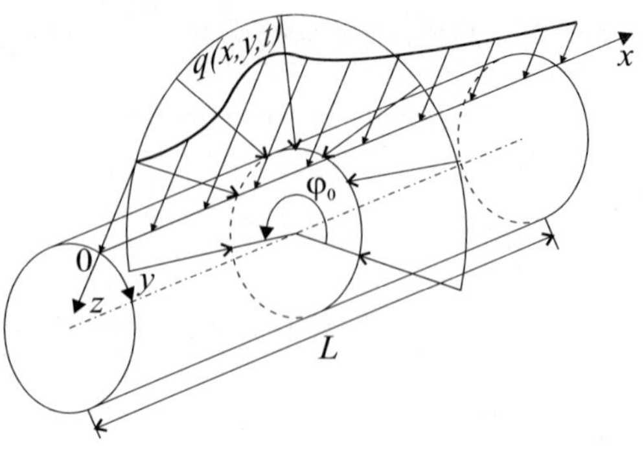

Fig. 10.1 Computational scheme

along a longitudinal coordinate; axis y goes along a circled coordinate; axis z is associated with a mean surface normal to the shell (see Fig. 10.1).

The cylindrical shell treated as the three-dimensional space Ω in the given coordinate system is defined as follows:

$$\Omega = \{x, y, z | (x, y) \in [0; L] \times [0; 2\pi], -h \le z \le h\}.$$

The following differential equations in the non-dimensional form are studied:

$$\frac{1}{12(1-\mu^2)}\left(\frac{1}{\lambda^2}\frac{\partial^4 w}{\partial x^4} + \lambda^2\frac{\partial^4 w}{\partial y^4} + 2\frac{\partial^4 w}{\partial x^2 \partial y^2}\right) - k_y\frac{\partial^2 F}{\partial x^2} - L(w, F)$$
$$-\frac{\partial^2 w}{\partial t^2} - \varepsilon\frac{\partial w}{\partial t} + k_y^2 q(x, y, t) = 0,$$
$$\frac{1}{\lambda^2}\frac{\partial^4 F}{\partial x^4} + \lambda^2\frac{\partial^4 F}{\partial y^4} + 2\frac{\partial^4 F}{\partial x^2 \partial y^2} + k_y\frac{\partial^2 w}{\partial x^2} + \frac{1}{2}L(w, w) = 0, \qquad (10.1)$$

where:

$$L(w, F) = \frac{\partial^2 w}{\partial x^2}\frac{\partial^2 F}{\partial y^2} + \frac{\partial^2 w}{\partial y^2}\frac{\partial^2 F}{\partial x^2} - 2\frac{\partial^2 w}{\partial x \partial y}\frac{\partial^2 F}{\partial x \partial y}$$

is the known nonlinear operator.

The system of Eq. (10.1) has already been reduced to a non-dimensional form with the use of the following non-dimensional parameters (bars over non-dimensional quantities are omitted for simplicity):

$$k_y = \frac{2h}{R^2}\bar{k}_y, \quad q = \bar{k}_y^2\frac{E_0(2h)^4}{L^2R^2}\bar{q}, \quad t = \frac{RL}{2h\sqrt{gE_0}}\bar{t},$$
$$w = 2h\bar{w}, \quad x = L\bar{x}, \quad y = R\bar{y}, \quad F = E_0(2h)^3\bar{F}.$$

The following boundary conditions with their geometrical schemes are applied (see Table 10.1):

Table 10.1 Applied boundary conditions

Type	Eq.
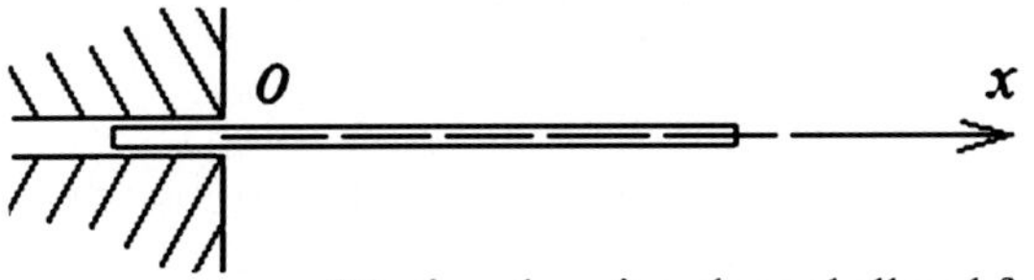 Moving clamping along shell end faces	(10.2)
 Free support along shell faces	(10.3)
 Movable clamping along shell end faces with additional ribs	(10.4)
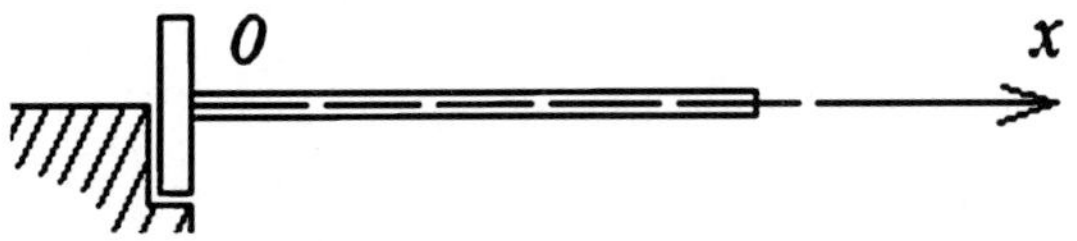 Ball-type support along shell end faces with additional flexible ribs	(10.5)

1. Moving clamping along shell end faces:

$$w = 0; \quad \frac{\partial w}{\partial x} = 0; \quad F = 0; \quad \frac{\partial F}{\partial x} = 0 \text{ for } x = 0; 1. \tag{10.2}$$

2. Free support along shell faces:

$$w = 0; \quad \frac{\partial^2 w}{\partial x^2} = 0; \; F = 0; \quad \frac{\partial F}{\partial x} = 0 \text{ dla } x = 0; 1. \tag{10.3}$$

3. Movable clamping along shell end faces with additional ribs

$$w = 0; \quad \frac{\partial w}{\partial x} = 0; \quad F = 0; \quad \frac{\partial^2 F}{\partial x^2} = 0 \text{ dla } x = 0; 1. \tag{10.4}$$

4. Ball-type support along shell end faces with additional flexible ribs:

$$w = 0; \quad \frac{\partial^2 w}{\partial x^2} = 0; \quad F = 0; \quad \frac{\partial^2 F}{\partial x^2} = 0 \text{ for } x = 0; 1. \tag{10.5}$$

It should be emphasized that the physical interpretation and mathematical mode of the boundary condition (10.5) were first introduced in monograph [144].

The following initial conditions are attached:

$$w(x,y)|_{t=0} = \varphi_1(x,y), \quad \frac{\partial w}{\partial t} = \varphi_2(x,y). \tag{10.6}$$

In the above the following notation is used: μ is the Poisson coefficient, $\lambda = L/R$ where L and $R = R_y$ is the length and radius of a cylindrical shell with a circular intersection; $k_y = 1/R_y$ is the curvature of a shell along y (shell curvature in relation to x is $k_x = 1/R_x = 0$), $q(x,y,t)$ is the outer pressure, ε is the damping coefficient of a medium.

We consider the transversal external load action in the zone of $0 \leq \varphi \leq \varphi_0$ and harmonically changing $q(t) = q_0 \sin(\omega_p t)$, where ω_p is the frequency of the exciting force, q_0 is the amplitude of the exciting force, $\mu = 0.3$, $\varepsilon = 9$.

10.3 Bubnov-Galerkin Method and Fourier Representation

The sought functions w and F are approximated by an analytical expression consisting of a finite number of arbitrary parameters in the form of products of functions dependent on time and spatial coordinates of the following forms:

$$w = \sum_{i=1}^{M_x} \sum_{j=0}^{M_y} A_{ij}(t) \varphi_{ij}(x,y),$$

$$F = \sum_{i=1}^{M_x} \sum_{j=0}^{M_y} B_{ij}(t) \psi_{ij}(x,y). \tag{10.7}$$

The coordinate systems $\{\varphi_{ij}(x,y),\ \psi_{ij}(x,y)\}$ are chosen in such a way as to keep functions $\varphi_{ij}(x,y)$, $\psi_{ij}(x,y)$ for $\forall i, j$: (i) linearly independent assuming their continuity together with their partial derivatives up to the fourth order in the space Ω; (ii) satisfying one of the corresponding boundary conditions (10.2)–(10.5); (iii) with compact properties.

Notice that in some papers (see for instance [82]) a question of neglecting the homogeneous solution for F is discussed.

We address this question using an example of solutions to a system of partial differential equations of flexible shells in the following hybrid form:

$$\nabla^4 w = L(w,F) + \nabla_k^2 F + q - \frac{\partial^2 w}{\partial t^2} - \varepsilon \frac{\partial w}{\partial t}, \tag{10.8}$$

$$\nabla^4 F = -\frac{1}{2} L(w,w) - \nabla_k^2 w. \tag{10.9}$$

In the above w denotes a deflection function, F denotes the Airy function, and $L(w,F)$ is a known nonlinear operator.

Now, in order to solve Eqs. (10.8) and (10.9) by the Bubnov-Galerkin method the following approach is widely applied, sometimes referred to as the Papkovitch method [241]. Namely, a deflection is assumed in the form $w = \sum_{i,j=1}^{N} A_{ij}(t)w_{ij}(x,y)$ and then it is substituted into the right-hand side of deformation compatibility Eqs. (10.9)

$$\nabla^4 F = -\frac{1}{2}\left(\sum_{i,j=1}^{N} A_{ij}(t)w_{ij}(x,y), \sum_{i,j=1}^{N} A_{ij}(t)w_{ij}(x,y)\right)$$
$$-\nabla_k^2\left(\sum_{i,j=1}^{N} A_{ij}(t)w_{ij}(x,y)\right) = f(A_{ij}(t), w_{ij}).$$

As a result one obtains a linear PDE of the fourth order of the form $\nabla^4 F = f(A_{ij}(t), w_{ij})$, which is solved by a method of successive iterations. In general, usually this equation is solved for $N = 1$, i.e., applying the first approximation only and then the function $F(x,y,t)$ is defined with respect to this first approximation. The value of F obtained so far is substituted into the right-hand side of equilibrium Eq. (10.8) and then the Bubnov-Galerkin procedure associated with Eq. (10.8) is applied. It is assumed then that the compatibility Eq. (10.9) is satisfied exactly, but in the first approximation, and the equilibrium Eq. (10.8) is solved in an averaged sense applying the Bubnov-Galerkin procedure.

Notice that in our work we apply another approach developed by Vlasov. Namely, we assume the deflection and Airy functions in the following forms:

$$w = \sum_{i,j=1}^{N} A_{ij}(t)w_{ij}(x,y), \; F = \sum_{i,j=1}^{N} B_{ij}(t)F_{ij}(x,y). \tag{10.10}$$

Then Eq. (10.10) is substituted into Eqs. (10.8), (10.9), and then simultaneously the Bubnov-Galerkin procedure is applied to both equations. In other words, both deformation and stress functions of the governing equations are satisfied simultaneously in some sense in the averaged meaning, but a solution is sought in higher order approximations and the solution convergence is studied with respect to N in (10.10). This approach is expected to have larger accuracy in comparison with the first one, and in addition is supported by theorems yielding convergence of the applied numerical procedures.

There exists also a third approach. Namely, the Airy function governed by the second relation of (iii) is substituted into equilibrium Eq. (10.8), and in the stationary case a linear fourth-order differential equation with periodic coefficients is obtained. However, in general finding a solution to this problem is more difficult than to apply the Papkovitch approach. There exist some exceptional cases, when a solution can be found. Then the Bubnov-Galerkin procedure is applied for the compatibility Eq. (10.9). The latter approach can be applied only for stationary problems. In the case of non-stationary problems like ours governed by Eqs. (10.8), (10.9) this procedure cannot be applied.

In other words, the Bubnov-Galerkin procedure applied by us includes a homogeneous solution, since both equations are satisfied in an averaged sense for higher approximations, as it has been explained earlier.

Let us introduce (for simplicity) the left-hand sides of Eqs. (10.1) in the following way:

$$\begin{gathered}\Phi_1(w, F, \frac{\partial^2 w}{\partial x^2}, \frac{\partial^2 F}{\partial x^2}, \ldots) + Kq(x,y,t) = 0, \\ \Phi_2(w, F, \frac{\partial^2 w}{\partial x^2}, \frac{\partial^2 F}{\partial x^2}, \ldots) = 0, \quad K = k_y^2. \end{gathered} \tag{10.11}$$

Applying the Bubnov-Galerkin procedure to Eq. (10.2) the following equations are obtained:

$$\begin{gathered}\int_0^1\int_0^{2\pi} \Phi_1 \varphi_{rs}(x,y)dx\,dy + \int_0^1\int_{-\frac{\varphi_0}{2}}^{\frac{\varphi_0}{2}} Kq(x,y,t)\varphi_{rs}(x,y)dx\,dy = 0, \\ \int_0^1\int_0^{2\pi} \Phi_2 \psi_{rs}(x,y)dx\,dy = 0, \quad r = 0,1,\ldots,M_x; \quad s = 0,1,\ldots M_y. \end{gathered} \tag{10.12}$$

Owing to Eqs. (10.3), Eqs. (10.3) are given in the forms

$$\begin{gathered}\sum_{rs}\Bigg[\sum_{ij} A_{ij}\sum_{kl} H_{ijklrs} + \sum_{ij} B_{ij}C_{1,ijrs} + \sum_{ij} A_{ij}W_{ijrs} \\ + MqQ_{rs} + \sum_{ij} A_{ij}\sum_{kl} B_{kl}D_{1,ijklrs} + \sum_{ij}\left[\frac{d^2A_{ij}}{dt^2} + \varepsilon\frac{dA_{ij}}{dt}\right]G_{ijrs}\Bigg] = 0, \\ \sum_{rs}\left[\sum_{ij} A_{ij}C_{2,ijrs} + \sum_{ij} B_{ij}\sum_{kl} P_{ijklrs} + \sum_{ij} A_{ij}\sum_{kl} A_{rs}D_{2,ijklrs}\right] = 0. \end{gathered} \tag{10.13}$$

Note that the operator $\sum_{rs}[*]$ before each equation of system (10.13) means that instead of the given equation the system of rs equations is taken, and integrals of the Bubnov-Galerkin procedure have the following form:

$$\begin{aligned} H_{ijklrs} = \int_0^1\int_0^{2\pi} \frac{1}{12(1-\mu^2)}&\left[\frac{1}{\lambda^2}\frac{\partial^2\varphi_{ij}}{\partial x^2}\frac{\partial^2\varphi_{kl}}{\partial x^2}\right. \\ &\left. + \lambda^2\frac{\partial^2\varphi_{ij}}{\partial y^2}\frac{\partial^2\varphi_{kl}}{\partial y^2} + 2\frac{\partial^2\varphi_{ij}}{\partial x\partial y}\frac{\partial^2\varphi_{kl}}{\partial x\partial x}\right]\varphi_{rs}dx\,dy, \end{aligned} \tag{10.14}$$

$$C_{1,ijrs} = \int_0^1\int_0^{2\pi}\left[-k_y\frac{\partial^2\Psi_{ij}}{\partial x^2}\right]\varphi_{rs}dx\,dy, \quad C_{2,ijrs} = \int_0^1\int_0^{2\pi}\left[k_y\frac{\partial^2\varphi_{ij}}{\partial x^2}\right]\Psi_{rs}dx\,dy,$$

$$D_{1,ijklrs} = \int_0^1 \int_0^{2\pi} [-L(\varphi_{ij}, \Psi_{kl})]\, \varphi_{rs} dx\, dy,$$

$$D_{2,ijklrs} = \int_0^1 \int_0^{2\pi} \left[\frac{1}{2} L(\varphi_{ij}, \varphi_{kl})\right] \Psi_{rs} dx\, dy,$$

$$P_{ijklrs} = \int_0^1 \int_0^{2\pi} \left[\frac{1}{\lambda^2}\frac{\partial^2 \Psi_{ij}}{\partial x^2}\frac{\partial^2 \Psi_{kl}}{\partial x^2} + \lambda^2 \frac{\partial^2 \Psi_{ij}}{\partial y^2}\frac{\partial^2 \Psi_{kl}}{\partial y^2} + 2\frac{\partial^2 \Psi_{ij}}{\partial x \partial y}\frac{\partial^2 \Psi_{kl}}{\partial x \partial x}\right] \Psi_{rs} dx\, dy,$$

$$G_{ijrs} = \int_0^1 \int_0^{2\pi} [-\varphi_{ij}\varphi_{rs}]\, dx\, dy, \quad Q_{rs} = \int_0^1 \int_{-\frac{\varphi_0}{2}}^{\frac{\varphi_0}{2}} Kq(x,y,t)\varphi_{rs} dx\, dy.$$

Integrals (10.14), perhaps in spite of Q_{rs}, and in spite of the case when the transversal load is only on part of the shell surface, are computed with respect to the whole mean shell surface.

As a result of the application of the Bubnov-Galerkin procedure and applying integrals (10.14) the following system of nonlinear second-order ODEs with respect to the coefficients A_{ij} and B_{ij}, and to the system of linear algebraic equations (LE) also with respect to the coefficients A_{ij} and B_{ij} is obtained:

$$\bar{G}(\ddot{\bar{A}} + \varepsilon\dot{\bar{A}}) + \bar{H}\bar{A} + \bar{W}\bar{A} + \bar{C}_1\bar{B} + \bar{D}_1\bar{A}\bar{B} = \bar{Q}q(t), \quad (10.15)$$

$$\bar{C}_2\bar{A} + \bar{P}\bar{B} + \bar{D}_2\bar{A}\bar{A} = 0, \quad (10.16)$$

where $\bar{H} = \|H_{ijrs}\|$, $\bar{G} = \|G_{ijrs}\|$, $\bar{C}_1 = \|C_{1ijrs}\|$, $\bar{C}_2 = \|C_{2ijrs}\|$, $\bar{D}_1 = \|D_{1ijklrs}\|$, $\bar{D}_2 = \|D_{2ijklrs}\|$, $\bar{W} = \|W_{ijrs}\|$, $\bar{P} = \|P_{ijrs}\|$ are the quadratic matrices with dimensions $2N_1N_2 \times 2N_1N_2$, and $\bar{A} = \|A_{ij}\|$, $\bar{B} = \|B_{ij}\|$, $\bar{Q} = \|Q_{ij}\|$ are the matrices with dimensions $2N_1N_2 \times 1$.

Equation (10.16) is solved on every time step and yields the matrix

$$\bar{B} = \left[-\bar{P}^{-1}\bar{D}_2\bar{A} - \bar{P}^{-1}\bar{C}_2\right]\bar{A}. \quad (10.17)$$

Multiplying (10.15) by $\bar{G}^{-1}$, and denoting $\dot{\bar{A}} = \bar{R}$ we get the following nonlinear system of first-order ODEs

$$\begin{cases} \dot{\bar{A}} = \bar{R}, \\ \dot{\bar{R}} = -\bar{\varepsilon}\bar{R} - \left[\bar{G}^{-1}\bar{C}_1 + \bar{G}^{-1}\bar{D}_1\bar{A}\right]\bar{B} - \bar{G}^{-1}\bar{H}\bar{A} \\ -\bar{G}^{-1}\bar{W}\bar{A} + \bar{G}^{-1}\bar{Q}q(\bar{t}). \end{cases} \quad (10.18)$$

Equations (10.18) are supplemented by one of the boundary conditions (10.2)–(10.5), the initial conditions (10.6), and the obtained Cauchy problem is solved using the fourth-order Runge-Kutta method.

Let us consider now a ball-type supported cylindrical shell along a curvilinear circle with homogeneous boundary conditions (10.5) and initial conditions (10.6).

For this purpose φ_{ij}, Ψ_{ij} in (10.10) are represented by a product of two functions, and each of them depends only on one argument that satisfies the boundary conditions (10.2) of the form

$$w = \sum_{i=1}^{M_x} \sum_{j=0}^{M_y} A_{ij}(t)\,\sin(i\pi x)\,\cos(jy),$$

$$F = \sum_{i=1}^{M_x} \sum_{j=0}^{M_y} B_{ij}(t)\,\sin(i\pi x)\,\cos(jy). \tag{10.19}$$

In the Bubnov-Galerkin procedure (in its Fourier form) the following integral representation is used:

$$I_{1,r} = \int_{x_1}^{x2} \sin(r\pi x)\,dx = \frac{\cos(r\pi x_1) - \cos(r\pi x_2)}{r\pi},$$

$$I_{2,s} = \int_{y_1}^{y2} \cos(s\pi y)\,dy = \frac{\sin(s\pi y_2) - \sin(s\pi y_1)}{s\pi},$$

$$I_{3,ir} = \int_0^1 \sin(i\pi x)\,\sin(r\pi x)\,dx = \begin{cases} \frac{1}{2}, \ i = r, \\ 0, \ i \neq r, \end{cases}$$

$$I_{4,js} = \int_0^{2\pi} \cos(j\pi y)\,\cos(s\pi y)\,dy = \begin{cases} 2\pi, & j = s = 0, \\ \pi, & j = s \neq 0, \\ 0, & j \neq s, \end{cases}$$

$$I_{5,ikr} = \int_0^1 \sin(i\pi x)\,\sin(r\pi x)\,\sin(k\pi x)\,dx$$

$$= \begin{cases} \frac{1}{4\pi}\Big[-\frac{\cos(\alpha_1\pi)}{\alpha_1} - \frac{\cos(\alpha_2\pi)}{\alpha_2} - \frac{\cos(\alpha_3\pi)}{\alpha_3} - \frac{\cos(\alpha_4\pi)}{\alpha_4} \\ \quad + \frac{1}{\alpha_1} + \frac{1}{\alpha_2} + \frac{1}{\alpha_3} - \frac{1}{\alpha_4}\Big], \quad \alpha_\nu \neq 0, \\ \frac{\cos(\alpha_\nu\pi)}{\alpha_\nu} \approx 0, \quad \frac{1}{\alpha_\nu} \approx 0, \quad \nu = 1,2,3, \quad \alpha_\nu = 0, \end{cases}$$

$$I_{7,ikr} = \int_0^1 \cos(i\pi x)\,\cos(k\pi x)\,\sin(r\pi x)\,dx$$

$$= \begin{cases} \frac{1}{4\pi}\Big[\frac{\cos(\alpha_1\pi)}{\alpha_1} - \frac{\cos(\alpha_2\pi)}{\alpha_2} - \frac{\cos(\alpha_3\pi)}{\alpha_3} - \frac{\cos(\alpha_4\pi)}{\alpha_4} \\ \quad - \frac{1}{\alpha_1} + \frac{1}{\alpha_2} + \frac{1}{\alpha_3} + \frac{1}{\alpha_4}\Big], \quad \alpha_\nu \neq 0, \\ \frac{\cos(\alpha_\nu\pi)}{\alpha_\nu} \approx 0, \quad \frac{1}{\alpha_\nu} \approx 0, \quad \nu = 1,2,3, \quad \alpha_\nu = 0, \end{cases}$$

$$I_{8,jls} = \int_0^{2\pi} \sin(jy)\,\sin(ly)\,\cos(sy)\,dy$$

$$= \begin{cases} \frac{1}{4}\Big[-\frac{\sin(\beta_1\pi)}{\beta_1} + \frac{\sin(\beta_2\pi)}{\beta_2} + \frac{\sin(\beta_3\pi)}{\beta_3} + \frac{\cos(\beta_4\pi)}{\beta_4} \\ \quad +\frac{1}{\beta_1} - \frac{1}{\beta_2} - \frac{1}{\beta_3} + \frac{1}{\alpha_4}\Big], \quad \beta_\nu \neq 0, \\ \frac{\sin(\beta_\nu\pi)}{\beta_\nu} \approx 0, \quad \frac{1}{\beta_\nu} \approx 0, \quad \nu = 1,2,3,4, \qquad \beta_\nu = 0, \end{cases}$$

$$I_{6,jls} = \int_0^{2\pi} \cos(jy)\,\cos(ly)\,\sin(sy)\,dy = 0.$$

where:

$$\alpha_1 = j+k-r, \quad \alpha_2 = k+r-i, \quad \alpha_3 = r+i-k, \quad \alpha_4 = i+k+r,$$
$$\beta_1 = j+l-s, \quad \beta_2 = l+s-j, \quad \beta_3 = s+j-l, \quad \beta_4 = j+l+s.$$
$$I_Q^{rs} = MI_{1r}I_{2s}, \quad I_P^{rs} = \left(s^2 p_x(t) + r^2 p_y(t)\right)\pi^2 I_{3,ir}I_{4,js},$$
$$I_{AB}^{rs} = \left(s^2 k_x + r^2 k_y\right)\pi^2 I_{3,ir}I_{4,js}, \quad I_{rs}^t = I_{3,ir}I_{4,js}, \quad M = k_y^2,$$
$$I_{ijklrs} = \pi^2\left[(i^2l^2 + j^2k^2)I_{5,ikr}I_{6,jls} - 2ijlkI_7, ikrI_{8,jls}\right],$$
$$J_{1,ijkl}^{rs} = \frac{\pi^2}{12(1-\mu^2)}\left[\frac{r^4}{\lambda^2} + 2r^2s^2 + \lambda^2 s^4\right] I_{3,ir}I_{4,js},$$
$$J_{2,ijkl}^{rs} = \left[\frac{r^4}{\lambda^2} + 2r^2s^2 + \lambda^2 s^4\right]\pi^2 I_{3,ir}I_{4,js}.$$

Then in view of the written integrals, system (10.13) is given in the form

$$\sum_{rs}\Bigg\{\sum_{ij}\sum_{kl}\Big[J_{1,ijkl}^{rs}A_{ij} + I_{AB}^{rs}B_{rs} + I_P^{rs}A_{ij} + I_Q^{rs}q(t)$$
$$+ A_{ij}B_{kl}I_{ijklrs} + \left(\frac{d^2A_{ij}}{dt^2} + \varepsilon\frac{dA_{ij}}{dt}\right)I_{rs}^t\Big]\Bigg\} = 0,$$
$$\sum_{rs}\Bigg\{\sum_{ij}\sum_{kl}\left[J_{2,ijkl}^{rs}B_{ij} + I_{AB}^{rs}A_{rs} + \frac{1}{2}A_{ij}A_{kl} + I_{ijklrs}\right]\Bigg\} = 0. \tag{10.20}$$

10.4 Static Problems of Closed Cylindrical Shell Theory

Consider a static loading of closed cylindrical shell consisting of uniform external load $q = q_1$ distributed within the zone with central angle φ_0 shown in Fig. 10.1.

We are going to study the dependence of critical loads on width of a loading pressure zone. A static solution is obtained from dynamic one with a help of the setup method first introduced by Feodosev [95] and associated with critical damping $\varepsilon = \varepsilon_{cr}$. In 1963 Feodosev proposed a dynamic approach to solve a problem related to the stability of shells. From a mathematical point of view, this method is called the "set-up" method. The main idea of this method is that solution of the nonlinear partial differential equations is reduced to a Cauchy problem of ODEs that is linear in time. This means that this method linearizes the nonlinear equations and decreases their dimension.

In what follows, we discuss briefly an advantage of this method. From a mathematical point of view, the "set-up" method can be treated as an iterative method to solve nonlinear algebraic equations, where each time step provides a new approximation to the exact solution. Like all iterative methods, this one is characterized by a high accuracy of computation. In addition, is does not have the common disadvantage of iterative methods of a high sensitivity to the choice of the initial approximation. In addition, the "set-up" method not only gives a very simple rule for obtaining nonunique solutions of static problems, but also allows one to find the stable and unstable branches of the equilibrium position of the system under consideration and to capture all process of the jumping behavior of a shell.

In the process of solution of homogeneous equations via traditional methods, in order to obtain a nontrivial solution one needs to introduce an artificial excitation (in the theory of shells this corresponds, for instance, to a small transverse load, a small curvature or some other initial imperfection). However, this influences (sometimes significantly) the results obtained. In the case of the "set-up" method, the initial conditions play the role of the initial excitations, and small changes of these conditions do not influence the static solution obtained. Another advantage of the method is related to its simple realization, because today there are many effective algorithms and programs devoted to solution of the Cauchy problem.

In order to obtain $q_1(\varphi_0)$ one needs to construct for $\forall \varphi_0 \in [0; 2\pi]$ the set $\{q_{1,i}, w_i\}$, due to which $q_{1,cr}$ (critical load) is defined.

Let us discuss the results obtained for different approximations. Since the load is applied along the whole length of cylindrical shell, the number of series terms with respect to coordinate x does not play any important role. Let us investigate the obtained results versus the number of series terms associated with circle coordinate $y - M_y$. The mentioned relations are shown in Fig. 10.2 for $M_y = 4$ and $M_y = 5$.

One may observe that the relation is non-monotonous. For different values of M_y the curves are convergent, but an increase of approximation number of one unit yields more accurate results (Fig. 10.2). Hence, for a non-homogeneous load the use of a small number of terms of series (10.10) yields its convergence beginning from $M_y = 13$ (see Fig. 10.3). Note that the mentioned results have been obtained for large values of the loading angles ($\varphi_0 = 6$).

Below, we investigate the method's convergence for different loading angles. For this purpose a dependence of critical load $q_{1,cr}$ on the number of series terms (10.10) for different angles of loading actions $q_{1,cr}(M_y)$ (see Fig. 10.4) has been determined. In a general case, for different angles, the convergence begins with $M_y = 15 - 20$.

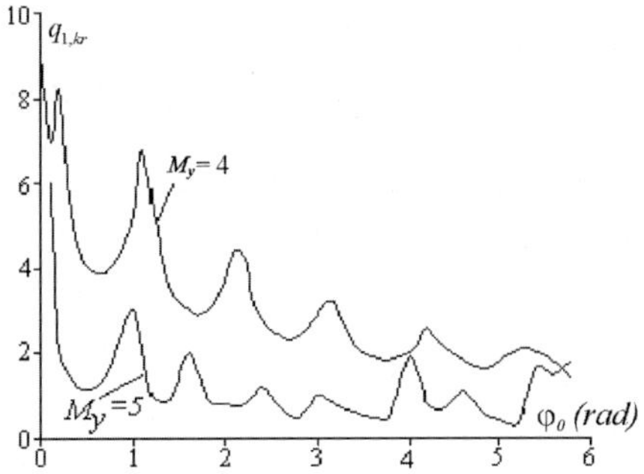

Fig. 10.2 Relation $q_{1,cr}(\varphi_0)$ for different approximations, $M_y = 4, 5$

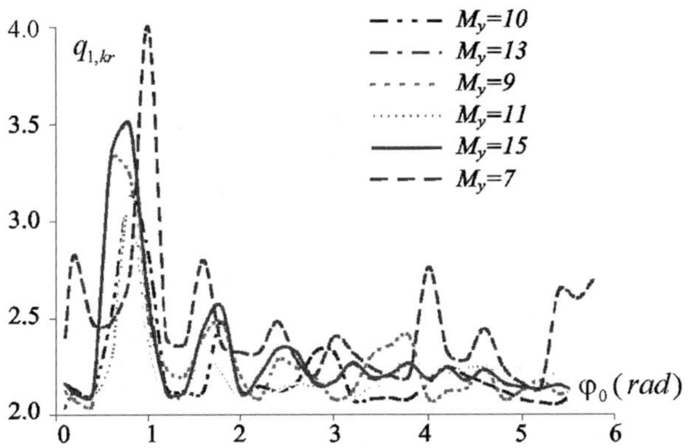

Fig. 10.3 Relation $q_{1,cr}(\varphi_0)$ for different approximations, $M_y > 6$

In order to verify the reliability of results, let us draw the dependence of $q_{1,cr}$ on the pressure zone width $\bar{q}_{1,cr}(\varphi_0)$ for $M_x = 1$, $M_y = 13$ (Fig. 10.5). In the above, we have taken $\bar{q}_{1,cr} = \frac{q_{1,cr}}{\tilde{q}_{cr}}$, where $\tilde{q}_{cr}$ is the classical critical value of the uniform external pressure computed using the Mizes-Papkovitch formula [241], i.e., $\bar{q}_{cr} = 0.92\frac{R}{L}\left(\frac{h}{R}\right)^{5/2}$. Note that the dependence $\bar{q}_{1,cr}(\varphi_0)$ has a non-monotonous vibrational

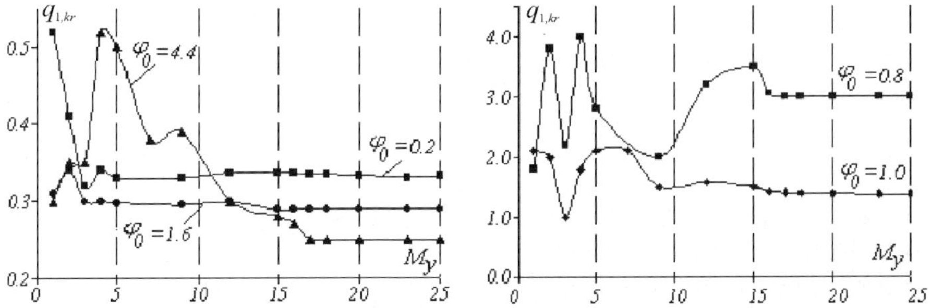

Fig. 10.4 Dependence of critical load on the number of series terms 10.10 for different loading angles $q_{1,kr}(M_y)$

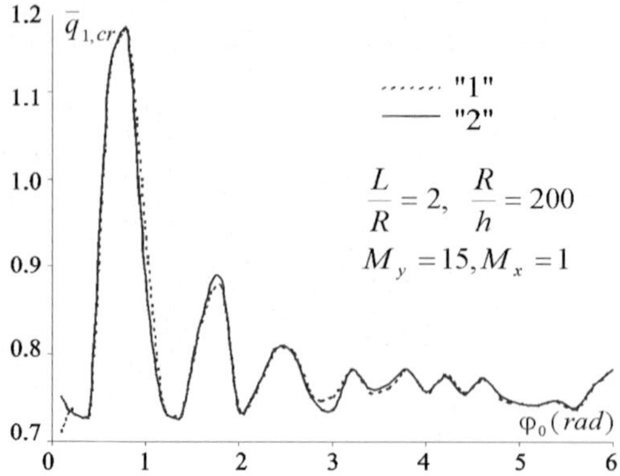

Fig. 10.5 Critical loads versus width of the pressure zone ("1", curve obtained via "set-up" method; "2", curve given in reference [11])

character. An increase of φ_0 from zero is associated with a series of maxima and minima, and beginning from $\varphi_0 \approx 4$ the critical load oscillates on the level of $\bar{q}_{1,cr} \approx 0.75$. In order to verify reliability of the results, the data given in reference [11] will be applied.

On the basis of a comparison of the two graphs shown in Fig. 10.5, one may conclude that the results obtained fully coincide with those reported in reference [11]. Although we show only a comparison with the literature concerning static critical loads, in all further computations we have used the Runge principle to control the results reliability.

10.5 Dynamics of Closed Cylindrical Shells

In this section we are going to analyze dynamic problems in the case of external harmonic excitation of the form $q(t) = q_0 \sin(\omega_p t)$. Analysis of complex nonlinear shell vibrations is carried out with respect to two control parameters in both pre- and post-critical states, i.e., width of the pressure zone φ_0 and linear dimension of the shell $\lambda = \frac{L}{R}$ are taken into account.

10.5.1 Convergence of the Fourier Representation for a Non-stationary Problem

We study a convergence of the Bubnov-Galerkin approach versus M_y in (10.19) for a cylindrical shell in the case of a non-stationary problem. Following Poincaré's ideas [247] that instead of studying one particular orbit it is more convenient to analyze the entire orbit manifold, various vibrational charts are constructed for the control parameters $\{q_0, \omega_p\}$ for $L = \frac{\lambda}{R} = 2$. First, the problem of solution convergence with an increase of control parameters plane partition is studied (see Table 10.2). The introduced figure notation is valid also in further considerations.

Table 10.2 Investigation of a solution convergence

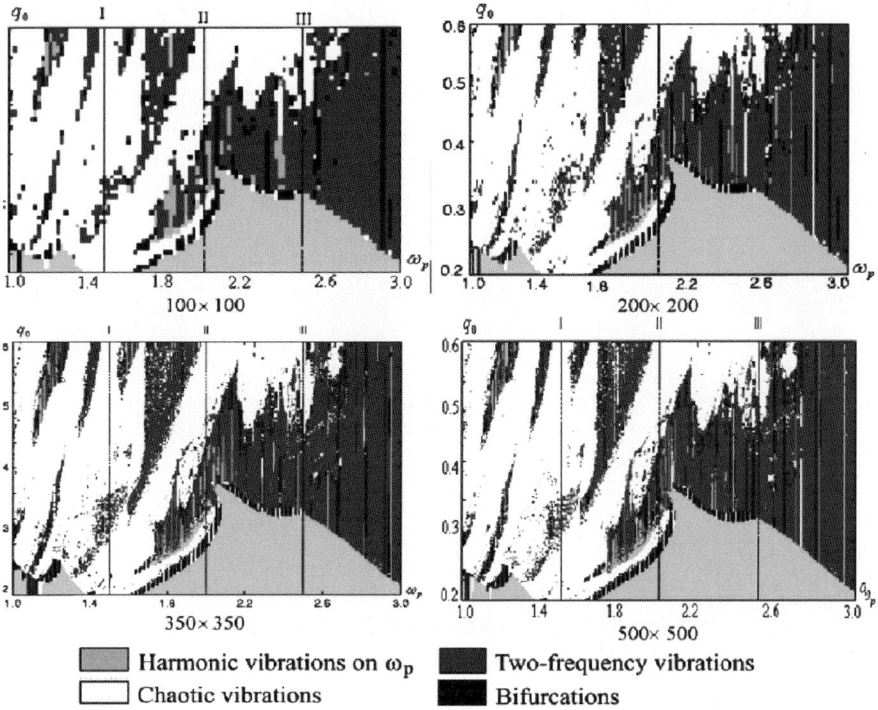

For construction of such maps the grid in which unit identification of vibration character was made has been imposed on the area of space $\{q_0, \omega_p\}$. Hence, it is necessary to solve a dynamic problem regarding the construction and analysis of a power spectrum for each unit of a grid, i.e., each set of parameters $\{q_0, \omega_p\}$.

Computations have shown that a convergence is realized for $N \times N \geq 350 \times 350$ (see Table 10.2). The introduced charts allow us to study the entire manifold of the shell's behavior.

Identification of the type of cylindrical shell vibrations during construction of the chart $\{q_0, \omega_p\}$ for each signal (time history) $w(t)$ is supported by analysis of the power spectrum $S(\omega)$ and Lyapunov exponents. The chart $\{q_0, \omega_p\}$ is divided into 350×350 parts.

A convergence of the Bubnov-Galerkin method in higher approximations and with the application of Fourier transform with respect to M_y is analyzed in the bifurcation zone (point A in chart $\{q_0, \omega_p\}$, Fig. 10.6), and in the chaotic zone (point B in the chart $\{q_0, \omega_p\}$).

Note that for small values $M_y \leq 9$ basically harmonic vibrations appear and zones of chaos are absent at all investigated frequencies (only small bifurcations areas are visible). At $M_y = 10$ the picture of vibrations varies; bifurcation areas at high frequencies also vanish, but they arise at the frequencies close to fundamental (natural) frequency of vibrations. Furthermore, increasing the parameter $M_y = 11$ causes an

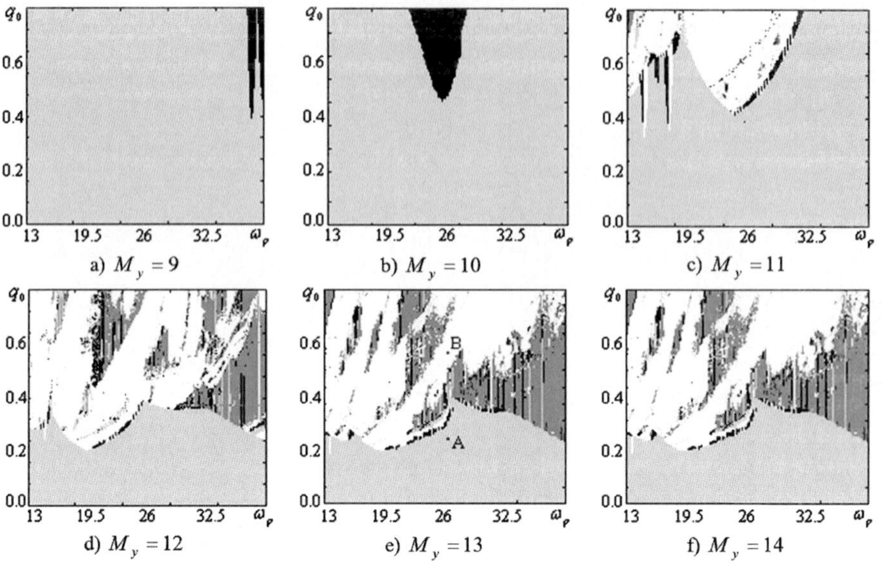

Fig. 10.6 Maps of vibrations character depending on number of M_y in (10.19)

occurrence of extensive zones of chaotic vibrations at low frequencies being close to fundamental vibrations frequency (the areas of bifurcations observed previously vanish at $M_y = 9$). A further increase of M_y, i.e., $M_y = 12$, results in changes associated with high-frequency zones of chaos that at low frequencies remain practically unchanged. Finally, one may observe that the maps constructed for $M_y = 13$ and $M_y = 14$ almost coincide, and hence the converging process begins at $M_y = 13$. It should also be emphasized that convergence at low frequencies is better than that at high frequencies, being close to the fundamental frequency of vibrations (≈ 26.1).

In Figs. 10.7 and 10.8 signals $w(0.5;0;t)$ and power spectra $S(\omega)$ associated with the mentioned points are reported. Analysis of the obtained results (Fig 10.7) shows that for $M_y \geq 12$ power spectra completely coincide, and signals are very close to each other. However, the results obtained for $M_y = 9, 10, 11$ essentially differ from the case of $M_y \geq 12$. On the basis of this observation one may conclude that for $M_y \geq 12$, the bifurcation process is truly, described, and a converging sequence of data is observed. Our computations allowed us also to establish the following conditions:

$$\left[w - \sum_{i=0}^{N_1} \sum_{j=0}^{N_2} A_{ij}(t)\varphi_{ij}(x,y) \right] = \min\{t \in [15;15.7]\},$$

$$\left[F - \sum_{i=0}^{N_1} \sum_{j=0}^{N_2} B_{ij}(t)\Psi_{ij}(x,y) \right] = \min\{t \in [15;15.7]\},$$

yielding the best approximation for both w and F.

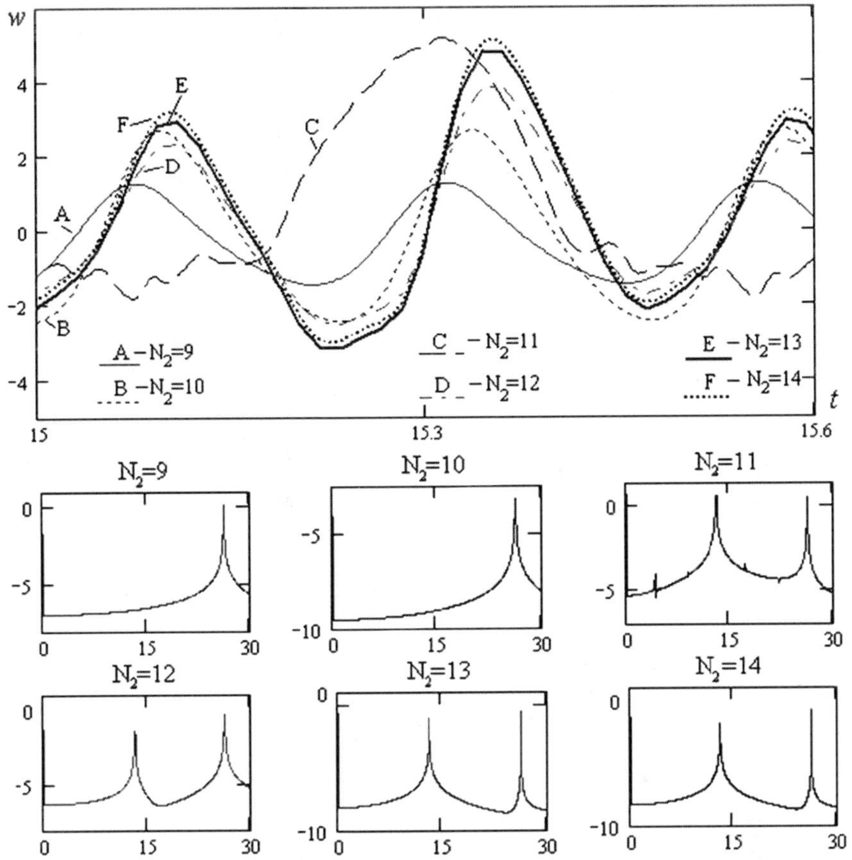

Fig. 10.7 Time history $w(0.5;0;t)$, $t = [15;15.7]$ and power spectra $S(\omega)$ versus M_y (bifurcation zone)

In Fig. 10.8 the same characteristics as in Fig. 10.7 are reported. Unfortunately, in this case we do not notice such a uniform convergence as in the previous case. The value of $M_y = 9$ corresponds to a period doubling bifurcation (the birth of orbit "2"); $M_y = 11$ is associated with two bifurcations of doubling of the period and trebling of the period; for $M_y = 10\ldots14$ in general chaotic vibrations are observed, but the character of chaos is essentially various. Namely, for $M_y = 14$ chaos at the basic frequency of excitations occurs; for $M_y = 10, 13$ chaos at the frequencies associated with the trebled period takes place, and for $M_y = 12$ chaos at the frequencies with a 12-fold increase in the period of vibrations appears. In other words, each member of the decomposition introduces its own contribution to the picture of chaos.

In order to investigate the convergence in more detail, it is required to analyze bending shell forms for fixed loading and width zone pressure action in various approximations. Table 10.3 gives shell cross sections in defined time instants and

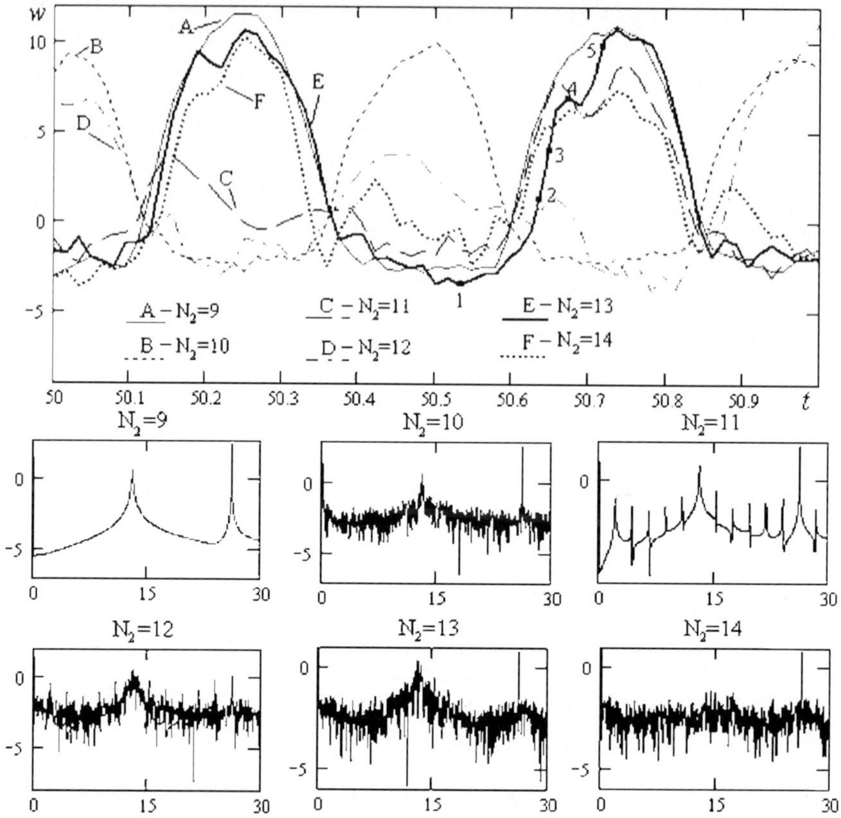

Fig. 10.8 Time history $w(0.5;0;t)$, $t = [15;15.7]$ and power spectra $S(\omega)$ versus M_y (chaotic zone)

Table 10.3 Shell cross-sections

M_y	Signal w(t)				
	1	2	3	4	5
13	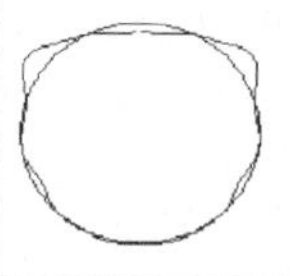				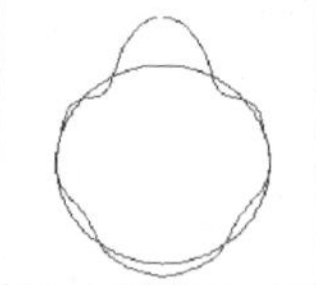
17				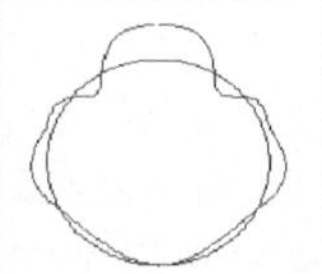	

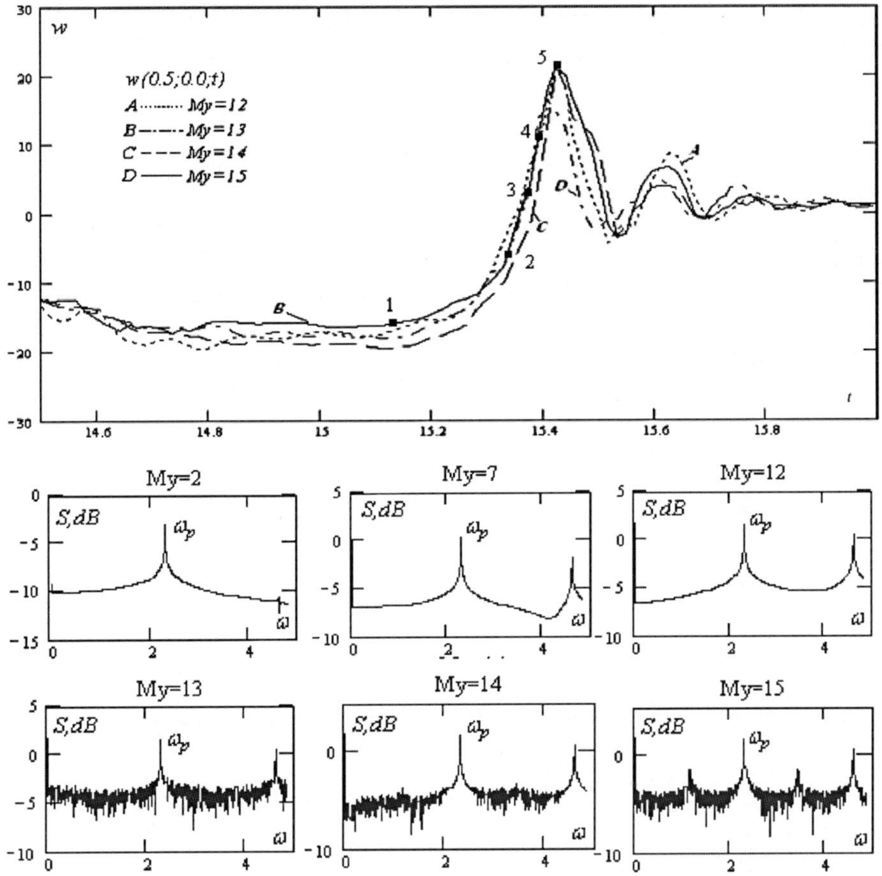

Fig. 10.9 Convergence of the Bubnov-Galerkin method in the chaotic zone

for the fixed value of $x = 0.5$, and $0 \leq y \leq 2\pi$ of point B in the chart $\{q_0, \omega_p\}$, i.e. in the chaotic zone. Points 1–5 correspond to the values reported in Fig. 10.9. It is clear that beginning from $M_y = 13$ one may observe the same picture as in both forms of transversal cross section and $w(t)$, i.e., a convergence with respect to signal (time history) and power spectrum is observed. An increase of the series terms of (10.19) does not improve the accuracy of the results.

Here it is necessary to note that in the considered class of problems an increase in the number of degrees of freedom does not result in a simplification of the character of vibrations as in classical Lorenz model, but leads to their serious complexity.

Furthermore, we have considered one of the most difficult cases by taking $M_x = 1$ and we have shown that by increasing only M_y the numerical process is convergent and hence the obtained results are reliable. However, it should be emphasized that another problem may appear regarding the accuracy of the Donnell shallow shell theory for $\frac{w}{h} \gg 1$.

10.5.2 Vibrations of Closed Cylindrical Shells Subjected to Transversal Sinusoidal Load

In this section we study the cylindrical shell with the attached boundary (10.2) and initial conditions (10.6) driven by external harmonic load distributed in a zone with central angle ϕ_0 (Fig. 10.10). The loading coordinates are: $0 \leq x \leq 1$; $0 \leq y \leq \varphi_0$.

For essentially small loading angles, and when the loading distribution is close to the loading curve, relatively large deflections are developed. The bending shell form is not being changed qualitatively during the loading process.

Note that pre-critical deflection is mostly expressed in zones being in the vicinity of loading part and inside it. For small values of angles φ_0 one concavity occurs in the loading zone (Fig. 10.10c).

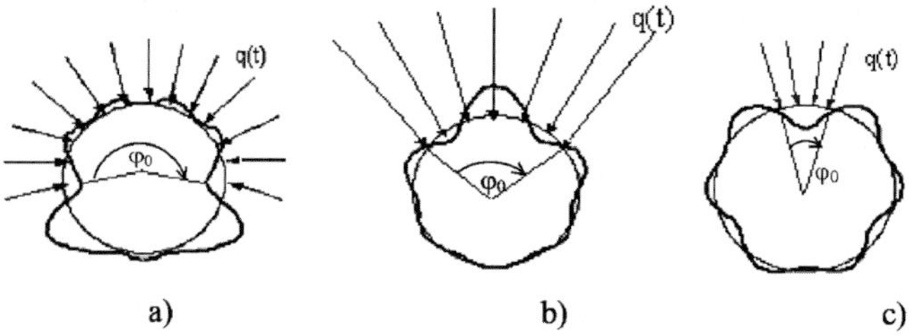

Fig. 10.10 Deflection forms of the shell for different "opening" angles of the loading zone

An increase of φ_0 causes the occurrence of two concavities lying on the boundaries of the loading zone (Fig. 10.10b). For $\varphi_0 \to 0$, and $\varphi_0 \to 2\pi$, the bending process is slightly manifested (Fig. 10.10a, c). For some discrete values of the loading angle φ_0 and during the process of loading increase, a bifurcation of the deflection form occurs and the number of half-waves increases. For larger loading angles the mentioned qualitative change is exhibited only locally, and it is visible in the center of the loading zone (see Fig. 10.11, where q_0 is the critical load).

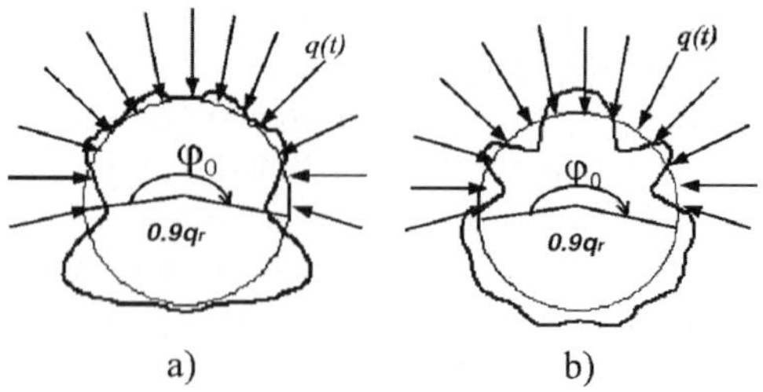

Fig. 10.11 Deflection forms for different loading bends

A stability loss of the shell is observed for loading angles larger than $\varphi_{\min}$ for which the bifurcation point occurs on the "load-deflection" diagram. Such diagrams for different loading angles are shown in Fig. 10.12.

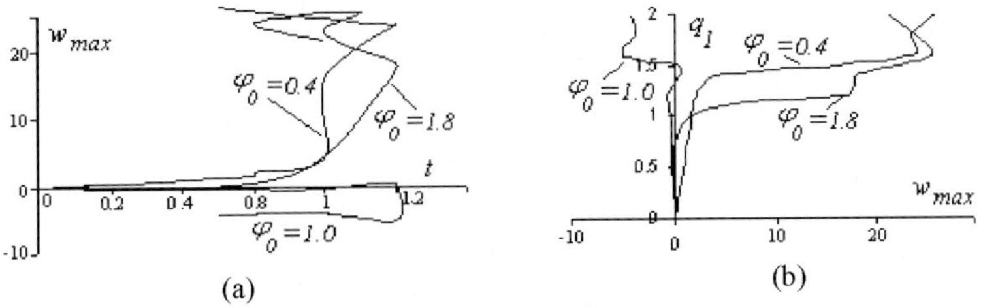

Fig. 10.12 "Maximum deflection–time" relation (a) and dependence of q_r vs. maximum deflection (b) for a variety of loading angles

In order to define the critical load, two earlier described criteria are applied: the dynamic Volmir [304] criterion (Fig. 10.12b) and the Shian et al. [270] criterion (Fig. 10.12a). Based on the analysis of the obtained results, one may conclude that critical loads obtained using the two criteria coincide with accuracy of 10^{-3}. In order to apply a more detailed analysis, let us consider forms of the shell deflection and forms of transversal deflection of the shell in both pre-critical and post-critical states for the series of values φ_0 (Table 10.4). Forms of transversal cross-section $(x = 0.5,\ y \in [0;\ 2\pi])$ in the same time instants as well as the characteristic wave forms correspond to point A in the signal.

Observe that the fast increase of deflection owing to a small change of the load yields an increase of half-wave number. During transition into the post-critical state also large deflections are transmitted into shell unloaded zones.

10.5.3 Dependence of Vibration Character on Width of the Pressure Zone

For some values of the pressure zone φ_0, the characteristic features of vibrations versus control parameters $\{q_0, \omega_p\}$ are specified (Fig. 10.13). Owing to these charts the whole shell behavior can be monitored.

It is evident that the character of vibrations depends on the loading angle significantly. For small values of φ_0 the sum of chaotic zones is reasonably high and consists of two subspaces corresponding to frequency values $\omega_p < \omega_0$ and $\omega_p > \omega_0$. Recall that ω_0 is the fundamental (dimensionless) shell natural frequency, whereas ω_p is the dimensionless frequency of excitation. Owing to an increase of loading of the cylindrical shell, the surface at the chaotic space is decreased into that of low and average frequencies. For a large surface of external pressure the chaotic windows are distributed over the whole chart. The largest part of the chaotic zone is concentrated in the vicinity of $\omega_p < \omega_0$, and the summed chaotic surface zone is relatively high. Also remarkably large are the zones of Andronov-Hopf bifurcations, which appear close to $\omega_p > \omega_0$.

Table 10.4 Pre- and post-critical shell states for various values of q_0

φ_0	Pre-critical states		Post-critical states	
0.4 rad = 23°	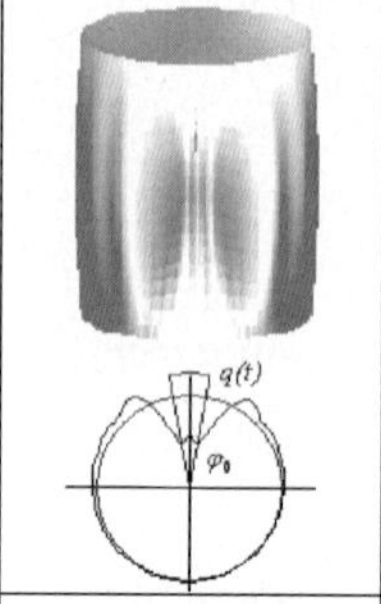	$q_1=0.5$ 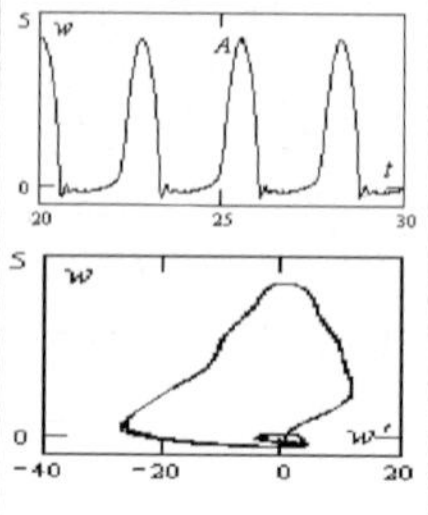	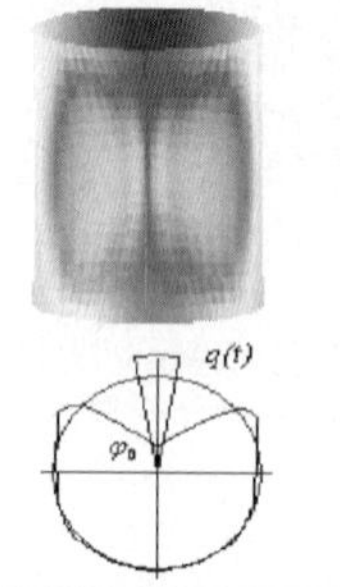	$q_1=1.5$ 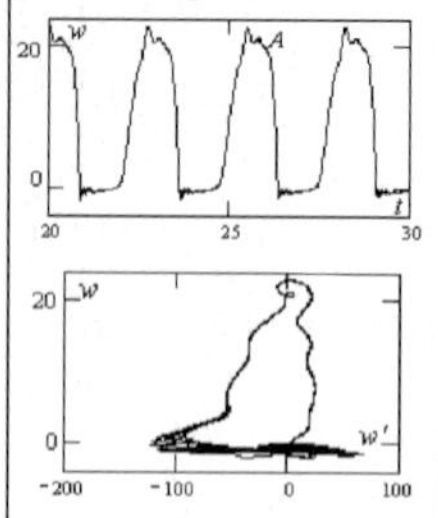
1.0 rad = 57°	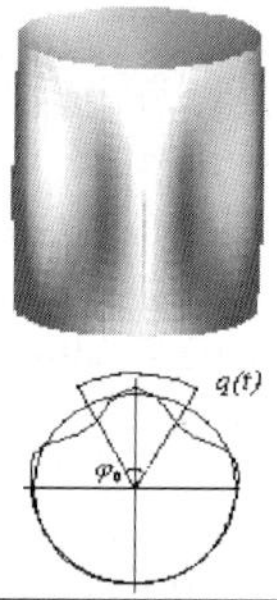	$q_1=0.5$ 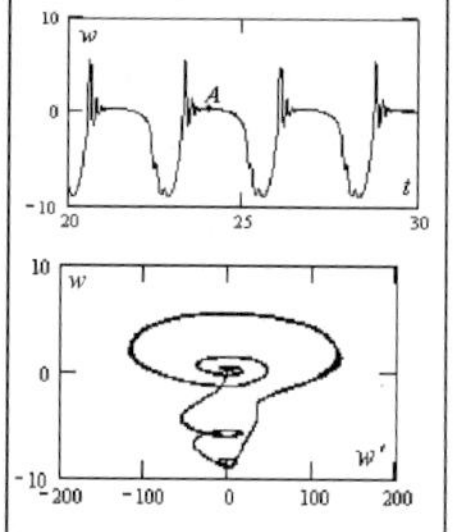	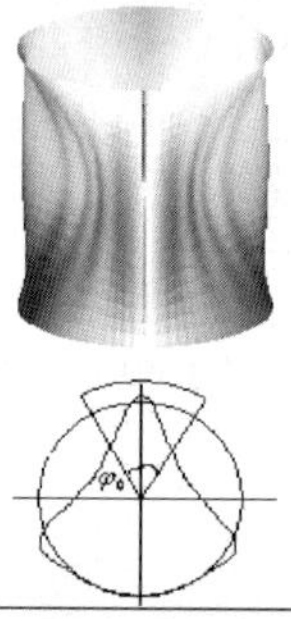	$q_1=1.5$ 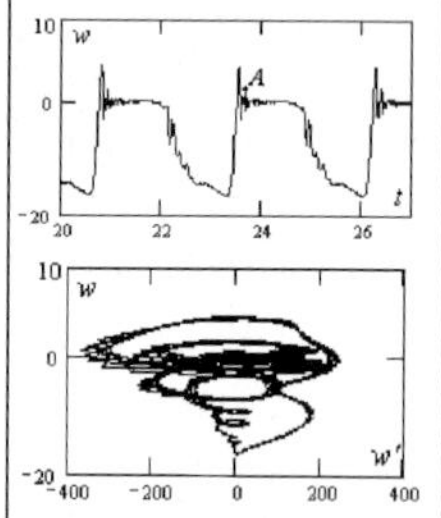
1.8 rad = 103°	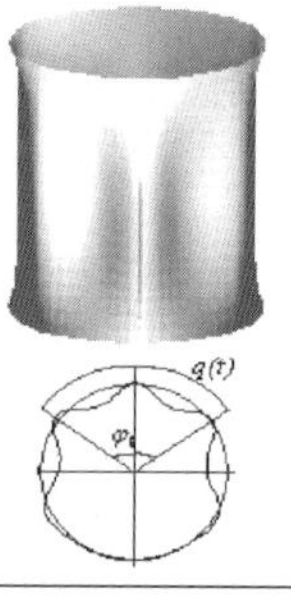	$q_1=0.5$ 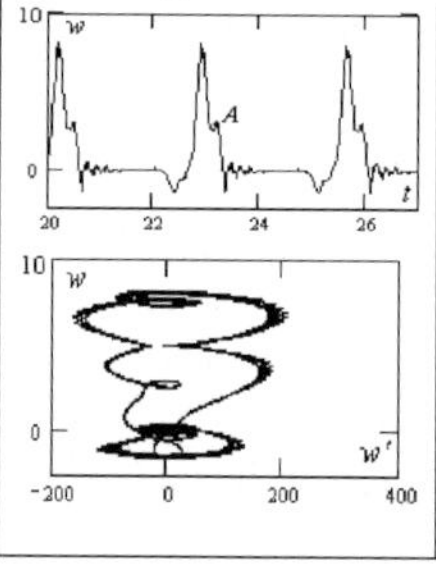	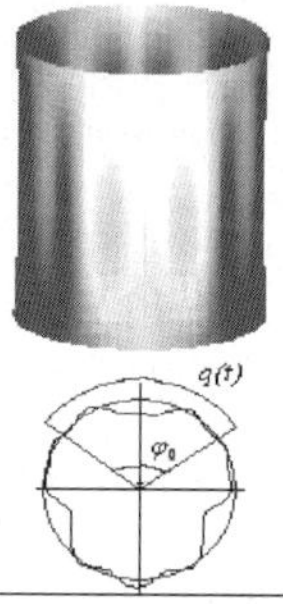	$q_1=1.7$ 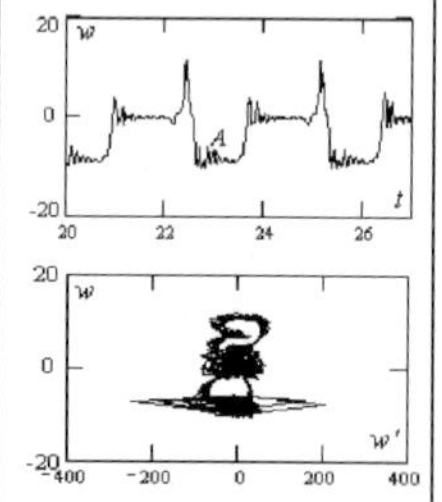

10.5.4 Dependence of Vibration Character on the Linear Shell Dimension

Now, we consider the action of harmonic load in the zone of width $\varphi_0 = 6\,\text{rad} = 343°$ (see Fig. 10.1). For each of fixed parameters of λ the following characteristics are constructed: time history $x(t;x_0;y_0)$, phase portrait $w(w')$, power spectrum $S(\omega)$, and Poincaré map $w_t(w_{t+T})$, where T is the period of excitation subject in the central shell point $(x_0,y_0) = (0.5;\pi)$. In addition, the relation $w_{\max}(q_0)$ for fixed values of the frequency of excitation $\omega_p = \omega_0$ (ω_0 is the fundamental natural frequency of linear vibrations) and vibration character zones (scales) are monitored. In order to investigate spatial vibrations, both forms of the waving shapes of the shell for

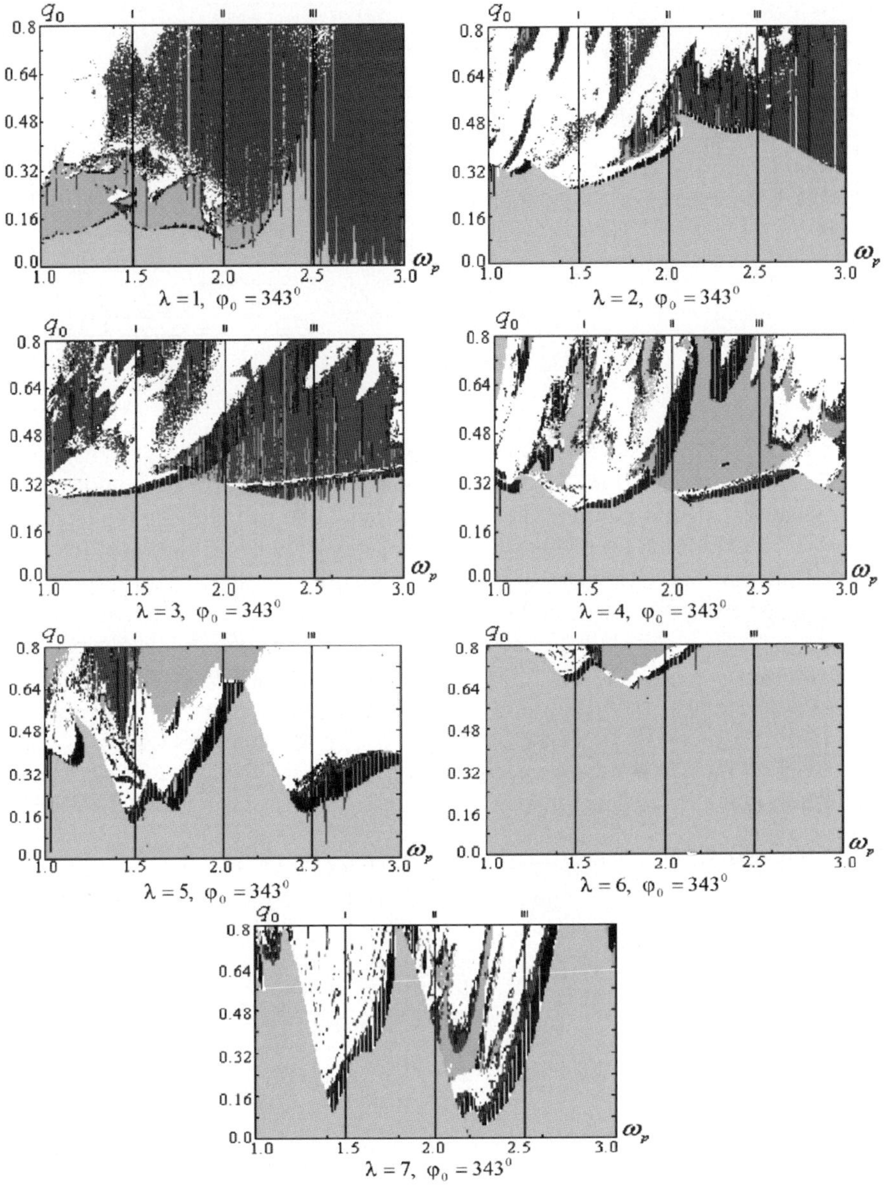

Fig. 10.13 Charts of the shell vibrations

$0 \leq x \leq 1$; $0 \leq y \leq 2\pi$ and forms of transversal cross sections $x = 0.5$; $0 \leq y \leq 2\pi$ are studied in pre- and post-critical states.

Owing to the investigation of relations $w_{\max}(q_0)$ for each λ, zones of stiff stability loss (Fig. 10.14) are detected, i.e., the criterion of stability loss for the studied types of shells is established. Observation of the scales of shell vibration character yields a transition of vibrations from harmonic to chaotic ones allowing us to de-

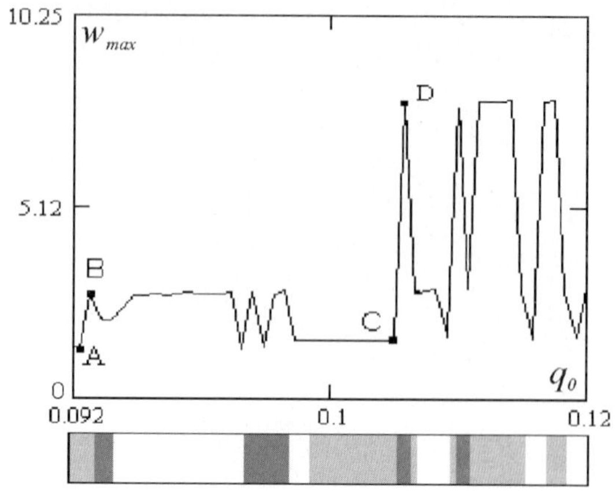

Fig. 10.14 Relation $w_{\max}(q_0)$ for $\lambda = 0.5$

fine the scenarios of such transitions. Let us analyze the occurrence of stability loss of the cylindrical shell depending on the parameter λ. For this purpose the relation $w_{\max}(q_0)$ is used for each fixed value of λ. In addition, the zones of stiff stability loss as well as fundamental characteristics in some controlling points that correspond to the shell state before and after stability loss are constructed. The characteristic relation $w_{\max}(q_0)$ for $\lambda = 0.5$ is illustrated in Fig. 10.14. Four control points are marked corresponding to local (A–B) and global points (C–D) of stability loss (a definition of the local and global stability follows). The analogous relations for other values of parameter λ are shown in Figs. 10.15–10.17. Fig. 10.15 shows how $w_{\max}$ increases with an increase of the excitation amplitude q_0. Note that the largest value of deflection in the investigated interval $q_0 \in [0;\ 0.8]$ is associated with the shell's chaotic motion.

A similar investigation is reported in Fig. 10.16 for $\lambda = 1$. However, in this case two-frequency vibrations (quasi-periodic motion) dominate within the considered interval $q_0 \in [0;\ 0.8]$.

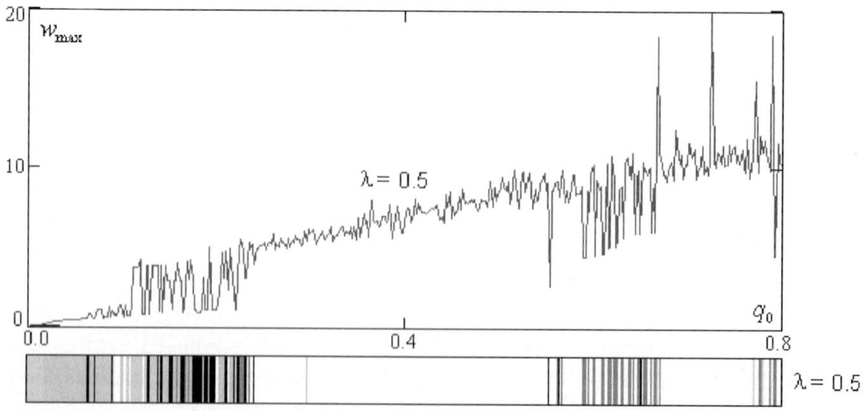

Fig. 10.15 Relations $w_{\max}(q_0)$ for $\lambda = 0.5$ and the scale of vibration character

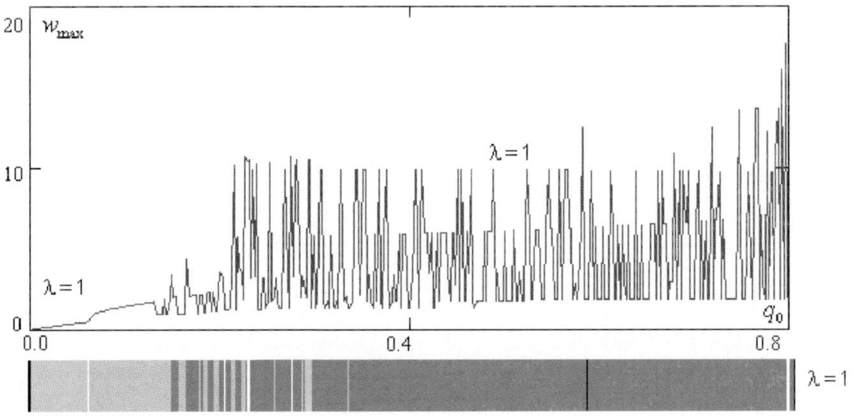

Fig. 10.16 Relations $w_{max}(q_0)$ for control parameter value $\lambda = 1$ and scale of vibrational character

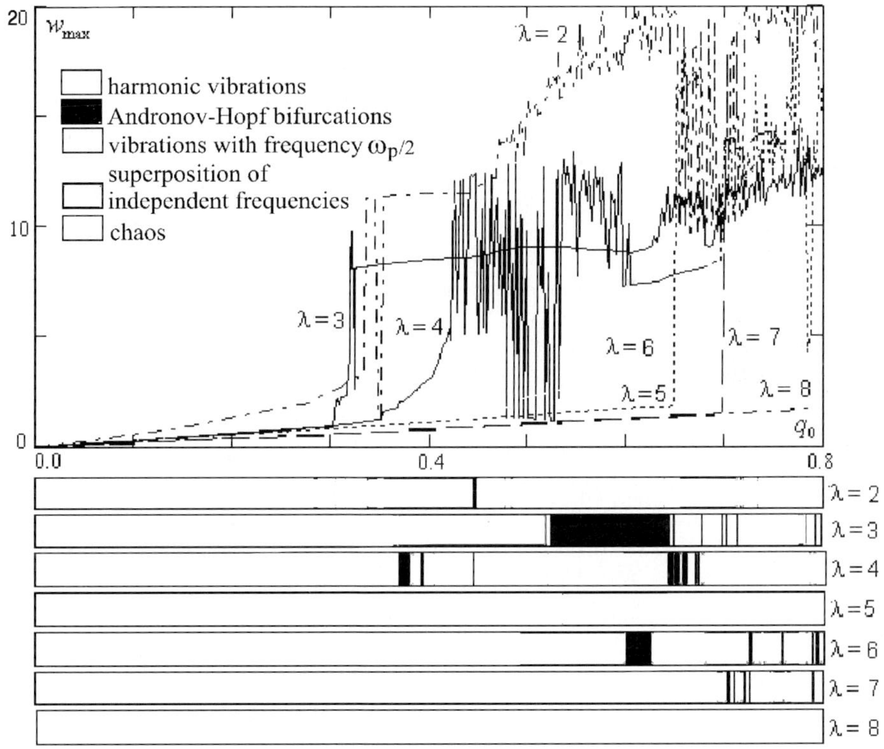

Fig. 10.17 Relations $w_{max}(q_0)$ for control values of λ and scale of vibrational character

Finally, in Fig. 10.17 a collection of similar investigations for $\lambda = 2$–8 is reported. It is seen that the largest shell deflection w_{max} is achieved for $\lambda = 2$, with the widest "length of chaos." Only harmonic and chaotic vibrations appear for $\lambda = 8$. To sum up, the reported results shown in Figs. 10.15–10.17 allow us to realize the required dynamics of the shell (harmonic, quasi-periodic and chaotic) by taking only two control parameters q_0 and λ.

Let us analyze the shell behavior during the transition through (A–B) and (C–D) for different λ. Determination of local and global stability loss for closed cylindrical shells subject to local harmonic load action is further carried out.

Local stability loss will be characterized by a qualitative change of vibration character from harmonic one (associated with Andronov-Hopf bifurcations) and accompanied by an increase of deflections by 5–6 shell thicknesses. The following definition of the local stability loss is applied further: Local stability loss is manifested by the change of vibration character. Harmonic vibrations are substituted by the vibrations with the frequency associated with the first Andronov-Hopf bifurcation ω_0, and vibrations with the frequency $\frac{\omega_0}{2}$ are substituted by quasi-periodic two-frequency vibrations. The number of waves along a circled coordinate is fixed and does not change in time.

Let us investigate longitudinal waves with respect to the linear shell dimensions $\lambda = \frac{L}{R}$ in chosen time instants corresponding to maximum, minimum, and averaged deflection values. In the point of time history maximum, the number of signal half-waves for all parameter values $\lambda = \frac{L}{R}$ remains constant (Fig. 10.18), whereas in the minimum and averaged point, the relation is non-monotonous. Minima of the graphics $n_{\min}(\lambda)$ correspond to maxima $n_{\text{average}}(\lambda)$ taken in the averaged sense. During transition through the point of local stability loss (Fig. 10.19), the relation $n(\lambda)$ coincides for all points in time and has one minimum (for $\lambda = 2$) and one maximum (for $\lambda = 3$). Therefore, it may be concluded that the number of half-waves associated with circled coordinate after the local stability loss is fixed and does not change in time.

By the global stability loss of a shell we will understand a fast increase of deflections (up to 15–20 shell thicknesses) corresponding to a small change of excitation force amplitude. The following definition for global stability loss is applied: Global stability loss is characterized by a change of vibration character. Our computations have shown that for an arbitrary value of λ in the frequency spectrum (after the global stability loss) first Andronov-Hopf bifurcation occurs, although the vibration character can be either periodic or chaotic. The number of half-waves is decreased and beginning with $\lambda = 4$ it becomes constant in time.

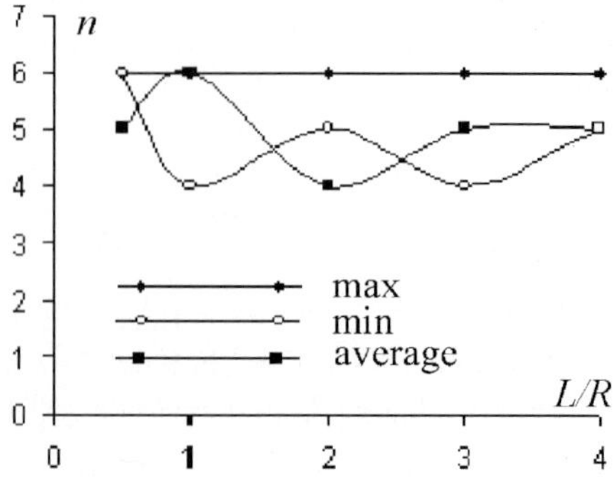

Fig. 10.18 Relation $n(\lambda)$ in the pre-critical state

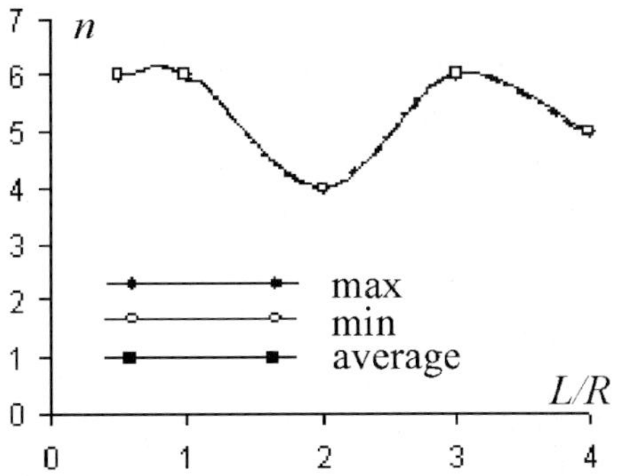

Fig. 10.19 Relation $n(\lambda)$ in the post-critical state

Although in the shell response both longitudinal (number of half waves) and circumferential (number of modal diameters) wave numbers are important, in this study we focused only on one of them. During transition through the point of global stability loss, a change of shell vibration character is obtained, and chaotic behavior associated with first Andronov-Hopf bifurcation appears. In the pre-critical state the system can exhibit either harmonic ($\lambda = 0.5,\ 2,\ 5,\ 7$) or chaotic ($\lambda = 3,\ 4,\ 6$) vibrations. The exceptional case is that of $\lambda = 1$, where the system exhibits quasi-periodic two-frequency vibrations with frequencies $\frac{\omega_0}{2}$ and ω_1.

Owing to an increase of the amplitude of external load the stiff stability loss occurs. Next, we will study the system behavior in the post-critical state.

For all values of the parameter λ the change of vibration character is observed. The following rule is detected for any λ in the spectrum (after a global stability loss): first, Andronov-Hopf bifurcation occurs, and then vibrations are either periodic or chaotic.

Also in the case of local stability loss one may trace the shell's behavior in space during transition through the critical point. Let us investigate a dependence of half-wave numbers with respect to linear shell dimensions in various time instants, corresponding to maximum, minimum and averaged deflection value (global stability loss).

Therefore, in the point of maximum, minimum, and averaged value of the signal, the number of half-waves is changing for all values of the parameter $\lambda = \frac{L}{R}$ (Fig. 10.20). Although for all relations $n(\lambda)$ there are local minima and maxima, but globally the number of half-waves along the circled coordinate decreases for all $n_{\text{average}}(\lambda)$, $n_{\min}(\lambda)$ and $n_{\max}(\lambda)$. The same holds for the post-critical state (Fig. 10.21), i.e., the number of half-waves along the circled coordinate decreases and beginning with $\lambda = 4$ it becomes constant for all $n_{\text{average}}(\lambda)$, $n_{\min}(\lambda)$ and $n_{\max}(\lambda)$.

To conclude, the number of half-waves along the circled coordinate does not depend on time instant for all $\lambda \geq 4$, i.e., it holds for all shells for which there is a lack of local stability loss.

Based on the obtained results, one may construct a graph of dependence of the critical load on the parameter λ, $q_0^+(\lambda)$ and $q_{0\,\text{loc}}^+(\lambda)$, where q_0^+ and $q_{0\,\text{loc}}^+$ denote the global and local critical loads, respectively (Fig. 10.22). The relation is non-

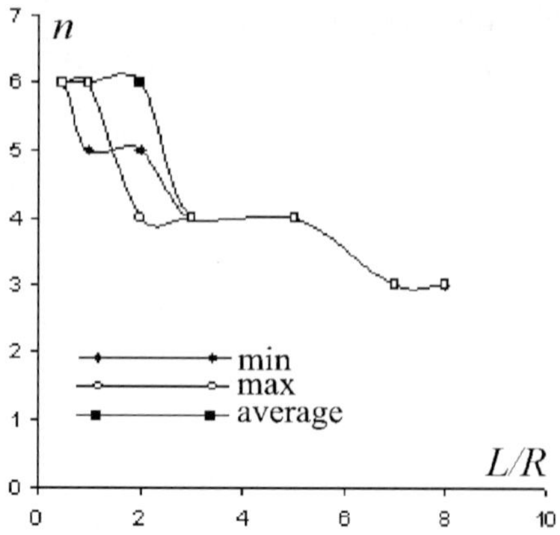

Fig. 10.20 Relation $n(\lambda)$ in the pre-critical state

monotonous, and there are two local minima ($\lambda = 3, 6$), whereas for the last values of λ this relation increases monotonically.

Therefore, an increase of a relative length of the shell causes an increase of critical load values. The analogous rule holds also for the local critical loads.

10.5.5 Scenarios of Shell Vibration Transition into Chaos Versus λ

In nonlinear dynamics there are a few scenarios of the transition of mechanical systems from harmonic to chaotic states (see, for instance, Landau-Hopf scenario [179], Ruelle-Takens-Newhouse scenario [262], Feigenbaum scenario [92], and Pomeau-Manneville scenario [197]).

In order to analyze the scenarios of transition into chaotic vibrations of the shell, the relations $w_{\max}(q_0)$ for different λ together with vibrational character scales are analyzed (they are shown in Figs. 10.15–10.17).

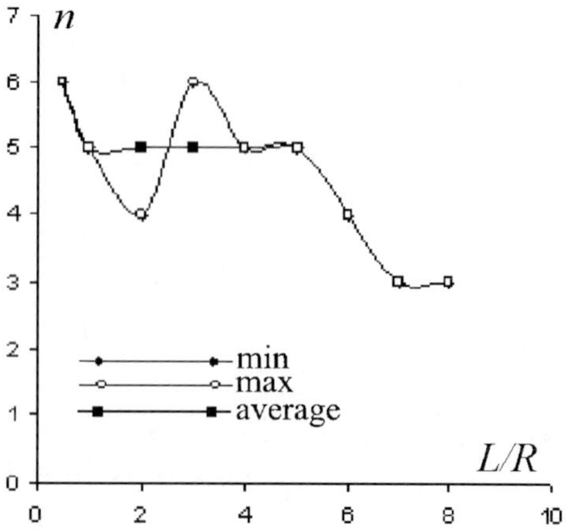

Fig. 10.21 Relation $n(\lambda)$ in the post-critical state

Fig. 10.22 Relation $q^+_{0\,loc}(\lambda)$ and $q^+_0(\lambda)$

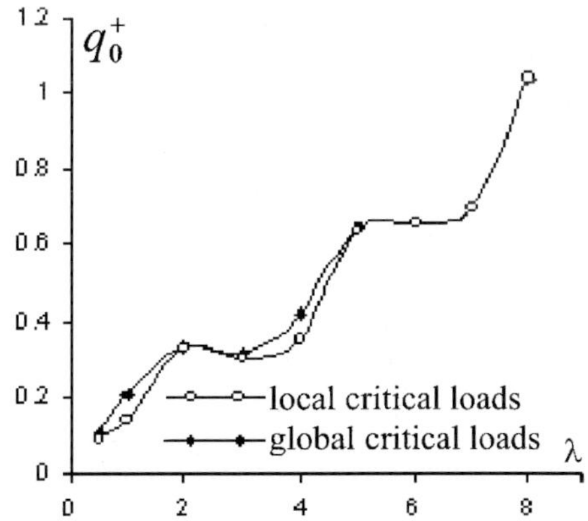

For all parameters λ one may find common properties of the dynamic shell behavior. A stiff stability loss corresponds to a change of vibration character, shivering of graphs $w_{\max}(q_0)$ corresponds to the chaotic zones in vibration scales. Finally, an increase of q_0 from 0 corresponds to a large zone of harmonic vibrations, i.e., on the graph of $w_{\max}(q_0)$ one observes a smooth increase of deflections.

In order to define these λ values for which the cylindrical shell exhibits slightly chaotic vibrations, the dependence of chaotic zone length ℓ on the parameter λ, i.e., the function $\ell(\lambda)$ for other conditions being fixed is analyzed further. The mentioned dependence is studied together with critical loads for each λ. The length of chaotic zones is computed using vibration character scales for each fixed value of λ.

The mentioned graphs are shown in Figs. 10.22 and 10.23.

The relation $\ell(\lambda)$ has non-monotonous vibrational character. While increasing λ, two local minima and two local maxima are found. After $\lambda \geq 4$ a monotonously decreasing relation is observed (Fig. 10.23), which corresponds to monotonously increasing part in the graph $q^+_0(\lambda)$. Therefore, one may conclude that owing to the increase of critical load, chaotic zones are decreased, i.e., low critical load is associated with a larger surface of chaotic vibrations. Furthermore, for lengthy

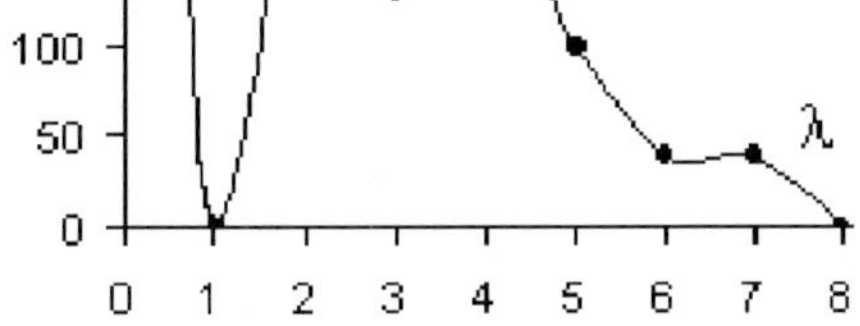

Fig. 10.23 Relation $\ell(\lambda)$, where ℓ denotes the length of chaotic zone for each λ

shells ($\lambda \geq 4$) the relation ($q_0^+(\lambda)$) monotonously increases, whereas it has a non-monotonous character for short and average shells. Next, we will study how a relative shell length influences the occurrence or lack of bifurcations in the frequency spectrum, the occurrence of independent frequencies in vibrations with the frequencies $\frac{\omega_0}{2}$, $\frac{\omega_0}{3}$, and so on. For this purpose the corresponding relations for various values of λ using vibration character scales are studied (see Figs. 10.24–10.26). Analyzing the obtained relations one may conclude that shells averaged in length ($1 \geq \lambda \geq 3$) behave similarly, whereas in the behavior of long shells some differences are observed.

The following general conclusions can be formulated. The stiff stability loss (i.e., both local and global) is characterized by the occurrence of the first Andronov-Hopf bifurcation and transition into vibrations with the frequency $\frac{\omega_0}{2}$, and they can be either periodic or chaotic. In other words, the mechanism of transition through the point of stability loss is the same as in both local and global cases. Exceptional cases include $\lambda = 1$ (owing to stiff stability loss the vibrations are transmitted from one frequency harmonic to two-frequency quasi-periodic ones) and $\lambda = 6$ (vibrations are transmitted from two-frequency to chaotic ones being associated with the frequency of excitations).

It is worth noting that analyzing the forms of transversal cross sections of the cylindrical shell (beginning from $\lambda = 4$), the number of half-waves along circled coordinate becomes constant and does not undergo changes in time. Furthermore, for $\lambda \geq 4$ local stability loss does not appear. When investigating scenarios of the transition of shell vibrations from harmonic to chaotic ones, it may be concluded that $\lambda = 4$ is responsible for the scenario type definition (see Figs. 10.24–10.26).

To conclude, the scenarios of transition from harmonic to chaotic vibrations essentially depend on the relative shell length and they are different for short ($\lambda < 4$) and long ($\lambda \geq 4$) shells.

Table 10.5 gives charts of vibration character $\{q_0, \omega_p\}$ depending on control parameters for different values of shell length λ. Analysis of the mentioned charts evidently supports our earlier conclusions that vibrational type essentially depends on the length of the cylindrical shell. Therefore, to control vibrations of our mechanical system it is worth changing the linear dimensions of our shell, keeping other parameters fixed.

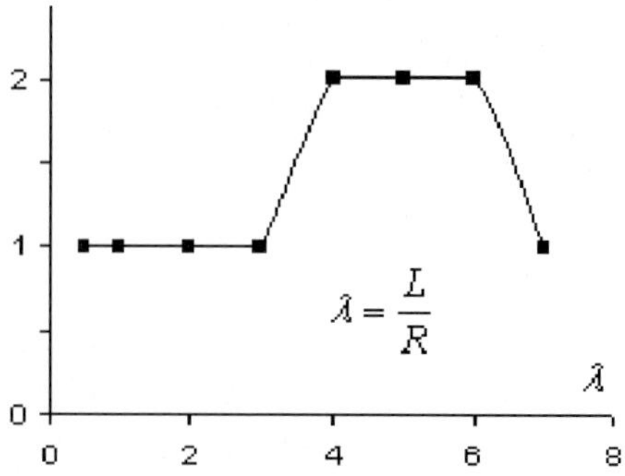

Fig. 10.24 Occurrence of vibrations at frequencies $\omega_0/2$: (1, yes; 2, no)

Table 10.5 Vibration types in the plane $\{q_0, \omega_p\}$ for different λ

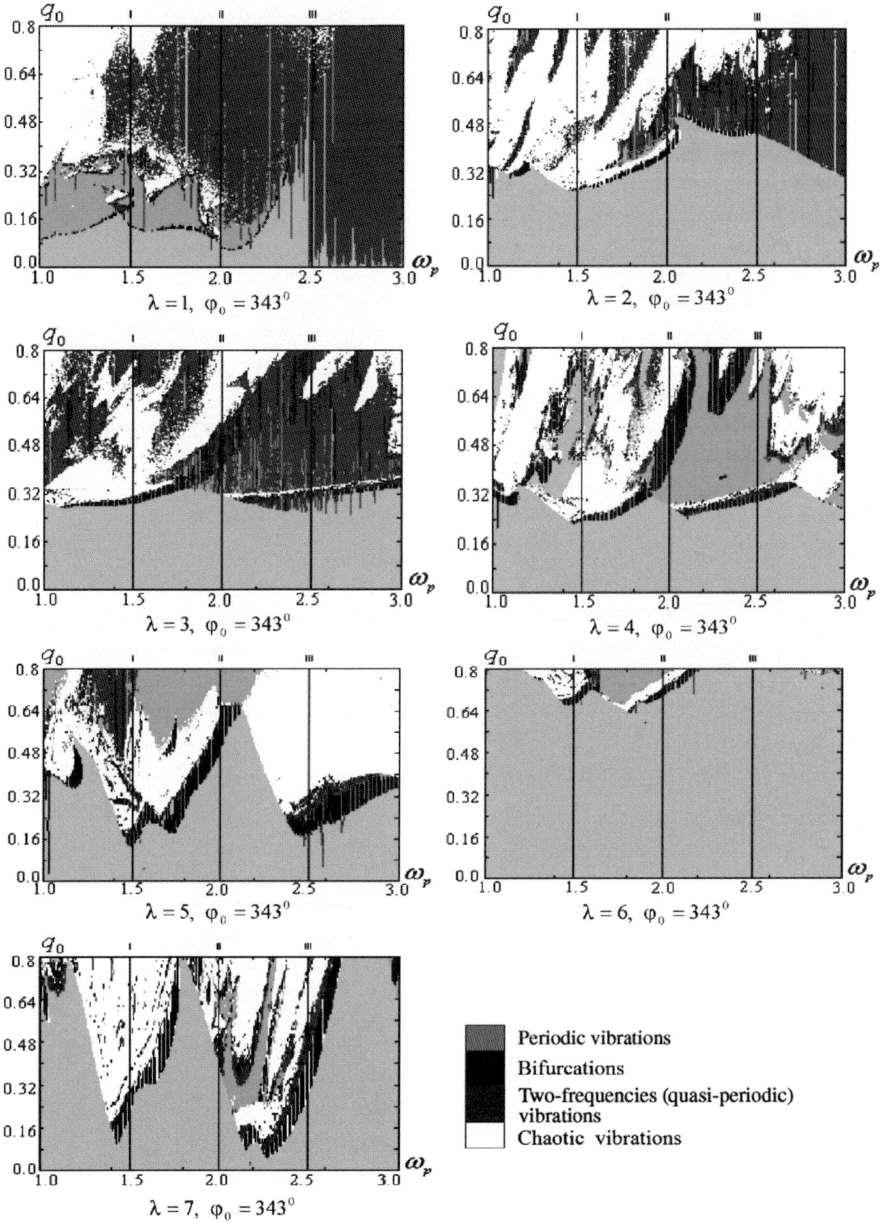

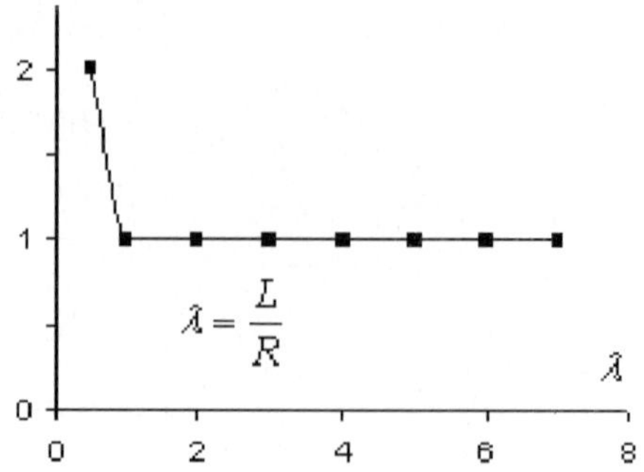

Fig. 10.25 Occurrence of independent frequencies (1, yes; 2, no)

10.5.6 Feigenbaum Scenario

The chart of vibration character versus control parameters $\{q_0, \omega_p\}$, constructed for a cylindrical shell of circled cross-section for $\lambda = 3$ and subject to non-symmetric external load applied to a zone of width $\varphi_0 = 343°$ (see Table 10.5), indicates the existence of a few zones where a transition into a chaotic state has been realized using the Feigenbaum scenario [92]. Namely, both the so-called Feigenbaum sequence and Feigenbaum constant $d = 4.6625$ have been numerically detected. The obtained value differs from its theoretical prediction by 0.14%. In this work the process of period doubling bifurcations is obtained owing to solution of the system of partial differential equations (PDEs) governing dynamics of the closed cylindrical shell using the Kirchhoff-Love kinematic model, and not the one associated with considerations of the logistic curves behavior.

In order to illustrate the described process of transition into chaos the following characteristics are monitored (see Table 10.6) for the Feigenbaum scenario [92] associated with 2–6 bifurcations: signal, phase portrait, power spectrum, Poincaré map, and modal portrait (in order to properly analyze space vibrations).

If we consider the deflection of a plate $w(x,y)$ then we have physical interpretation to the fourth derivative. The first derivatives $w_x(x,y)$ and $w_y(x,y)$ correspond to tangents of angles between the deflection slopes and the corresponding axes at the points where the derivatives are calculated. The second order derivatives

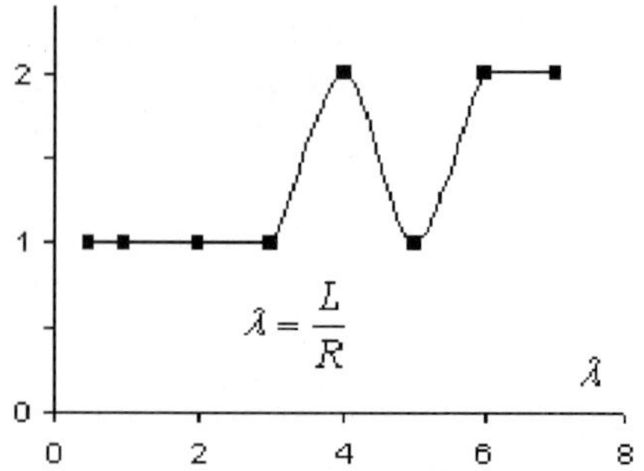

Fig. 10.26 Occurrence of Hopf bifurcations (1, yes; 2, no)

Table 10.6 Various indicators of the analysed shell dynamics

k	signal and phase portrait	Frequency spectrum	Poincaré section	Modal portrait
2	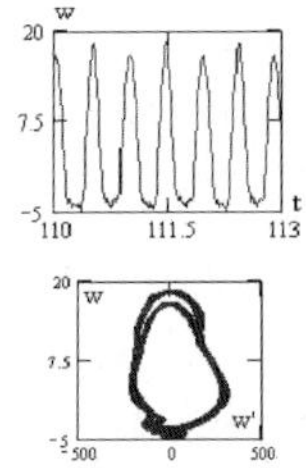	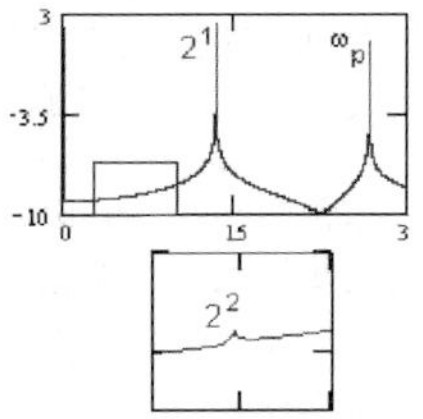	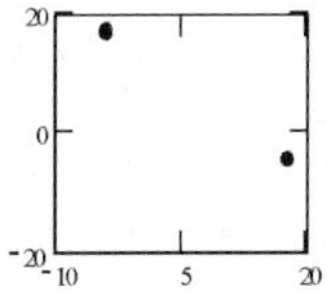	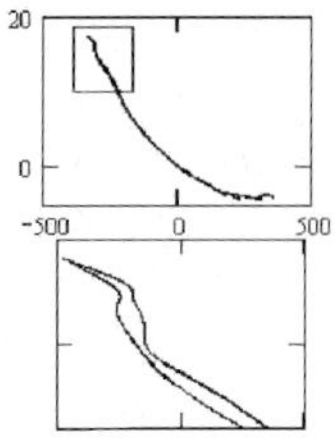
3	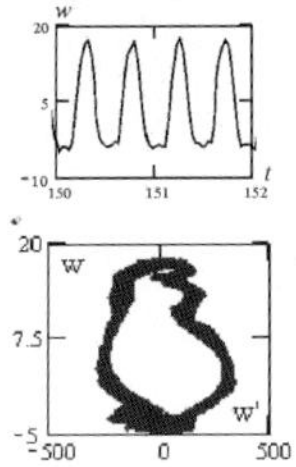	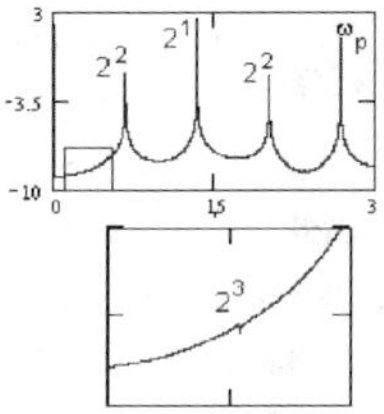	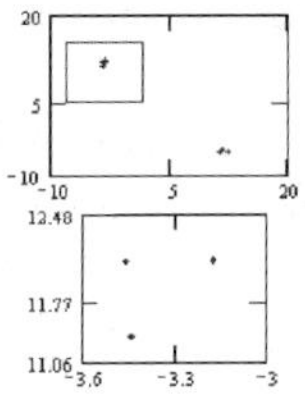	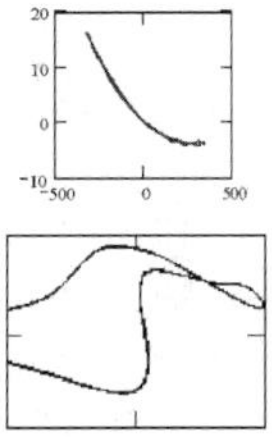
4	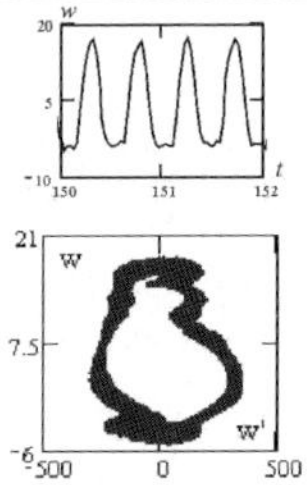	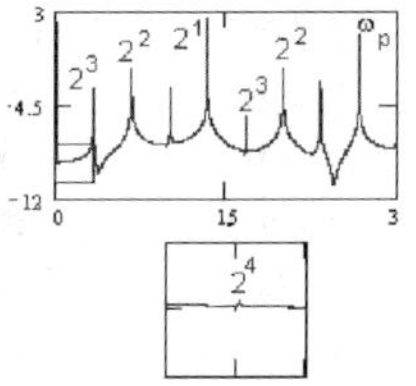	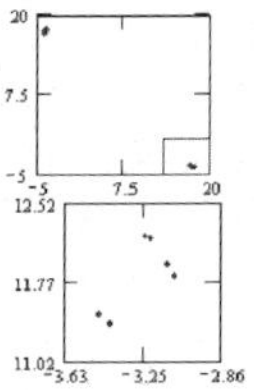	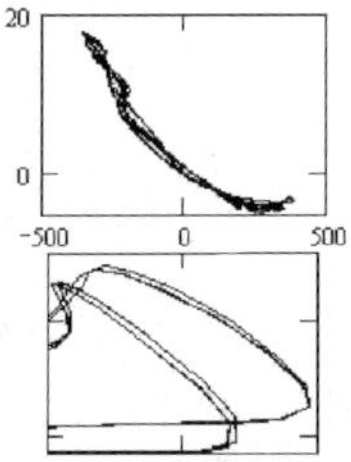
5	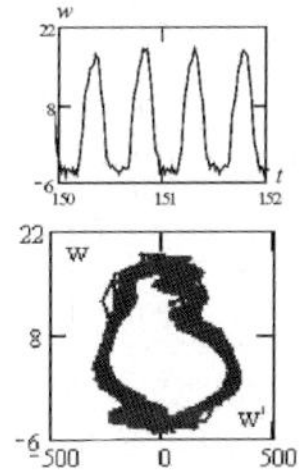	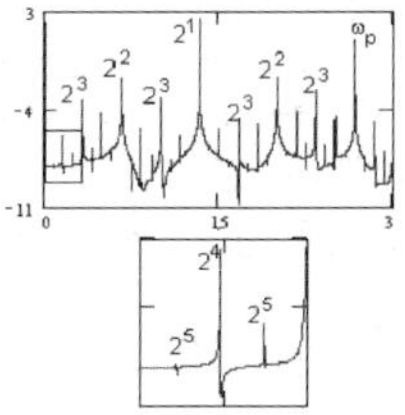	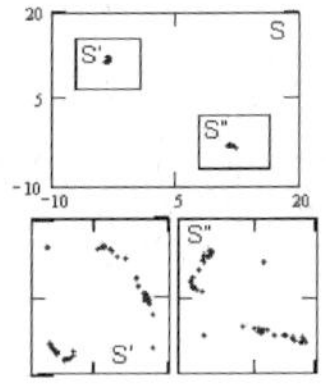	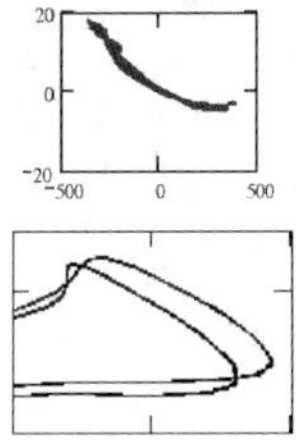
6	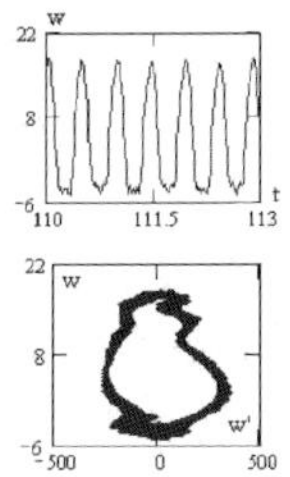	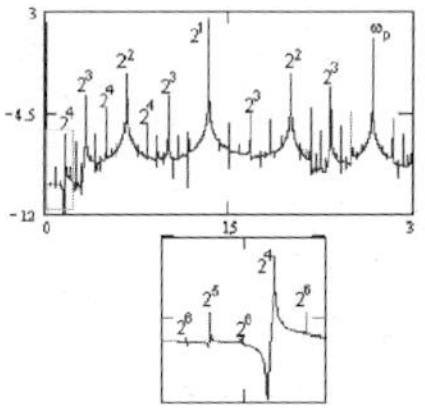	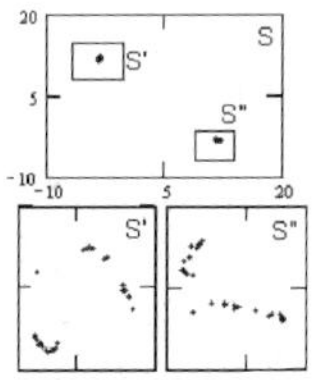	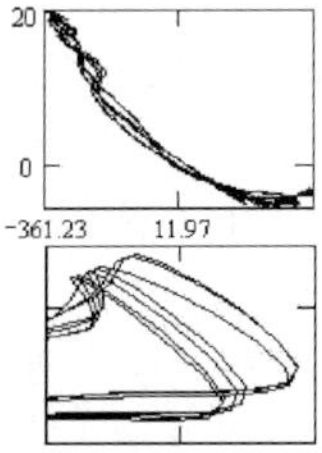

$w_{xx}(x,y)$, $w_{yy}(x,y)$ and $w_{xy}(x,y)$ correspond to the curvatures of the deflection functions at the points where the derivatives are calculated. If they are multiplied by the corresponding constants and summed up, then the bending and torsion moments are obtained. Therefore, on analyzing those characteristics in time one can conclude about a spatial change of a plate surface. We refer further to them as the characteristics of "modal portraits" while solving problems of plates using the method of finite differences. In a phase portrait the dependence between a deflection and its velocity is reported for each time moment, whereas in a modal portrait we have a relation between deflection and a tangent of its slope to the corresponding coordinate axis in a plane. If we consider the phase portraits in space, then for each time moment we get a dependence between deflection, velocity and acceleration. If we consider the modal portraits in space then each plate point gives information about deflection, tangents of its slope and curvatures. As a result we can get all required information about the character of deflection of the plate surface (see also [36] for application of modal portraits).

Note that in Table 10.6 k defines the number of bifurcations.

The detected scenario follows:

1. The second bifurcation yields period doubling and occurrence of two points in the Poincaré map. A modal portrait also exhibits two orbits.
2. The third bifurcation yields tripling of the phase portrait trajectories, and each of the previous two points of the Poincaré section bifurcates into three points. In the modal portrait interlacing of trajectories takes place.
3. The fourth bifurcation yields matching of the trajectories in the phase portrait. In the Poincaré section each of the three points splits into two points. In the modal portrait an arc is observed, and with an increase of the scale two loops are formed, each consisting of two curves.
4. The fifth bifurcation gives rise to a fractal structure in the Poincaré section, i.e., two arcs appear. In the modal portrait trajectories in each of two fundamental orbits are matched.
5. After six bifurcations have occurred, the fractal structure collapses into a few analogous ones. The modal portrait consists of six orbits.

Therefore, the behavior of our mechanical system can be traced with respect to both power spectrum and modal portraits, which means that in order to study the spatial chaos one has to apply an analysis in the modal plane.

As a result of the study, also the value of Feigenbaum constant is obtained. Note that the obtained value differs from its theoretical prediction only by 0.14% (Table 10.7).

Table 10.7 The Feigenbaum scenario

n	2.	3.	4.	5.	6.
$q_{0,n}$	0.45605	0.5663	0.58527	0.589	0.5898
d_n	-	5.811808...	5.085791...	4.6625...	-

10.5.7 The Ruelle-Takens-Feigenbaum Scenarios

As already mentioned, in various zones of the chart of control parameters $\{q_0, \omega_p\}$ different scenarios of the transition of our mechanical system into chaotic state are found. For example, the Ruelle-Takens scenario for the cylindrical shell for $\lambda = \frac{L}{R} = 2$ has been found. It is worth noting, however, that the detected scenario is not in full agreement with the classical one proposed by Ruelle and Takens. This is rather the so-called modified Ruelle-Takens scenario. Namely, as in the classical Ruelle-Takens scenario, one observes here initially the independent frequency ω_1, and then formation of two-dimensional orbits with two independent frequencies. An increase of q_0 and fixing of ω_p results in a sequence of soft Hopf bifurcations, i.e., the Feigenbaum process is observed and the system dynamics is transmitted into chaos with the excitation frequency ω_p. Note that in this case the novel type of transition into chaos is illustrated owing to the combined scenarios of Ruelle-Takens [262] and Feigenbaum [92], which are further referred to as the Ruelle-Takens-Feigenbaum scenario (Table 10.8).

The occurrence of each of the successive Hopf bifurcations in a two-dimensional orbit induces essential changes only in the Poincaré section. Note that both modal and phase portraits remain almost the same when a new Hopf bifurcation appears. In the Poincaré section a successive doubling of points is observed.

To conclude this section, the series of Hopf bifurcations on the two dimensional orbit has been detected and the Feigenbaum constant has been found ($d = 4.6685$), which differs from its theoretical value by 0.01%.

Table 10.8 Dynamics indicators

q_{0k}	Power spectrum $S(\omega)$	Signal $w(t;0.5;0.0)$ and phase portrait $w(w')$	Poincaré section (w_t, w_{t+T})	Modal portrait $w(w_{yy})$
1	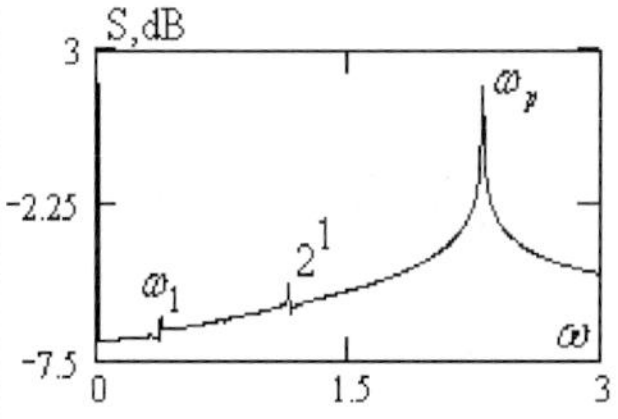	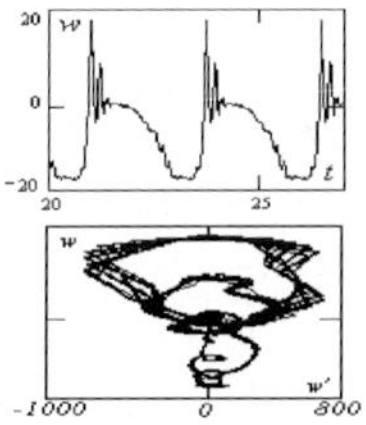	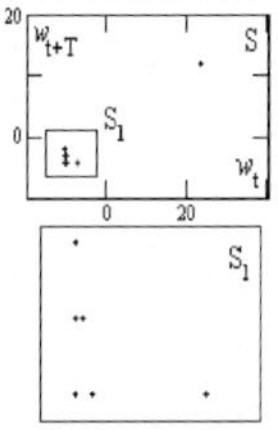	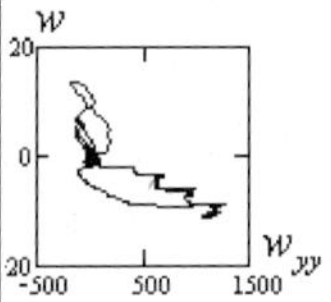
4	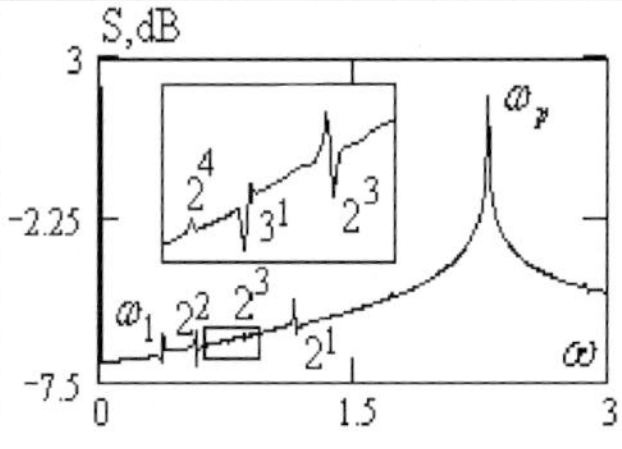	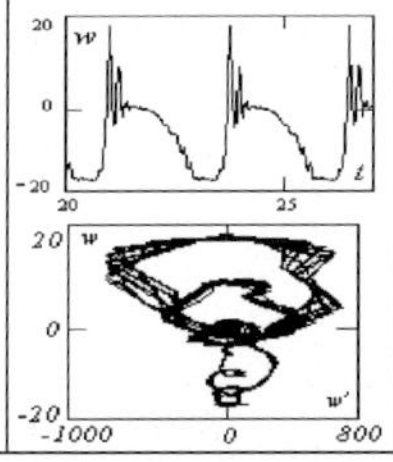	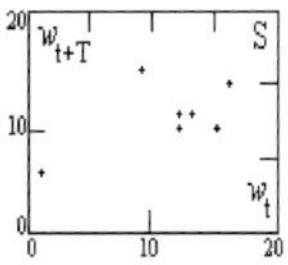	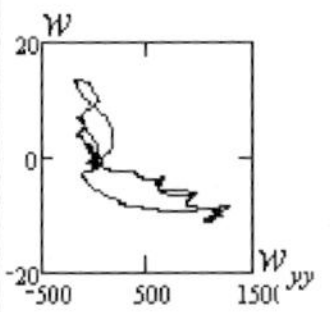

10.6 Conclusions

We have presented both a novel approach to study bifurcation and chaos exhibited by vibrated flexible cylindrical shells as well as some new results associated with stability, bifurcation, and chaos of the analyzed shells.

Particular attention has been paid to verification of the result reliability while using a higher order Bubnov-Galerkin approach and Fourier representation (Sects. 10.4 and 10.5).

Dynamic stability loss of cylindrical shells is widely described with emphasis on associated and not appropriately solved so far problems related to this important question for both pure and applied scientists.

In the case of investigation of closed cylindrical shells subjected to transversal sinusoidal loading it has been found that the character of the investigated shell vibrations depends essentially on loading angle φ_0. For small values of φ_0 the total surface of chaos is high, whereas an increase of loading of the cylindrical shell yields a decrease of the space of chaos which is shifted into low and averaged frequencies. For a large surface of external pressure the chaotic areas are concentrated on the whole chart, but the largest part is located in the vicinity of $\omega_p < \omega_0$. The corresponding total surface of the chaotic zone is essentially high. Also the areas associated with Hopf bifurcations located in the vicinity of $\omega_p > \omega_0$ are relatively high.

Both local and global stability losses of the investigated shells have been illustrated using numerous figures and tables, and discussed. It has been shown, among others, that stiff stability loss is associated with occurrence of the first Andronov-Hopf bifurcation and transition into vibrations with the frequency $\frac{\omega_0}{2}$, which can be either harmonic or chaotic. New scenarios of the transition of our shell vibrations from harmonic to chaotic ones have been detected and illustrated.

Chapter 11
Controlling Time-Spatial Chaos of Cylindrical Shells

In this chapter an analysis of controlling the chaos occurring in closed cylindrical shells is presented. The process of controlling chaos is understood as the transformation of chaotic dynamics into regular, or the other way around, but of different characteristics with the use of small external periodic input functions and by the influence of a transverse load applied in antiphase.

11.1 Introduction

Chaotic vibrations are connected with the occurrence of disordered movements in highly nonlinear physical systems. Such movements were already observed in mechanics of fluids, but have also been recently found in the mechanics of plates and shells [21, 25, 31, 46, 170]. The problems of the existence and uniqueness of the solutions for Timoshenko shells were analyzed.

A mathematical model of cylindrical shells was constructed on the basis of Kirchhoff-Love kinematic model with a nonlinear relation between deflection and dislocations taken into account. The problem is reduced to the analysis of partial differential equations in relation to the functions of stress and strain. In order to reduce the continuous system to a discrete one, the Bubnov-Galerkin method with higher approximations was used, which allows one to consider a cylindrical shell as a mechanical system with an infinite number of degrees of freedom. The obtained system of ordinary differential equations is solved with the use of the Runge-Kutta method of the order of four, and the system of algebraic equations is solved with the use of one of many numerical procedures available.

Controlling chaos was the subject of works that mostly referred to simple spatial systems modeled by the chains of mappings. The first remarks about chaos control appeared in studies Jackson of [126, 127] and Otto et al. [236]. In the review work [271] are found references to even earlier sources. The problems of controlling chaos were studied in hydrodynamics [272], chemistry [244], biology, and medicine [266]. This chapter presents ways of controlling chaos in such complex objects as cylindrical shells under a nonuniform external load.

J. Awrejcewicz, V.A. Krysko, *Chaos in Structural Mechanics*,
DOI: 10.1007/978-3-540-77676-5_12, © Springer-Verlag Berlin Heidelberg 2008

11.2 Mathematical Model

Within a classical nonlinear theory of shells, we shall consider a closed cylindrical shell with a circular intersection and finite length and constant rigidity and density under nonuniform and variable load.

We shall introduce the coordinate system in the following way: axis x is directed along the length of a surface, axis y is directed along the circular coordinate, and axis z is directed along the perpendicular of its central surface (see Fig. 10.1).

A cylindrical shell as a three-dimensional object Ω in a given coordinate system is characterized in the following way: $\Omega = \{x, y, z | (x_y) \in [0; l] \times [0; 2\pi], \; -h \le z \le h\}$. The following form of the equations of the theory of shells are the basis for further analysis [315]:

$$\frac{1}{12(1-\mu^2)}(\nabla^4 w) - k_y \frac{\partial^2 F}{\partial x^2} - L(w,F) - \frac{\partial^2 w}{\partial t^2} - \varepsilon \frac{\partial w}{\partial t} + k_y^2 q(x,y,t) - p_x(x,y,t)\frac{\partial^2 w}{\partial x^2} = 0,$$

$$\nabla^4 F + k_y \frac{\partial^2 w}{\partial x^2} + \frac{1}{2} L(w,w) = 0 \,, \tag{11.1}$$

where $L(w,\; F)$ is a known nonlinear operator.

The system (11.1) has been reduced to a non-dimensional form with the use of known nondimensional parameters. In Eq. (11.1) μ denotes Poisson coefficient, ε is the damping coefficient, $\lambda = L/R$ where L and $R = R_y$ denote the length and radius of a circular cylindrical shell, $k_y = 1/R_y$ is the curvature of a shell in relation to coordinate y, and $q(x,y,t)$ denotes the external load.

The boundary conditions have the following form:

$$w = \frac{\partial^2 w}{\partial x^2} = 0; \quad F = \frac{\partial^2 F}{\partial x^2} = 0 \text{ for } x = 0, 1,$$

$$w = \frac{\partial^2 w}{\partial y^2} = 0; \quad F = \frac{\partial^2 F}{\partial y^2} = 0 \text{ for } y = 0, 2\pi, \tag{11.2}$$

and the initial values follow:

$$w(x,y)|_{t=0} = 0, \quad \text{for } w|_{t=0} = 0. \tag{11.3}$$

Let us consider our dissipative system under an external transverse load applied in the sector of surface characterized by angle $0 \le \varphi \le \varphi_0$, $0 \le x \le 1$ and harmonically changing $q(t) = q_0 \sin(\omega_p t)$, where q_0 and ω_p are, respectively, the amplitude and frequency of the exciting force. Moreover, the following was assumed: $\mu = 0.3$, $\varepsilon = 9$, $\lambda = L/R = 2$.

11.3 Bubnov-Galerkin Method and Fourier Transformation

Let us analyze the dynamics of a closed cylindrical shell freely supported along the curved side with uniform boundary conditions (11.2) and initial conditions (11.3). We are looking for the solutions of Eq. (11.1) in the form of a product of two functions in relation to spatial coordinates, each of which depends on one argument only, and the Eq. (11.2) are satisfied. Test functions have the following form here:

$$\varphi_{ij}(x,y) = \sin(i\pi x)\cos(jy). \tag{11.4}$$

The solution is based on such test functions that are energetically normalized, i.e., such that

$$(\nabla^4(\varphi_{ij}), \varphi_{nm}) = \begin{cases} 0, & \text{for} \quad i,j \neq n,m, \\ 1, & \text{for} \quad i,j = n,m, \end{cases} \tag{11.5}$$

and then

$$\begin{aligned} w &= \sum_{i=1}^{M_x}\sum_{j=0}^{M_y} A_{ij}(t)\sin(i\pi x)\cos(jy), \\ F &= \sum_{i=1}^{M_x}\sum_{j=0}^{M_y} B_{ij}(t)\sin(i\pi x)\cos(jy). \end{aligned} \tag{11.6}$$

After the application of Bubnov-Galerkin method with higher approximations, we obtain a system of linear algebraic equations in relation to coefficients (that can be solved by inversion of the matrix) and the system of differential equations of the order of two in relation to coefficients, which is later reduced to an ordinary form and solved with the use of the Runge-Kutta method of the order of four:

$$K_{ij}B_{ij} = F_1(A_{ij}), \tag{11.7}$$

$$\begin{cases} \frac{dA_{ij}}{dt} = X_{ij}, \\ \frac{dX_{ij}}{dt} + \varepsilon X_{ij} = F_2(A_{ij}, B_{ij}, t), \quad i = \overline{1, M_x}, \quad j = \overline{0, M_y}, \end{cases} \tag{11.8}$$

where K_{ij} in (11.7) is the matrix of coefficients of the system of algebraic equations in relation to unknown B_{ij}, $F_1(A_{ij})$ is the column of elements depending on parameters A_{ij}, (11.8) is an ordinary system of differential equations of the order of 1 in relation to unknown A_{ij} and X_{ij}. Step in time is selected from the condition of maintaining the stability of solution ($\Delta t = 1.9531 \cdot 10^{-3}$).

The convergence of the Bubnov-Galerkin method was tested earlier in relation to the number of elements of series (11.6).

As the load is applied along the whole length of a cylindrical shell, the number of elements of a series in relation to coordinate x is not very significant and in (11.6) only one element ($M_x = 1$) was assumed. During numerical tests it occurred that the optimal number of elements in series (11.6) along the circular coordinate is $M_y = 15$ (more details related to the convergence of this method are given in [169]).

It is important to note that in the case of a stationary problem, the numerical results obtained in this chapter are fully convergent with the results obtained in [11], which confirms the feasibility of the Bubnov-Galerkin method with higher approximations [169].

11.4 Control of Chaos

The process of control of chaos is understood as the transformation of chaotic dynamics into a regular motion or other chaotic manifold with the use of small longitudinal periodic input functions with the form of $p_x(x,t) = p_0(x)\sin(\omega_p,\ t)$, used on purpose, as well as by the influence of a transverse input function in antiphase.

In our case, steering of chaotic vibrations in a cylindrical shell involved steering a set $\{q_0,\ \omega_p\}$ with the application of a load distributed along the surface of a shell, in accordance with the equation $q(x,y,t) = q_0(x,y)\sin(\omega_p t)$, where $q_0(x,y)$ is an exciting force depending on coordinates (x,y) and ω_p is the frequency of this exciting force.

The analysis is conducted by monitoring phase portraits of spectral power frequency, the spectrum of Lyapunov exponents, and Poincaré maps. This allows for finding and analyzing multi-frequency and chaotic structures of vibrations, and the analysis of the mechanism of transition between different vibration regimes and steering them.

Let us analyze the case of input function with a transverse harmonic force in antiphase, i.e., $q(x,y,t) = q_0(x,y)\sin(\omega_p t + 2\pi)$, where $q_0(x,y) = const$ (Fig. 10.1). The classification of vibrations will be made on the basis of the relation between $w_{\max}(q_0)$ and $w_{\max,1}(q_0)$ for two cases of load distribution:

(i) Load influences one area characterized by angle $\varphi_0 = 180°$;
(ii) Load is applied on two areas, each of which has the width of angle $\varphi_{0,1} = \varphi_{0,2} = 90°$.

Above, $w_{\max}(q_0)$ denotes maximum deflections in relation to amplitude q_0 for the case of $q(x,y,t) = q_0(x,y)\sin(\omega_p t)$, and $w_{\max,1}(q_0)$ denote maximum deflections in relation to amplitude q_0 for the case of $q(x,y,t) = q_0(x,y)\sin(\omega_p t + 2\pi)$.

Let us note that the character of the relation between $w_{\max}(q_0)$ and $w_{\max,1}(q_0)$ in both studied cases depends significantly on antiphase, but for the small amplitude of exciting force ($q_0 \leq 0.4$ for $\varphi_{0,1} = \varphi_{0,2} = 90°$ and $q_0 \leq 0.52$ for $\varphi_0 = 180°$) maximum deflections overlap.

To sum up, changing the character of a load enables steering the vibrations of shells, and it offers the possibility of a significant increase of the area of the occurrence of harmonic vibrations.

By applying the influence of a transverse load in antiphase, the possibility occurred of increasing a critical load for which rigid stability loss occurs, and the value of over-critical deflection was significantly reduced (Table 11.1; $\varphi_0 = 180°$).

Table 11.1 Shell load and load values

External load	w_{subcr}	q^0_{cr}	w_{postcr}
$q(x,y,t) = q_0(x,y)\sin(\omega_p t)$	1.2671	0.384	12.963
$q(x,y,t) = q_0(x,y)\sin(\omega_p t + 2\pi)$	1.2887	0.415	3.4796

Let us analyze the scale of vibration character distribution for $w_{\max}(q_0)$ and $w_{\max,1}(q_0)$ (Fig. 11.1). The identification of the type of vibration of a cylindrical shell in the process of constructing the aforementioned scales $\{q_0,\ w_0\}$ for every time course $w(t)$ was conducted with the use of the analysis of power spectrum $S(\omega)$ and Lyapunov exponents. The notation assumed is given in Fig. 11.1. Chaotic vibrations will be understood as the occurrence of a series of disordered movements in deterministic systems.

It can be observed that the application of a load in antiphase significantly influences the type of vibrations of a cylindrical shell, which corresponds with large values of the amplitude of external input function q_0. Research has shown that the application of a load in antiphase can be treated as an effective way of steering the chaotic motion.

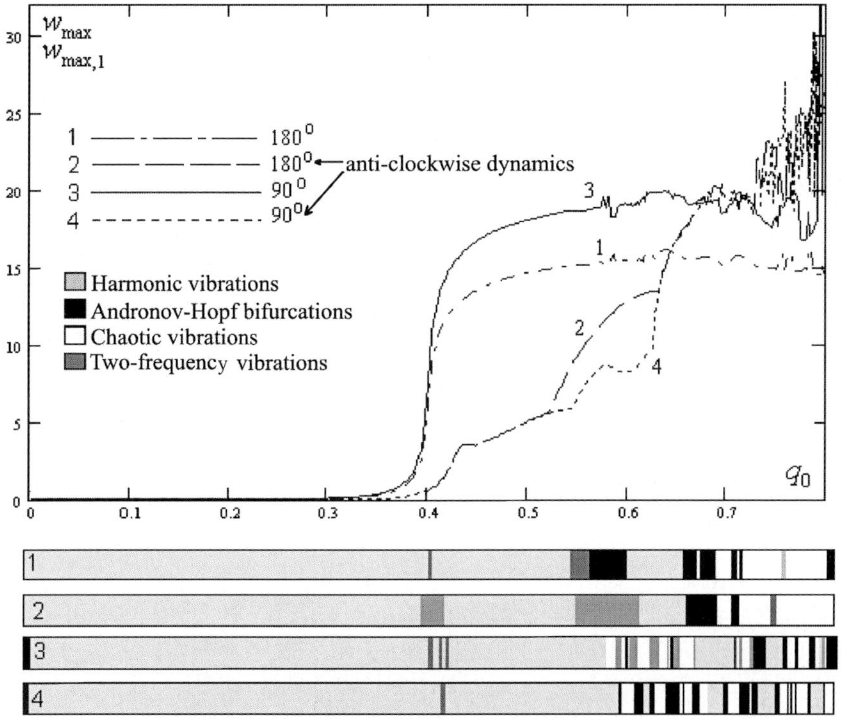

Fig. 11.1 Relation between $w_{\max}(q_0)$ and $w_{\max,1}(q_0)$ and the scales of character of vibration

Our next step will be the analysis of simultaneous harmonic input function $q(x,y,t) = q_0 \sin(\omega_p t)$ and longitudinal harmonic input function $p_x(x,t) = p_0(x)\sin(\omega_p t)$, where $\omega_p = \omega_0$ is the frequency of linear vibrations of a cylindrical shell.

Fig. 11.2 shows the relations $w_{\max}(q_0)$ for the set values of p_0.

The analysis of the scales of character of vibrations proves that the application of a longitudinal load leads to the change of the character of vibrations of a mechanical system. Both the transition from chaotic to harmonic (or to the creation of Andronov-Hopf bifurcations) and from harmonic to chaotic vibration has been observed.

In this way, for certain load values, we can either lead the system out of chaos, or cause the occurrence of chaos.

Let us now establish the amplitude of transverse load on the value of $q_0 = 0.71$ (intermittent on the plane of the character of vibrations, Fig. 11.4a). For lack of longitudinal vibrations, the mechanical system is (for such a value of q_0) in the regime of chaotic vibrations. Here, we will in addition force our cylindrical shell longitudinally.

Fig. 11.3 presents relation $w_{\max}(p_0)$. The analysis of both this relation and of the scale of vibration character leads us to the conclusion that the analyzed mechanical system represented by point $\{q_0,\ w_0\} = \{0.71,\ \omega_p\}$ has left the state of chaotic vibration and transformed to harmonic vibration with mild Andronov-Hopf bifurcations.

Let us now discuss the influence of a longitudinal harmonic load with the frequency from the set of $\{\omega_0 - \omega_0/2,\ \omega_0 + \omega_{/}2\}$ (Fig. 11.4a,b). For further analysis,

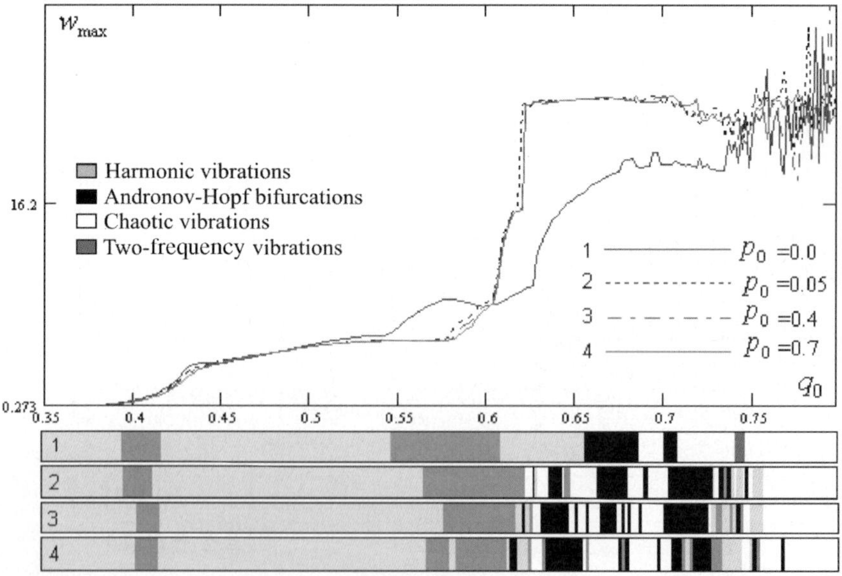

Fig. 11.2 Relations $w_{\max}(q_0)$ and vibration character scales

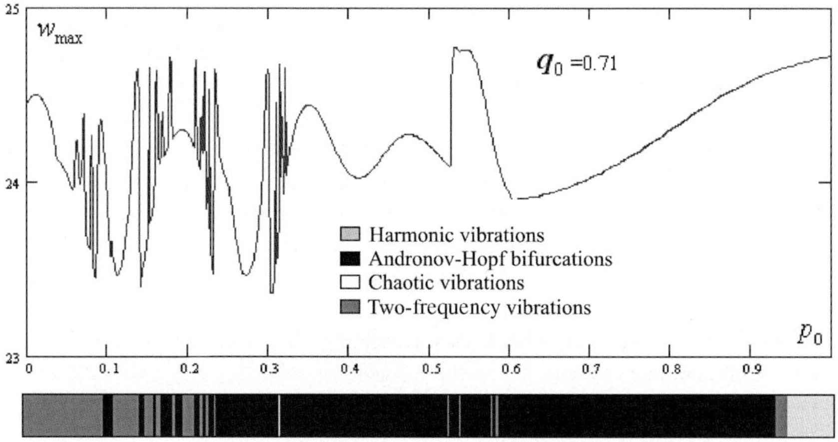

Fig. 11.3 Relation $w_{\max}(q_0)$ and a vibration character scale

flat images of the character of vibrations have been constructed for the set of control parameters $\{q_0,\ \omega_p\}$, where q_0 changes within the range of [0.32, 0.8], and the frequency of exciting force within $\omega_p \in [\omega_0 - \omega_0/2,\ \omega_0 + \omega_0/2]$.

It can be observed that a general image of dynamics is maintained also for the simultaneous influence of transverse and longitudinal load, i.e., we observe the existence of a relatively large area of harmonic vibrations on all frequencies of input function ($q_0 \leq 0.47$ under transverse load only and $q_0 \leq 0.52$ for simultaneous influence of transverse and longitudinal load), and then small "spots" appear that consist of the areas of Andronov-Hopf bifurcations, the areas of vibrations with two incommensurable frequencies, and the areas of chaos.

It can be easily observed that such phenomena resulting from the influence of longitudinal load occur for a larger amplitude of external transverse input functions. A significant reduction of the areas of mild bifurcations was also noticed in

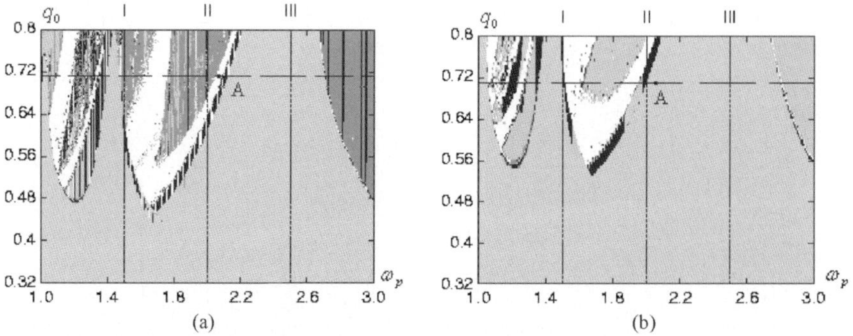

Fig. 11.4 Control parameter planes $\{q_0,\ w_0\}$: **(a)** the influence of a transverse load $q = q_0 \sin(\omega_p t)$; **(b)** the influence of a transverse $q = q_0 \sin\,(\omega_p t)$ and longitudinal load $p = 20 \sin(\omega_0 t)$

the regime of chaotic vibrations. The points corresponding to the vibration with two frequencies (see the first control curve), however, transformed into the area of chaotic vibration after the application of a longitudinal load.

Eventually, we can conclude that the combination of longitudinal and transverse input functions leads to a change of the character of vibrations of a mechanical system in the range of all the analyzed frequencies, and the change may take place both in the direction from chaotic vibration and from periodic to chaotic. In other words, for some parameters of input functions we have chaos, and for other ones, regular motion.

Let us now consider the behavior of the system in space. Therefore, we shall trace the influence of transverse load on various degrees of loading on the behavior of a shell for changes of x and y in the ranges $x = 0.5$, $0 \leq y \leq 2\pi$. Table 11.2 presents such wavy shapes and the corresponding transverse intersections of a cylindrical shell.

Together with the increase of the amplitude of longitudinal input function p_0 the number of wave halves does not change and it is 7 ($p_0 \leq 5.0$), but a dislocation of maximum deflections takes place.

For such small values of $p_0 = 0.05$, maximum deflections are concentrated within the area of application of transverse load, and then together with a small increase of

Table 11.2 Shapes of a deflected cylindrical shell

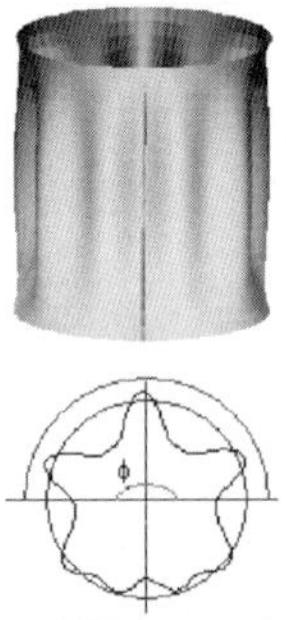 $q_0 = 0.71,\ p_0 = 0.05$	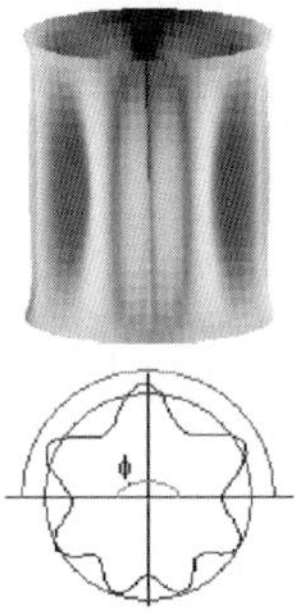 $q_0 = 0.71,\ p_0 = 0.4$	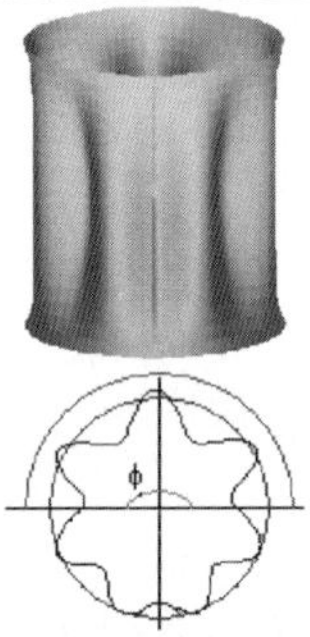 $q_0 = 0.71,\ p_0 = 0.7$	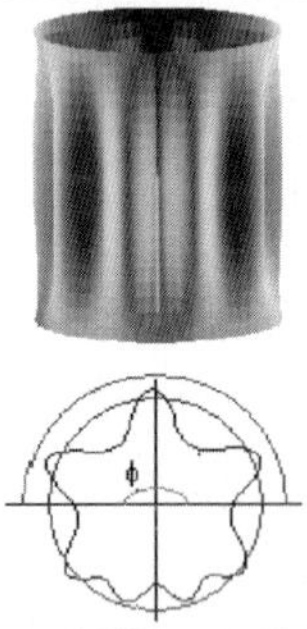 $q_0 = 0.71,\ p_0 = 2.5$
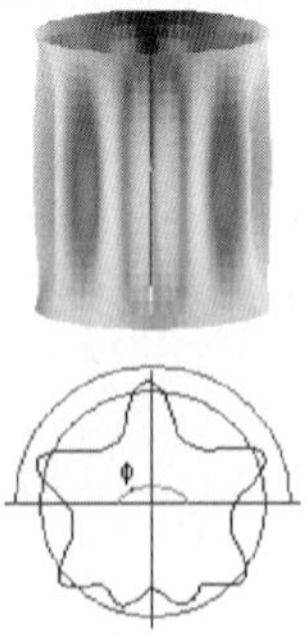 $q_0 = 0.71,\ p_0 = 5.0$	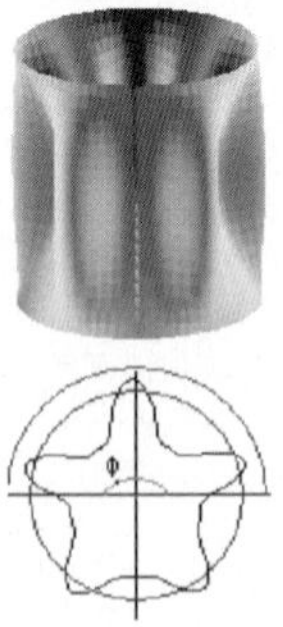 $q_0 = 0.71,\ p_0 = 7.0$	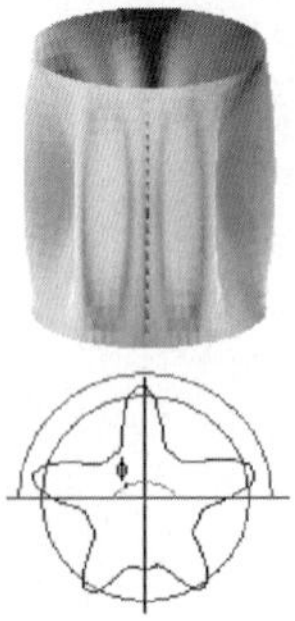 $q_0 = 0.71,\ p_0 = 20.0$	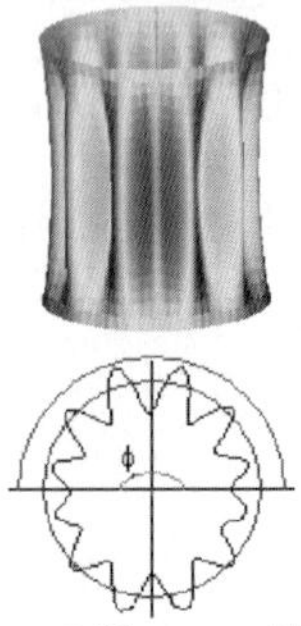 $q_0 = 0.71,\ p_0 = 50.0$

the amplitude of longitudinal input function p_0 maximum values also change to the areas around the transverse load.

Then, for a certain control value $p_0 = 7.0$ the number of wave halves created along the circular coordinate is reduced to 5, and significant changes take place only in the area beyond the point of application of the transverse load, and within this area the number of wave halves and the character of vibration do not change.

The image of spatial dynamics of a cylindrical shell described in this way remains the same up to the value of $p_0 = 50$. For such a critical longitudinal load, the number of wave halves suddenly increases to the value of 12 and their distribution becomes asymmetrical in relation to the curve of transverse load application.

The above consideration leads to the conclusion that one of the effective ways of steering time-spatial chaos in mechanical systems represented by closed cylindrical shells is purposeful application of small longitudinal periodic input functions.

Qualitative change of chaotic dynamics of cylindrical shells can be realized with the use of small periodic steering input functions. As the regime of chaotic vibrations includes a countable set of regular unstable states, the possibility of realization of various regimes of the work of a mechanical system increases in an unlimited way.

The approach described here is connected with the problem of controllable synchronization. This enables the determination and isolation of chaotic subsets with the use of the procedure mentioned, as well as to transform synchronic motion of identical systems into stable motion along some selected directions, and into a non-stable one along other directions. In this way it is possible to realize the transition from asynchronic chaotic vibration to the regime of full chaos synchronization.

11.4.1 Conclusions

The chapter has presented the method of controlling the time-spatial chaos of a cylindrical shell by transforming chaotic dynamics of a system into regular dynamics with other topological characteristics with the purposeful use of periodic input functions as well as transverse load in antiphase.

A longitudinal input function leads to the change of vibration character of the analyzed mechanical system, and these can be connected with the transition from chaotic to periodic vibration (or to the occurrence of Andronov-Hopf bifurcations), or the other way around, i.e., with the transition from periodic to chaotic vibration.

It is possible to lead the system out of one state of chaos and at the same time into a topologically different state of chaos. In the latter case, we managed to reduce critical loads for the analyzed shell and reduce the state of over-critical deflection, as well as to reduce the areas of chaotic vibrations by leading them into a periodic regime of system vibrations.

Chapter 12
Chaotic Vibrations of Flexible Rectangular Shells

In this chapter, chaotic vibrations of flexible rectangular shells forced by transversal harmonic load are analyzed. The study is conducted with the application of the qualitative theory of differential equations and nonlinear dynamics. An infinitely dimensional problem is reduced to a finitely dimensional one with the application of the Bubnov-Galerkin method with higher approximations and the method of finite differences with approximation $O(\Delta^2)$. An initial problem has been solved with the use of the fourth-order Runge-Kutta method. It has been shown that within the range of harmonic vibrations, the results obtained from both methods are fully convergent, whereas in the range of chaos such convergence can be obtained only in relation to the character of vibrations, i.e., in relation to the frequency spectrum. The increase in the number of element partitions in the method of finite differences and the number of approximations in the Bubnov-Galerkin method leads to better results, but there is some threshold value beyond which further calculations are impossible.

12.1 Fundamental Equations

Within a classical theory of nonlinear equations, we shall analyze a rectangular spherical shell (Fig. 12.1) with constant stiffness and density and harmonically forced. In an assumed system of rectangular coordinates, a three-dimensional area fulfilled with a shell has the following form:

$$\Omega = \{x_1, x_2, x_3 | (x_1, x_2) \in [0;a] \times [0;b],\ \ x_3 \in [-h;h]\} \quad 0 \le t < \infty.$$

We shall assume the following non-dimensional form of equations of the theory of shallow shells as the basis for further research [21]:

$$\frac{1}{12(1-\mu^2)}\left(\frac{1}{\lambda^2}\frac{\partial^4 w}{\partial x_1^4} + \lambda^2\frac{\partial^4 w}{\partial x_2^4} + 2\frac{\partial^4 w}{\partial x_1^2 \partial x_2^2}\right) - k_{x_2}\frac{\partial^2 F}{\partial x_1^2} - k_{x_1}\frac{\partial^2 F}{\partial x_2^2}$$
$$- L(w,F) - \frac{\partial^2 w}{\partial t^2} - \varepsilon\frac{\partial w}{\partial t} - q(x_1,x_2,t) = 0,$$

J. Awrejcewicz, V.A. Krysko, *Chaos in Structural Mechanics*,
DOI: 10.1007/978-3-540-77676-5_13, © Springer-Verlag Berlin Heidelberg 2008

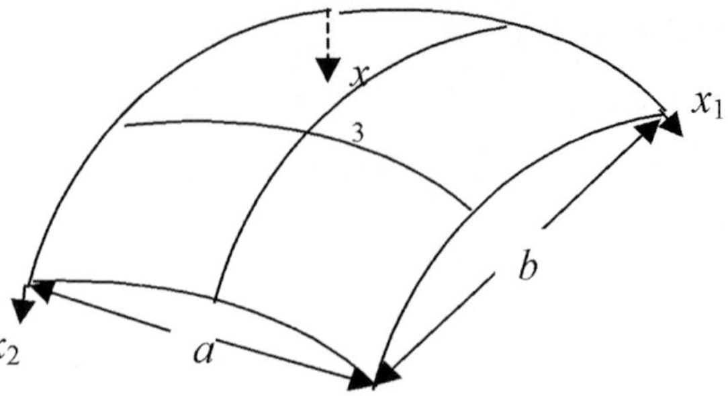

Fig. 12.1 Scheme of the spherical rectangular shell under analysis

$$\frac{1}{\lambda^2}\frac{\partial^4 F}{\partial x_1^4}+\lambda^2\frac{\partial^4 F}{\partial x_2^4}+2\frac{\partial^4 F}{\partial x_1^2\partial x_2^2}+k_{x_2}\frac{\partial^2 w}{\partial x_1^2}+k_{x_1}\frac{\partial^2 w}{\partial x_2^2}+\frac{1}{2}L(w,w)=0, \qquad (12.1)$$

where

$$L(w,F)=\frac{\partial^2 w}{\partial x_1^2}\frac{\partial^2 F}{\partial x_2^2}+\frac{\partial^2 w}{\partial x_2^2}\frac{\partial^2 F}{\partial x_1^2}-2\frac{\partial^2 w}{\partial x_1\partial x_2}\frac{\partial^2 F}{\partial x_1\partial x_2}$$

is a known nonlinear operator.

System (12.1) has been reduced to a non-dimensional form with the use of the following nondimensional parameters. Parameters: $\lambda = a/b$; $x_1 = a\bar{x}_1$, $x_1 = a\bar{x}_2$; $k_{x_1} = a^2/R_{x_1}h$, $k_{x_2} = b^2/R_{x_2}h$ denote the curvatures of a shell with relation to, respectively, x_1 and x_2, $w = 2h\bar{w}$ is the deflection; $F = E(2h)^3\bar{F}$ is the function of strains; $t = t_0\bar{t}$ is time; $q = \frac{E(2h)^4}{a^2b^2}\bar{q}$ is the external load; $\varepsilon = (2h)\bar{\varepsilon}$ is the coefficient of damping.

The bars over nondimensional values have been omitted in the equations. The following denotation has also been introduced: a, b are shell dimensions, respectively toward coordinates x_1 and x_2; μ is the Poisson coefficient.

We shall attach the following boundary conditions to Eqs. (12.1):

1. Support on a flexible incompressible ribs

$$w=0;\quad \frac{\partial^2 w}{\partial x_1^2}=0;\quad F=0;\quad \frac{\partial^2 F}{\partial x_1^2}=0\quad \text{for } x_1=0,1;$$

$$w=0;\quad \frac{\partial^2 w}{\partial x_2^2}=0;\quad F=0;\quad \frac{\partial^2 F}{\partial x_2^2}=0\quad \text{for } x_2=0,1. \qquad (12.2)$$

2. Edge clamping

$$w=0;\quad \frac{\partial w}{\partial x_1}=0;\quad F=0;\quad \frac{\partial^2 F}{\partial x_1^2}=0\quad \text{for } x_1=0,1;$$

$$w=0;\quad \frac{\partial w}{\partial x_2}=0;\quad F=0;\quad \frac{\partial^2 F}{\partial x_2^2}=0\quad \text{for } x_2=0,1. \qquad (12.3)$$

3. Free edge support

$$w = 0; \quad \frac{\partial^2 w}{\partial x_1^2} = 0; \quad F = 0; \quad \frac{\partial F}{\partial x_1} = 0 \quad \text{for } x_1 = 0, 1;$$

$$w = 0; \quad \frac{\partial^2 w}{\partial x_2^2} = 0; \quad F = 0; \quad \frac{\partial F}{\partial x_2} = 0 \quad \text{for } x_2 = 0, 1. \tag{12.4}$$

4. Clamping with respect to flexible incompressible ribs:

$$w = 0; \quad \frac{\partial w}{\partial x_1} = 0; \quad F = 0; \quad \frac{\partial^2 F}{\partial x_1^2} = 0 \quad \text{for } x_1 = 0, 1;$$

$$w = 0; \quad \frac{\partial w}{\partial x_2} = 0; \quad F = 0; \quad \frac{\partial^2 F}{\partial {x_2}^2} = 0 \quad \text{for } x_2 = 0, 1. \tag{12.5}$$

The initial conditions follow:

$$w(x_1, x_2)|_{t=0} = \varphi_1(x_1, x_2), \quad \frac{\partial w}{\partial t} = \varphi_2(x_1, x_2). \tag{12.6}$$

We shall still consider a case of the application of a transverse harmonic load $q(t) = q_0 \sin(\omega_p t)$, where ω_p is the frequency of exciting force and q_0 is the amplitude of exciting force, and we will search for the solution to differential equations (12.1)–(12.6) with the application of the Bubnov-Galerkin method with higher approximations and the method of finite differences with a different partition of spatial coordinates. Our aim is to find, with the use of these computational methods, the so-called "real" chaos.

12.2 Bubnov-Galerkin Method with Higher Approximations

We approximate the functions sought, the solutions to Eqs. (12.1), with the use of expressions with a finite number of random parameters, and we shall present them as the product of two functions satisfying one of the boundary conditions (12.2)–(12.5) in the following form:

$$w = \sum_{i=1}^{N} \sum_{j=1}^{N} A_{ij}(t) \varphi_{ij}(x_1, x_2),,$$

$$F = \sum_{i=1}^{N} \sum_{j=1}^{N} B_{ij}(t) \psi_{ij}(x_1, x_2). \tag{12.7}$$

This solution is based on test functions that are energetically orthonormalized, i.e.:

$$(\nabla^4(\varphi_{ij}), \varphi_{nm}) = \begin{cases} 0, & \text{for } i,j \neq n,m, \\ 1, & \text{for } i,j = n,m. \end{cases} \tag{12.8}$$

We shall choose such coordinates of system $\{\varphi_{ij}(x_1,x_2),\ \psi_{ij}(x_1,\ x_2)\}$ that functions $\varphi_{ij}(x_1,\ x_2),\ \psi_{ij}(x_1,\ x_2)$ for $\forall i,j$ are linearly independent, continuous together with their first derivatives up to the order of four within the range of Ω, and that $\varphi_{ij}(x_1,\ x_2),\ \psi_{ij}(x_1,\ x_2)$ satisfy one of the respective boundary conditions (12.2)–(12.5). In addition, $\varphi_{ij}(x_1,\ x_2),\ \psi_{ij}(x_1,\ x_2)$ are supposed to be complete.

For convenience, we shall denote the left sides of the equations of system (12.1) respectively by Φ_1 and Φ_2:

$$\Phi_1\left(w,F,\frac{\partial^2 w}{\partial x_1^2},\frac{\partial^2 F}{\partial x_1^2},\ldots\right)+q(x_1,x_2,t)=0,$$
$$\Phi_2(w,F,\frac{\partial^2 w}{\partial x_1^2},\frac{\partial^2 F}{\partial x_1^2},\ldots)=0. \tag{12.9}$$

By applying the Bubnov-Galerkin procedure in system (12.9), we obtain:

$$\int_0^1\int_0^1 \Phi_1\varphi_{kl}(x_1,x_2)dx_1\ dx_2+\int_0^1\int_0^1 q(x_1,x_2,t)\varphi_{kl}(x_1,x_2)dx_1\ dx_2=0,$$
$$\int_0^1\int_1^1 \Phi_2\psi_{kl}(x_1,x_2)dx_1dx_2=0,\quad k=1,2,\ldots,N,\quad l=1,2,\ldots N \tag{12.10}$$

Taking (12.10) into account Eq. (12.9), we obtain

$$\sum_{kl}\Bigg[\sum_{ij}A_{ij}I_{1,klij}-\sum_{ij}B_{ij}I_{2,klij}+\sum_{ij}qI_{3,klij}-\sum_{ij}A_{ij}\sum_{rd}B_{rs}I_{4,klijrs}$$
$$-\sum_{ij}\left[\frac{d^2A_{ij}}{dt^2}+\varepsilon\frac{dA_{ij}}{dt}\right]I_{8,klij}\Bigg]=0,$$
$$\sum_{kl}\left[\sum_{ij}A_{ij}I_{7,klij}+\sum_{ij}B_{ij}I_{5,klij}-\sum_{ij}A_{ij}\sum_{rs}A_{rs}I_{6,klijrs}\right]=0,$$
$$k,i,r=1,2,\ldots,N,\quad l,j,s=1,2,\ldots,N. \tag{12.11}$$

In the preceding equations, $\sum_{kl}[*]$ before all the equations of system (12.11) means that each equation is for us a complex system with kl of this type of equation, and integrals from the Bubnov-Galerkin method have the following form:

$$I_{1,klij}=\int_0^1\int_0^1\frac{1}{12(1-\mu^2)}\left[\frac{1}{\lambda^2}\frac{\partial^2\varphi_{ij}}{\partial x_1^2}\frac{\partial^2\varphi_{kl}}{\partial x_1^2}+\lambda^2\frac{\partial^2\varphi_{ij}}{\partial x_2^2}\frac{\partial^2\varphi_{kl}}{\partial x_2^2}\right.$$
$$\left.+2\frac{\partial^2\varphi_{ij}}{\partial x_1\partial x_2}\frac{\partial^2\varphi_{kl}}{\partial x_1\partial x_2}\right]\psi_{kl}dx_1\ dx_2,$$

$$I_{2,klij} = \int_0^1\int_0^1 \left[-k_{x_2}\frac{\partial^2\psi_{ij}}{\partial x_1^2} - k_{x_1}\frac{\partial^2\psi_{ij}}{\partial x_2^2}\right]\varphi_{kl}dx_1\, dx_2,$$

$$I_{3,klij} = \int_0^1\int_0^1 \varphi_{kl}q(x,y,t)dx_1\, dx_2,$$

$$I_{4,klirsj} = \int_0^1\int_0^1 L(\varphi_{ij},\psi_{rs})\varphi_{kl}dx_1\, dx_2,$$

$$I_{5,klij} = \int_0^1\int_0^1 \left[\frac{1}{\lambda^2}\frac{\partial^2\psi_{ij}}{\partial x_1^2}\frac{\partial^2\psi_{kl}}{\partial x_1^2} + \lambda^2\frac{\partial^2\psi_{ij}}{\partial x_2^2}\frac{\partial^2\psi_{kl}}{\partial x_2^2}\right.$$

$$\left. + 2\frac{\partial^2\psi_{ij}}{\partial x_1\partial x_2}\frac{\partial^2\psi_{kl}}{\partial x_1\partial x_2}\right]\varphi_{kl}dx_1\, dx_2,$$

$$I_{6,klijrs} = \int_0^1\int_0^1 \frac{1}{2}L(\varphi_{ij},\varphi_{rs})\psi_{kl}dx_1\, dx_2,$$

$$I_{7,klij} = \int_0^1\int_0^1 \left[-k_{x_2}\frac{\partial^2\varphi_{ij}}{\partial x_1^2} - k_{x_1}\frac{\partial^2\varphi_{ij}}{\partial x_2^2}\right]\psi_{kl}dx_1\, dx_2,$$

$$I_{8,klij} = \int_0^1\int_0^1 \varphi_{ij}\psi_{kl}dx_1\, dx_2. \tag{12.12}$$

Integrals (12.12), apart from $I_{3,klij}$ in some cases, i.e., when the transverse load is not applied to the whole surface of a shell, are calculated along the whole central surface of a shell.

After applying Bubnov-Galerkin method, with respect to higher approximations describing spatial coordinates, we obtain a system of algebraic linear equations with respect to coefficients B_{ij}, that is solved by reversing the matrix, and the system of differential equations of the order of four with respect to coefficients A_{ij}, that is later reduced to a normal form and solved with the use of fourth-order Runge-Kutta method:

$$K_{ij}B_{ij} = F_1(A_{ij}), \tag{12.13}$$

$$\begin{cases} \dfrac{dA_{ij}}{dt} = X_{ij}, \\ \dfrac{dX_{ij}}{dt} + \varepsilon X_{ij} = F_2(A_{ij},B_{ij},t), \quad i=\overline{1,M_x}, \quad j=\overline{0,M_y}, \end{cases} \tag{12.14}$$

where K_{ij} in (12.13) is the matrix of coefficients of a linear algebraic system with respect to unknown parameters B_{ij}, $F_1(A_{ij})$ is a column of elements depending on parameters A_{ij}, and (12.14) is a normal system of differential equations of the order of four with respect to A_{ij} and X_{ij}. The step of integration in time is selected from the condition of maintaining the stability of the solution.

As an example, we shall analyze a spherical rectangular shell with uniform boundary conditions (12.2) and zero initial conditions (12.6) $\varphi_1(x_1,x_2) = 0$, $\varphi_2(x_1, x_2) = 0$. Functions φ_{ij}, ψ_{ij} in Eq. (12.7) will be presented as a product of two functions, each of which depends on one argument only (such functions should satisfy boundary conditions (12.2)):

$$w = \sum_{i=1}^{N}\sum_{j=1}^{N} A_{ij}(t)\sin(i\pi x_1)\sin(j\pi x_2),$$

$$F = \sum_{i=1}^{N}\sum_{j=1}^{N} B_{ij}(t)\sin(i\pi x_1)\sin(j\pi x_2). \tag{12.15}$$

Let us analyze numerically the convergence of the Runge-Kutta method in relation to N in Eq. (12.15) or flexible rectangular shell within the range of harmonic vibrations and within the area of chaos. Let us analyze point $\{q_0, \omega_p\} = \{5, 25\}$, situated in the area of harmonic vibrations (Fig. 12.2).

For all values of N we observe one-frequency harmonic vibrations that for small values of N have small vibration amplitudes that are similar to one another.

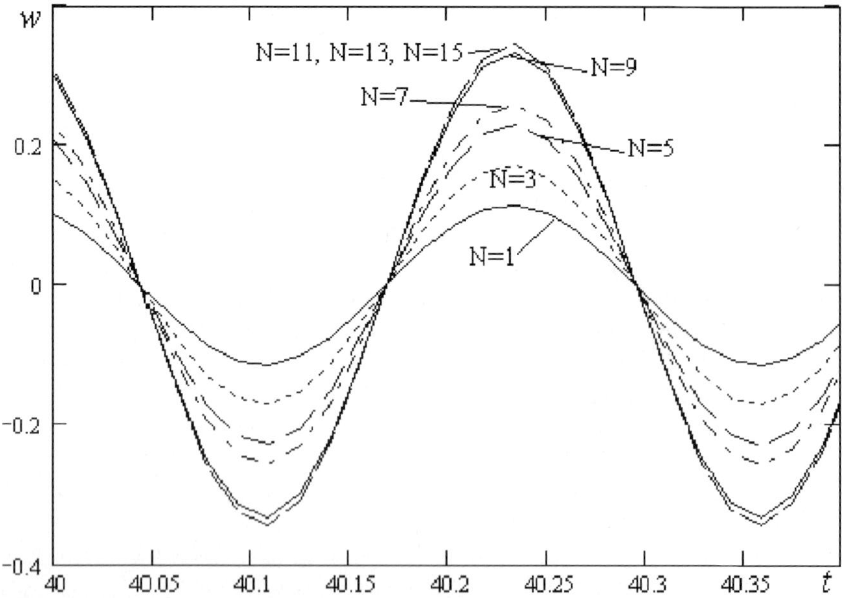

Fig. 12.2 Dependence of $w(t)$ for different approximations of the Bubnov-Galerkin method in the range of harmonic vibrations

With larger number of approximations $N = 11, 13, 15$ we obtain the consistency of results also with respect to an amplitude, i.e., time histories fully overlap.

Further, we shall analyze the convergence of the Bubnov-Galerkin method in the range of chaotic vibrations. Let us analyze point $\{q_0, w_0\} = \{151.8, \omega_0\}$. After more detailed analysis, we can see that the convergence of vibrations in the area of chaotic vibrations is much worse than in the area of harmonic vibrations.

Table 12.1 presents the following characteristics: N is the number of elements of the order of (12.15), time history $w(0.5,\ 0.5,\ t)$ in the central point of a spherical shell, phase portrait $w(w')$, frequency spectrum $S(\omega)$, Poincaré map $w_t(w_{t+T})$, where $T = 2\pi/\omega_p$ is the period of an excitation force.

The analyzed problem is significantly different from the Lorenz problem. For $N = 1, 3$ periodic vibrations are observed, as well as unirotational cycle in a phase portrait, whereas the Poincaré map consists of one point. For $N = 5$ the system transforms into the state of chaos with a dominant frequency of an exciting force, and the Poincaré intersection describes a chaotic set of points. For $N = 7$ we deal with periodic vibrations again, there is a point in Poincaré map, and a phase portrait presents a unirotational cycle. Apart from that, we can see that time courses $N = 3,\ 7$ practically overlap, but they differ with respect to the amplitude from a time course for $N = 1$. All three courses are harmonically characterized by the frequency of an exciting force, whereas a signal is chaotic for $N = 5$. Such a process can be described the transition from harmonic motion to chaotic vibrations, as further for $N = 9,\ 11,\ 13,\ 15$ we can observe chaotic vibrations in time course and in frequency spectrum, i.e., the convergence starts only for $N \geq 9$.

Figure 12.3 shows time history $w(0.5,\ 0.5,\ t)$ for $t \in [40;\ 41]$ and for various elements of series (12.15) in the area of chaos (see scales of the character of vibrations in Fig. 12.4). On the other hand, convergence of $w(0,\ 5,\ 0.5,\ t)$ vs. is reported in Fig. 12.5.

We do not observe chaos during the analysis of time course of convergence, which implies checking the possibility of convergence in an average sense, i.e., in relation to a frequency spectrum. Therefore, we shall analyze a frequency spectrum and Poincaré map for different values of N in (12.15) and for point $q_0 = 151.8$, $\omega_p = \omega_0$.

Let check the relations of $w_{max}(q_0)$ for these approximations of the Bubnov-Galerkin method for which convergence with respect to the average spectrum in an average sense has been detected (Fig. 12.4). Apart from monitoring the relations of $w_{max}(q_0)$ we shall also observe the scales of the character of vibrations in relation to control parameters $\{q_0, \omega_p\}$, where ω_p is the frequency of excitement.

Identification of the type of vibrations of a spherical rectangular shell on the basis of construction of the scales of character of vibrations $\{q_0, \omega_p\}$ for each time course $w(t)$ has been conducted by the analysis of frequency spectrum $S(\omega)$ of Lyapunov exponents, and the denotation applied is presented in Fig. 12.4.

At this point, let us emphasize that full convergence of the Bubnov-Galerkin method could not be achieved in relation to time courses, but the convergence in an average sense in relation to frequency spectrum has been achieved.

Table 12.1 Shell dynamics characteristics

N	Time history $w(0.5;0.5;t)$	Phase portrait $w(w')$	Power spectrum $S(\omega)$	Poincaré section $w_t(w_{t+T})$
N – 1	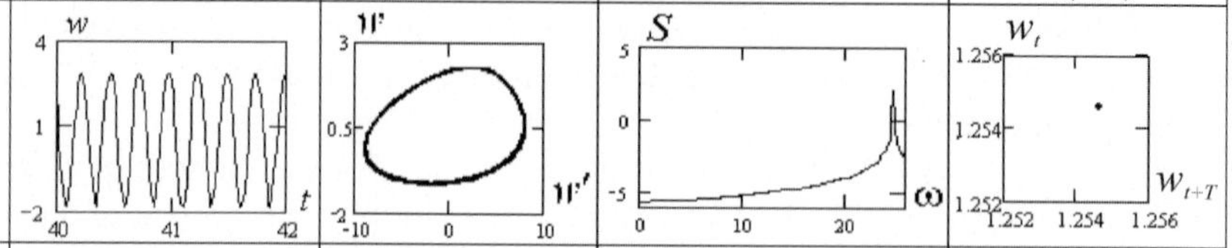			
N – 3	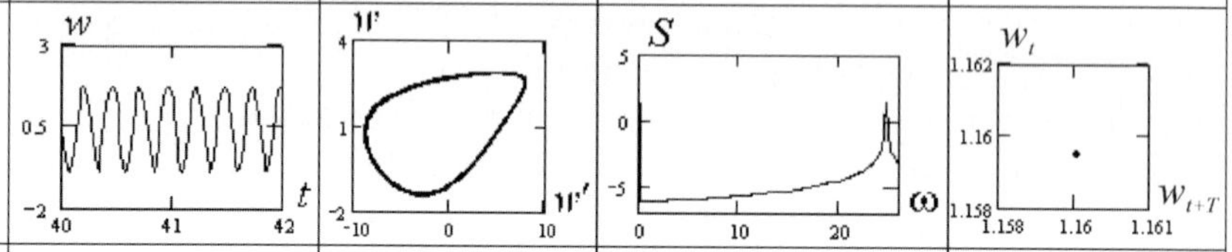			
N – 5	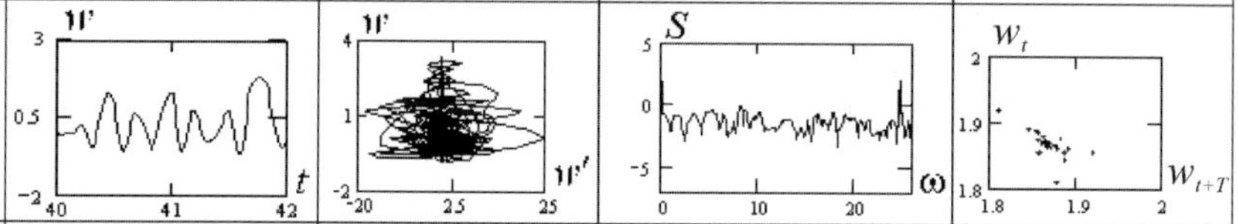			
N – 7	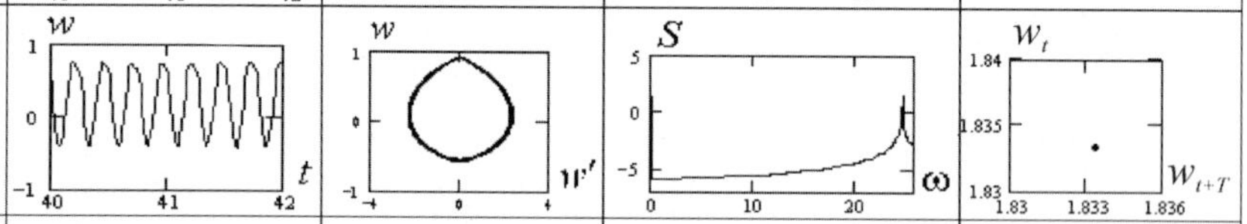			
N – 9	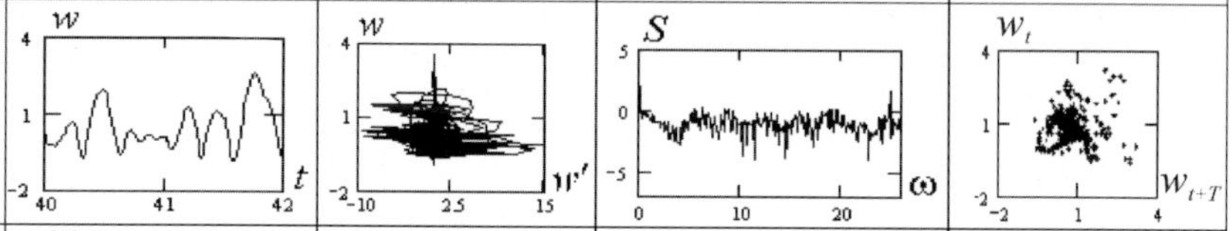			
N – 11	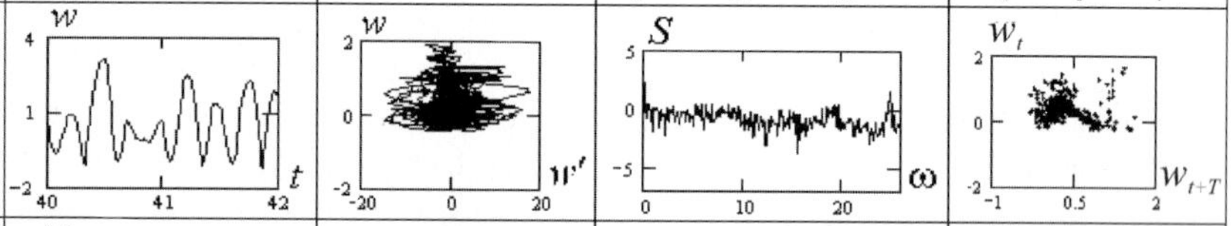			
N – 13	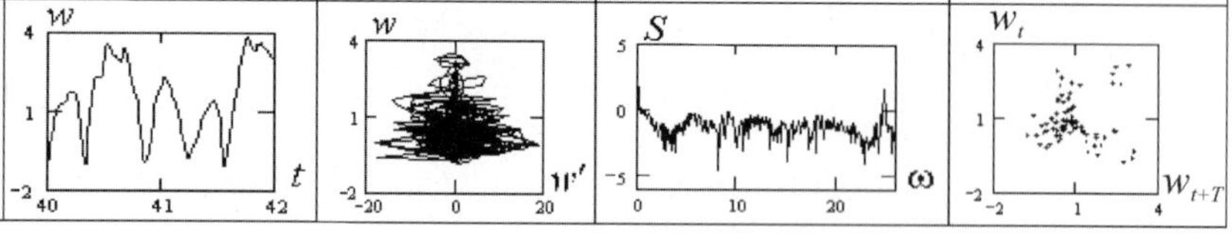			

12.3 Method of Finite Differences

Let us now analyze the possibility of application of some other method, namely the method of finite differences with approximation $O(\Delta^2)$ in relation to spatial coordinates x_1 and x_2. In this case, after applying it in continuous system (12.1)–(12.5) we obtain a discrete system that has the following operational form

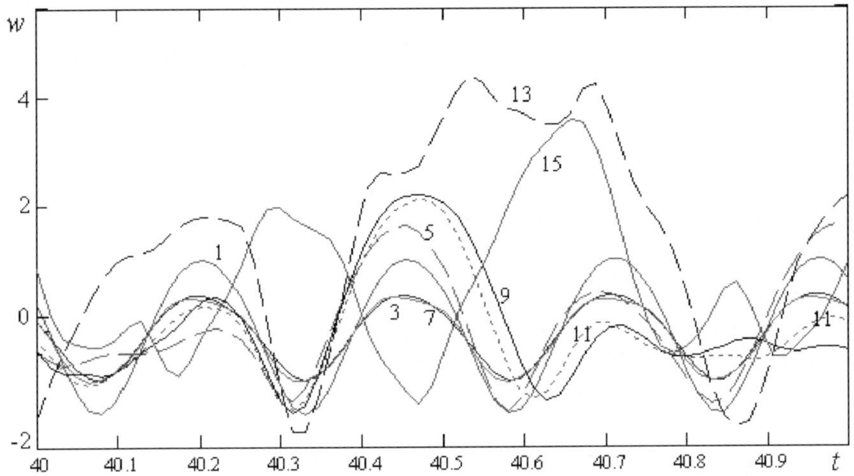

Fig. 12.3 Convergence of the Bubnov-Galerkin method in relation to time history $w(0.5,\ 0.5,\ t)$ in the range of chaos

$$-\frac{1}{12(1-\nu^2)}\left(\lambda^{-2}\Lambda_1^2 w_{ij}+2\Lambda_{12}^2 w_{ij}+\lambda^2\Lambda_2^2 w_{ij}\right)-K_x\Lambda_2 F_{ij}-K_y\Lambda_1 F_{ij}$$
$$-\Lambda_1 w_{ij}\cdot\Lambda_2 F_{ij}-\Lambda_2 w_{ij}\cdot\Lambda_1 F_{ij}+\Lambda_{12}w_{ij}\cdot\Lambda_{12}F_{ij}+q_i=(w_{tt}+\varepsilon w_t)_{i,j}$$
$$\left(\lambda^{-2}\Lambda_1^2 F_{ij}+2\Lambda_{12}^2 F_{ij}+\lambda^2\Lambda_2^2 F_{ij}\right)$$
$$=-K_x\Lambda_2 w_{ij}-K_y\Lambda_1 w_{ij}-\Lambda_1 w_{ij}\cdot\Lambda_2 w_{ij}+(\Lambda_{12}w_{ij})^2,$$

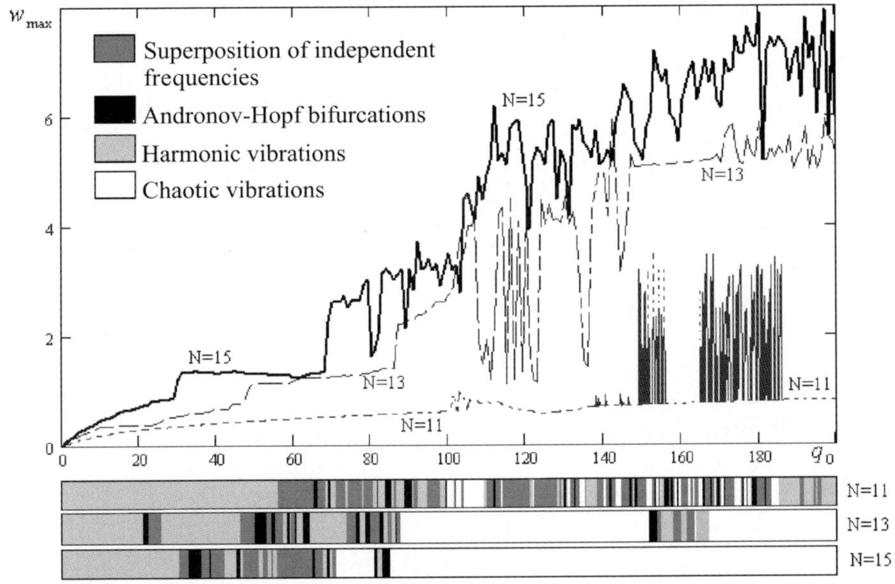

Fig. 12.4 Dependence $w_{max}(q_0)$ for various values of N

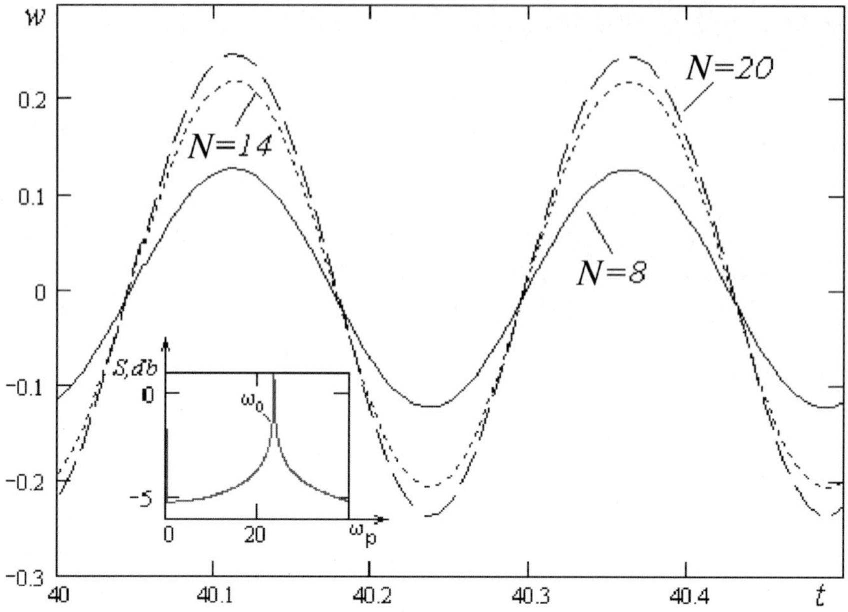

Fig. 12.5 $w(t)$ and $S, db(\omega_p)$ in the area of periodic vibrations in relation to N

where

$$\Lambda_i y = \frac{1}{h_i^2}\left[y(x_i - h_i) - 2\cdot y(x_i) + y(x_i + h_i)\right], \quad i = 1,2,$$

$$\Lambda_{12} y = \frac{1}{4h_1h_2}[y(x_1+h_1, x_2+h_2) + y(x_1-h_1, x_2-h_2)$$
$$- (x_1+h_1, x_2-h_2) - (x_1-h_1, x_2+h_2)],$$

$$\Lambda_i^2 y = \frac{1}{h_i^4}[y(x_i - 2h_i) - 4y(x_i - h_i) + 6y(x_i)$$
$$- 4y(x_i + h_i) + y(x_i + 2h_i)], \quad i = 1,2,$$

$$\Lambda_{12}^2 y = \frac{1}{h_1^2h_2^2}[y(x_1-h_1, x_2-h_2) - 2y(x_1-h_1, x_2) + y(x_1-h_1, x_2-h_2)$$
$$- 2(x_1, x_2-h_2) + 4y(x_1, x_2) - 2y(x_1, x_2+h_2)$$
$$+ y(x_1+h_1, x_2-h_2) - 2(x_1+h_1, x_2) + y(x_1+h_1, x_2+h_2)]$$

$$\Lambda_{12}^2 y = \frac{1}{h_1^2h_2^2}[y(x_1-h_1, x_2-h_2) - 2y(x_1-h_1, x_2) + y(x_1-h_1, x_2-h_2)$$
$$- 2(x_1, x_2-h_2) + 4y(x_1, x_2) - 2y(x_1, x_2+h_2)$$
$$+ y(x_1+h_1, x_2-h_2) - 2(x_1+h_1, x_2) + y(x_1+h_1, x_2+h_2)]. \quad (12.16)$$

We must attach the following initial and boundary conditions to Eqs. (12.16):

$$w_{ij} = f_1(r_k, \theta_k), \quad w'_n = f_2(r_k, \theta_k), \quad (0 \le k \le n), \quad 0 \le t \le \infty,$$

$$w_{N,j} = 0, \quad \Lambda_1 w_{N,j} = 0, \quad F_{N,j} = 0, \quad \Lambda_1 F_{N,j} = 0, \quad j = 1, M-1,$$

$$w_{i,M} = 0, \quad \Lambda_1 w_{i,M} = 0, \quad F_{i,M} = 0, \quad \Lambda_1 F_{i,M} = 0, \quad i = 1, N-1. \tag{12.17}$$

The first equation of a system is solved with the application of the fourth-order Runge-Kutta method, and the second one with the application of the method of inverse matrix. The time threshold has been selected in accordance with the Runge law.

Let us also numerically analyze convergence of the method of finite differences in relation to n, i.e., the number of elements of division of segments $[0;a]$ and $[0;b]$ for a rectangular spherical shell. Let us analyze the point with parameters used in the previous paragraph, i.e., $\{q_0,\ \omega_p\} = \{151.8, \omega_0\}$. Table 12.2 provides the following characteristics: time course $w(0.5, 0.5t)$ for $40 \le t \le 42$, phase portrait $w(w'_t)$, frequency power spectrum S, $db(\omega_p)$, and Poincaré map $w_t(w_{t+T})$ for a different number of division of segments.

For all n signals are different (Fig. 12.6), but the convergence of characteristics such as frequency spectrum, the Poincaré map or a phase portrait is significant, beginning with $n > 10$.

For a more detailed analysis of relationships of the convergence of the method of finite differences, the diagrams of relation $w_{\max}(q_0)$ have been prepared for $n = 8; 14; 20$ (Fig. 12.7). Under these diagrams, scales characterizing the type of vibrations have been provided that were prepared on the basis of the analysis of the power spectrum. Conventional denotation used there is provided under Fig. 12.4. Together with the increase of the number of segment divisions, the area of chaotic vibrations also increases ($n = 20$ shall be our model value). Together with the increase of the number of segment divisions, the diagrams approach one another and their shape comes closer to the shape of a model diagram.

12.4 Comparison of Results Obtained with the Use of the Bubnov-Galerkin Method and the Method of Finite Differences

Figure 12.8 shows two relationships $w_{\max}(q_0)$ corresponding with the results obtained with the application of the Bubnov-Galerkin method for $N = 11$ and $N = 15$, and with the application of the method of finite differences for the division of $n = 8$ and $n = 20$. It can be observed that approximations for $N = 11$ and the number of division $n = 8$, as well as for $N = 15$ and $n = 20$ qualitatively overlap.

Let us now focus on the first case ($N = 11$ and $n = 8$).

For small values of the amplitude of exciting force, when the system undergoes harmonic vibrations, we observe complete overlapping of diagrams $w_{\max}(q_0)$

Table 12.2 Characteristics of shell dynamics

n	Time history $w(0.5;0.5;t)$	Phase portrait $w(w')$	Power spectrum $S,\ db(\omega_p)$	Poincaré section $w_t(w_{t+T})$
n - 6	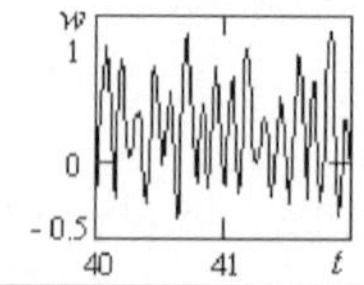	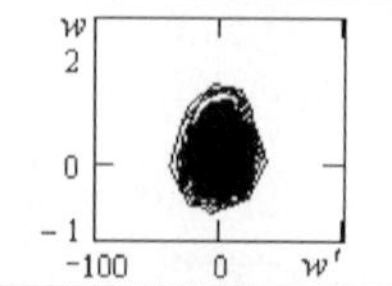	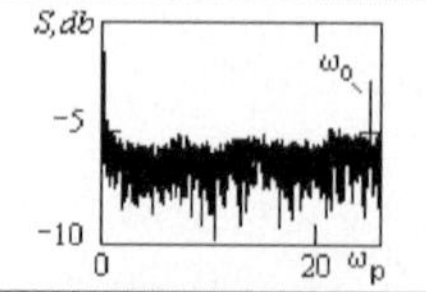	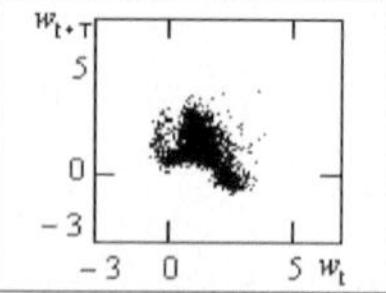
n - 8	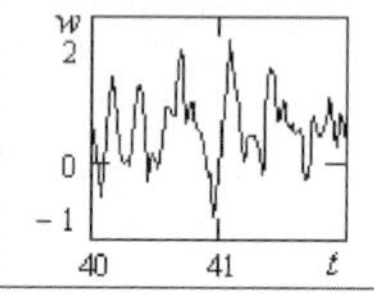	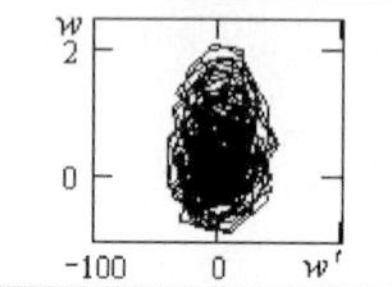	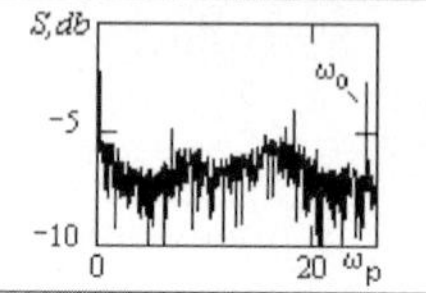	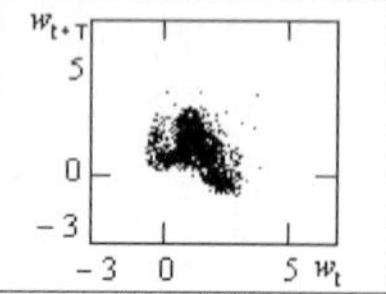
n - 10	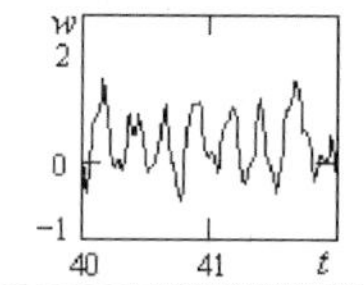	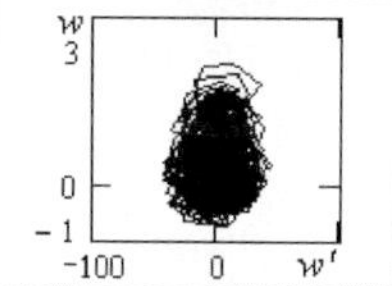	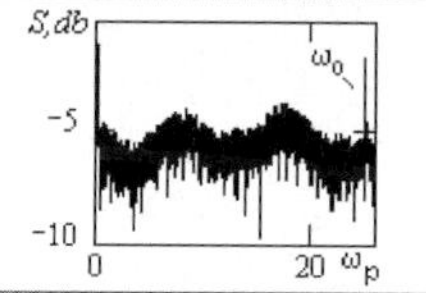	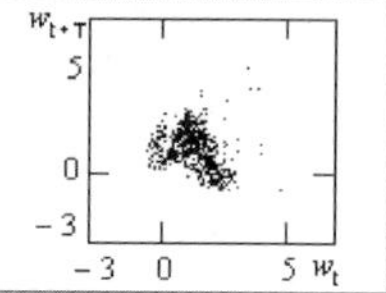
n - 12	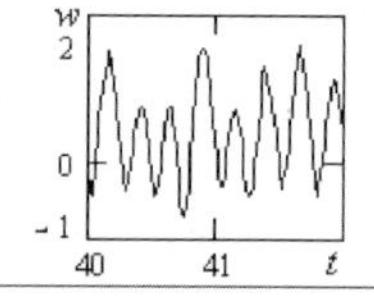	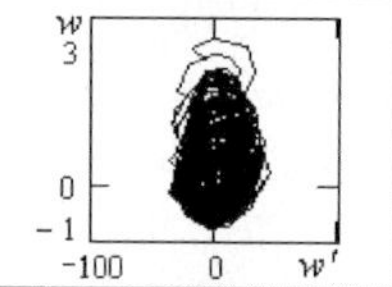	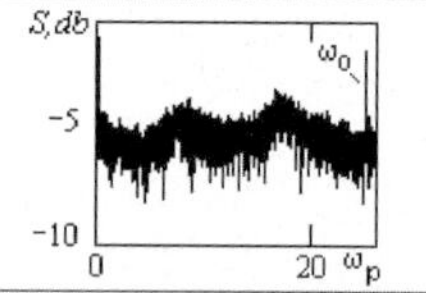	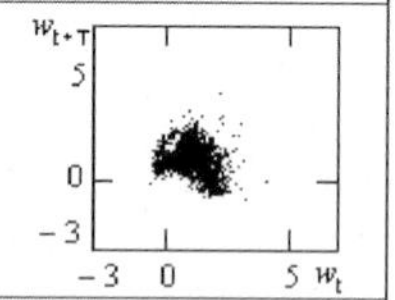
n - 14	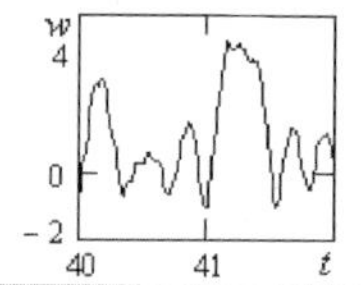	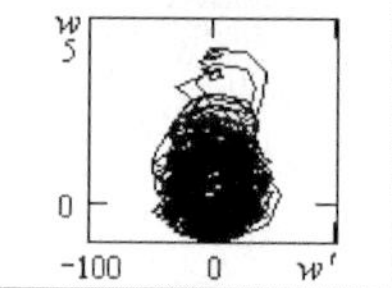	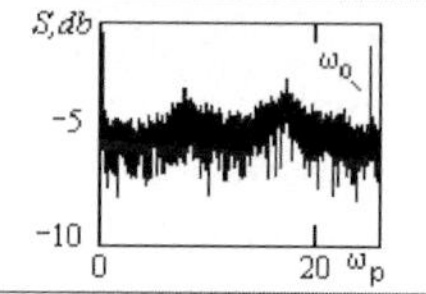	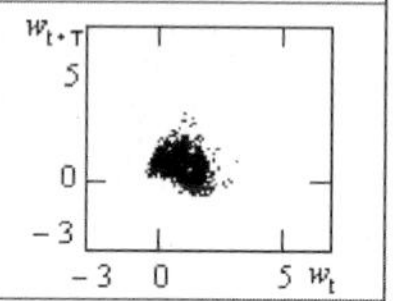
n – 16	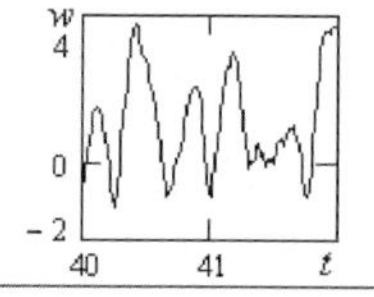	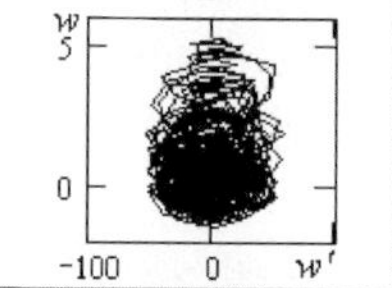	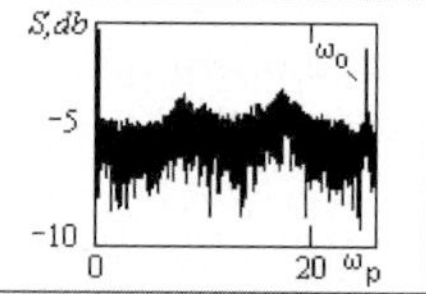	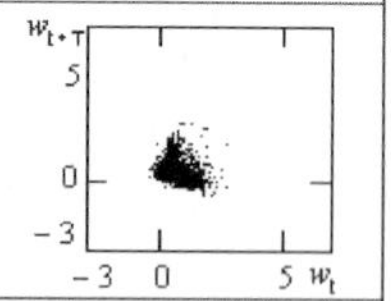
n - 18	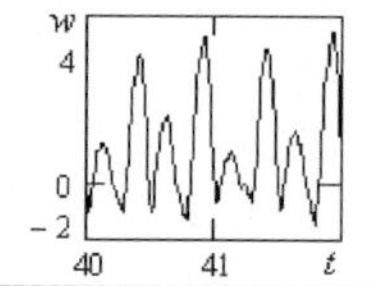	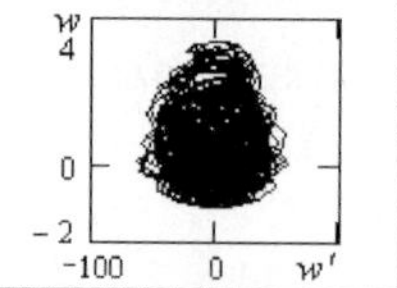	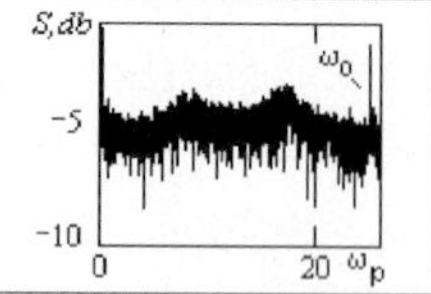	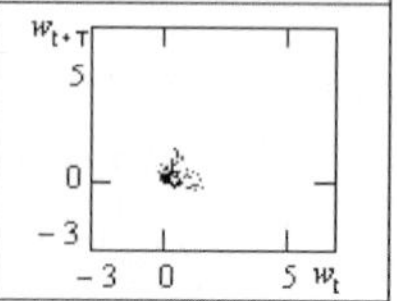
n - 20	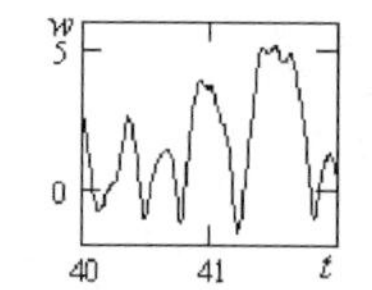	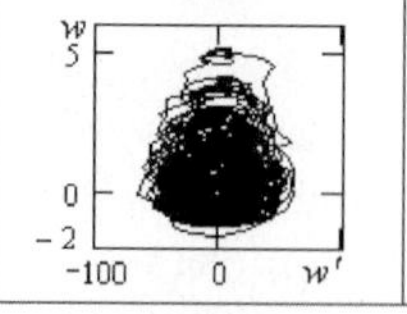	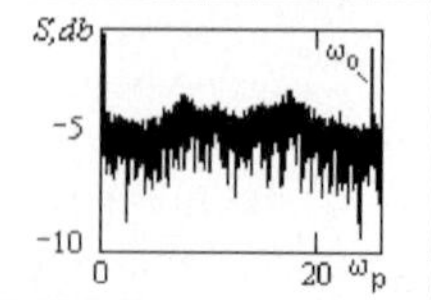	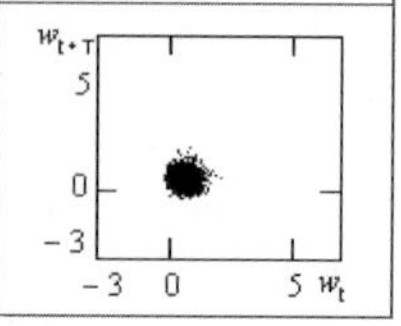

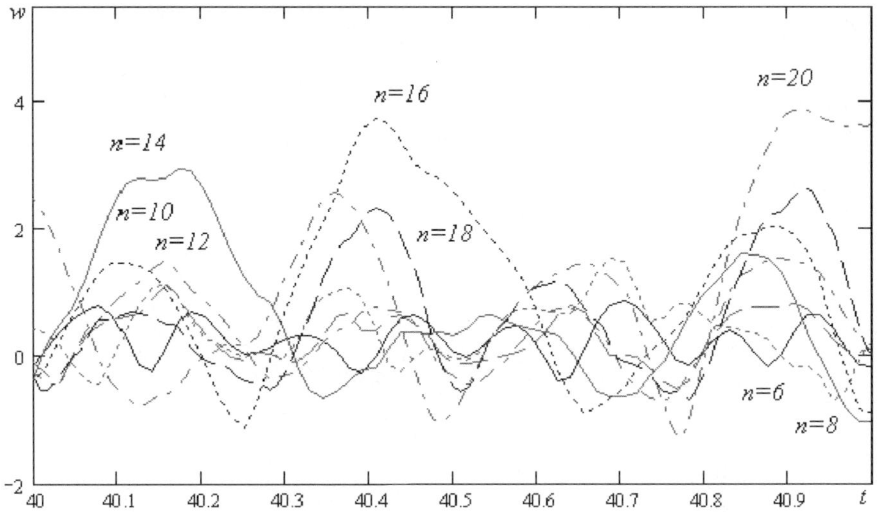

Fig. 12.6 Convergence of the method of finite differences in relation to time history $w(0.5, 0.5, t)$

for both cases, i.e., the convergence of results is obtained with respect to both the amplitude and the scales of the character of vibrations. When the character of vibrations changes, slight differences can be observed, but the values of critical load overlap, and the type of vibrations is the same for both methods. However, for large values of the amplitude of external excitation, significant differences in the obtained type of vibrations are observed (harmonic vibrations for $N = 11$ and chaotic vibrations for $n = 8$).

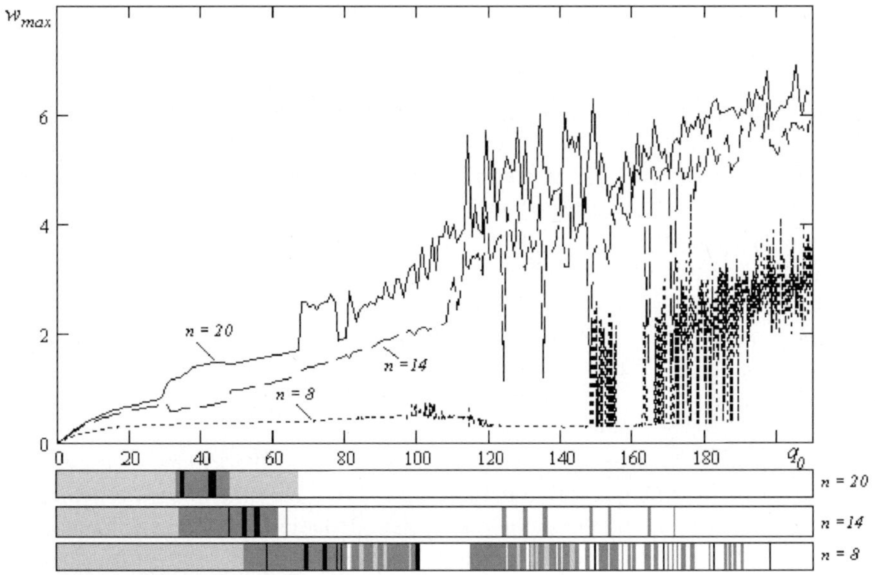

Fig. 12.7 Dependence $w_{max}(q_0)$ for various values of n

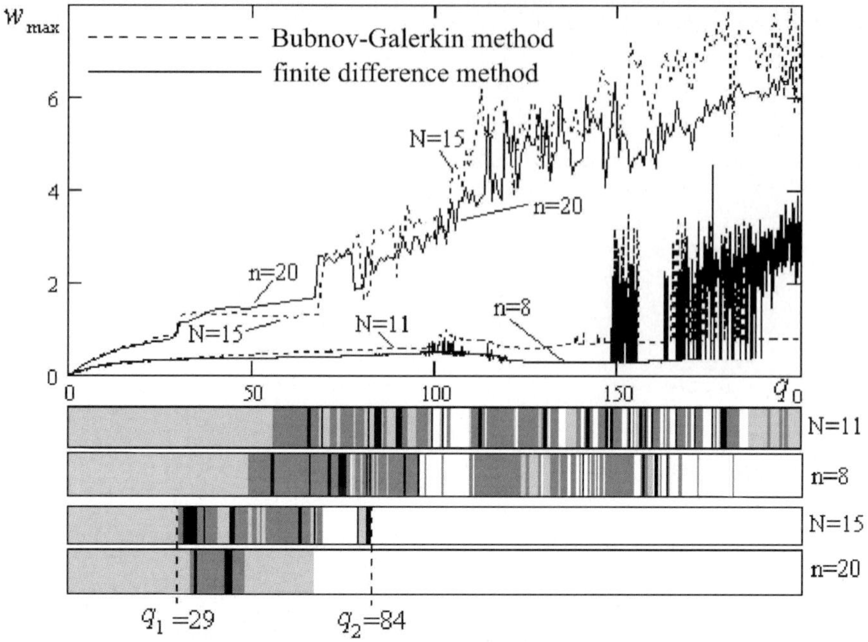

Fig. 12.8 Comparison of the method of finite differences and the Bubnov-Galerkin method

By increasing the number of division of spatial coordinates in the method of finite differences ($n = 20$) and the number of approximations in the Bubnov-Galerkin method ($N = 15$) we managed to obtain the convergence of results. In this way, complete overlap of the results obtained for the small amplitude of exciting force ($q_0 \leq q_1$), and the convergence with respect to the type of vibrations for a larger amplitude of exciting force ($q_0 \geq q_2$) has also been achieved. Segment $\{q_1, q_2\}$ is a transformation element between harmonic vibrations and the state of complete chaos of a material system. We can observe complex vibrations here, e.g., two-frequency vibrations and subharmonic vibrations and the areas of Andronov-Hopf bifurcation. Obtaining complete overlap of the results of both methods is not possible, but such convergence has been achieved at some isolated points. Generally, we can state that it is possible to obtain such qualitative convergence of the results from the two methods mentioned before. We should note here that the increase of the number of range divisions in the method of finite differences and the number of approximations in the Bubnov-Galerkin method leads to better results.

The approximation $N = 15$ in the case of Bubnov-Galerkin method and approximation $n = 20$ for the method of finite differences turn out to be threshold values for the time of calculations, as increasing them leads to the sudden prolongation of work of computational algorithm. Table 12.3 shows the time needed for calculations in the analysis of chaotic vibrations of flexible and rectangular shells with the application of the Bubnov-Galerkin method and the method of finite differences on a PC computer with Pentium 4 processor ($2GHz$).

Table 12.3 Comparison of effectiveness of the Bubnov-Galerkin method and the method of finite differences

Bubnov-Galerkin method	$N = 11$	$N = 13$	$N = 15$
Time of time history analysis	1 min	2.2 min	4 min
Time of vibration scale construction	3.5 hours	7.7 hours	14 hours
Time of control parameters plane construction	30 days	68 days	122 days
Finite difference method	$n = 8$	$n = 14$	$n = 20$
Time of time history analysis	0.66 min	3 min	9 min
Time of vibration scale construction	2.25 hours	10.5 hours	30 hours
Time of control parameters plane construction	20 days	81.2 days	261 days

12.5 Conclusions

On the basis of cases analyzed here, it has been shown that the convergence of the results obtained with the application of both methods can be achieved with respect to the type of vibrations, i.e., by the power spectrum. As in this case, unlike in static problems, the convergence on the basis of time courses cannot be obtained, we still managed to obtain such convergence in an average aspect, i.e., with respect to the character of vibrations. It should be emphasized here that for small amplitudes of excitement, we obtained the convergence of the results for both methods also in a traditional way, i.e., on the basis of a time course. The increase of the number of division elements in the method of finite differences and of the number of approximations in the Bubnov-Galerkin method lead to significantly better results, although there is some threshold value beyond which reliable results cannot be obtained for the time being.

Chapter 13
Determination of Three-layered Nonlinear Uncoupled Beam Dynamics with Constraints

In this chapter both regular and chaotic vibrations, various bifurcations, and scenarios exhibited by three-layered non-linear uncoupled beams with constraints are illustrated and discussed. The finite difference approximation is applied and numerical results reliability is first rigorously discussed. New scenarios of transition to chaos and synchronization phenomena are reported, and the essential influence of four various boundary conditions on various nonlinear behaviors is outlined.

13.1 Introduction

Investigation of multi-layered beams with constraints is one of the challenging tasks of today's nonlinear dynamics. Classical approaches of modeling three-layer beams and plates are reported in references [79, 80, 202, 204, 205]. Experimental results of compressional damping measurement and transverse vibrations study in an elastic-viscoelastic-elastic sandwich beam are presented in references [273, 275]. Free vibration analysis of non-uniform beams with an arbitrary number of cracks and concentrated masses is carried in Li's paper [187]. The eigenvalue equation of a non-uniform beam with any two kinds of end support, any finite number of cracks, and with concentrated masses is determined from a second-order determinant, which significantly saves computational effort. Three identification methods of nonlinear model behavior of an externally excited cantilever beam are proposed and discussed in reference [81]. The propagation of structural waves on an infinitely long periodically supported Timoshenko beam is studied in reference [120]. The power series expansion of displacement components method, yielding a set of fundamental dynamic equations of a one-dimensional higher order theory for a laminated composite beam subjected to axial stress derived via the Hamilton principle, is applied by Matsunaga [201]. The approach introduced is used for the analysis of natural frequencies and buckling stresses of laminated composite beams, taking into account the complete effects of transverse shear, normal stresses, and rotary inertia. A vast number of papers is devoted to control of beams. Here

J. Awrejcewicz, V.A. Krysko, *Chaos in Structural Mechanics*,
DOI: 10.1007/978-3-540-77676-5_14, © Springer-Verlag Berlin Heidelberg 2008

only a few recent ones are mentioned, but readers may use the references cited in those papers. Active vibration control of beams with smart constrained layer damping treatment is proposed by Balamurugan and Narayanan [50]. The sensor piezoelectric layer measures the vibration response of the structure, and a feedback controller is used to regulate the axial deformation of the piezoelectric actuator and hence to provide adjustable damping in the structure. The piezoelectric control of damped sandwich beams is proposed in reference [296]. The analysis of the proposed linear quadratic regulator, and the real representation of complex modal reduced models for hybrid piezoelectric-active viscoelastic-passive vibration control, are reported. In addition, the parametric analysis and the comparative study of control strategies of damped sandwich beams are given. Investigation of chaos exhibited by beams with different boundary conditions is reported in a series of references [146, 223, 224, 226, 320, 323]. In this chapter contact complex vibrations of three-layered nonlinear uncoupled beams are studied. A gap (constraint) occurs between three-layered beams. It is assumed that in a contact zone the beams may freely slip, which characterizes a contact condition violation due to occurrence of a transversal load action. In contact zones where stretching occurs, a sudden lack of contact may appear. A pressure in a contact zone is defined via the hypothesis given in reference [131]. Each of the beams is governed by the Euler-Bernoulli kinematics hypothesis. This chapter addresses the problems of regular and chaotic vibrations of three uncoupled beams in a dissipative surrounding medium. In general, the beams material is considered a nonlinear and elastic one. In particular, problems with the following types of nonlinearity are considered: (i) material nonlinearity, where beam material is elastic and (ii) material kinematic and geometric nonlinearities (beam material deformation is governed by one of the given deformation diagrams). Each of these nonlinearities is illustrated and discussed separately. Our continuous system consisting of the package of three-layered beams is solved using a finite difference method with error $O(h^2)$. The various scenarios of transition of harmonic vibrations of a three-layered package into a chaotic state for different boundary conditions (four boundary conditions, 13.10–13.12) are studied. In all cases considered, zero values of the initial conditions are applied.

13.2 Fundamental Relations

In order to outline a need for theoretical analysis of a sandwich beam, two different classes of problems will be briefly discussed. The first class consists of coupled beams with separate contact zones, which occur due to either technological defects or special exploration conditions. The second class includes uncoupled beams with constraints and interactions through boundary conditions. Note that such a sandwich beam may include many layers with different thickness and mechanical properties, as well as clearances. This class is studied within this chapter. The mentioned layers may either slip freely or with friction. The contact conditions between layers may

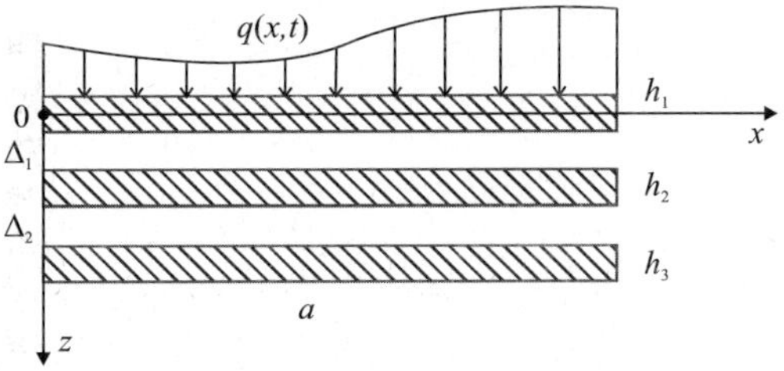

Fig. 13.1 The analyzed beam sandwich

depend on time and coordinates, and may include all possible non-ideal one-sided contacts between layers.

However, weld conditions in both normal and horizontal directions are not considered. Layer behavior coincides with one of the nonlinear beam theories given later. The functions of contact pressure between layers are excluded from the unknowns number [131].

The analyzed beam sandwich is reported in Fig. 13.1, which includes three beams with thickness h_l $(l = 1,\ 2,\ 3)$.

Observe that the origin of the coordinates is situated left in the upper beam in its mean curvature, and constraints are denoted by δ_m $(m = 1,\ 2)$.

The coordinate x is measured beginning from the left beam and is extended along the beam axis, whereas the z axis is measured from the middle beam curve down. The sandwich length is denoted by a. Deformations occurring in the middle beam curvatures are assumed to be small and are neglected. A beam curvature deformation is characterized through the parameter χ_l. The beam material is assumed to be isotropic, and hence the Young modulus E_l, the shear modulus G_l, the volume elasticity modulus K_l, the transversal deformation coefficient ν_l, and the plastic flow coefficient σ_{sl} are functions of (x,y). In other words, we consider the beam material as non-homogeneous until a deformation takes place, but is physically nonlinear. In what follows we assume that physical material parameters $E_l,\ G_l,\ K_l,\ \nu_l$ are unique functions of a material point and its deformation state. This assumption is within the frame of nonlinear theory of elastic deformations, or the theory of small elastic-plastic deformations. A material point deformation state is characterized by the volume deformation ε_{0l} and intensity of deformation ε_{il}. The theory introduced below is based on the hypothesis of straight normals, i.e., the Euler-Bernoulli hypothesis.

For the case of two beams made from the same material and having the same thickness, the following Winkler relation between clamping and contact pressure holds (see Fig. 13.2):

$$q_k = k\frac{E}{h}(w_1 - \delta_1 - w_2), \tag{13.1}$$

where k is the Winkler proportionality coefficient, δ_1 denotes beams distance, whereas $w_1,\ w_2$ describe deflections of first and second beams, respectively. Otherwise (i.e., for two different beam materials and different distances) one obtains

$$q_k = k\left(1 + \frac{E_1 h_2}{E_2 h_1}\right)\frac{E_1}{h_1}(w_1 - \delta_1 - w_2). \tag{13.2}$$

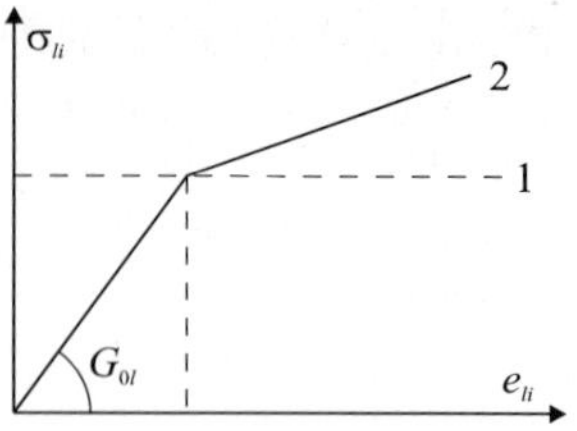

Fig. 13.2 Scheme of ideally (13.1) and with linear strain hardening (13.2) elastic-plastic material

Since in real systems very often $E_1 h_2 \simeq E_2 h_1$, the further formula (13.1) is applied assuming E and h to be characteristic quantities. Note that the function $q_k(w)$ defined through formula (13.1) is linear with respect to a measure of transversal beam damping in a contact zone. However, use of nonlinear relation between q_k and w does not lead to serious complexity of the further introduced method to solve the considered contact problems, i.e., instead of q_k one may substitute $q_k(w)$ in the equilibrium equations, which is widely used in equations of nonlinear beams theory. It is worth noting that deriving the beam motion, the dependencies $E = E(x,\ z,\ \varepsilon_0,\ \varepsilon_i)$, $\nu = \nu(x,\ z,\ \varepsilon_0,\ \varepsilon_i)$ are assumed to be given through theory of small elastic-plastic deformations [57, 64, 83, 124, 255].

13.3 Formulation of the Problem and Computational Algorithm

It is assumed that one of the beams is transversally and periodically driven. Let the investigated beams occupy the following domains in R^3:

$$\Omega_l = \{(x,y,z)|\quad 0 \le x \le a, \quad 0 \le y \le b, \quad a_l \le z \le b_l\}, \quad (l = 1,2,3),$$

where:

$$\begin{aligned} &\alpha_1 = -h_1/2, \quad \alpha_2 = (h_1/2 + \delta_1), \qquad \alpha_3 = (h_1/2 + \delta_1 + \delta_2 + h_2), \\ &\beta_1 = h_1/2, \qquad \beta_2 = (h_1/2 + \delta_1 + h_2), \quad \beta_3 = (h_1/2 + \delta_1 + \delta_2 + h_2 + h_3) \end{aligned}$$

for the first, second, and third beam, respectively.

Following the mentioned theory of small elastic-plastic deformations, the differential equation modeling an equilibrium state of the beam with bending stiffness $J_l(x)$ and with an account of the kinematic Euler-Bernoulli model and subject to an action of the transversal load $q_l^*(x,t)$ reads

$$\frac{\partial^2}{\partial x^2}\left(J_l(x)\frac{\partial^2 w_l}{\partial x^2}\right) = q_l^*(x,t), \tag{13.3}$$

where due to the d'Alembert principle:

$$q_l^*(x,t) = q_l + q_{kl} - \frac{\gamma_l}{g}h_l\frac{\partial^2 w_l}{\partial t^2} - \varepsilon_l\frac{\partial w_l}{\partial t}.$$

The following notation is used: $q_l(x_l,t)$ is the transversal load; h_{0l} l-th beam thickness with respect to its center $x = a/2$; w_l is the beam deflection; E_l is the Young modulus; b_l is the beam width; t is the time; ε_l is the damping coefficient; γ_l is the weight density; g is the acceleration due to gravity; and l is the beam number.

Equation (13.3) can be presented in its equivalent form

$$h_l \frac{\partial^2 w_l}{\partial t^2} + \varepsilon_l \frac{\partial w_l}{\partial t} = -L_l(w_l) - q_{kl} - q_l, \quad l = (1,2,3). \tag{13.4}$$

Here and later indices $l = 1,\ 2,\ 3$ are associated with upper, middle, and below beam, respectively.

The introduced nonlinear operators in (13.4)

$$L_l(w_l) = \frac{\partial^2}{\partial x^2}\left(J_l(x)\frac{\partial^2 w_l}{\partial x^2}\right), \tag{13.5}$$

where

$$J_l(x) = b_l \int_{d_l}^{b_l} E_l z^2 dz. \tag{13.6}$$

A lack of contact between beams is accounted for through formulas analogous to formula (13.1), i.e.:

$$q_{kn} = (-1)^n k_1 \frac{E_1}{h_1}\left(w_1 - d_1 - w_2 \frac{h_{02}}{h_{01}}\right)\psi_1, \quad (n = 1,2), \tag{13.7}$$

$$q_{km} = (-1)^{m+1} k_2 \frac{E_2}{h_2}\left(w_2 - d_2 - w_3 \frac{h_{03}}{h_{02}}\right)\psi_2, \quad (m = 1,2), \tag{13.8}$$

where k_l is the proportionality coefficient between contact pressure and clamping. The function ψ_l characterizes contact zone dimension, and possesses the following form:

$$\psi_n = \left(1 + sgn\left(w_n - \delta_n - wn + 1\frac{h_{0n+1}}{h_{0n}}\right)\right)/2, \quad (n = 1,2). \tag{13.9}$$

Recall that an occurrence of the multiplier ψ_n in equations governing beams dynamics transforms them to a nonlinear form, and the new problem obtained is both physically and structurally a nonlinear one. By a structural nonlinearity we mean nonlinearity causing variation of a computational scheme within the process of deformation. The integration space holds for $X_l(x = (0,a))$ with the boundary ∂X_l.

Equations (13.4) are solved together with one option of the following boundary conditions on ∂X_l:

(i) Stiff clamping

$$\frac{\partial w_l(x,t)}{\partial x} = w_l(x,t) = 0\,; \tag{13.10}$$

(ii) Hinged clamping

$$\frac{\partial^2 w_l(x,t)}{\partial x^2} = w_l(x,t) = 0\,; \tag{13.11}$$

(iii) Free boundary

$$\frac{\partial^2 w_l(x,t)}{\partial x^2} = \frac{\partial}{\partial x}\left(J_l(x)\frac{\partial^2 w_l(x,t)}{\partial x^2}\right) = 0\,; \tag{13.12}$$

Furthermore, initial conditions may take the form

$$\frac{\partial w_l(x,0)}{\partial x} = F_l(x), \quad w_l(x,0) = f_l(x), \quad (0 \le x \le a),\ l = 1,2,3, \tag{13.13}$$

where F_l and f_l are functions characterizing a distribution of velocities and deflections of beams in the initial time instant.

The following non-dimensional parameters are introduced:

$$x = \bar{x}a,\ z = \bar{z}h_{0l},\ h_l = \bar{h}_l h_{0l},\ w_l = \bar{w}_l h_{0l},$$
$$E_l = \bar{E}_l G_{0l},\ b_l = \bar{b}_l b_{0l},\ d_l = \bar{d}_l h_{0l},$$

$$t = \bar{t}\frac{a^2}{h_{0l}}\sqrt{\frac{\gamma_l}{G_{0l}b_{0l}}},\ \varepsilon_l = \bar{\varepsilon}_l\frac{a^2}{h_{0l}^2}\sqrt{\frac{\gamma_l}{gG_{0l}b_{0l}}},\ k_l = \bar{k}_l\frac{G_{02}h_{02}^4 b_{0l}}{G_{0l}a^4}.$$

Beam dynamics is governed by the following non-dimensional equations, where the overbars are omitted for the sake of simplicity:

$$h_l\frac{\partial^2 w_l}{\partial t^2} + \varepsilon_l\frac{\partial w_l}{\partial t} = -L_l(w_l) - q_{kl} - q_l, \quad l = (1,2,3). \tag{13.14}$$

An account of physical beam material nonlinearity is realized through the method of variated elasticity parameters. Owing to this method shear and Young moduli are coupled via the following relation

$$E_l = \frac{9K_lG_l}{3K_l + G_l}. \tag{13.15}$$

K_l is treated as a constant and equal to 1.94 G_{0l}. Recall that in theory of plasticity, shear modulus is defined through the formula

$$G_l = \frac{1}{3}\frac{\sigma_{li}(e_{li})}{e_{li}}. \tag{13.16}$$

For example, one of the following cases can be applied [25, 26]:

1. Ideally elastic-plastic material:

$$\sigma_{li} = 3G_{0l}e_{sl} \quad \text{for} \quad e_{li} < e_{ls},$$
$$\sigma_{li} = \sigma_{ls} \quad \text{for} \quad e_{li} \ge e_{ls}\,; \tag{13.17}$$

Fig. 13.3 Scheme of pure aluminum behavior

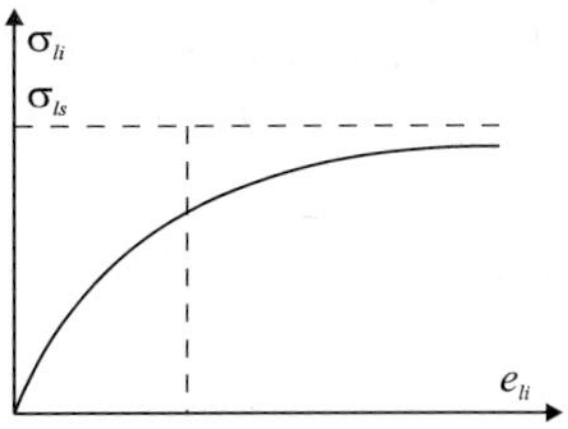

2. Elastic-plastic material with linear strain hardening:

$$\sigma_{li} = 3G_{0l}e_{li} \quad \text{for} \quad e_{li} < e_{ls},$$
$$s_i = 3G_{0l}e_{ls} + 3G_{1l}(e_{li} - e_{ls}) \quad \text{for} \quad e_{li} \geq e_{ls}\,; \tag{13.18}$$

3. Diagram for pure aluminum (Fig. 13.3):

$$\sigma_{li} = \sigma_{ls}\left(1 - \exp\left(-\frac{e_{li}}{e_{ls}}\right)\right). \tag{13.19}$$

In the above the following notation is applied: G_{0l}, G_{1l} is the shear moduli; e_{il} is the deformation intensity; σ_{il} is the stress intensity; e_{sl} is the flow deformation intensity; and K_l is the volume elasticity modulus.

Since for a beam we have $e_{yyl} = e_{zzl} = e_{xyl} = e_{xzl} = e_{zyl} = 0$, the deformation intensity reads

$$e_{li} = \frac{2}{3}\left| z\frac{\partial^2 w_l}{\partial x^2}\right|. \tag{13.20}$$

In our proposed model the introduced transversal load may act either on all beams simultaneously, or on each of them separately. Variations of the load along beam length and in time can be realized in an arbitrary manner.

In order to integrate Eq. (13.1) the finite differences method with error $O(h^2)$ is applied. For this purpose the space $D = \{(x,t)|0 \leq x \leq 1,\ 0 \leq t \leq T\}$ is covered by the rectangular mesh $x_i = ih_x$, $t_j = jh_t$ $(i = 0,\ 1,\ 2,\ldots n;\ j = 0,\ 1,\ 2,\ldots)$, where $\triangle\, x_i = x_{i+1} - x_i = h_x = 1/n_x$ (n_x is integer) and $h_t = t_{j+1} - t_j$. In the mesh x_i, t_j, the differential eqs. (13.4), (13.10)–(13.13) are substituted by the corresponding finite differences. In order to increase accuracy of the obtained results symmetric formulas for derivatives have been applied (for a comparison see Tables 13.1–13.3).

Table 13.1 Comparison of beam center deflection computations for various time and spatial steps

	$h_t = 0.001$	0.0001	0.00004	0.00002
$n_x = 10$	0.02362	0.02362	0.02362	0.02362
16	0.02351	0.02351	0.02351	0.02351
20	0.02348	0.02348	0.02348	0.02348
24	0.02347	0.02347	0.02347	0.02347
28	0.02346	0.02346	0.02346	0.02346

Table 13.2 Magnitudes of the fundamental frequency for different time and spatial steps

	$h_t = 0.001$	0.0001	0.00004	0.00002
$n_x = 10$	32.8657	32.8651	32.8651	32.8650
16	32.4956	32.4949	32.4949	32.4949
20	32.4109	32.4103	32.4102	32.4102
24	32.3651	32.3644	32.3644	32.3644
28	32.3375	32.3368	32.3268	32.3368

Table 13.3 Absolute and relative errors of fundamental amplitude estimation

	10	16	20	24	28
Amplitude	32.8650	32.4949	32.4102	32.3644	32.3368
δ	0.5282	0.158	0.0734	0.0276	0
$\delta\%$	1.63	0.488	0.226	0.085	0

For example, one may derive the following relations for the beam with number l:

$$w_{li,j+1} = \frac{1}{1+\frac{\varepsilon_l h_t}{2h_{li}}}\left[2w_{li,j} + \left(\frac{\varepsilon_l h_t}{2h_{li}} - 1\right) w_{li,j} + \frac{h_t^2}{h_{li}}\left((-1)^l q_{kl} + L(w_l)_{i,j}\right)\right],$$

where

$$L(w_l)_{i,j} = \left[J_{li+1}w_{li+2,j} - 2(J_{li+1} + J_{li})w_{li+1,j} + (J_{li+1} + 4J_{li} + J_{li-1})w_{li,j}\right.$$
$$\left. -2(J_{li} + J_{li-1})w_{li-1,j} + J_{li-1}w_{li-2,j}\right]/h_x^4, \quad (l = 1,2,3).$$

In the above $i(j)$ corresponds to a spatial (time) $(l = 1,\ 2,\ 3)$ coordinate.

Finally, a three-layered scheme can be obtained. In order to compute $w_l(x,t)$ on the layer $(j+1)$, the values $w_l(x,t)$ of two previous layers j-th and $(j-1)$-th are accounted for. The computation begins from $w_l(x,t)$ in a fictitious layer with the number $j = -1$. Furthermore, a derivative in the initial conditions (13.13) is also substituted by finite difference relations using the non-symmetric difference.

In order to apply the method of variated parameters, a beam is divided into n_2 layers. In each time step the following quantities are computed for a node x_j; deformation intensity through formula (13.20); E_l and G_l using one of relations (13.17)–(13.19) (depending on the chosen deformation diagram); integral (13.6) is estimated through the Simpson rule.

An example where upper and below beams are hinged on their ends is further illustrated and discussed.

The following initial conditions are introduced:

$$F_l(x) = f_l(x) = 0, \quad i = 1,2,3 \quad (x \le x \le 1). \tag{13.21}$$

Note that a transversal load is applied only to the top beam and possesses the following harmonic form:

$$q_1 = q_{10}\cos(\omega t). \tag{13.22}$$

The given algorithm has been tested for its convergence with respect to both spatial and time meshes in static and dynamic problems. Applying a set-up method [95], a deflection of a hinged beam center with a uniformly distributed load along its length within a static problem has been calculated.

Owing to analysis of the obtained results, a beam deflection does not depend on a time step magnitude at least for $2 \cdot 10^{-5} \leq h_t \leq 10^{-3}$. A change of spatial distribution from $n_x = 28$ to $n_x = 10$ results in an error only of 1%. Note that for $h_t = 2 \cdot 10^{-5}$ and $n_x = 28$, the deflection achieved in the beam center is equal to 0.02346, which is in very good agreement with results obtained through an analytical method in statics (see [54]), i.e., 0.02343. The same values of h_t and n_x have been applied to solve dynamical problems. Namely, a hinged beam subject to transversal load (13.22) is studied, assuming $q_{10} = 1.5$ and $\omega = 0.5$. Analysis of the obtained results yielded the conclusion that a variation from $n_x = 28$ to $n_x = 10$ caused a change of the fundamental amplitude of only 1.6% in a frequency spectrum. Owing to the discussed preliminary computations, the spatial step with respect to the x coordinate equals 0.05 ($n_x = 20$), whereas the time step $t = 2 \cdot 10^{-5}$ and the number of layers along thickness $n_z = 20$.

13.4 Structurally Nonlinear Problems

Problem 13.4.1 Computations are carried out for $k_1 = k_2 = 2000$; $0 \leq t \leq 120$; $\delta_1 = \delta_2 = 0.05$; beam material is linear; upper ($l = 1$) and lower ($l = 3$) beams are hinged on their left edges ($x = 0$) (see (13.11)) whereas they are clamped on their right edges (see (13.10)) and free on their right ends (see (13.12)). The whole sandwich is governed by Hook's law.

Recall that the top beam is always subjected to the harmonic excitation $q_1 = q_{10}\cos(\omega t)$. Let us fix the excitation frequency $\omega = 1.0$ and treat the excitation amplitude q_{10} as the control parameter. Note that now and later excitation frequency is chosen to be close to free vibration frequencies of both upper and lower beams. Observe that the top beam vibrations are harmonic unless its contact with the middle beam is achieved. For each value of q_{10} the following dependencies are constructed: deflection variation ($w_1(0.5, t)$, $w_2(0.5, t)$, $w_3(0.5, t)$), power spectrum (note that the mentioned characteristics are qualitatively similar for other beams and they are not shown), deflection surface ($w_1(x, t)$, $w_2(x, t)$, $w_3(x, t)$) and the contact pressure between first and second, as well as second and third beams ($q_{12}(x, t)$, $q_{23}(x, t)$) for the time and space intervals attached to figures.

Now a scenario of a three-layered package from harmonic to chaotic vibrations will be briefly described.

For $q_{10} = 0.5$ the first contact between upper and middle beams occurs, then a contact loss (jump) between the second and third beams takes place. Since our system is deterministic, the middle beam vibrations are damped owing to occurrence of damped waves. Both jump and successive damped vibrations effects are clearly expressed by the contact surface pressure, which occupies a narrow zone

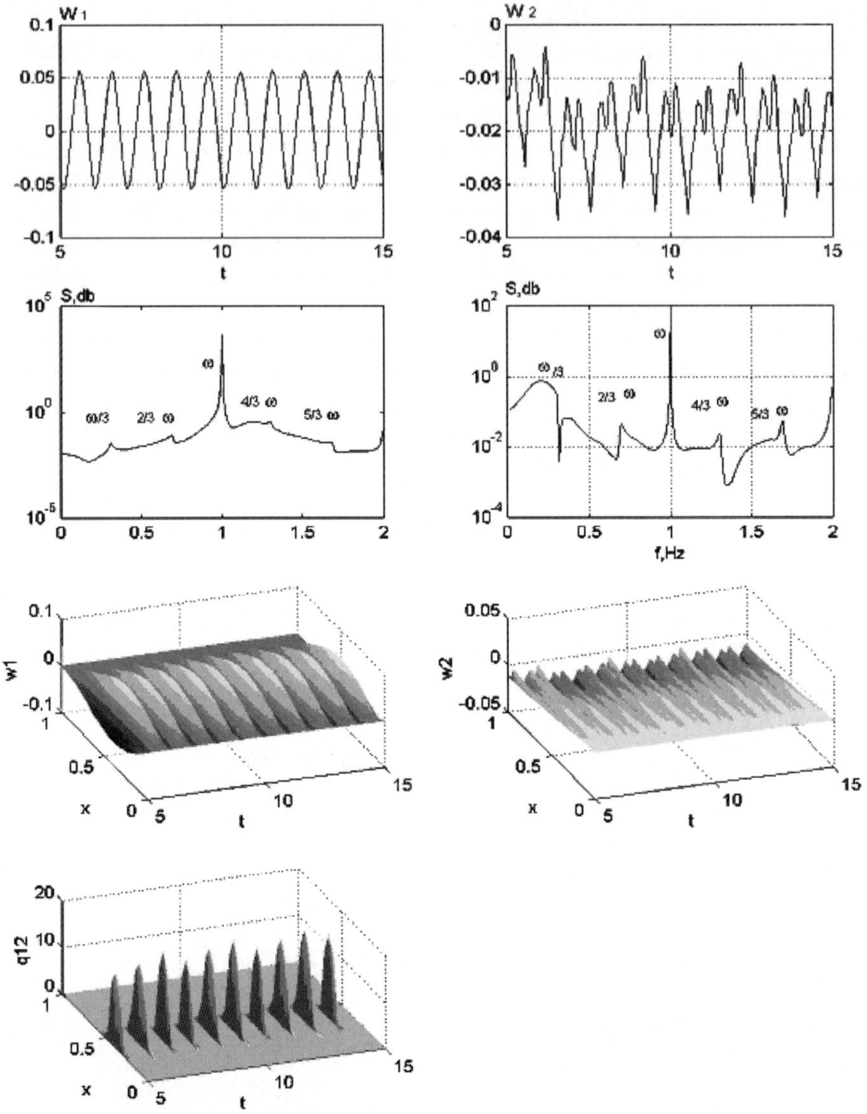

Fig. 13.4 Time histories of deflection $w_i(0.5,\ t)$ $i = 1,2$, the corresponding power spectra, deflection surfaces $w_i(x,\ t)$ $i = 1,\ 2$, and the contact pressure $q_{12}(x,\ t)$ between the first two beams for $q_{10} = 0.8$

in space ($x = 0.4$) and time. Observe that when the beam inertia wins a challenge with the dissipative forces ($q_{10} = 0.8$), synchronization occurs associated with period tripling bifurcations in both beams simultaneously (Fig. 13.4). The latter phenomenon is also exhibited via a periodic variation with the same amplitude of the contact pressure. It should be emphasized that the reported period tripling exhibited by both beams simultaneously has not been found yet during investigation of either one-layered beams, plates, or shells. Nowadays chaos exhibited by discrete dynamical systems is relatively well understood. Many appropriate theorems are formulated for such systems including the very important one, the so-called window 3, which is associated with analysis of the map $f(x) = x^2 + C$. A natural question arises: do other periodic orbits exist? The answer is given by the Sharkovsky theorem. However, in the beginning of the Sharkovsky ordered sequence period doubling bifurcation appears almost everywhere (one or few of them), and then period tripling bifurcations follow. In contrast, in our case already in the beginning a period tripling bifurcation occurs.

It is worth noting that for constructively nonlinear beams treated within the theory of contact problems of continuous systems, period tripling bifurcations similar to that governed by the mentioned one-dimensional map have been detected. Observe that the described chart of synchronized vibrations of two beams can be traced on a relatively long interval of the load until a contact between second and third beam is reached. The contact pressure between the first and second beams is uniform in the sense of excitation amplitude and frequency, but occurring locally. In other words the full synchronization of vibrations between two beams is observed. Owing to increase of q_{10}, the second beam starts to touch the third beam, it starts to jump, and the damped vibrations appear. Although synchronization of the first two beams for the frequency associated with the period tripling holds, the second beam end exhibits a beating phenomenon. A minor increase of q_{10} yields an occurrence of a new synchronization type. Namely, two upper beams vibrate with the period 4. However, when q_{10} achieves the new critical value ($q_{10} = 1.0$) all three beam dynamics are synchronized for the frequency corresponding to a doubled period of the two upper beams. Note that the third beam vibrates with the frequency associated with two doubled periods. Further increase of the load up to ($q_{10} = 1.2$) yields a collapse of synchronized motion of the two top beams (Fig. 13.5), and the whole sandwich is transited to chaos on the frequencies associated with the doubled period. In particular, chaotic vibrations are exhibited by the third beam with the accompanied vibration break (see $w_3(0.5,\ t)$ and contact pressure charts).

Problem 13.4.2 Now upper ($l = 1$) and lower ($l = 3$) beams are clamped on both ends (see (13.10)), whereas the middle beam is fixed as in Problem 13.4.1.

In this case the frequency $\omega = 1.54$ is fixed. First of all, the vibrations are qualitatively similar to those reported in the previous case until a contact between the second and third beams occurs. Namely, the first two beams vibrate in a synchronized manner, but the synchronization follows one Hopf-Andronov bifurcation. Owing to a successive increase of the transversal load amplitude, the vibration character of two synchronized top beams dynamics is changed (after period doubling bifurcation, a period tripling bifurcation takes place, i.e., Sharkovsky diagram is manifested (Fig. 13.6)). The third beam exhibits the vibration break. In other words, a synchronization of a whole three-layered package is not observed. The increase of q_{10} successfully transits the vibrations to chaos, which is clearly outlined by contact pressure distribution between first and second, as well as second and third beams.

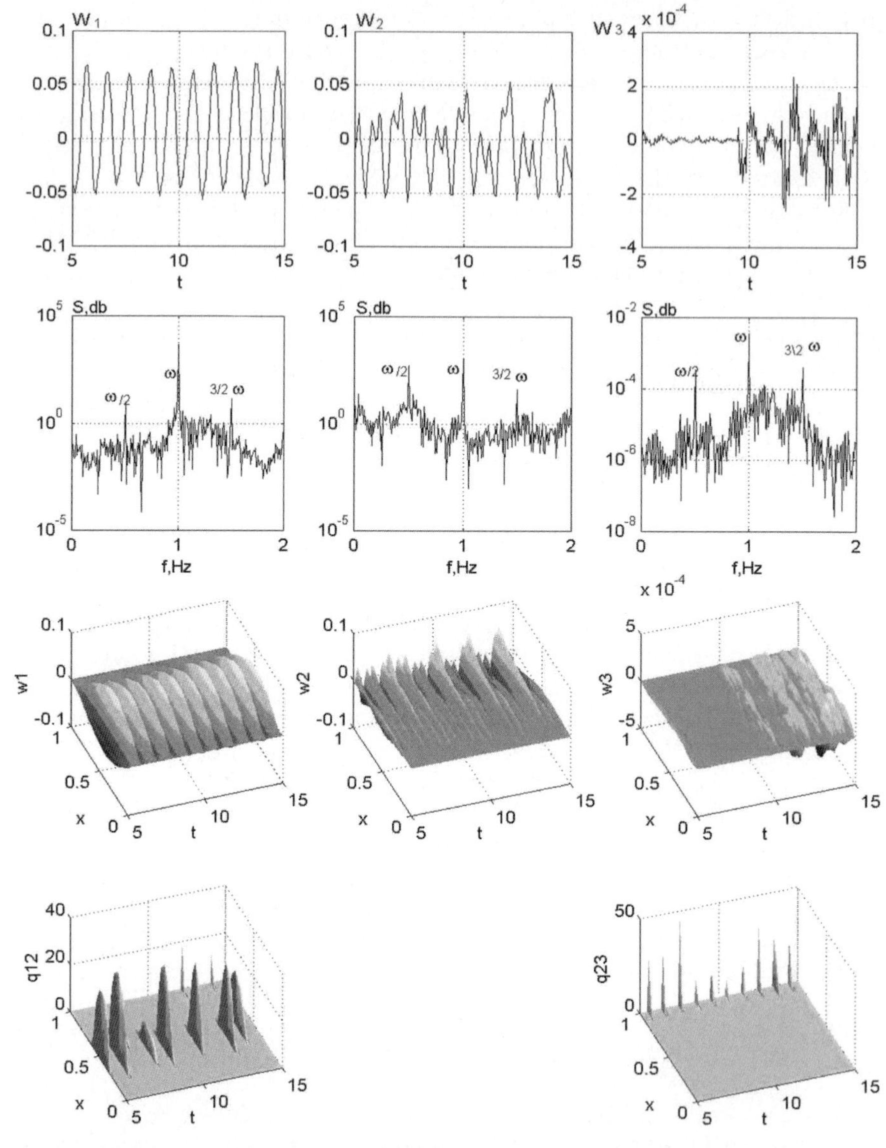

Fig. 13.5 Time histories of deflection $w_i(0.5, t)$ $i = 1, 2$, the corresponding power spectra, deflection surfaces $w_i(x, t)$ $i = 1, 2, 3$, and the contact pressure $q_{12}(x, t)$ between the first two beams for $q_{10} = 1.2$

Problem 13.4.3 Here upper ($l = 1$) and lower ($l = 3$) beams are clamped on their left ends (see (13.10)), whereas they are hinged on their right ends (see (13.11)); middle beam is fixed as in previous problem.

In this case the external frequency $\omega = 1.0$ is fixed. In this problem, similarly to the discussed Problems 13.4.1 and 13.4.2, the dynamic behavior is qualitatively similar until a contact associated with the third beam occurs. Synchronization

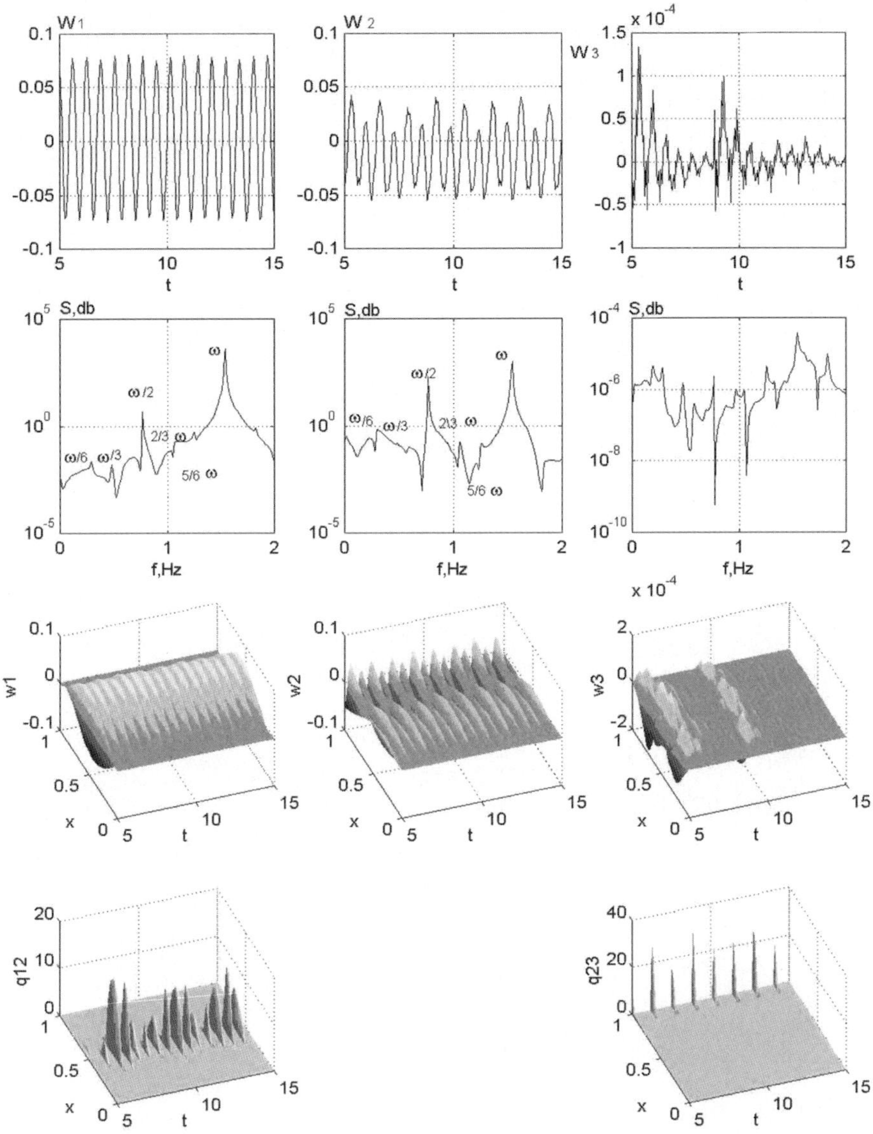

Fig. 13.6 Time histories of deflection $w_i(0.5,\ t)$ $i = 1,\ 2,\ 3$, the corresponding power spectra, deflection surfaces $w_i(x,\ t)$ $i = 1,\ 2,\ 3$, and the contact pressure $q_{12}(x,\ t)$ and $q_{23}(x,t)$, $(q_{10} = 1.5,\ \omega = 1.54)$

occurrence of the first and second beams on the frequency associated with period tripling (Fig. 13.7) is followed by a transition into excitation period, i.e., here the full phase locking phenomenon with respect to one excitation frequency occurs. The latter observation is exhibited by periodic behavior of contact pressure, which has not appeared in the beginning of synchronization (Fig. 13.7). For $q_{10} = 0.65$ a strange chaotic attractor is exhibited, which is associated with period doubling of the

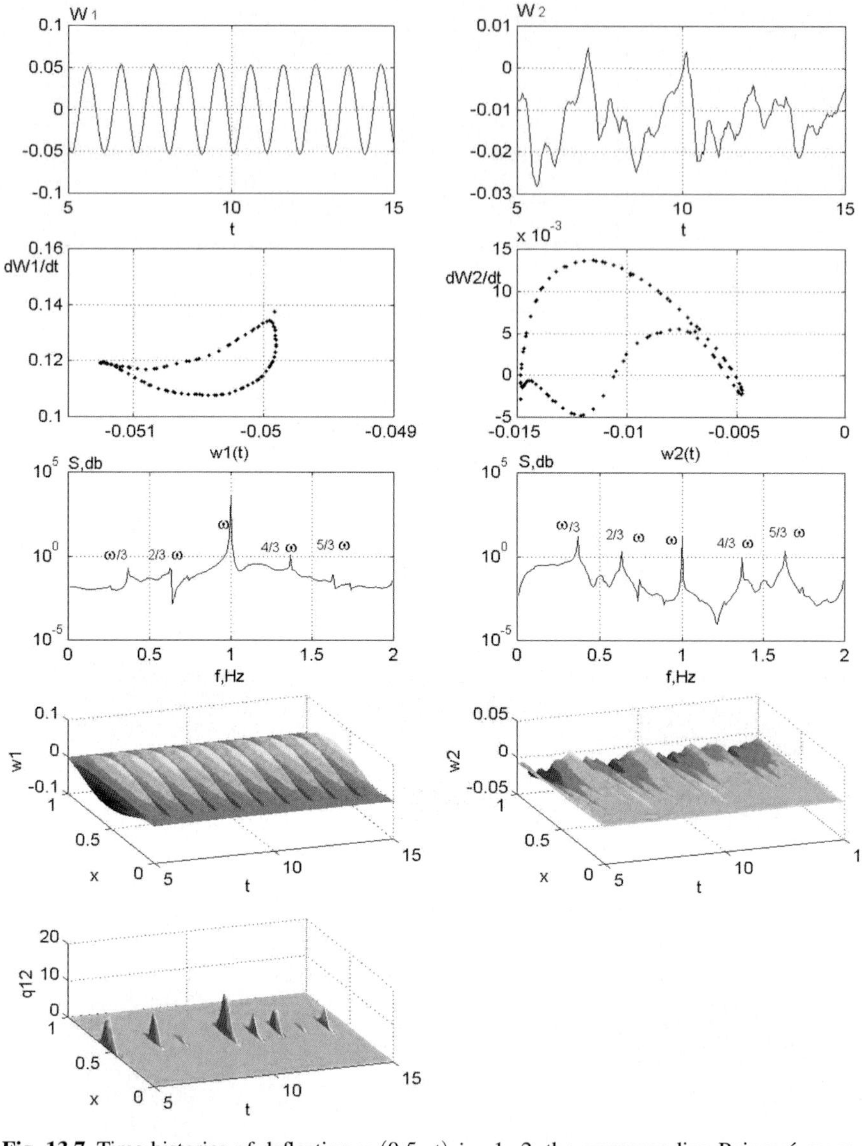

Fig. 13.7 Time histories of deflection $w_i(0.5,\ t)$ $i = 1,\ 2$, the corresponding Poincaré maps and power spectra, deflection surfaces $w_i(x,t)$ $i = 1,\ 2$, and the contact pressure $q_{12}(x,\ t)$, $(q_{10} = 0.65,\ \omega = 1.0)$

fundamental excitation period. The contact between third and second beams forces the system to change the synchronized regime of two first beams, i.e., vibrations with two Andronov-Hopf bifurcations occur (Fig. 13.8). A successive increase of the transversal load amplitude causes again a change of the first two beam vibrations (period tripling) with the successive system transition into a chaotic state on tripled frequencies of the first two beams (the third beam vibrates with the fundamental frequency).

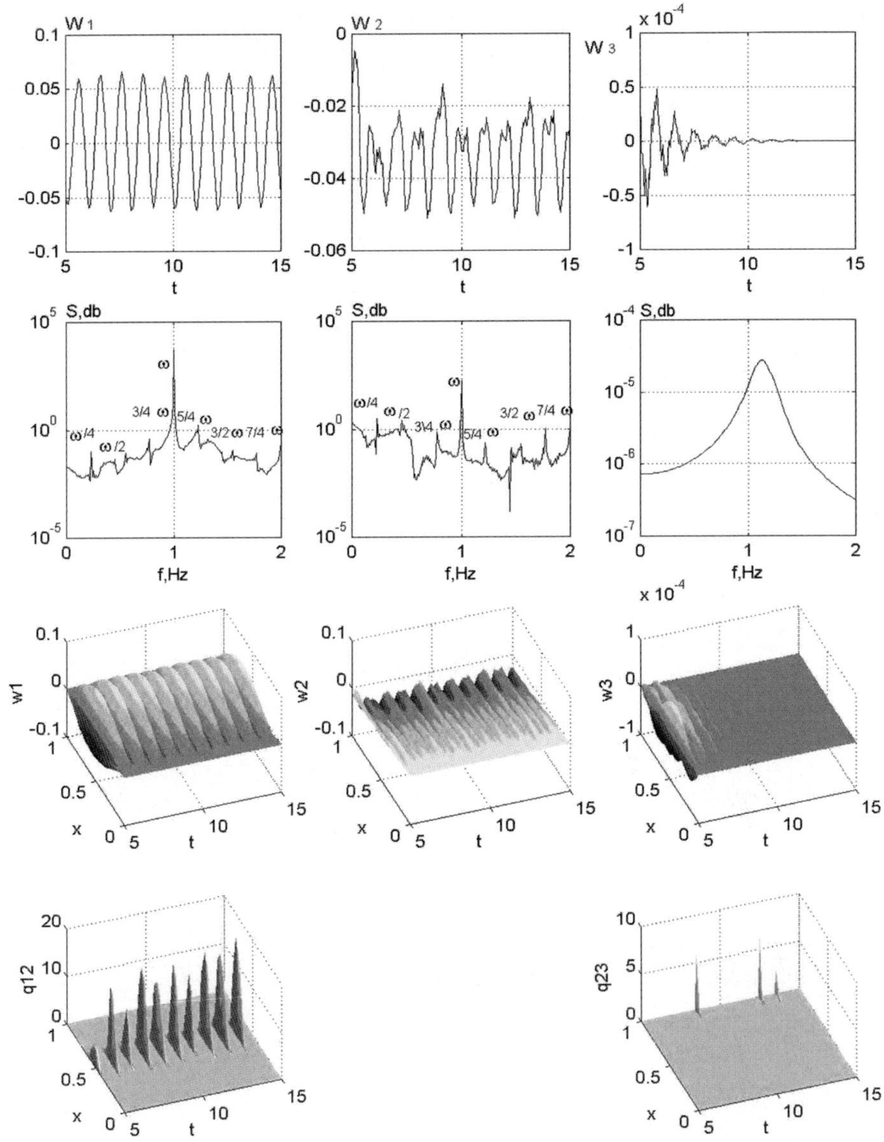

Fig. 13.8 Time histories of deflection $w_i(0.5,\ t)$ $i = 1,\ 2,\ 3$, the corresponding power spectra, deflection surfaces $w_i(x,\ t)$ $i = 1,\ 2,\ 3$, and the contact pressure $q_{12}(x,\ t)$ and $q_{23}(x,t)$, ($q_{10} = 0.9,\ \omega = 1.0$)

Problem 13.4.4 Now in the considered package the left-hand side beam ($l = 1$) end is clamped (see (13.10)), whereas the right one is hinged (see (13.11)). The boundary conditions of the lower beam are applied in an inverse manner. The middle beam ($l = 2$) is hinged on both ends. The top beam vibrates with the frequency $\omega = 1.0$ first. Then, after an impact between upper and middle beams, the second beam vibrates with the given frequency within a transitional time interval, and then its vibrations are damped. Owing to an increase of the external load amplitude on the third beam, the synchronized vibrations of two-layer package with the excitation frequency appear. This feature is represented by a variation of the contact pressure between beams, i.e., it is synchronized on the excitation frequency. Further increase of the excitation amplitude forces the system to exhibit its synchronized state after the period doubling Hopf-Andronov bifurcations (Fig. 13.9). Observe that for

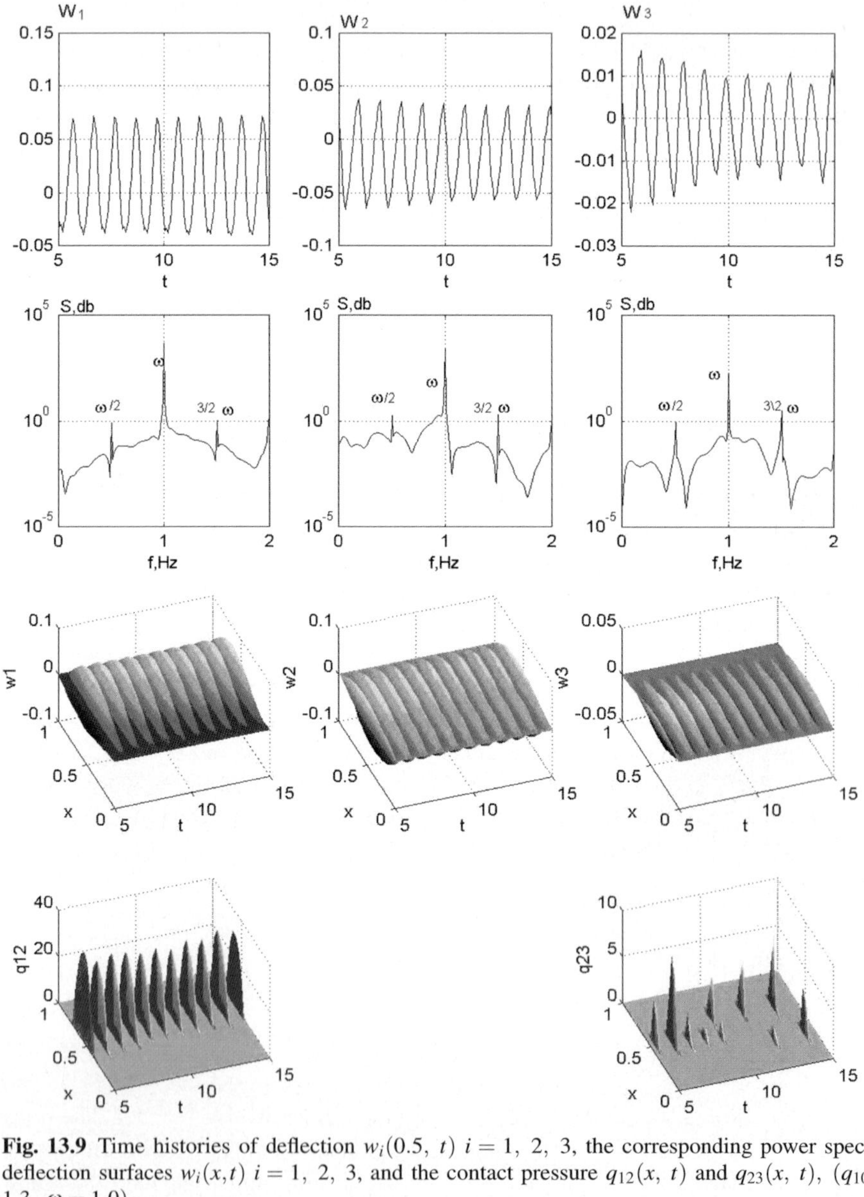

Fig. 13.9 Time histories of deflection $w_i(0.5,\ t)$ $i = 1,\ 2,\ 3$, the corresponding power spectra, deflection surfaces $w_i(x,t)$ $i = 1,\ 2,\ 3$, and the contact pressure $q_{12}(x,\ t)$ and $q_{23}(x,\ t)$, $(q_{10} = 1.3,\ \omega = 1.0)$

two first beams the full synchronization is exhibited. The successive increase of the external load amplitude shifts the system to chaos on the fundamental frequency (see the associated diagrams of contact pressure between first–second and second–third beams).

Problem 13.4.5 In this case the boundary conditions differ from those given in problem 13.4.4 slightly, since only boundary conditions associated with the middle beam are changed, i.e., it is clamped on both ends (see (13.10)).

In general, the initial vibration regime, when the top and middle beams are in a contact (or a contact of two-layered package with the third beam occurs) is similar to the earlier reported cases. However, now already in the beginning period tripling bifurcation occurs (Fig. 13.10). Then a successive period doubling occurs, the system is qualitatively changed, collapse of period tripling occurs, and the package synchronization with the frequency associated with period doubling is exhibited. Then the three-layer package transits into chaotic vibrations state on the excitation frequency. The latter behavior is exhibited by all characteristics.

13.5 Structurally and Physically Nonlinear Problems

In this section same boundary conditions are applied as in Sect. 13.4. A beam material is considered as physically nonlinear with deformation diagram (13.17). Computations have been also carried out for $k_1 = k_2 = 2000$; $0 \leq t \leq 120$; $\delta_1 = \delta_2 = 0.05$. Flow deformation intensity for all beams $e_{1s} = e_{2s} = e_{3s} = 0.1$. It should be emphasized that instead of an elastic-plastic problem a simplified nonlinear-elastic one has been solved, when loading and relief create the same curve.

We are going to study the formulated problem when upper ($l = 1$) and lower ($l = 3$) beams are hinged on their left ends ($x = 0$) (see relation (13.11)), whereas they are clamped on their right ends ($x = 1$); see relation (13.12). The middle beam ($l = 2$) is clamped on its left end (see (13.10)), whereas it is free on its right end (see (13.12)). The excitation frequency $\omega = 1.0$.

It is worth noticing that a supplement of physical nonlinearities essentially changes the vibration chart as well as the scenario of three-layered uncoupled beams into chaotic state.

Although each of the beams of the three-layered package is not elastic now, but physically nonlinear, the upper beam vibrations are periodic due to small deflections. Furthermore, the physical nonlinearity is rather exhibited in the negligible manner until a contact between the upper and middle beams is achieved. Note that the first contact between these beams yields an occurrence of the broad band spectrum support typical for a chaotic state. In the Poincaré sections, distribution of chaotic points is visible. There is a lack of synchronization. All other fundamental characteristics exhibit chaos of two-beam dynamics on the excitation frequency. A

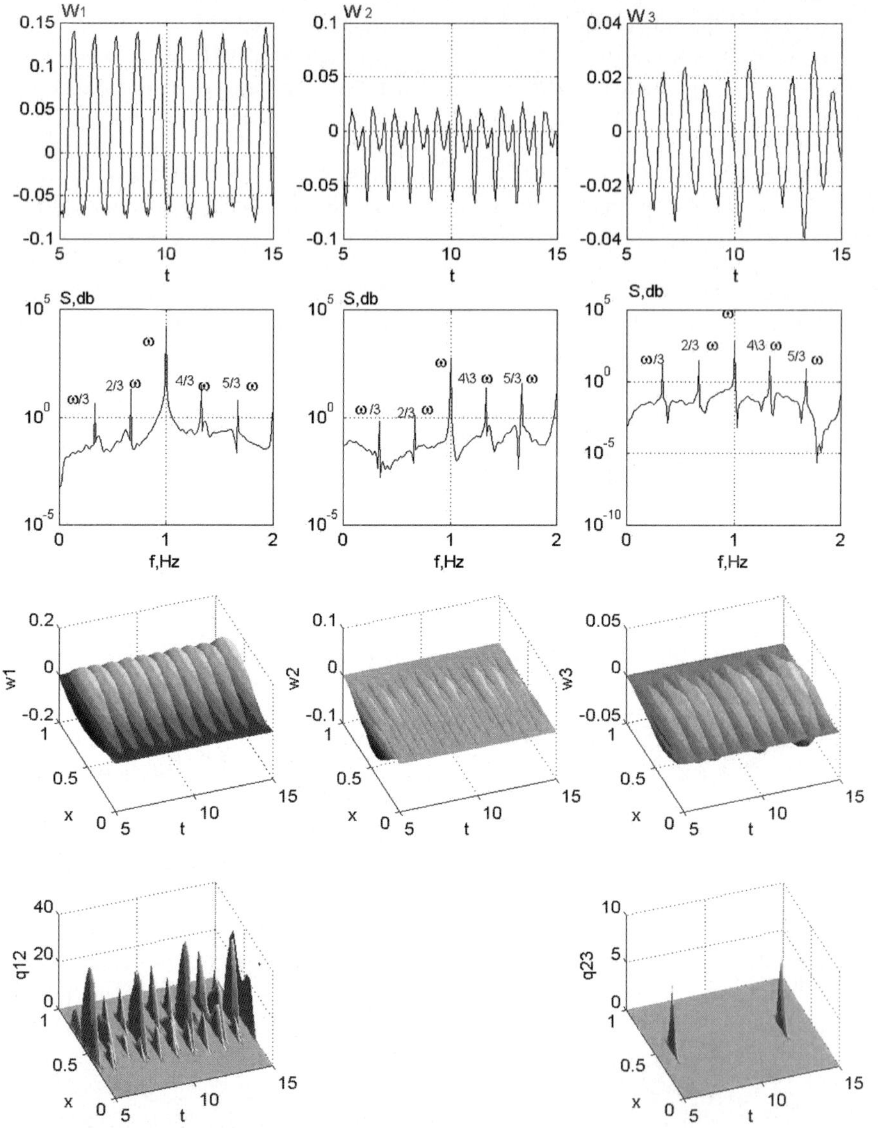

Fig. 13.10 Time histories of deflection $w_i(0.5,\ t)$ $i = 1,\ 2,\ 3$, the corresponding power spectra, deflection surfaces $w_i(x,t)$ $i = 1,\ 2,\ 3$, and the contact pressure $q_{12}(x,\ t)$ and $q_{23}(x,\ t)$, $(q_{10} = 2.4,\ \omega = 1.0)$

contact between second and third beams (Fig. 13.11) yields more developed chaotic dynamics into a general picture of a three-layered vibration sandwich. The power spectrum of the third beam is fully broad band.

Analysis of structurally and physically nonlinear three-layered beams package with the boundary conditions, where both upper and lower beams are clamped

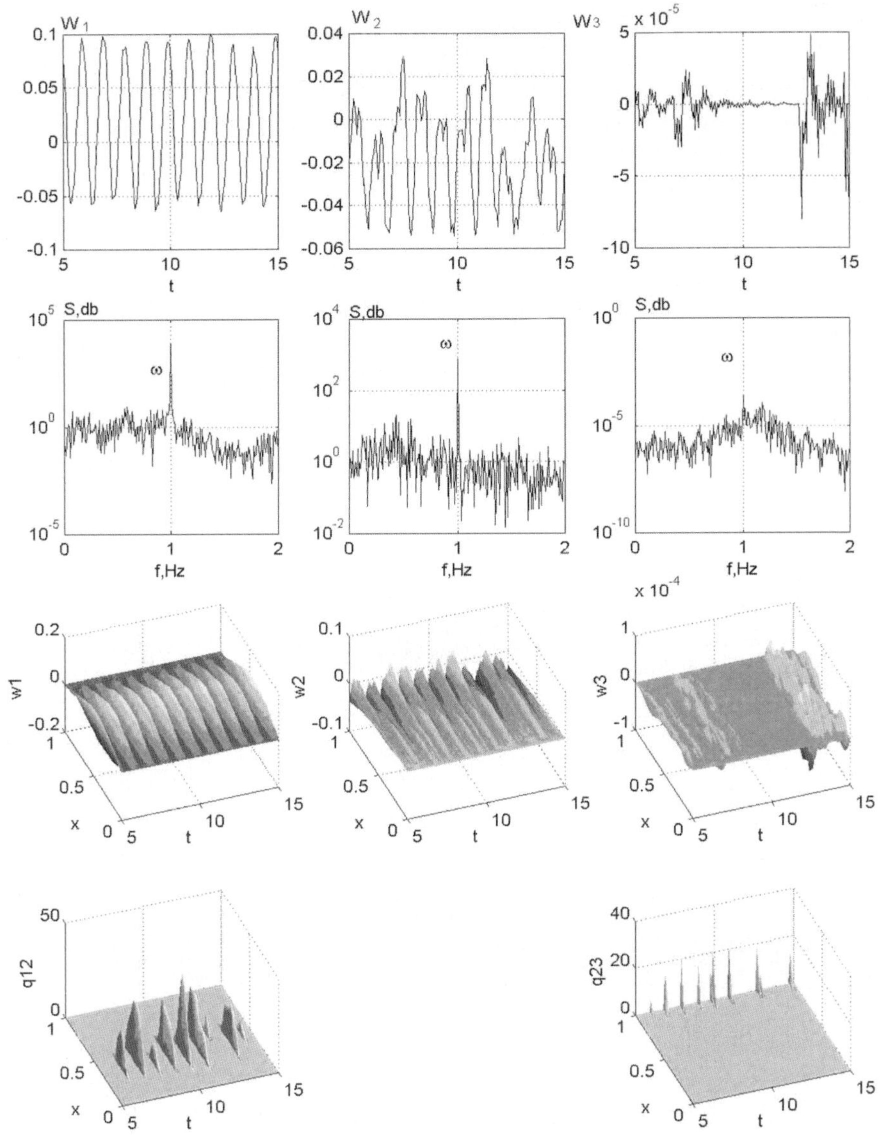

Fig. 13.11 Time histories of deflection $w_i(0.5,\ t)$ $i = 1,\ 2,\ 3$, the corresponding power spectra, deflection surfaces $w_i(x,\ t)$ $i = 1,\ 2,\ 3$, and the contact pressure $q_{12}(x,\ t)$ and $q_{23}(x,\ t)$, $(q_{10} = 1.2,\ \omega = 1.0)$

through (13.10) ($\omega = 1.54$), and when these beams are clamped on their left ends and hinged on their right ends ($\omega = 1.0$) (middle beams is clamped (free) on its left (right) end), outlines the same scenario of transition of three-layered package vibrations into chaos owing to increase of the amplitude of the excitation q_{10}.

13.6 Special Case

In this section more complicated problems are studied. Namely, we follow a way of study of three-layered beams sandwich with finite bending stiffness, when each of beams is linearly elastic.

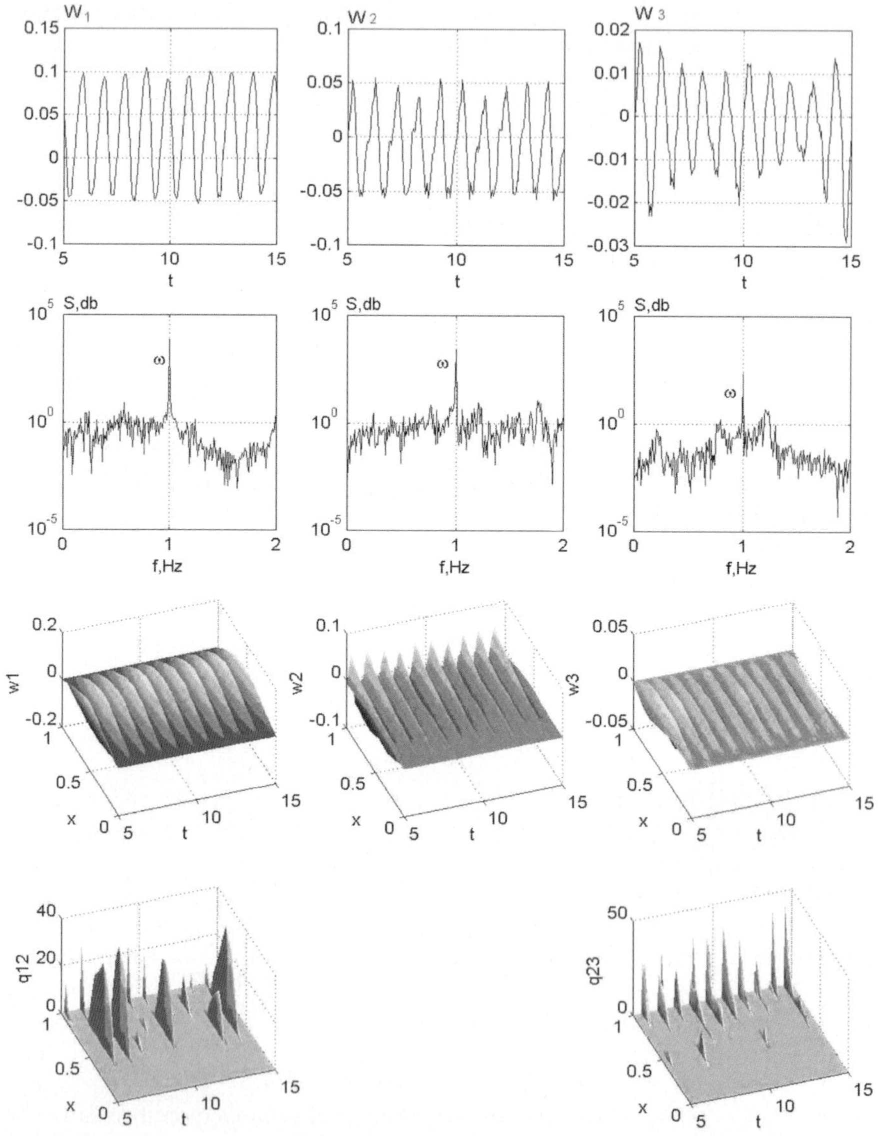

Fig. 13.12 Time histories of deflection $w_i(0.5,\ t)$ $i = 1,\ 2,\ 3$, the corresponding power spectra, deflection surfaces $w_i(x,\ t)$ $i = 1,\ 2,\ 3$, and the contact pressure $q_{12}(x,\ t)$ and $q_{23}(x,\ t)$, ($q_{10} = 2.4$, $\omega = 1.54$)

Problem 13.6.1 The middle ($l = 2$) beam possesses non-clamped ends (boundary conditions (13.12), $\omega = 1.0$), but in the initial time instant it is in an equilibrium state. The left end of the upper beam ($l = 1$) is clamped (see (13.10)), whereas its right end is hinged (see (13.11)). The mentioned conditions are applied in the reverse manner to the lower beam ($l = 3$).

Problem 13.6.2 The middle ($l = 2$) beam is free as in the previous case, whereas upper ($l = 1$) and lower ($l = 3$) beams are hinged on both ends. In both the problems, the third beam is subjected to sinusoidal load.

Consider first Problem 13.6.1. Note that a first contact between the top and middle beams is associated with their bending and a contact with the third beam, which is exhibited by the contact pressure diagrams. The top beam vibrations are harmonic with fundamental frequency, and vibrations of external beams are damped with time. Increase of initial amplitude of excitation transits whole package vibrations into chaos on the excitation frequency. The middle beam, owing to an impact within its averaged movement time, maintains contact with the top beam in a chaotic manner. Since it is bended, it touches the third beam along its ends also chaotically. Further increase of the external excitation amplitude shifts the package into a deep chaotic state (see the fundamental characteristics reported in Fig. 13.12). The change of the boundary conditions on external beams practically does not change a scenario to chaos for the considered three-layered package, but now in the power spectra a transition to chaos is observed on the fundamental excitation frequency and on the linearly dependent frequencies.

To conclude, a transition of three-layered package of elastic beams into chaos in practice does not depend on the boundary conditions associated with top and lower beams only if the third beam has unclamped ends.

13.7 Conclusions

A study of a three-layered uncoupled beams package subjected to harmonic excitation acting on the upper beam with an account of physical nonlinearities yields immediately chaotic dynamics only if a contact between beams is achieved (clearly exhibited peak in the power spectrum associated with the fundamental frequency occurs). In this case a synchronization phenomenon has not been detected, although its occurrence is typical for our three-layered package considered as a structurally nonlinear problem. In the case of structurally nonlinear three-layered beams, the period tripling phenomenon has been detected within a synchronization regime of the first two beams. Furthermore, the full agreement of vibration and bifurcation behavior with that predicted by the Sharkovsky theorem has been illustrated. Increase of number of layers in the package of uncoupled beams over two yields a violation of two-layered package synchronization, and the system transits into a chaotic state. However, the latter conclusion essentially depends on a choice of the boundary conditions.

Chapter 14
Bifurcation and Chaos of Dissipative Nonlinear Mechanical Systems of Multi-layer Sandwich Beams

The dynamics of a physically dissipative non-linear multi-layer sandwich of three beams is analyzed. The boundary conditions are arbitrary. The transversal load can be applied either simultaneously to all beams or separately to each of the beams. The finite difference method is used to solve the governing equations. Different types of beam material are considered: ideally elastic-plastic, elastic-plastic with linear straightness, and pure aluminum. Some new bifurcation and chaotic phenomena of the system are reported and discussed.

14.1 Introduction

Damping and dissipating energy of a beam (or of sandwich beams) have been studied by a large number of researchers. Mead and Markus [202, 203, 204] studied a sixth-order differential equation of motion in terms of the transverse displacement of a beam with arbitrary boundary conditions, and it serves as a classical method of modeling and describing damped three-layer beams and plates (see also [79, 80]). In references [273, 274] the experimental results from a cantilever beam under impact loading and with an operating four-bar mechanism are reported. Both experimental and analytical results for the compressional vibration of an elastic-viscoelastic-elastic three-layer sandwich beam are reported by Sisemore and Darveness [275].

Another research direction includes the free vibration of beams with concentrated masses (see Chen [69] and Goel [111]). The natural frequencies of a cantilevered beam with a slender tip mass are investigated in references [55, 56]. The Euler-Bernoulli and Timoshenko beam models are used in reference [66] to analyze the free vibration of simply supported and cantilever beams with distributed mass. Two distributed masses in-span attached to a beam are studied [67]. A mass carried by two different beam segments is studied by Kopmaz and Telli [141].

Free vibration analysis of non-uniform beams with an arbitrary number of cracks and concentrated masses is carried out in Li's paper [187]. The eigenvalue equation of a non-uniform beam with any two kinds of end support, any finite number of

J. Awrejcewicz, V.A. Krysko, *Chaos in Structural Mechanics*,
DOI: 10.1007/978-3-540-77676-5_15,

cracks, and with concentrated masses are determined from a second order determinant, which significantly saves the computational effort.

A nonlinear model analysis approach, based on invariant manifold theory, is proposed by Shaw and Pierre [264], and then used by Xie et al. [319].

Three identification methods of nonlinear model behavior of an externally excited cantilever beam are proposed and discussed [81]. The propagation of structural waves, on an infinitely long periodically supported Timoshenko beam, are studied in reference [120]. The power series expansion of displacement components method, yielding a set of fundamental dynamic equations of a one-dimensional higher order theory for a laminated composite beam subjected to axial stress derived through the Hamilton principle, is applied by Matsunaga [201]. The approach introduced is used for the analysis of natural frequencies and buckling stresses of laminated composite beams, taking into account the complete effects of transverse shear, normal stresses and rotary inertia.

The dynamic stability of a stepped beam subject to a moving mass is analyzed [2]. It is shown that the stability of certain beam models can be improved by providing the beam with appropriately spaced steps.

Two simple systems comprising straight uniform Euler-Bernoulli beams in which there are internal self-balancing axial loads are analyzed by Mead [203].

A vast number of papers is devoted to control of beams. Here only a few recent ones are mentioned, but a reader may use references cited in those papers. Active vibration control of beams with smart constrained layer damping treatment is proposed by Balamurgan and Narayanan [50]. The sensor piezoelectric layer measures the vibration response of the structure, and a feedback controller is used to regulate the axial deformation of the piezoelectric actuator and hence to provide adjustable damping in the structure. The piezoelectric control of damped sandwich beams is proposed in reference [296]. The analysis of the proposed linear quadratic regulator, and the real representation of complex modal reduced models for hybrid piezoelectric-active viscoelastic-passive vibration control are reported. In addition, the parametric analysis and the comparative study of control strategies of damped sandwich beams are given.

Investigation of chaos exhibited by beams with different boundary conditions is reported in series of references [146, 223, 225, 226, 320]. Different aspects of nonlinear longitudinal or transverse vibrations of beams subject to periodical longitudinal or transverse excitations are analyzed.

In this chapter a system consisting of three sandwich beams in a dissipative medium is analyzed. The beam material is assumed to be nonlinear and elastic.

14.2 Problem Formulation and Computational Algorithm

Dynamic interaction of three sandwich beams with clearance in their equilibrium state is analyzed. The kinematic J. Bernoulli hypothesis is assumed for the formulation of the governing equations. It is assumed that in a contact zone the beams slip freely. A contact pressure (load) is found using Winkler hypothesis [131]. Finally, one of the beams is subjected to periodic transverse load excitation.

The beam occupies the following domain in a three-dimensional space:

$$G = \left\{ (x,y,z) \middle| \quad 0 \le x \le a, \quad 0 \le y \le b, \quad -\frac{h}{2} \le z \le \frac{h}{2} \right\},$$

and the following notation is used: h is the beam thickness; h_{10} is the beam thickness in a center; w_l is the beam deflection; E_l is the elasticity modulus; b_l is the beam width; t is time; ε_l is the damping coefficient; u is beam length; ρ_l is material density; p_l is the clearance between beams $(1 = 1,\ 2,\ 3)$; σ_l is the Poisson coefficient; e_i is deformation intensity; σ_i is stress intensity; e_{sl} is intensity of flow deformation; σ_s is intensity of stress flow deformation; and K is the volume elasticity modulus.

The following non-dimensional parameters are introduced:

$$x = \bar{x}a, \quad y = \bar{y}b, \quad z = \bar{z}h_{0l}, \quad h_l = \bar{h}_l h_{0l},$$
$$w_l = \bar{w}_l h_{0l}, \quad E_l = \bar{E}_l G_{0l}, \quad b_l = \bar{b}_l b_{0l}, \quad p_l = \bar{p}_l h_{0l}.$$

The x coordinate is measured beginning from the left plate end and is extended along the beam axis, whereas the z axis is measured beginning from the mean curve down.

The beams motion is governed by the following non-dimensional equations:

$$h_l = \frac{\partial^2 w_l}{\partial t^2} + \varepsilon_l \frac{\partial w_l}{\partial t} = -L_l(w_l) - q_{kl} - q_l, \quad l = 1,2,3. \tag{14.1}$$

Now and later the index $l = 1$ corresponds to the top beam, $l = 2$ corresponds to the middle beam, and $l = 3$ refers to the bottom beam.

The operators $L_l(w_l)$ are defined via the formula

$$L_l(w_{l_i}) = \frac{\partial^2}{\partial x^2}\left[(b_1 P_l(x))\frac{\partial^2 w_l}{\partial x^2}\right], \tag{14.2}$$

where:

$$P_l(x) = \int_{-\frac{h}{2}}^{\frac{h}{2}} E_l z^2 dz. \tag{14.3}$$

The method of variation of the elasticity parameters is further applied in order to include physical material nonlinearities. In accordance with this method the elasticity modulus is coupled with the Poisson and deformation moduli via the relation

$$E = \frac{9KG}{3K + G}. \tag{14.4}$$

The modulus K is assumed to be constant and equal to $1.94G_{0l}$. Recall that in strain theory the shear modulus is defined via the formula

$$G = \frac{1}{3}\frac{\sigma_i(e_i)}{e_i}. \tag{14.5}$$

An arbitrary strain diagram of the beam material $\sigma_i(e_i)$ can be used. For example, it can be represented by one of the following choices [159]:

1. Ideally elastic–plastic material.

$$\begin{aligned} &\sigma_i = 3G_{0l}e_{sl}, && \text{for} \quad e_i < e_{sl}, \\ &\sigma_i = \sigma_s, && \text{for} \quad e_i \geq e_{sl}. \end{aligned} \tag{14.6}$$

2. Elastic–plastic material with linear strengthness.

$$\begin{aligned} &\sigma_i = 3G_{0l}e_{sl}, && \text{for} \quad e_i < e_{sl}, \\ &\sigma_i = 3G_{0l}e_{sl} + 3G_l(e_i - e_{sl}), && \text{for} \quad e_i \geq e_{sl}. \end{aligned} \tag{14.7}$$

3. Diagram for pure aluminum has the form:

$$\sigma_i = \sigma_s \left[1 - \exp\left(-\frac{e_i}{e_{sl}}\right)\right]. \tag{14.8}$$

The deformation intensity is governed by the following relation:

$$e_i = \frac{\sqrt{2}}{3}\left[(e_{xx} - e_{yy})^2 + (e_{yy} - e_{zz})^2 + (e_{xx} - e_{zz})^2 + \frac{3}{2}e_{xy}^2\right]^{\frac{1}{2}}.$$

Neglecting (for a beam) the components e_{yy} and e_{zz} one obtains

$$e_i = -\frac{2}{3}e_{xx}, \quad or \quad e_i = -\frac{2}{3}z\frac{\partial^2 w}{\partial x^2}. \tag{14.9}$$

The boundary conditions of the beam system are arbitrary. In particular, one may apply free, simple support or clamping conditions. The initial conditions have the form

$$\frac{\partial w_l(x,0)}{\partial t} = F_l(x), \quad w_l(x,0) = f_l(x), \quad l = 1,2,3, \tag{14.10}$$

where F_l and f_l are function of velocities and deflection distributions at the initial time instant, respectively. The transversal load can be applied either to all beams simultaneously or it can act on each of the beams separately. The load distributions along the beam and in time can be taken arbitrarily. In addition, a jump phenomenon (the beams separation phenomena) during interactions of the beams is included. The contact stresses are defined by the relations

$$q_{kl} = (-1)^l k_1 \frac{E_1}{h_1}\left(w_1 - p_1 - w_2\frac{h_{02}}{h_{01}}\right)\psi_1, \quad l = 1,2, \tag{14.11}$$

$$q_{kl} = (-1)^{i+1} k_2 \frac{E_2}{h_2}\left(w_2 - p_2 - w_3\frac{h_{03}}{h_{02}}\right)\psi_2, \quad l = 2,3, \tag{14.12}$$

where k_1 is the proportionality coefficient between contact pressure and clamping; the function ψ_1 defines a contact zone dimension and is found from the formula

$$\psi_l = \frac{1}{2}\left[1 + sign\left(w_l - p - w_{l+1}\frac{h_{0l+1}}{h_{0l}}\right)\right]. \tag{14.13}$$

In order to integrate Eq. (14.1) the finite differences method of approximation $O(h^2)$ is applied. The space $D = (x,t)|\, 0 \le x \le 1,\ 0 \le t \le T$ is covered by the rectangular mesh $x_i = ih_x$, $t_j = jh_t$ $(i = 0,\ 1,\ 2\ldots n;\ j = 0,\ 1,\ 2,\ldots)$, where $\Delta x_i = x_{i+1} - x_i = h_x = 1/n_x$ (n_x is an integer) and $h_t = t_{j+1} - t_j$. Between the nodes the differential equation (14.1) is approximated by finite-differences relations. In order to increase the accuracy, results of the symmetric formulas for derivative are applied. After a series of easy transformations for a beam with beam number l one obtains

$$w_{li,j+1} = \frac{1}{1 + \frac{\varepsilon_l h_t}{2h_{li}}}\left[2w_{li,j} + \left(\frac{e_l h_t}{2h_{li}} - 1\right) w_{li,j} + \frac{h_t^2}{h_{li}}((-1)^l q_{kl} + L(w_l)_{i,j})\right],$$

where

$$L(w)_{i,j} = [p_{li+1}w_{li+2,j} - 2(p_{li+1} + p_{li})w_{li,j} + (p_{li+1} + 4p_{li} - p_{li-1})w_{li,j}$$
$$-2(p_{li} + p_{li-1})w_{li-1,j} + p_{li-1}w_{li-2,j}]/h_x^4 + k_{fl}w_{li,j},$$
$$p(x) = b_l P(x), \quad l = 1,2.$$

Note that a three-layered system is obtained. In order to compute $w_l(x,t)$ on the layer $(j+1)$ the values $w_l(x,t)$ in two other layers, j-th and $(j-1)$-th, are used. First the $w_l(x,t)$ values of a fictive layer with number $j = -1$ are taken in order to initiate computations. Note also that a derivative in the initial conditions (14.10) is substituted by a finite difference relation.

An application of the variation of parameters method requires a splitting of the beam along its thickness into n_z layers.

Next, for the node x_j of each layer, the deformation intensity is found using the formula (14.9) on each time step of the computation. The elasticity modulus is defined via formulas (14.4), (14.5) and one of the expressions (14.6)–(14.8), and next the integral (14.3) is computed using the Simpson method.

14.3 Numerical Results

As an example the case when the top and lower beams are simply supported through balls on both ends is considered:

$$\frac{\partial w_1(0,t)}{\partial x} = w_1(0,t) = \frac{\partial w_1(1,t)}{\partial x} = w_1(1,t), \quad l = 1,3. \tag{14.14}$$

The middle beam is treated as the cantilever one:

$$\frac{\partial w_2(0,t)}{\partial x} = w(1,t) = 0,$$

$$\frac{\partial^2 w_2(1,t)}{\partial x^2} = \frac{\partial}{\partial x}\left(b_2 P(x)\frac{\partial^2 w_2(1,t)}{\partial x^2}\right) = 0. \tag{14.15}$$

The following initial conditions are applied:

$$F_i(x) = f_i(x) = 0, \quad i = 1,2,3 \quad (0 \le x \le 1), \tag{14.16}$$

and the transverse load drives only the top beam in a harmonic way:

$$q_1 = q_{10}\cos(\omega t). \tag{14.17}$$

The algorithm used to solve the equation governing beam vibrations is investigated from the point of view of both space and time meshes. The relaxation method is used to find the deflection in the center of a simply supported beam after the sudden application of uniformly distributed load along its length ($q_{10} = 0.15$).

The computational results, depending on both the mesh step in time h_t (along horizontal direction) and the number of beams splitting along its length n_x (along vertical direction), are reported in Table 14.1.

Note that the center deflection, found analytically for a static approach (14.9), is equal to 0.2343. Analysis of these results implies that a deflection does not depend on the computational time step, at least for the intervals investigated. However, it depends essentially on the number of splittings along the spatial coordinate. On the other hand, a solution to the problem leads to the conclusion that, for $h_t \le 0.0001$ and for a given n_x, one obtains the same behavior for the interaction of the beams. However, the results depend strongly on n_x, which may have the following physical interpretation. Observe that an application of a beam reduces the problem to the analysis of a multibody system dynamics with finite degrees of freedom instead of a continuous system governed by partial differential equations.

Hence, vibration of a number of nodes along the thickness direction yields in practice to the solution of another problem.

The results obtained suggest the following choice of computational steps. The spatial step along the x coordinate is set equal to $0.05 (n_x = 20)$, and the time step is taken as $2 \cdot 10^{-5}$. The number of layers along the z coordinate is taken equal to $n_z = 20$. The computations are carried out for $k_1 = k_2 = 2000$, $a \le t \le 120$; the clearances between beams are assumed to be $p_1 = p_2 = 0.05$. The material of the beams is assumed to be elastic-plastic with linear strain hardening (14.7). Flow

Table 14.1 Beam center deflection for different mesh steps in time h_t and along its length n_x

	$h_t = 0{,}001$	0,0001	0,00004	0,00002
$n_x = 10$	0,02362	0,02362	0,02362	0,02362
16	0,02351	0,02351	0,02351	0,02351
20	0,02348	0,02348	0,02348	0,02348
24	0,02347	0,02347	0,02347	0,02347
28	0,02346	0,02346	0,02346	0,02346

deformation intensity is assumed to be $e_{s1} = e_{s2} = e_{s3} = 0.1$ for all three beams, and the excitation frequency $\omega = 0.80$.

Note that a non-linear elastic problem is solved instead of the elastic-plastic one, when the load-relief curve is without hysteresis. Of course, this is a simplification of the real deformation process of the beams.

14.4 All Three Beams are Linearly Elastic

Beams will be linearly elastic if $G_1 = G_2 = G_3 = 1$. The computational results are shown in Figs. 14.1–14.24. In Figs. 14.1, 14.4, 14.7, 14.10, 14.13, 14.16, 14.19, 14.22, the amplitude characteristics $w(0.5;\ t)$, phase portraits, Poincaré sections, and the power spectrum of beams vibration are located vertically. The results corresponding to the top, middle, and lower beams are located beginning from the left to right. Time histories of the deformation of the beams surfaces are shown in Figs.14.2, 14.5, 14.8, 14.11, 14.14, 14.17, 14.20, 14.23. Finally, in Figs.14.3, 14.6, 14.9, 14.12, 14.15, 14.18, 14.21, 14.24 the time movement of the surfaces of contact pressure are reported. The damping coefficients $\varepsilon_1 = \varepsilon_2 = \varepsilon_3 = 1.4$. Note that all characteristics are taken for center points in the cases of freely supported beams (on balls), and for the end point of the cantilever beam.

For small load values $q_{10} < 0.34$ only the top beam vibrates, and the frequency spectrum includes only the external excitation frequency ω. With the load increase, after an impact between top and middle beams, the latter starts to vibrate (Fig. 14.1). The top beam exhibits quasi-periodic vibrations (view a phase portrait similar to an ellipse in shape and see the Poincaré sections). Although in the frequency spectrum the frequency 2ω is observed, but its amplitude is much more smaller than that corresponding to the fundamental frequency. A deflection surface of the top beam creates a regular waveform structure. The middle beam vibrates in a quasi-periodical manner, and in the spectrum the fundamental frequency ω and integral multiple frequencies 2ω, 3ω appear. Its deflection surface is much more complicated than that of the top beam (Fig. 14.2). The contact zone between them occurs in their central parts (Fig. 14.3).

The described qualitative behavior takes place until $q_{10} = 0.596$ when the middle beam touches the lower one. This event causes a qualitative change in the type of vibration of all three beams: the first period doubling appears. In the power spectrum the frequency components $\omega/2$, $3\omega/2$ and so on, are visible. Note that in the top beam the fundamental frequency ω dominates in its power spectrum. In the middle beam the amplitudes corresponding to the frequencies mentioned above are similar. Therefore, the phase portrait changes slightly for the first beam, whereas loops appear in the phase portrait of the second beam. Two sets of points are observed in the Poincaré sections. The sudden triple bifurcation, just after vibration occurrence, is observed in the third beam (see Fig. 14.4).

The deflection surface of the top beam remains practically unchanged, but this cannot be said of the second beam (Fig. 14.5). Characteristic properties of a contact zone between the two top beams are conserved; only its width is increased.

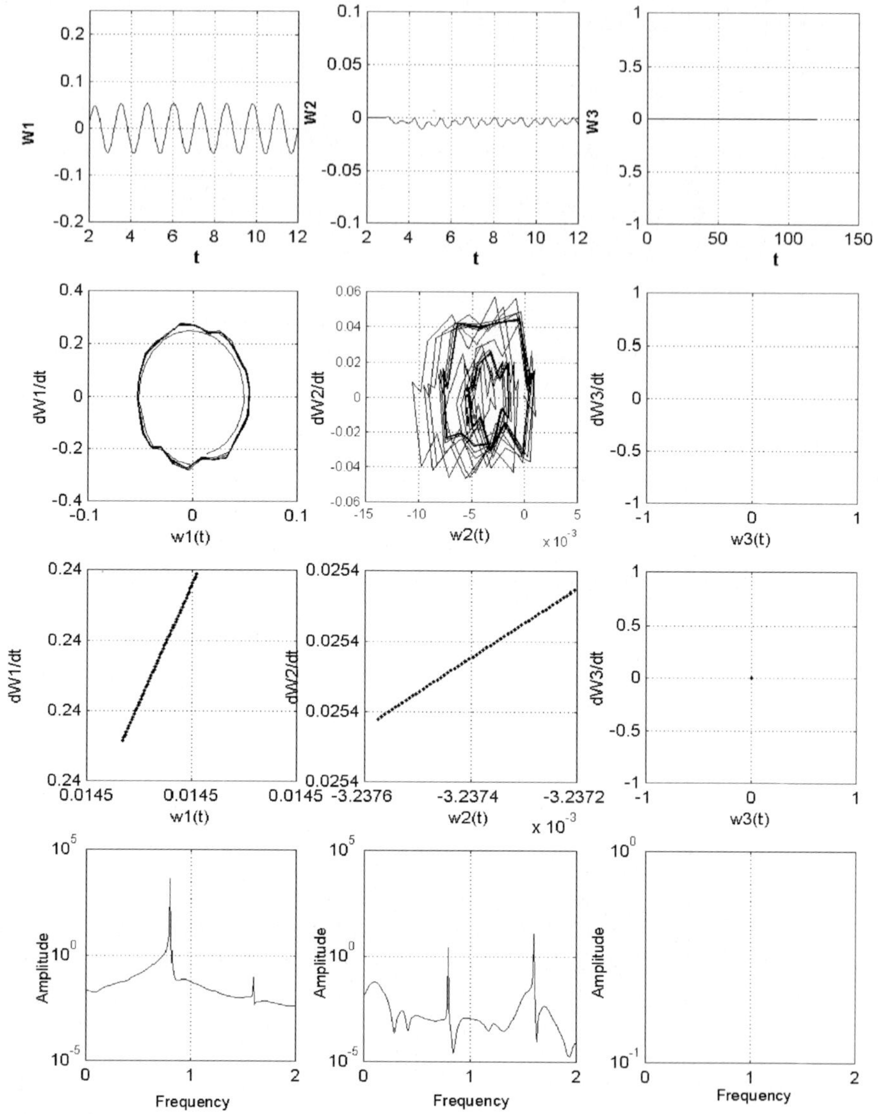

Fig. 14.1 Vibrations $w_i(0.5; t)$, velocities $\dot{w}_i(0.5; t)$, Poincaré sections $(\dot{w}_i, w_i)$, and power spectra for $q_1 = 0.34$

An interaction between middle and top beams appears at the end of the console beam (Fig. 14.6). The phase portrait and power spectrum prove that a slight degree of chaos appears. Increasing the load up to $q_{10} = 0.606$ (Fig. 14.7) the frequency components corresponding to a triple bifurcation appear also in both top and middle beams. The physical explanation follows: the amplitude of the lower vibrations increases, and hence it interacts more strongly with two remaining beams. Simultaneously, in the power spectra of all three beams the components corresponding to

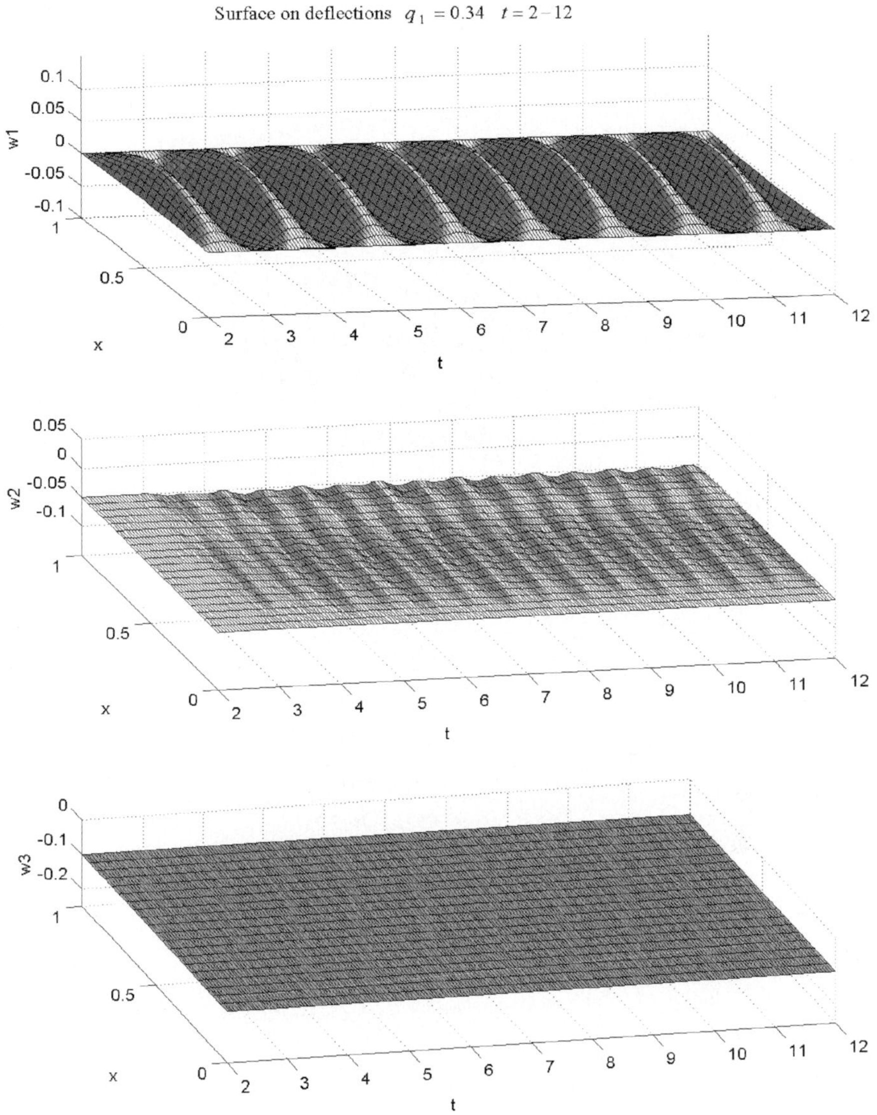

Fig. 14.2 Time histories of the beam surfaces for $q_1 = 0.34$

the first Hopf bifurcation appear. The deflection surfaces of the beams are practically conserved (Fig. 14.8). The contact zone between the top and middle beams changes, and contact takes place on some intervals along the beams length (Fig. 14.9). A collapse of the triple bifurcation with a simultaneous occurrence of period doubling (Hopf) bifurcation occurs for $q_{10} = 0.612$ (Fig. 14.10). However, the frequency components produced by this bifurcation are essentially smaller in comparison to the analogous components of the first bifurcation. Therefore, the fundamental

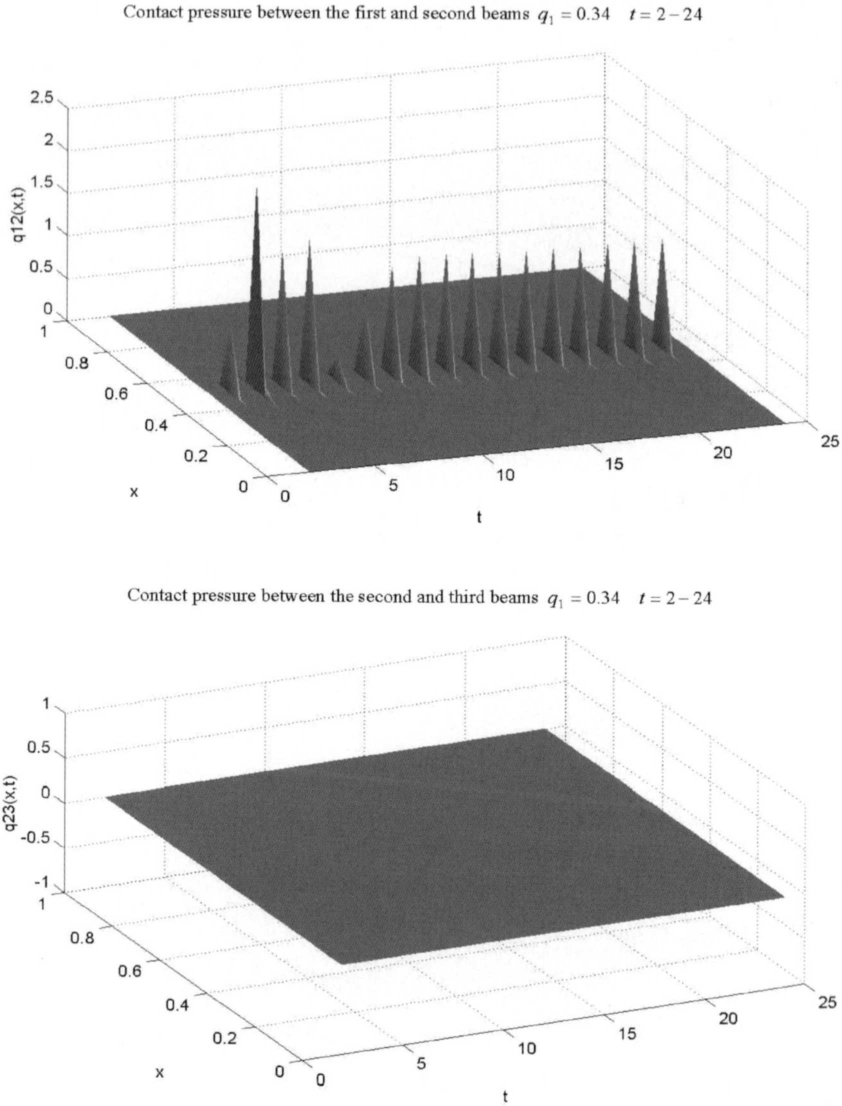

Fig. 14.3 Contact pressure between beams for $q_1 = 0.34$

associated characteristics, like amplitude characteristics, Poincaré section, phase portrait, and power spectra, remain practically unchanged (see Figs. 14.10, 14.12). Furthermore, for $q_{10} \geq 0.614$ (Fig. 14.13), in the spectra of second and lower beams, noisy components appear. All beams are in the synchronization regime within the frequencies of the secondary Hopf bifurcation. The points of the Poincaré sections display irregularity. The vibration picture is conserved up to $q_{10} = 0.65$ (see Fig. 14.16) with a simultaneous increase of noisy components. The deflection

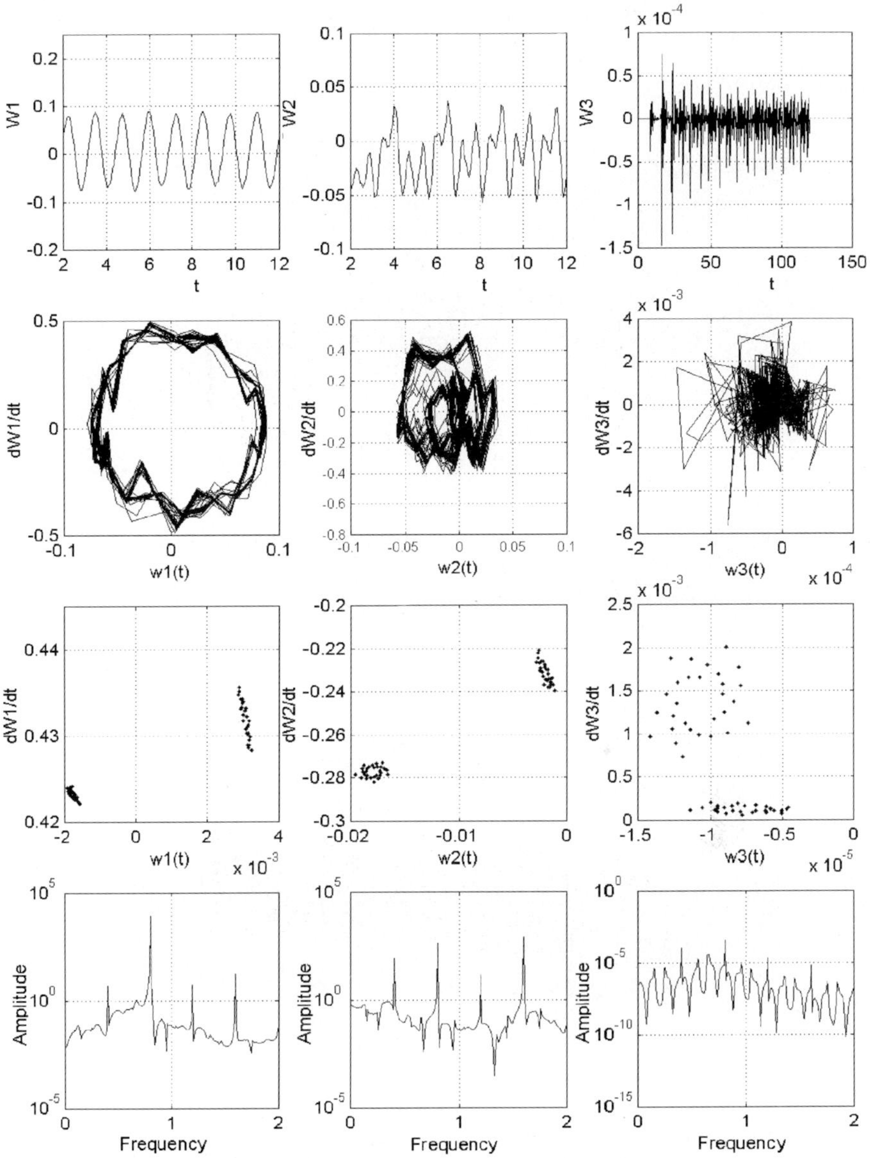

Fig. 14.4 Vibrations $w_i(0.5;\, t)$, velocities $\dot{w}_i(0.5;\, t)$, Poincaré sections $(\dot{w}_i,\, w_i)$, and power spectra for $q_1 = 0.596$ and $q_1 = 0.6$

character and interaction zones of the beams do not change (see Fig. 14.17 and Fig. 14.18). Next, the second Hopf bifurcation is collapsed, and vibration of all beams is synchronized on the fundamental frequency ω, and all other frequency components are integral multiples, i.e., 2ω, 3ω and so on (Fig. 14.19). Furthermore, when the load increases up to $q_{10} = 2$, the motion of the second beam becomes chaotic, which is displayed by the "wash-out" phase portraits and Poincaré sections

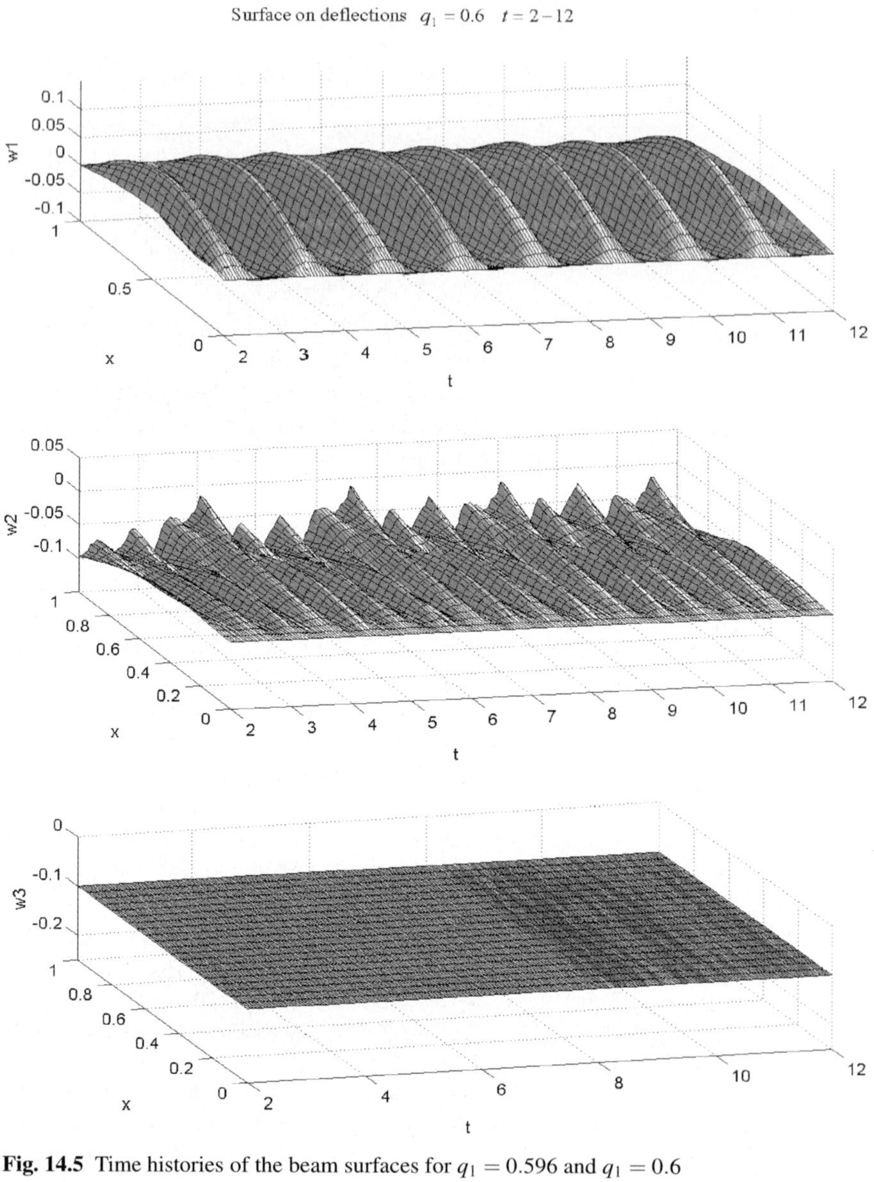

Fig. 14.5 Time histories of the beam surfaces for $q_1 = 0.596$ and $q_1 = 0.6$

(Fig. 14.22). On the other hand, those characteristics exhibit the regularization of vibrations of top and lower beams. The beams interaction changes qualitatively. The contact zones between all three beams are split into intervals (Fig. 14.23 and Fig. 14.24).

To conclude, the following scenarios leading to chaos are reported: one-frequency vibration (harmonic regime), first Hopf bifurcation, triple bifurcation and its collapse,

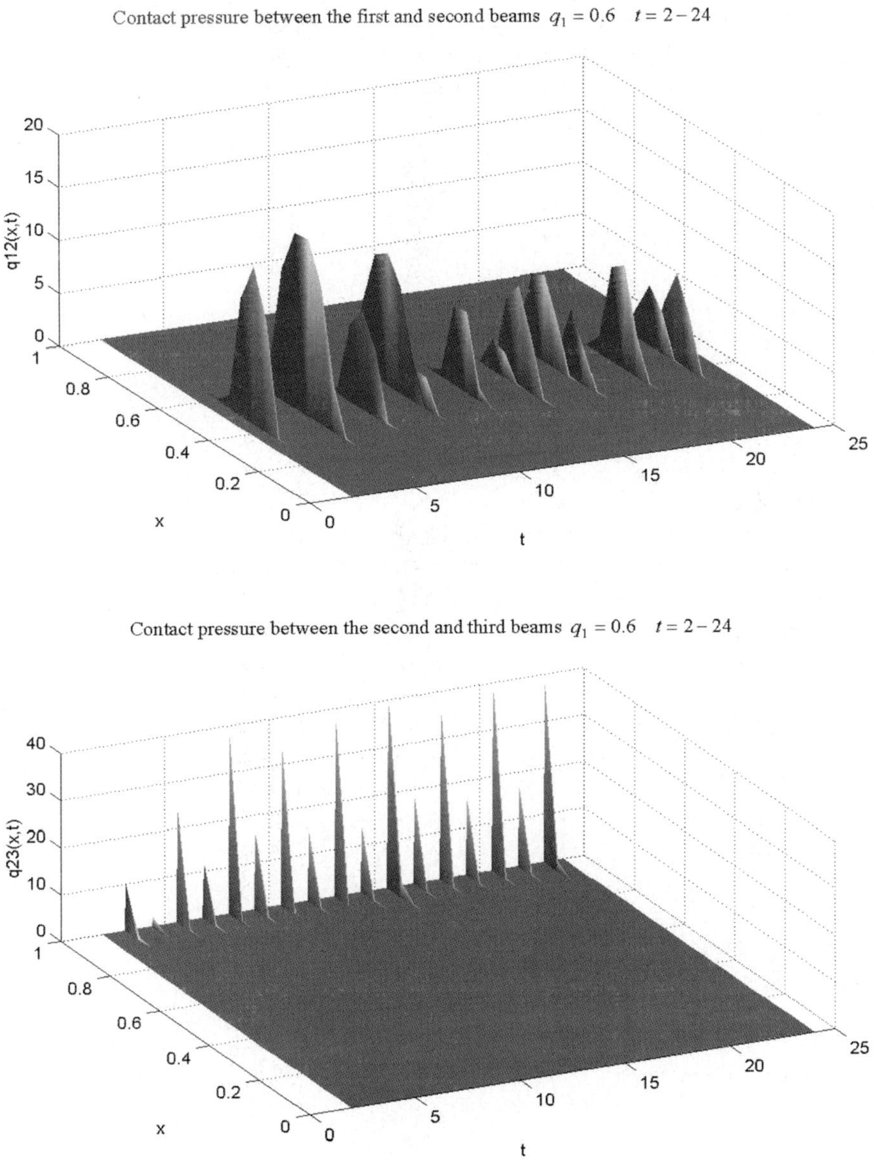

Fig. 14.6 Contact pressure between beams for $q_1 = 0.596$ and $q_1 = 0.6$

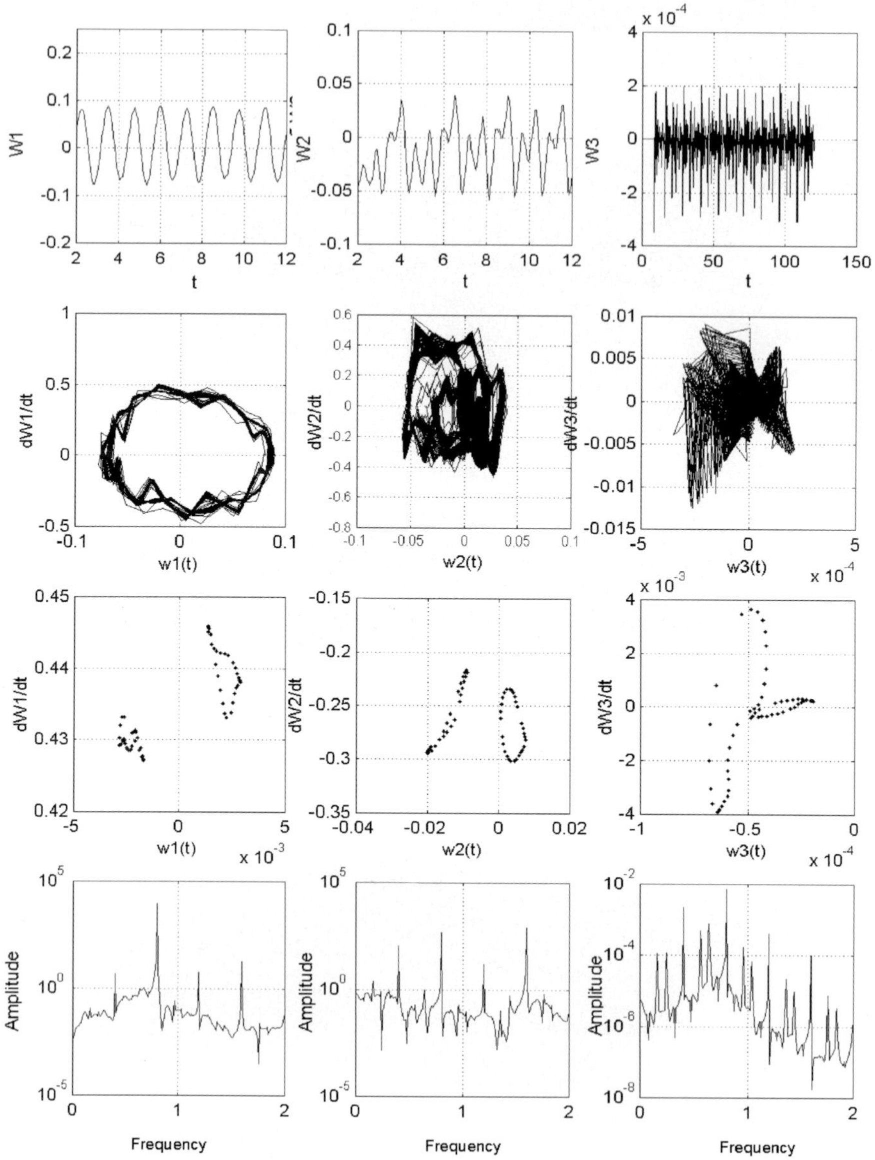

Fig. 14.7 Vibrations $w_i(0.5; t)$, velocities $\dot{w}_i(0.5; t)$, Poincaré sections $(\dot{w}_i, w_i)$, and power spectra for $q_1 = 0.606$ and $q_1 = 0.61$

second Hopf bifurcation and its collapse, transition to chaos. This scenario is not similar to that of the Feigenbaum transition: the bifurcation sequence is finite, and in addition in spite of a period doubling bifurcation, a triple bifurcation is observed.

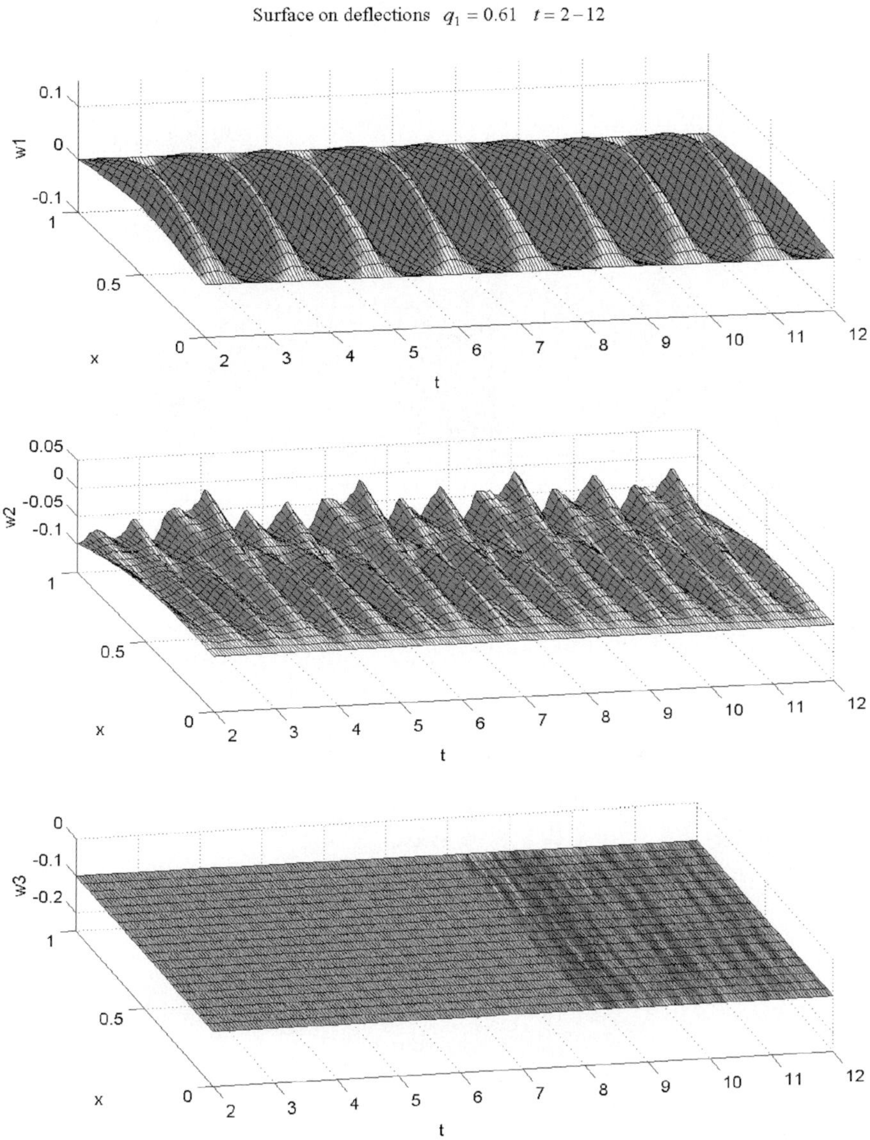

Fig. 14.8 Time histories of the beam surfaces for $q_1 = 0.606$ and $q_1 = 0.61$

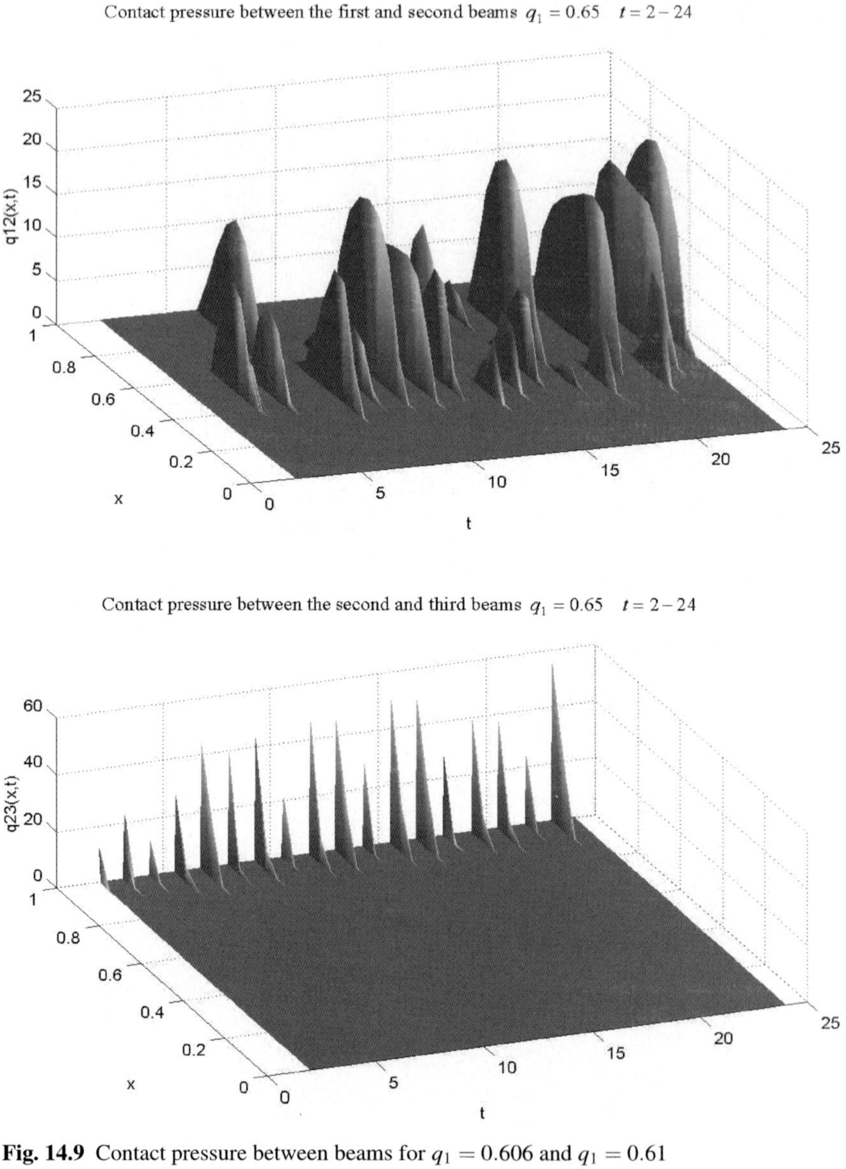

Fig. 14.9 Contact pressure between beams for $q_1 = 0.606$ and $q_1 = 0.61$

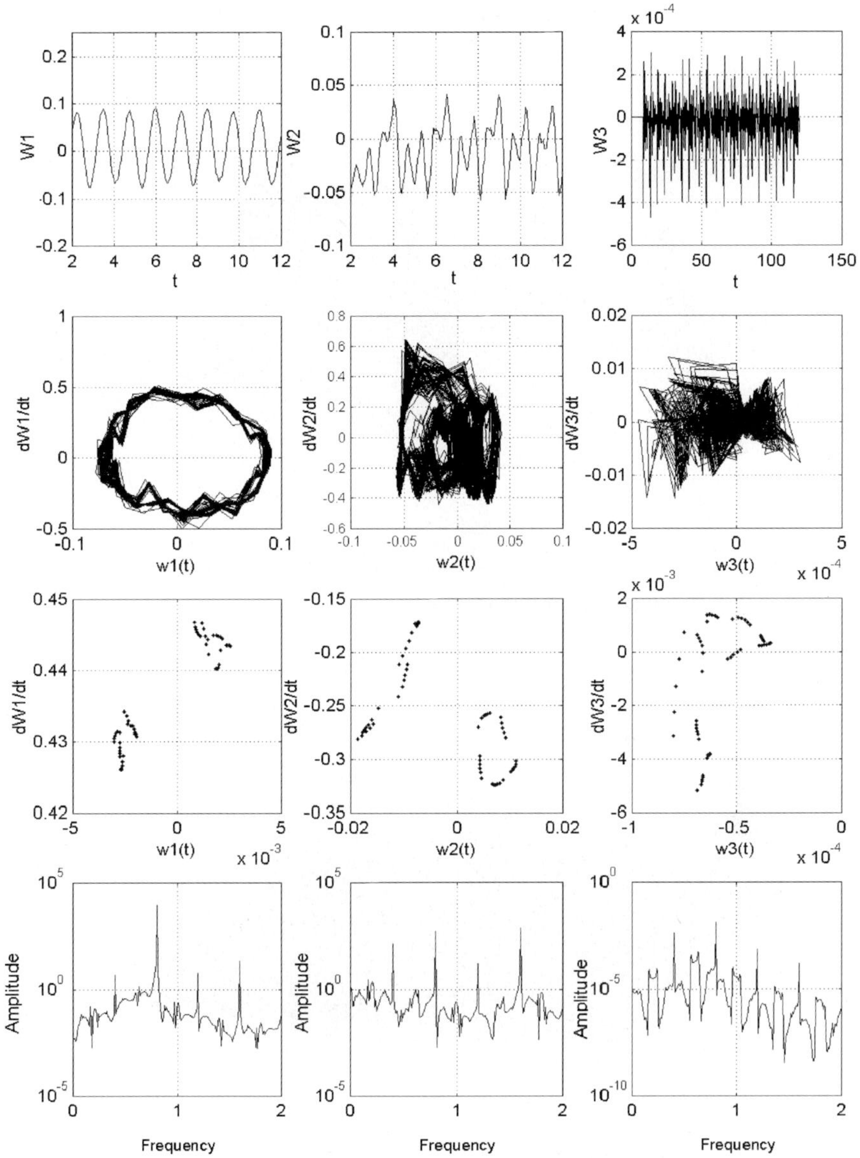

Fig. 14.10 Vibrations $w_i(0.5; t)$, velocities $\dot{w}_i(0.5; t)$, Poincaré sections $(\dot{w}_i, w_i)$, and power spectra for $q_1 = 0.612$ and $q_1 = 0.61$

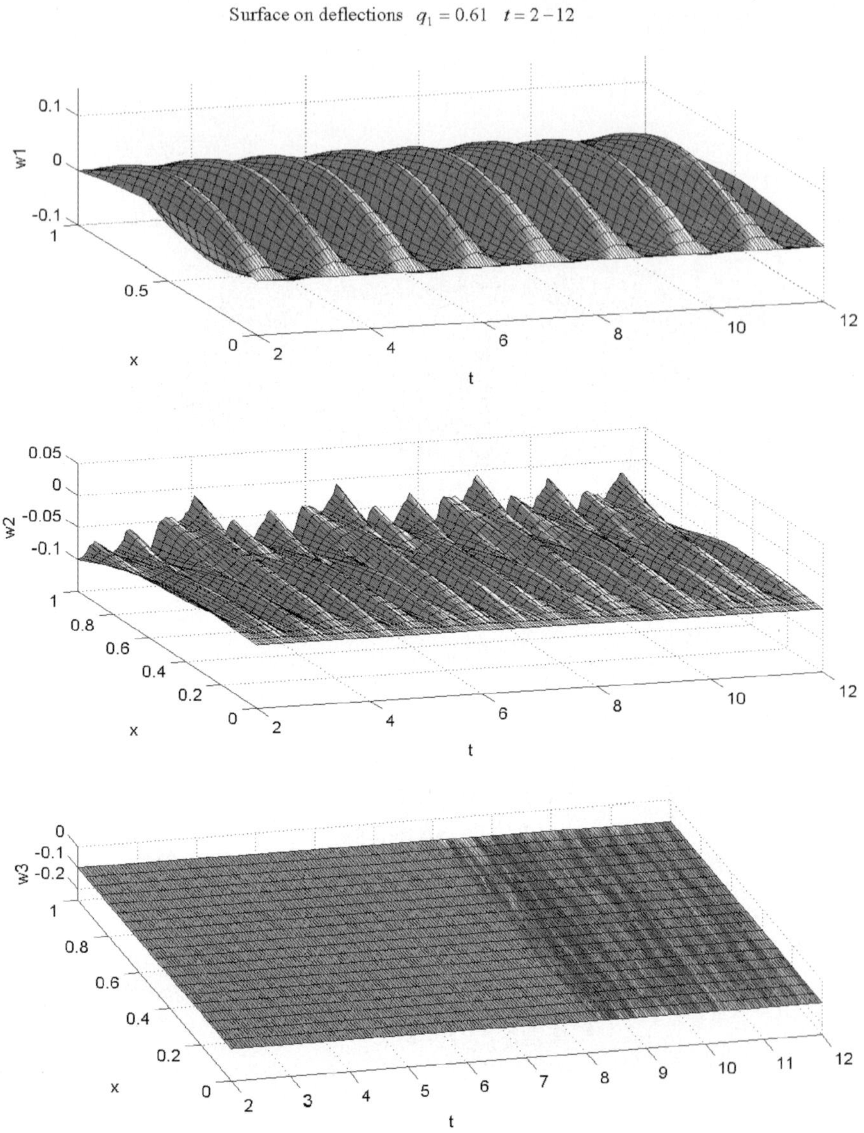

Fig. 14.11 Time histories of the beam surfaces for $q_1 = 0.612$ and $q_1 = 0.61$

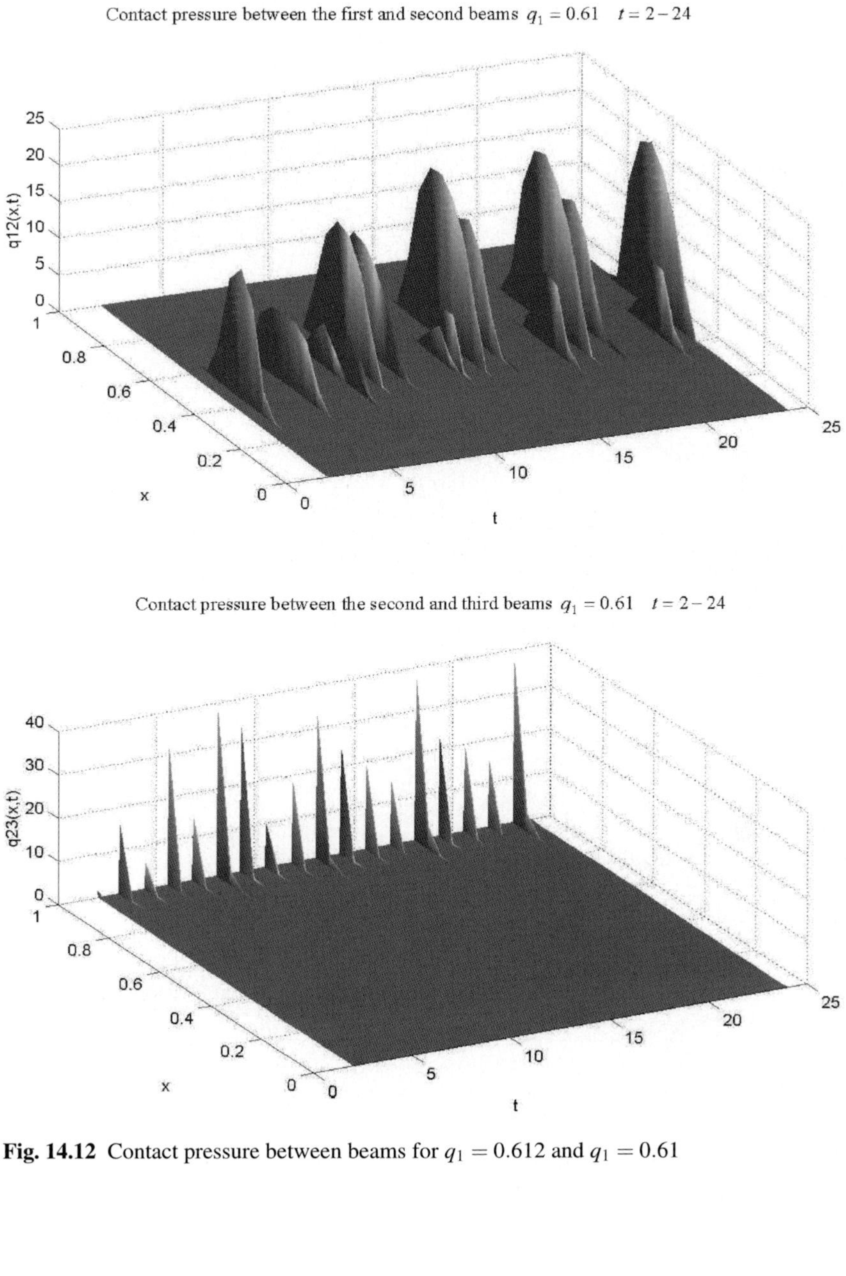

Fig. 14.12 Contact pressure between beams for $q_1 = 0.612$ and $q_1 = 0.61$

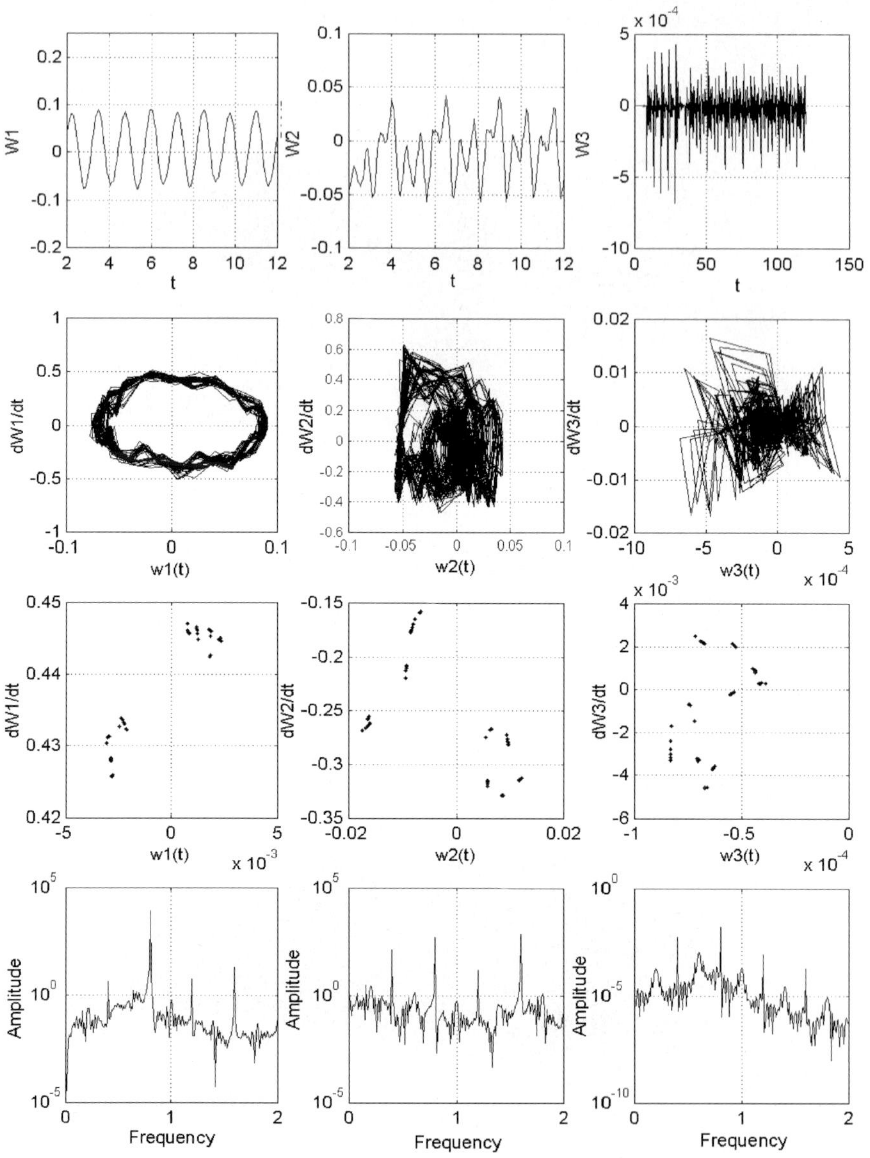

Fig. 14.13 Vibrations $w_i(0.5;\, t)$, velocities $\dot{w}_i(0.5;\, t)$, Poincaré sections $(\dot{w}_i,\ w_i)$, and power spectra for $q_1 = 0.614$ and $q_1 = 0.61$

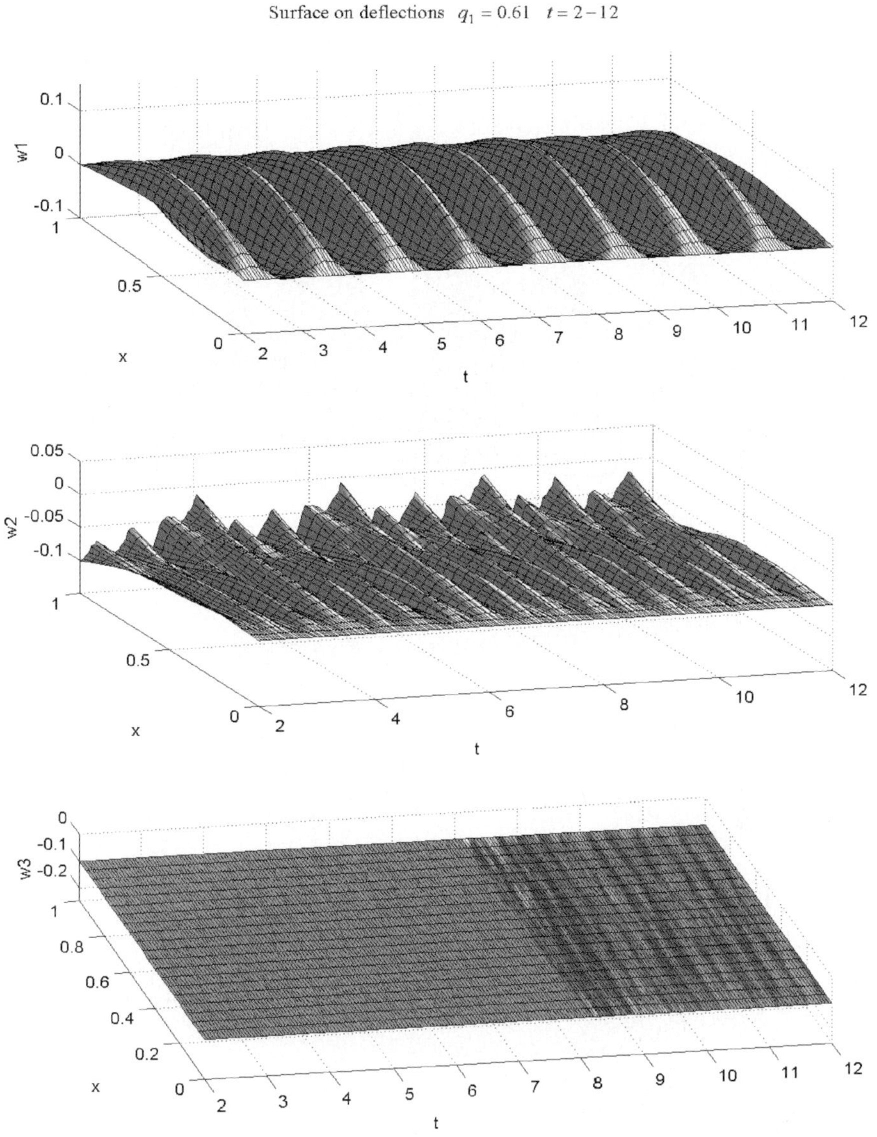

Fig. 14.14 Time histories of beam surfaces for $q_1 = 0.614$ and $q_1 = 0.61$

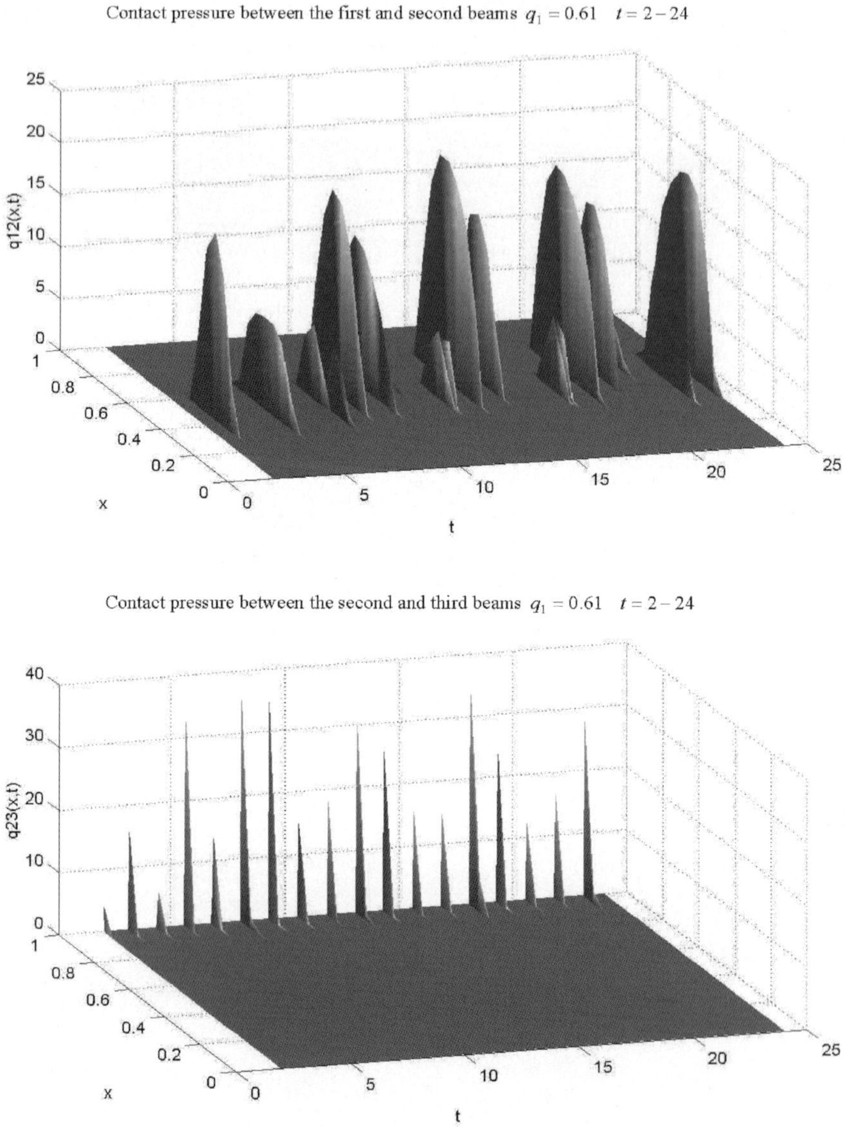

Fig. 14.15 Contact pressure between beams for $q_1 = 0.614$ and $q_1 = 0.61$

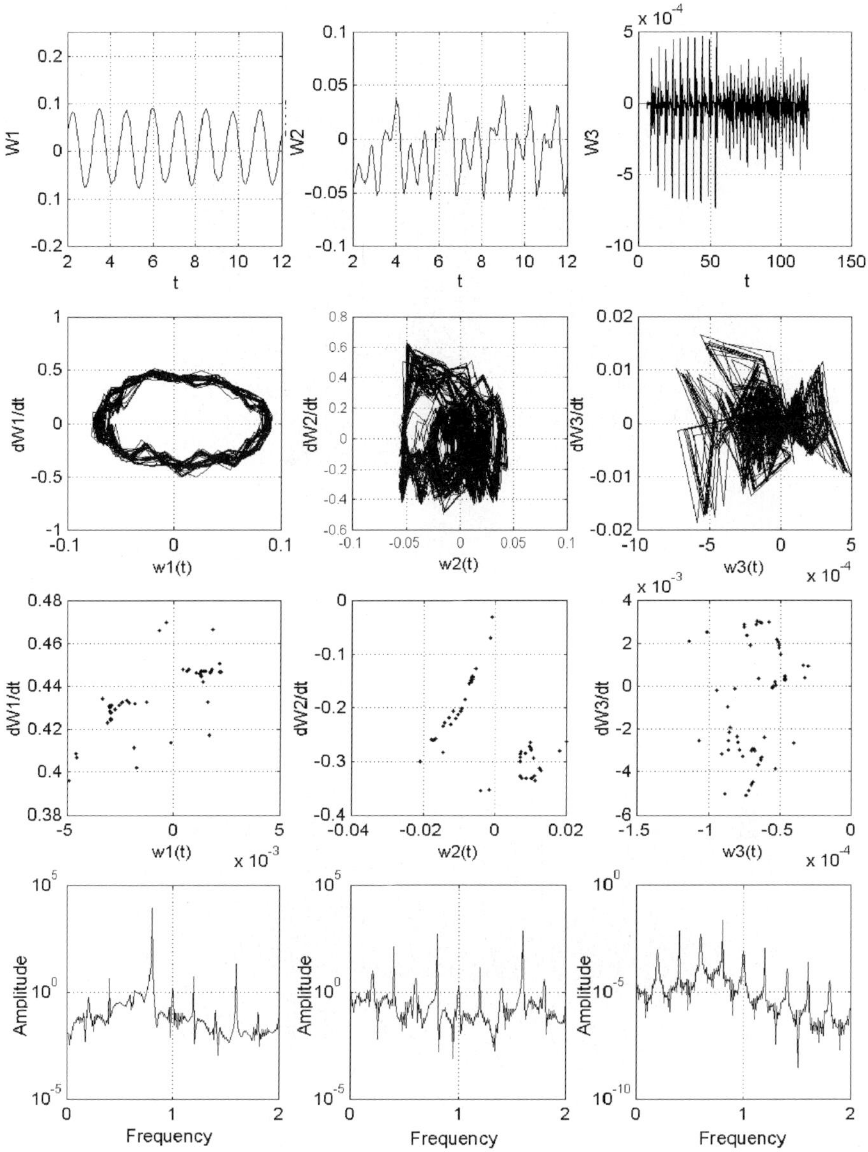

Fig. 14.16 Vibrations $w_i(0.5;\, t)$, velocities $\dot{w}_i(0.5;\, t)$, Poincaré sections $(\dot{w}_i,\, w_i)$, and power spectra for $q_1 = 0.616$ and $q_1 = 0.62$

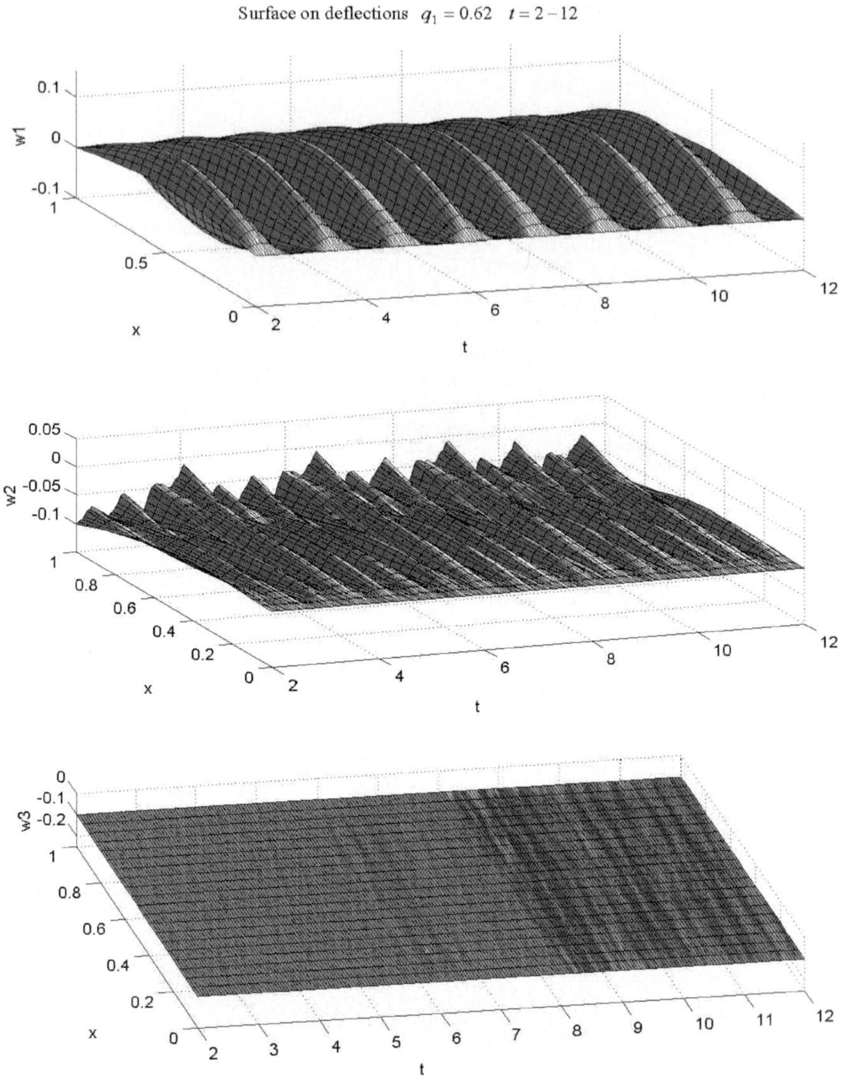

Fig. 14.17 Time histories of the beam surfaces for $q_1 = 0.616$ and $q_1 = 0.62$

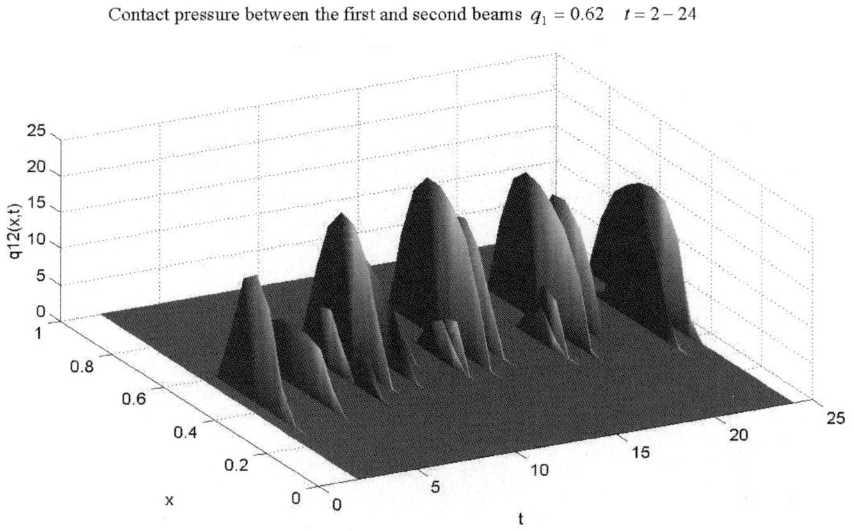

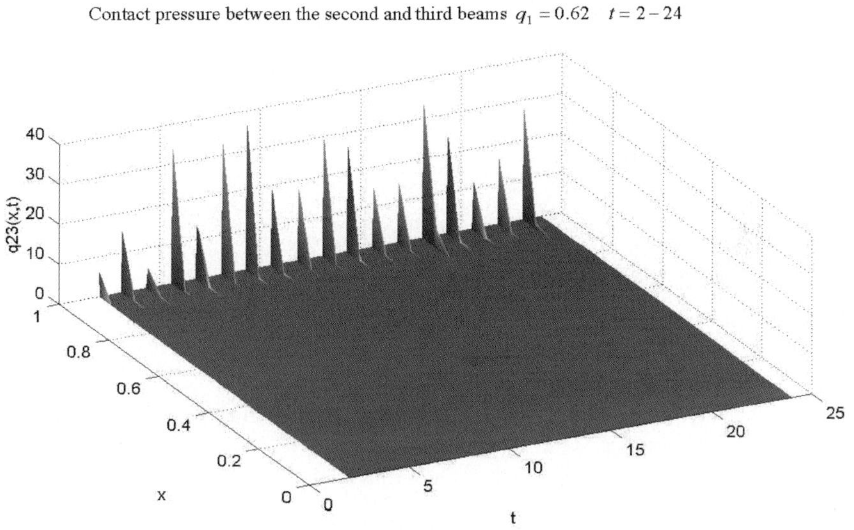

Fig. 14.18 Contact pressure between beams for $q_1 = 0.616$ and $q_1 = 0.62$

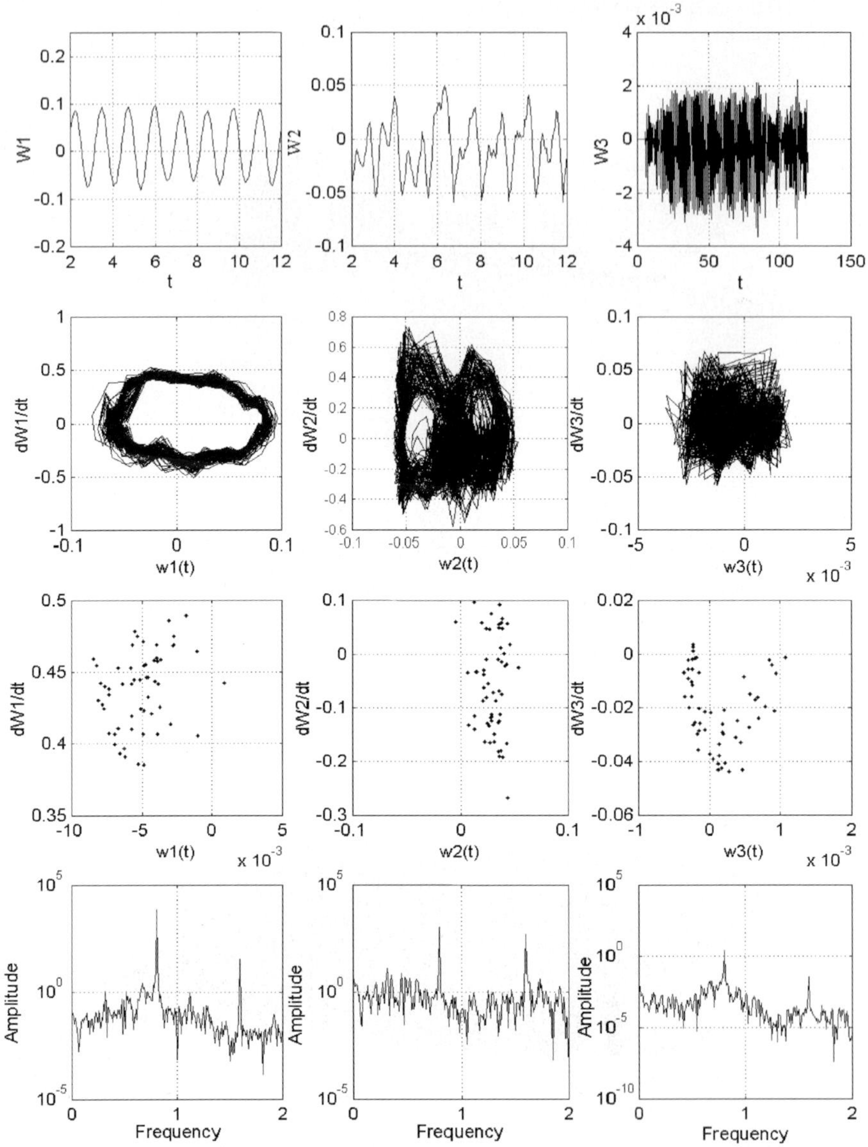

Fig. 14.19 Vibrations $w_i(0.5;\, t)$, velocities $\dot{w}_i(0.5;\, t)$, Poincaré sections $(\dot{w}_i,\, w_i)$, and power spectra for $q_1 = 0.65$

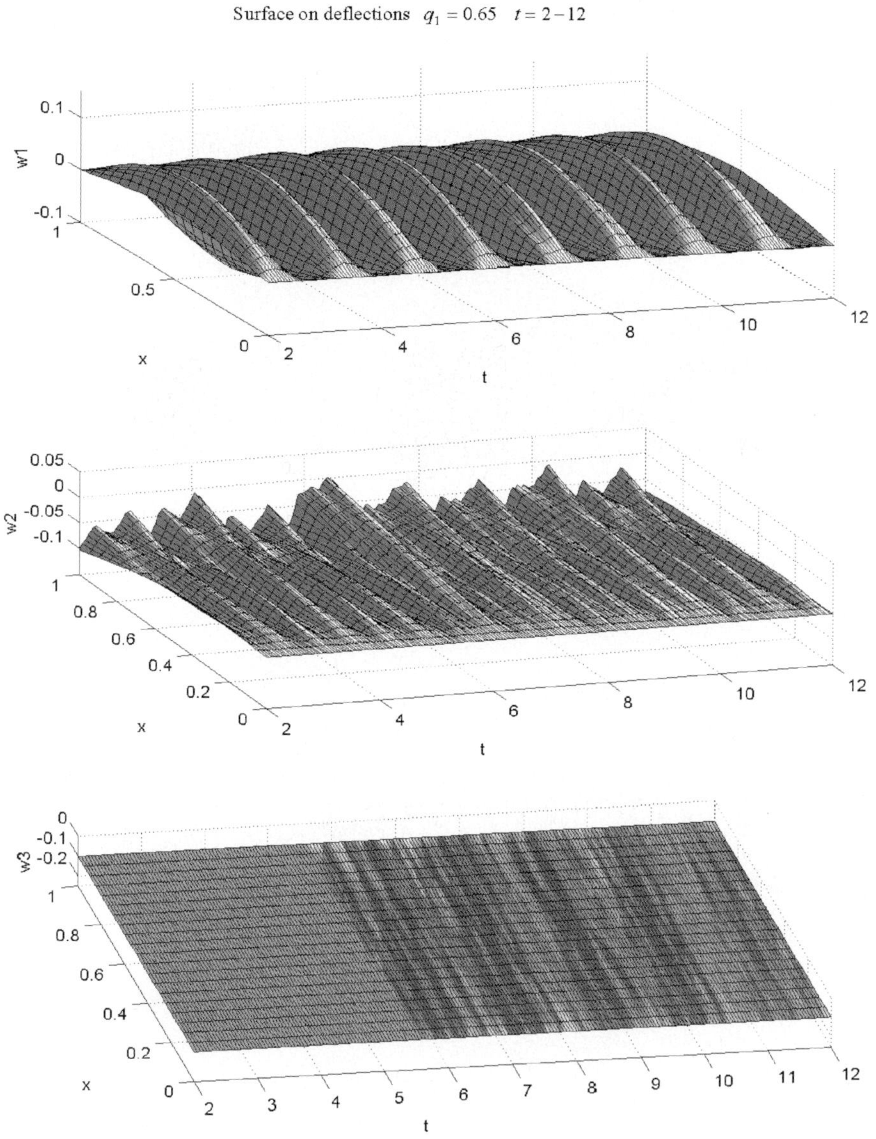

Fig. 14.20 Time histories of the beam surfaces for $q_1 = 0.65$

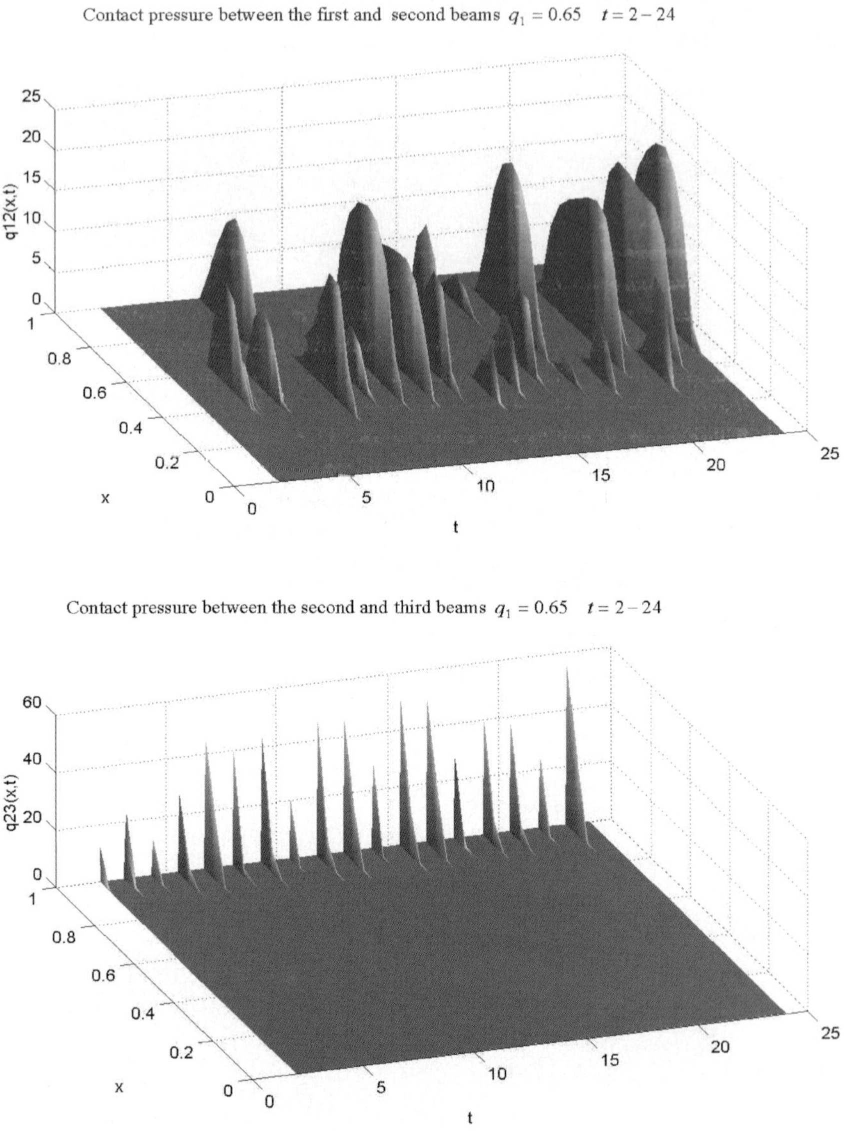

Fig. 14.21 Contact pressure between beams for $q_1 = 0.65$

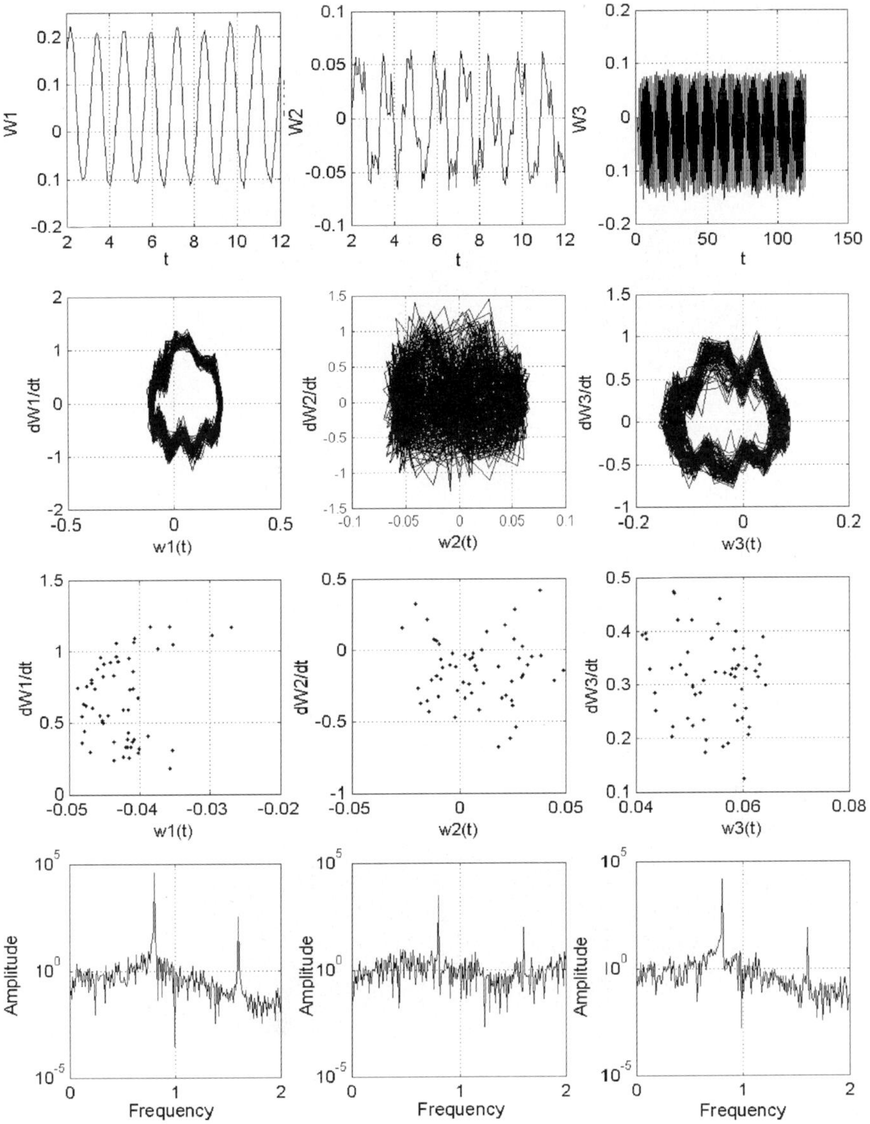

Fig. 14.22 Vibrations $w_i(0.5;\ t)$, velocities $\dot{w}_i(0.5;\ t)$, Poincaré sections $(\dot{w}_i,\ w_i)$ and power spectra for $q_1 = 2.0$

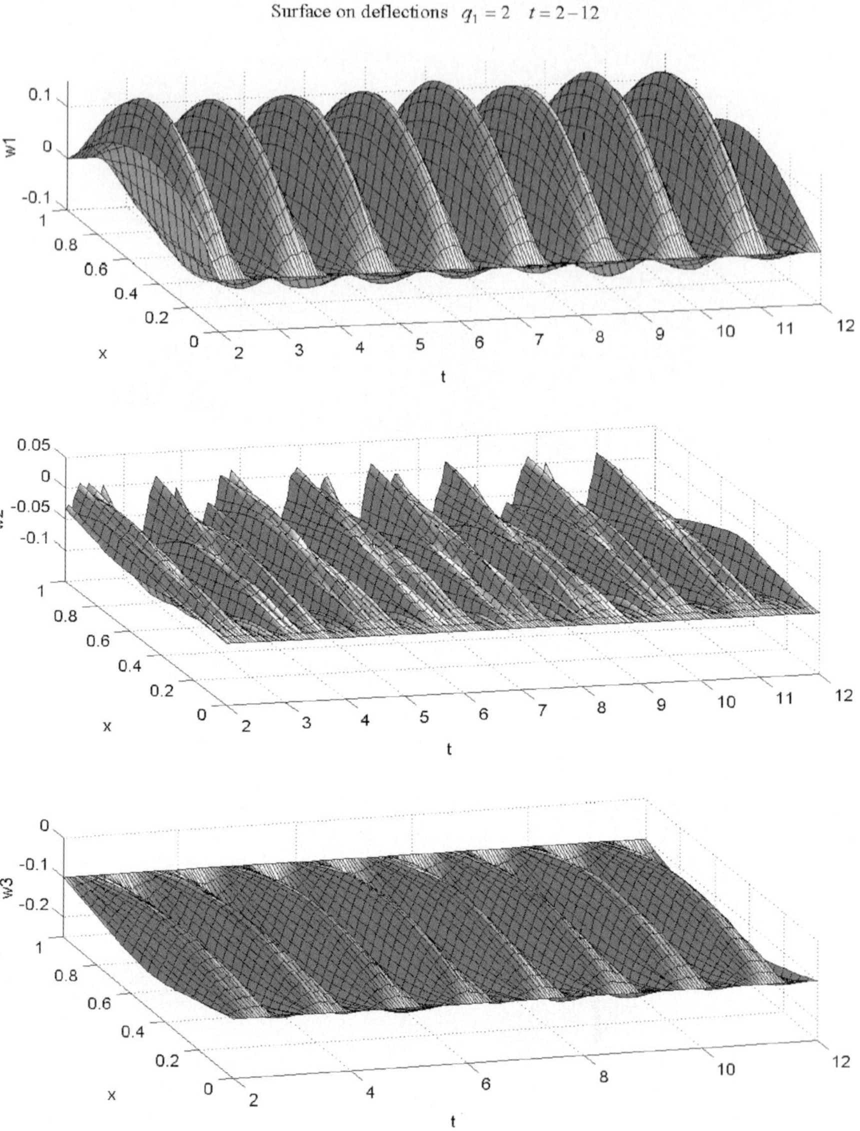

Fig. 14.23 Time histories of the beam surfaces for $q_1 = 2.0$

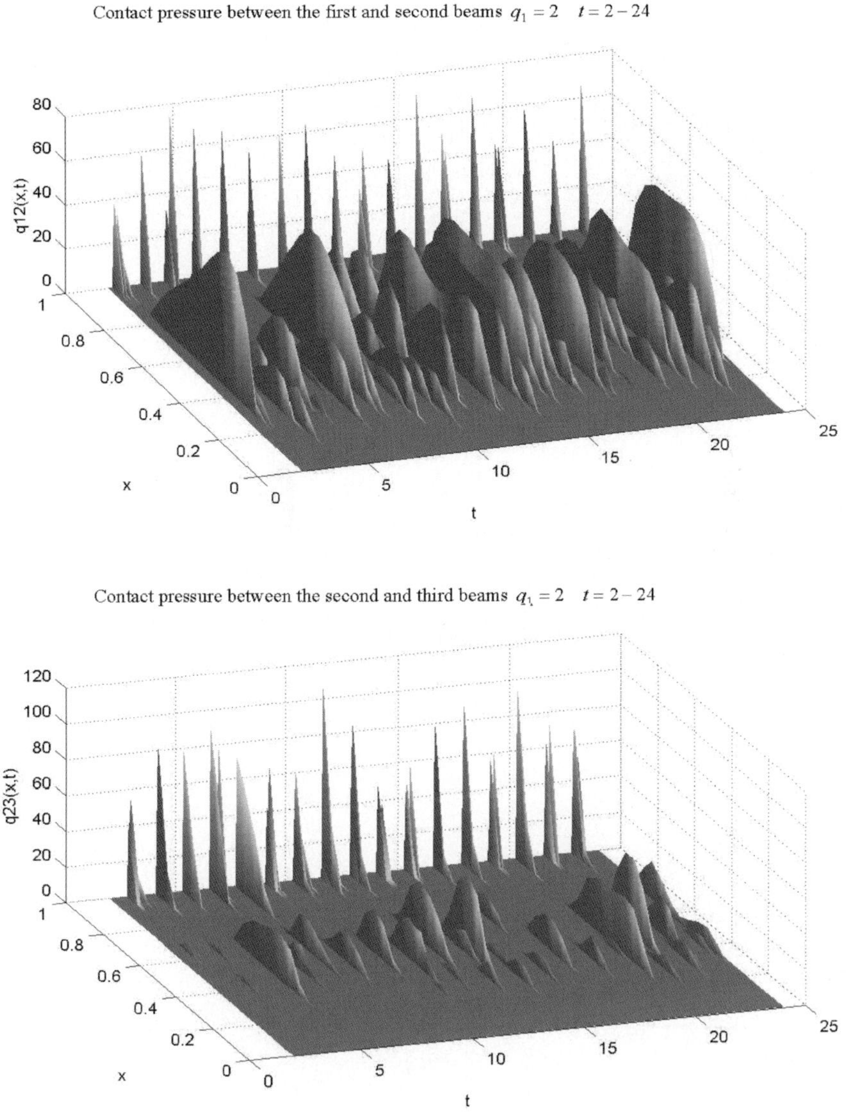

Fig. 14.24 Contact pressure between beams for $q_1 = 2.0$

14.5 All Three Beams are Nonlinearly Elastic

Beams will be absolutely nonlinearly elastic if $G_1 = G_2 = G_3 = 0$. The computational results are reported in Figs. 14.25–14.29. The vibration characteristics are located similarly to those in the linearly elastic case. In addition, the dependence on time of the deformation intensity is reported in $e_i(t)$. This is defined on the beam surfaces in the case of free (balls) support, and clamping for the cantilever beam.

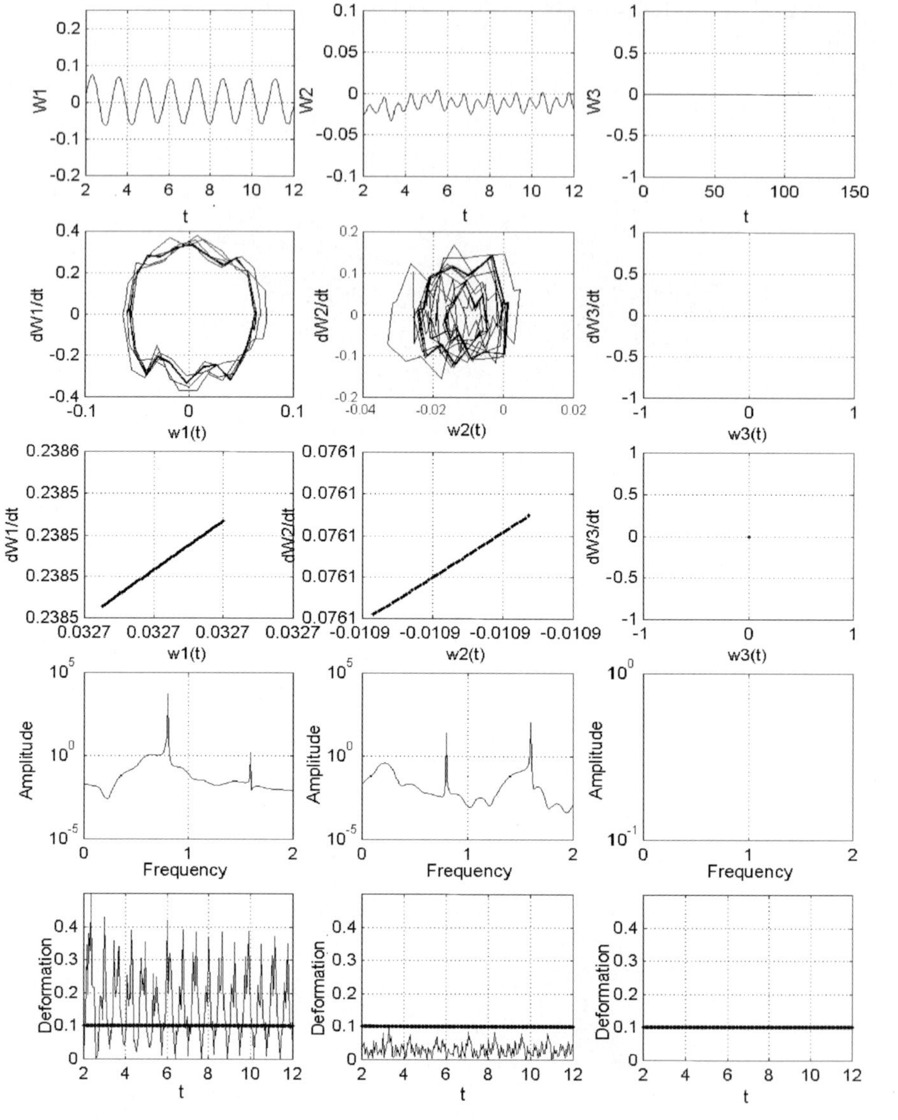

Fig. 14.25 Vibrations $w_i(0.5;\ t)$, velocities $\dot{w}_i(0.5;\ t)$, Poincaré sections $(\dot{w}_i,\ w_i)$, power spectra and $e_i(t)$ for $q_1 = 0.34$

The solid curve corresponds to the flow deformation intensity e_{s1}. Similar to the previous case, the top beam exhibits quasi-harmonic vibrations (Fig. 14.25). This can be easily observed in the phase portrait and the power spectrum. Note that deformation intensity exceeds e_{s1} i.e., the deformation intensity achieves a horizontal part of the deformation diagram (14.7). Increasing q_{10} more excites the second beam. In the power spectrum of the second beam, the components ω and 2ω are

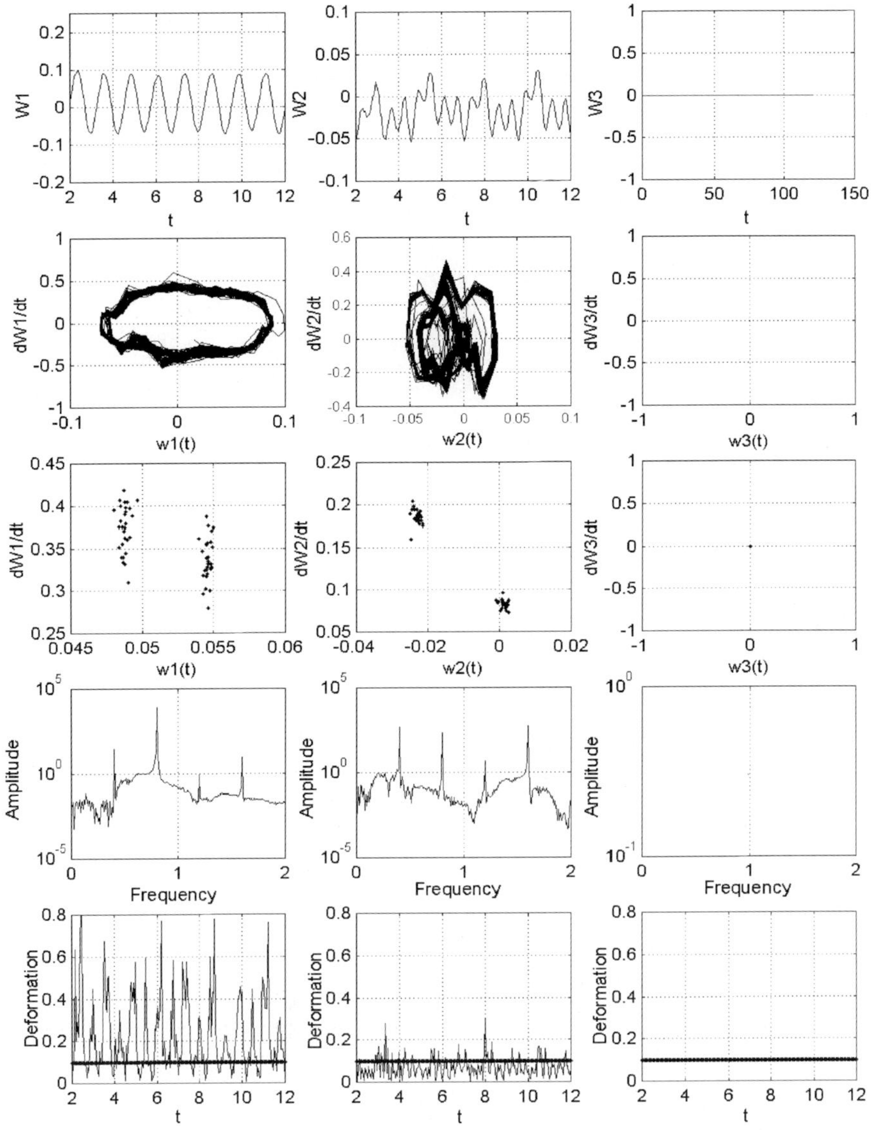

Fig. 14.26 Vibrations $w_i(0.5;\ t)$, velocities $\dot{w}_i(0.5;\ t)$, Poincaré sections $(\dot{w}_i,\ w_i)$, power spectra and $e_i(t)$ for $q_1 = 0.75$

observed. Since the second component is essentially larger then the first one, the amplitude characteristics are qualitatively different in comparison with the case of harmonic vibrations. For $q_{10} = 0.75$ the first period doubling (Fig. 14.26) occurs and the components $\omega/2$, $3\omega/2$, and so on, appear.

However, the quasi components are very small for the first beam, so its vibrations remain quasi-harmonic. For the second beam, those components are large and

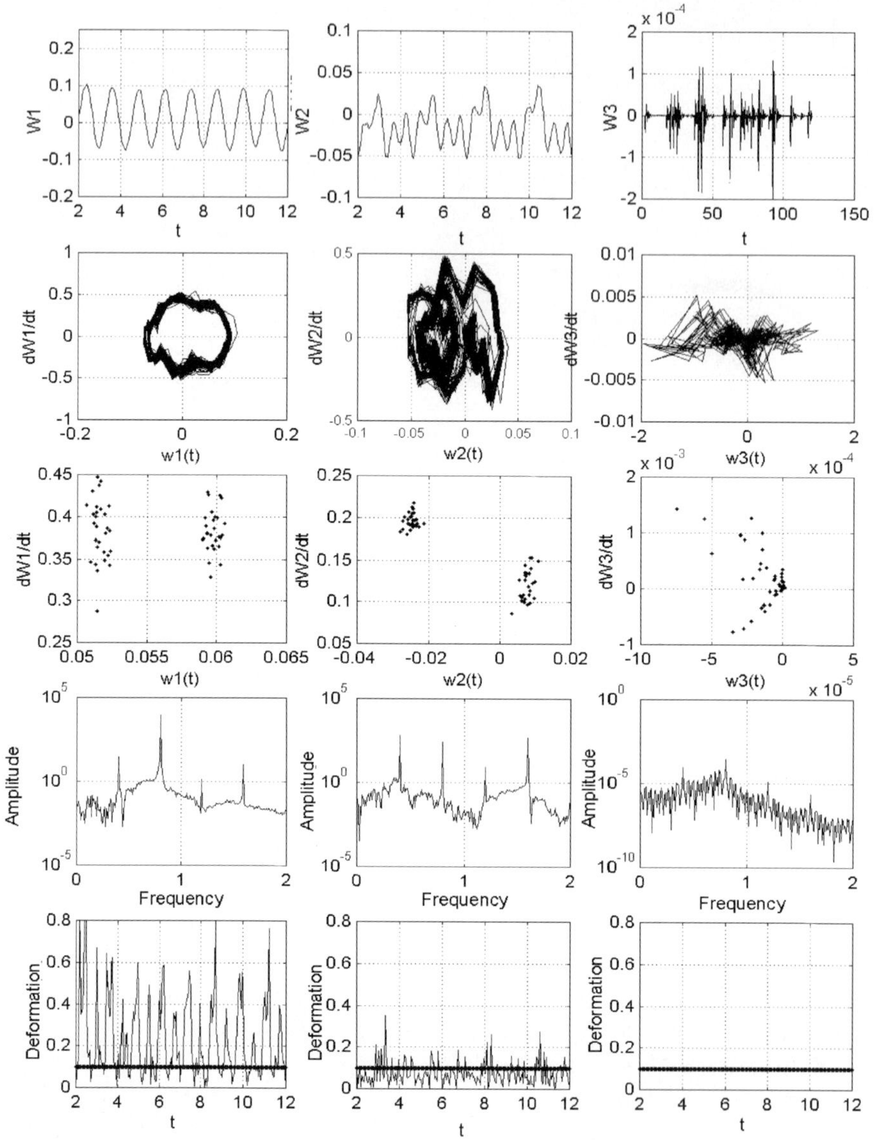

Fig. 14.27 Vibrations $w_i(0.5;\ t)$, velocities $\dot{w}_i(0.5;\ t)$, Poincaré sections $(\dot{w}_i,\ w_i)$, power spectra and $e_i(t)$ for $q_1 = 0.8$

the vibrations more complex, as shown by the amplitude characteristics and the phase portraits. In the Poincaré section, with respect to the period of exciting force, two sets of points occur, instead of one point observed for a smaller load. Both beams achieve the horizontal part of the deformation diagram, which is exhibited by the characteristics $e_i(t)$. This picture holds also during contact with the third

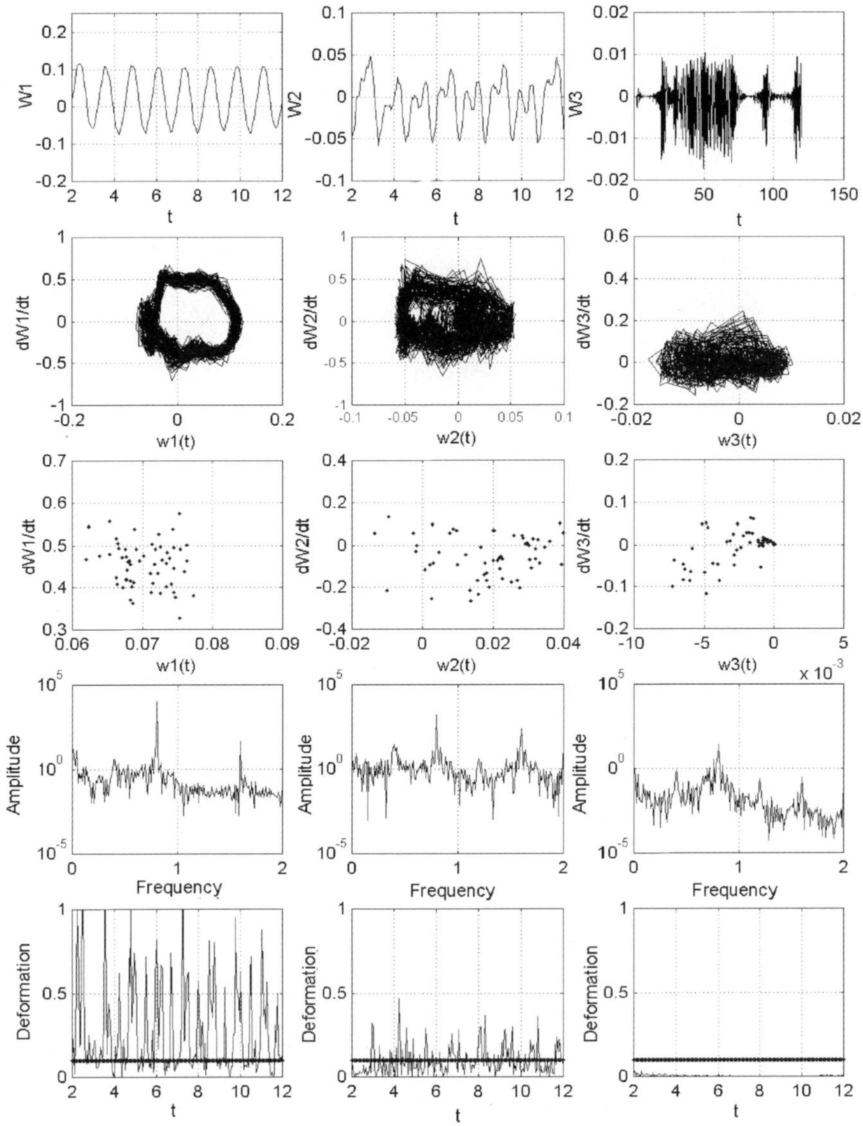

Fig. 14.28 Vibrations $w_i(0.5;\ t)$, velocities $\dot{w}_i(0.5;\ t)$, Poincaré sections $(\dot{w}_i,\ w_i)$, power spectra and $e_i(t)$ for $q_1 = 1.0$

beam (Fig. 14.27), which in practice immediately approaches a chaotic state. For $q_{10} = 1$ the first Hopf bifurcation is collapsed (Fig. 14.28); in the spectrum of the top and middle beams a strong increase of noisy components is observed, and then the system moves into chaos (Fig. 14.29).

To conclude, the increase of the parameter q_{10} is accompanied by a motion at low scales. The higher frequencies appear and the beams motion is more complex.

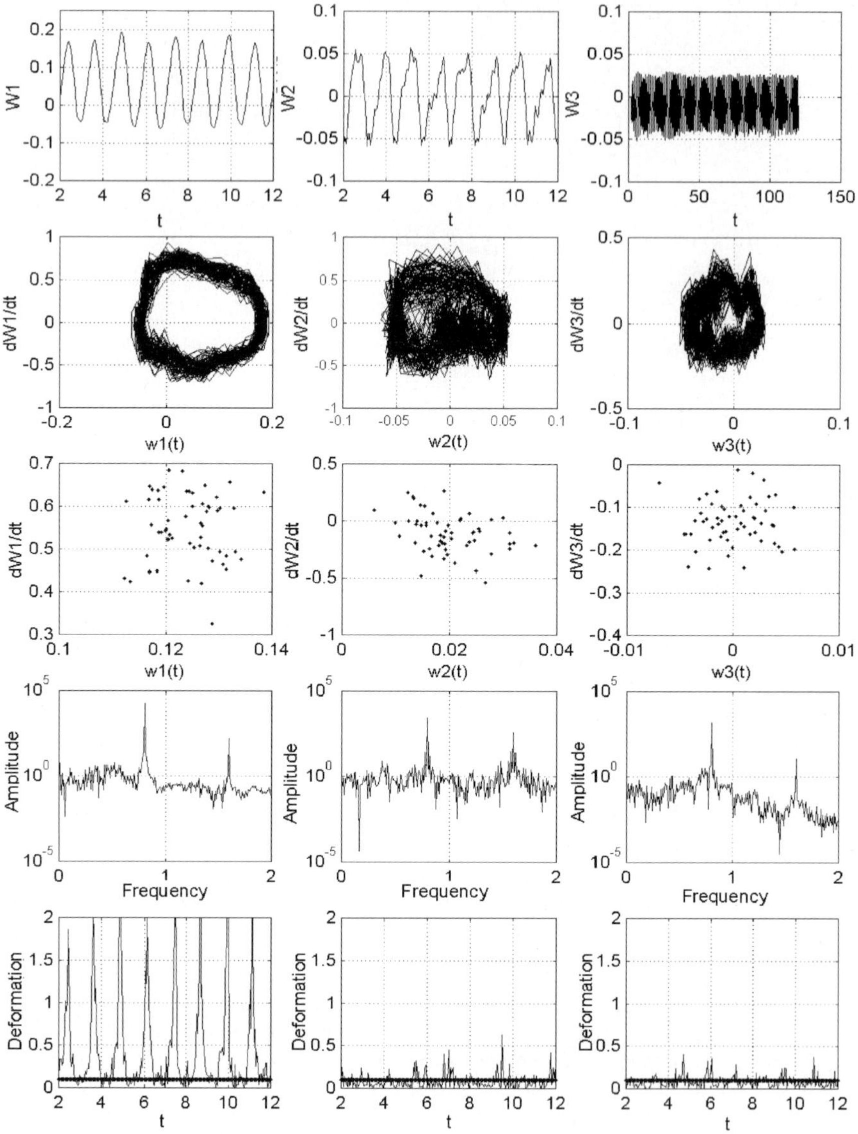

Fig. 14.29 Vibrations $w_i(0.5;\ t)$, velocities $\dot{w}_i(0.5;\ t)$, Poincaré sections $(\dot{w}_i,\ w_i)$, power spectra and $e_i(+)$ for $q_1 = 1.5$

In the power spectrum a transition from the discrete to the continuous spectrum is observed, and it is widened into a set of high frequencies. During a transition to chaos, and occurrence of large motions, the shape of the power spectrum recalls that of the processes where the energy cascades into the top of spectrum. In other words, a clearly expressed maximum, with mild decrease of the spectrum density in the direction of high frequencies, and a sudden decrease in the direction of low frequencies are observed (Figs. 14.27–14.29). The increase of q_{10} is accompanied by a shift of the spectrum density maximum in the direction of high frequencies.

It should be emphasized that energy cascade into the high part of the spectrum is typical for continuous systems exhibiting complex dynamics, including hydrodynamic objects.

Note that a route to chaos is different than in the case previously analyzed. The bifurcation series are not observed, and after the first Hopf bifurcation the system immediately transits to chaos. Therefore, the presence of nonlinearly elastic material qualitatively changes the vibrational process.

14.6 Conclusions

The general and high-accuracy method to trace bifurcation and chaos of physically dissipative nonlinear sandwich of three beams is proposed. First, the governing equations are derived and the computational finite difference algorithm is described. Three different materials of the beams are considered (ideal elastic-plastic, elastic-plastic with linear strengthens, and the pure aluminum). The amplitude characteristics, phase portraits, Poincaré sections, and the power spectrum are used to trace bifurcations and chaos of the analyzed system. Two main computational examples are reported. In the first one, the dynamics of linearly elastic beams are analyzed. The periodic, quasi-periodic, and chaotic dynamics of the beams is described in the text and is not repeated here. In addition, the contact zones between beams are traced. One frequency bifurcation, a first Hopf bifurcation, the triple bifurcation birth and its collapse, a second Hopf bifurcation and its collapse, and the bifurcation scenario leading to chaos are illustrated and discussed, among others.

The second example includes three beams that are nonlinearly elastic (detailed description is given in the latter section). In the beam's power spectrum, a transition from the discrete to continuous spectrum is observed with a clearly exhibited energy cascade into the high-spectrum component. The bifurcation series scenario leading to chaos is not observed in this case. After the first Hopf bifurcation the system suddenly jumps straight into chaos.

Chapter 15
Nonlinear Vibrations of the Euler-Bernoulli Beam Subjected to Transversal Load and Impact Actions

In this chapter, complex vibrations of a flexible Euler-Bernoulli type beam driven by dynamical load and with various type of inputs on its edge are studied. The governing equations include damping terms with damping coefficients ε_1, ε_2 associated with deflection w and displacement u, respectively. Damping coefficients ε_1, ε_2 and the transversal load coefficients (q_0, ω_p) serve as control parameters. The formulated infinite dimensional problem is reduced to that of final dimension applying the finite difference method with approximation $O(h^2)$ with regard to spatial coordinates and it is solved via the fourth-order Runge-Kutta method. This approach enabled identification of damping coefficients ε_1 and ε_2, as well as investigation of elastic waves generated by an impact introduced through a mass (lumped body) moving at constant velocity V. The introduced analysis is supported by applied achievements of the qualitative theory of differential equations and nonlinear dynamics.

15.1 Introduction

This chapter provides a novel approach to various topics of nonlinear vibration of a continuous system, impact, wave, bifurcation, and chaos.

It is well known that dynamical behavior of the Bernoulli-type beams with impacts has been studied theoretically, numerically, and experimentally for a long time. For example, the Bernoulli-type beam supported by springs and periodically forced has been studied using the finite elements method [152]. Dynamics of vibro-impacts are observed evaluating the impact velocity as a function of excitation frequencies.

In reference [75] an experimental study of a beam held in a fixed mount with clamped-free boundary conditions forced by a sinusoidal impactor was carried out. The authors show that a mean square response amplitude is captured by the first proper orthogonal beam mode.

J. Awrejcewicz, V.A. Krysko, *Chaos in Structural Mechanics*,
DOI: 10.1007/978-3-540-77676-5_16,

Periodic solutions of a periodically driven multi-DOF beam system possessing an elastic stop in the middle was studied in reference [306]. The steady-state beam responses were analyzed using two-point boundary value problems and multiple shooting applying the approximation of one, two, and four degrees of freedom of the beam.

Regular and chaotic impacting and non-impacting motion of a driven beam has been studied in [58]. One-, two- and three-impacts per motion of a vibro-impacting pinned beam are studied experimentally and the results are compared with those obtained from multi degree-of-freedom models. It is concluded that the latter models are necessary to estimate proper system responses for a high-frequency range [91].

Regular and chaotic dynamics of a long-term behavior of periodically excited linear beams with a one-sided spring has been studied in reference [307] both numerically and experimentally.

A rod subjected to both supercritical force and a time-dependent transverse loading has been studied numerically via the discretization method [283].

On the other hand, chaotic motions exhibited by deterministic systems are nowadays widely reported in the scientific literature. Although they have been observed in fluid mechanics, other mechanical objects governed by partial differential equations like plates and shells may also exhibit the so-called spatial-temporal deterministic chaos [21, 25, 31]. More rigorously the associated problem of uniqueness of solution to dynamical problems for shells of Timoshenko type has been investigated in references [160].

However, in spite of an increasing number of investigations of chaotic dynamics of such complex deterministic objects like plates, conical, spherical, and cylindrical shells the investigation of geometrically nonlinear Euler-Bernoulli type beams still await further study. Our work, among others, addresses the influence of damping terms ε_1, ε_2 and various geometrical beam parameters on the frequency characteristics of the mentioned objects, and we analyze the behavior of waves generated by the lumped body impact on the beam.

15.2 Problem Formulation

We mainly address the problem of mathematical modeling of the longitudinal vibrations of flexibly Euler-Bernoulli type beams with various boundary conditions (see Fig. 15.1).

Introducing the coordinates XOZ we consider the mentioned beam in the space $\Omega = \{x \in [0,a];\ -c \le z \le c;\ -\frac{b}{2} \le y \le \frac{b}{2}\}$ with its middle surface deformation $\varepsilon_x = \frac{\partial u}{\partial x} + \frac{1}{2}\left(\frac{\partial w}{\partial x}\right)^2$, where $w(x,t)$ denotes the beam deflection, and $u(x,t)$ describes displacement of its middle surface along the OX axis. It is assumed that owing to the Euler-Bernoulli hypothesis a normal (to the middle surface) remains all the time normal during the beam deformation process $\varepsilon_{xx} = \varepsilon_x - z\frac{\partial^2 w}{\partial x^2}$. The mentioned assumptions allow us to obtain the following dimensional governing equations:

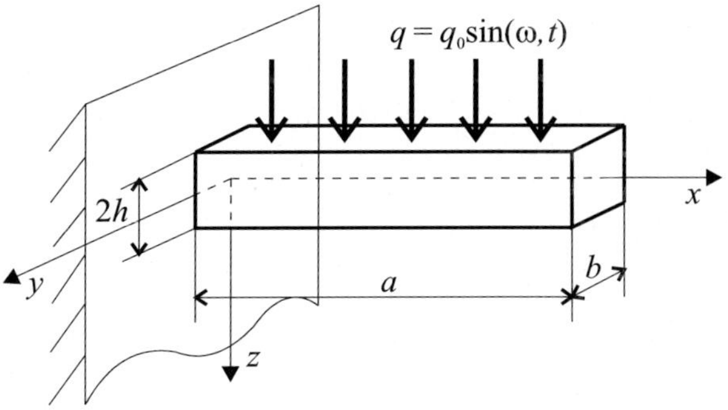

Fig. 15.1 Model of the studied problem

$$\rho\frac{\partial^2 u}{\partial t^2}+\rho\varepsilon_2\frac{\partial u}{\partial t}-\frac{E}{1-\nu^2}\left(\frac{\partial^2 u}{\partial x^2}+\frac{\partial w}{\partial x}\frac{\partial^2 w}{\partial x^2}\right)=0,$$

$$\rho\frac{\partial^2 w}{\partial t^2}+\rho\varepsilon_1\frac{\partial w}{\partial t}+\frac{2c^3}{3}\frac{E}{1-\nu^2}\frac{\partial^4 w}{\partial x^4}-\frac{E2c}{1-\nu^2}\left(\frac{\partial^2 u}{\partial x^2}+\frac{\partial w}{\partial x}\frac{\partial^2 w}{\partial x^2}\right)\frac{\partial w}{\partial x}$$

$$-\frac{E2h}{1-\nu^2}\left(\frac{\partial u}{\partial x}+\frac{1}{2}\left(\frac{\partial w}{\partial x}\right)^2\right)\frac{\partial^2 w}{\partial x^2}-q=0, \qquad (15.1)$$

where E is the Young modulus, ρ denotes beam material density, ε_1, ε_2 are the already mentioned damping coefficients, $2c$ is the beam height, a denotes the beam length, and q is the beam transversal load.

Let us apply new dimensionless parameters in the following form:

$$A=\frac{E}{1-\nu^2},\quad \bar{w}=\frac{w}{2c},\quad \bar{u}=\frac{ua}{(2c)^2},\quad \bar{x}=\frac{x}{a},\quad \lambda=\frac{a}{2c},$$

$$\bar{q}=q\frac{a^4}{(2c)^4A},\quad \bar{t}=\frac{t}{\tau},\quad \tau=\frac{a}{h},\quad h=\sqrt{\frac{Ag}{\rho}},\quad \bar{\varepsilon}=\varepsilon\frac{a}{h}, \qquad (15.2)$$

where $\bar{x}$, $\bar{u}$, $\bar{w}$, $\bar{t}$, $\bar{q}$, $\bar{\varepsilon}$ are the mentioned dimensionless parameters, and in the forthcoming system 15.3 bars are omitted. Now dimensionless parameters are as follows: u is the displacement along axis x, w is the deflection, t denotes time, q is the transversal load, ε is the dissipation coefficient, g is the gravitational acceleration, and a, $2c$ are the linear beam dimensions.

The dimensionless equations are given in the form

$$\frac{\partial^2 w}{\partial t^2}+\varepsilon_1\frac{\partial w}{\partial t}=\frac{1}{\lambda^2}\left(-\frac{1}{12}\frac{\partial^4 w}{\partial x^4}+l_1(w,u)+l_2(w,u)+q\right),$$

$$\frac{\partial^2 u}{\partial t^2}+\varepsilon_2\frac{\partial u}{\partial t}=\frac{\partial^2 u}{\partial x^2}+l_3(w,w), \qquad (15.3)$$

where

$$l_1(w,u) = \frac{\partial w}{\partial x}\frac{\partial^2 u}{\partial x^2} + \left(\frac{\partial w}{\partial x}\right)^2 \frac{\partial^2 w}{\partial x^2}, \quad l_2(w,u) = \left(\frac{\partial u}{\partial x} + \frac{1}{2}\left(\frac{\partial w}{\partial x}\right)^2\right)\frac{\partial^2 w}{\partial x^2},$$

$$l_3(w,w) = \frac{\partial w}{\partial x}\frac{\partial^2 w}{\partial x^2}$$

are the nonlinear operators.

The following boundary conditions are attached to Eqs. (15.3):

1. Clamping-clamping:

$$w(0) = w(1) = 0, \quad u(0) = u(1) = 0, \quad \frac{\partial w(0)}{\partial x} = \frac{\partial w(1)}{\partial x} = 0\,; \tag{15.4}$$

2. Simple-support:

$$w(0) = w(1) = 0, \quad u(0) = u(1) = 0, \quad \frac{\partial^2 w(0)}{\partial x^2} = \frac{\partial^2 w(1)}{\partial x^2} = 0\,; \tag{15.5}$$

3. Clamping-free edge:

$$w(0) = u(0) = 0, \quad M_x = N_x = 0, \tag{15.6}$$

where $N_x = \int_{-c}^{c} \sigma_{xx} dz$ is the longitudinal force, $M_x = \int_{-c}^{c} \sigma_{xx} z dz$ is the torque.

The following initial conditions are attached to system Eqs. (15.3), (15.4), (15.5), (15.6) in the considered problems 1–3:

$$w(x)_{|t=0} = u(x)_{|t=0} = 0, \quad \dot{w}(x)_{|t=0} = \dot{u}(x)_{|t=0} = 0. \tag{15.7}$$

15.3 Finite Differences Method

The infinite dimensional problem (15.3), (15.4), (15.5), (15.6), (15.7) is reduced to a system of ordinary differential equations (ODEs) using the finite differences method with approximation $O(h^2)$. In each mesh node the following system of ODEs is defined:

$$\begin{aligned} L_{1,h}(w_i, u_i) &= \varepsilon_i \dot{w}_i + \ddot{w}_i, \\ L_{2,h}(w_i, u_i) &= \varepsilon_2 \dot{u}_i + \ddot{u}_i, \quad i = 0, \ldots, n, \end{aligned} \tag{15.8}$$

where:

$$L_{1,h}(w_i, u_i) = \frac{1}{\lambda^2}\Bigg[-\frac{1}{12}\frac{1}{h^4}(w_{i-2} - 4w_{i-1} + 6w_i$$
$$-4w_{i+1} + w_{i-2}) + \frac{1}{2h}(w_i - 1 - w_{i+1})\frac{1}{h^2}(u_{i+1} - 2u_i + u_{i-1})$$

$$+\frac{1}{2h}(w_{i-1}-w_{i+1})\frac{1}{h^2}(u_{i+1}-2u_i+u_{i-1})$$
$$+\left(\frac{1}{2h}(w_{i-1}-w_{i+1})\right)\frac{1}{h^2}(w_{i+1}-2w_i+w_{i-1})$$
$$+\frac{1}{h^2}(w_{i+1}-2w_i+w_{i-1})\left(\frac{1}{2h}(u_{i+1}-u_{i-1})\right.$$
$$\left.\left.+\frac{1}{8h^2}(w_{i-1}-w_{i+1})(w_{i-1}-w_{i+1})\right)+q\right],$$
$$L_{2,h}(w_i,u_i)=\frac{1}{h^2}(u_{i+1}-2u_i+u_{i-1})$$
$$+\frac{1}{2h}(w_{i-1}-w_{i+1})\frac{1}{h^2}(w_{i+1}-2w_i+w_{i-1}).$$

The following boundary conditions are added to the considered cases:

(i) Problems 1 and 2:

$$w_0=w_n=0,\quad u_0=u_n=0\,; \tag{15.9}$$

(ii) Problem 3:

$$w_0=u_0=0,\quad M_x=N_x=0. \tag{15.10}$$

Initial conditions (15.7) in the difference form for problems 1–3 are:

$$w(x_i)_{|t=0}=u(x_i)_{|t=0}=0,$$
$$\dot{w}(x_i)_{|t=0}=\dot{u}(x_i)_{|t=0}=0,\quad i=0,\ldots,n. \tag{15.11}$$

The ODEs obtained for all problems are solved via fourth-order Runge-Kutta algorithm. The beam considered is subjected to the action of sign changeable load of the form

$$q=q_0\cos(\omega_p t), \tag{15.12}$$

where ω_p is the frequency of excitation, and q_0 is the excitation amplitude. Let us emphasize that the considered system is dissipative and damping coefficients $\varepsilon_1,\varepsilon_2$ correspond to deflection w and displacement u, respectively.

15.4 Influence of Damping Coefficients on the Frequency Characteristics

Let us consider the influence of damping ε_1 on the behavior of frequency characteristics along the beam length. The relative beam length $\lambda=\frac{a}{2c}=50$; it is sinusoidally driven with amplitude q_0. Note that both steps h (spatial) and Δt (time) have been chosen from stability conditions applying the Runge principle. We have taken $n=40,\ h=1/40$, and the time step $\Delta t=3.9052\times10^3$.

Below, we study the influence of ε_1 on variations of the fundamental frequency characteristics along the beam length applying the following indicators (for problems 1 and 3): Poincaré sections, modal portraits $w(w'_x)$, time histories $w(t)$, power spectra, and phase portraits $w(\dot{w})$.

15.4.1 Power Spectra

In Table 15.1 power spectra for problem 1 for various damping coefficients ε_1, $\varepsilon_2 = 0$ are reported. The amplitude of excitation q_0 for each ε_1 has been chosen to keep the condition $w(0.5) \approx 1.5$.

Analysis of the results reported in Table 15.1 shows that depending on ε_1 value, the number of frequencies along the beam length changes. Decreasing ε_1 induces a decrease of frequencies and already for $\varepsilon_1 = 5$ all beam points vibrate with one frequency along the whole beam length.

Below more detailed investigations are presented. With a decrease of parameter ε_1 one may observe a convergence of the frequencies on the graphs of power spectra depending on the beam length $(x = 0.25,\ x = 0.5,\ x = 0.75)$. For $\varepsilon_1 = 20$ the number of frequencies in the beam center $(x = 0.5)$ is significantly smaller in comparison to the beam quadrants $(x = 0.25;\ 0.75)$. In the center a linear combination of the free frequency $\omega_3 = \frac{\omega_p}{6}$ and the free frequencies $\omega_5 = 2.12$ and $\omega_1 \approx 10^{-6}$ occurs, whereas in the quadrants $(x = 0.25;\ 0.75)$ five more linearly independent frequencies and the linear combination $\omega_2 = \frac{\omega_p}{11}$ appear. The frequency ω_4 essentially depends on the length: for $x = 0.25$, we have $\omega_4 = 1.76$, and for $x = 0.75$, we have $\omega_4 = 1.22$. In the considered case the number and magnitude of frequencies in the beam quadrants fully overlap already for $\varepsilon_1 = 10$. The number of frequencies for $x = 0.5$ is less than for $x = 0.25;\ 0.75$, but the values of frequencies overlap their values for $x = 0.25$ and $x = 0.75$. If $\varepsilon_1 = 10$ then in points $x = 0.25$ and $x = 0.75$ a larger set of linearly dependent frequencies appears, whereas for $x = 0.5$ this number is smaller. Full overlapping of frequencies in the beam quadrants is caused by a symmetric clamping introduced from both beam sides. For $\varepsilon_1 = 5$ the number and magnitude of frequencies do not depend on the beam length. Finally, in general, the influence of the damping ε_1 on the frequency characteristics plays an essential role in the case of problem 1.

Investigation of problem 3 $(\lambda = 50)$ In this section we study beam vibrations for the series of fixed values of damping $\varepsilon_1 = 20,\ 10,\ 5,\ 1$ and use variation of damping ε_2. In Tables 15.2 and 15.3 power spectra for problem 3 and for various values of damping coefficients are reported. The amplitude q_0 has been chosen for each case in a way analogous to that in the previous problem.

The following main results have been obtained while analyzing this problem. Full overlapping of the number and magnitude of frequencies was observed only for $\varepsilon_1 = 1$ and $\varepsilon_2 = 0$. Only one linear combination $\omega_5 = \frac{\omega_p}{2}$ appears. Furthermore, increasing the value of ε_1 (for $\varepsilon_2 = 0$), and decreasing the distance between the

Table 15.1 Power spectra

$\varepsilon_1 = 20,\ q_0 = 3.5 \cdot 10^5$

$x = 0.25$	$x = 0.5$	$x = 0.75$
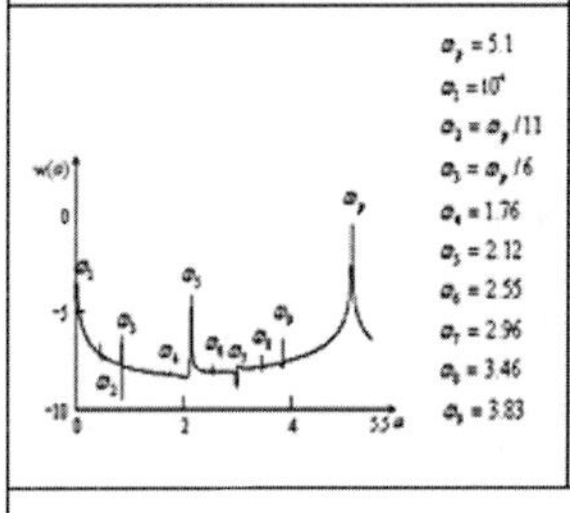	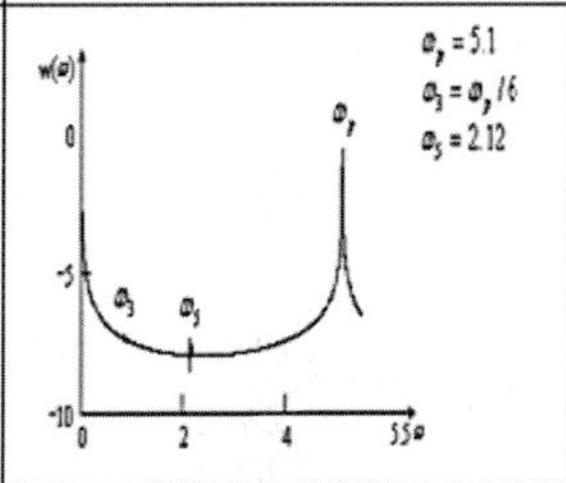	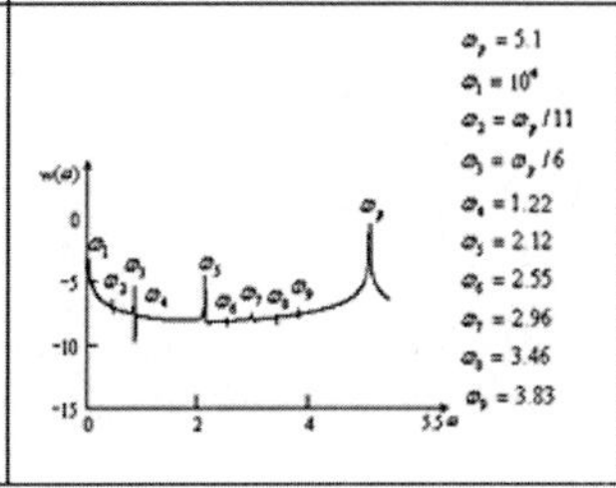

$\varepsilon_1 = 10,\ q_0 = 2.2 \cdot 10^5$

$x = 0.25$	$x = 0.5$	$x = 0.75$
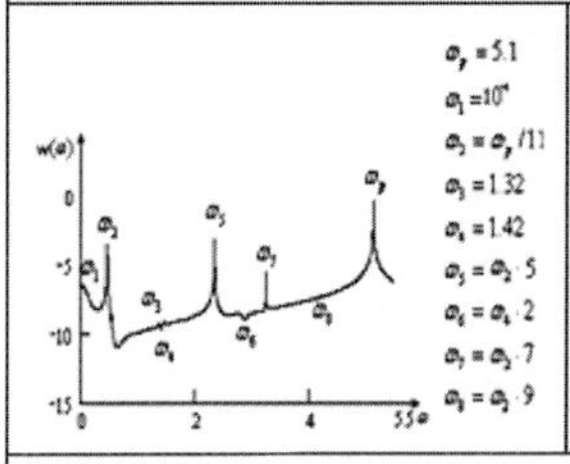	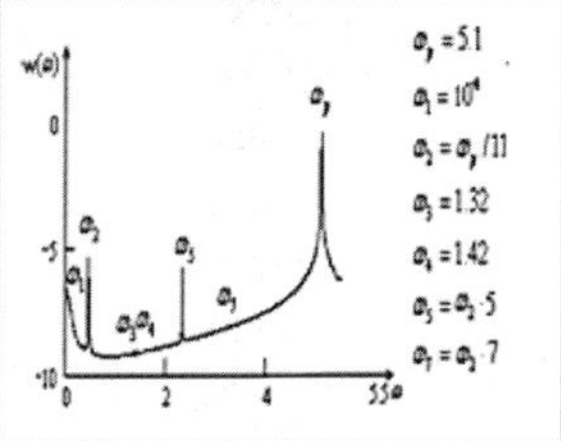	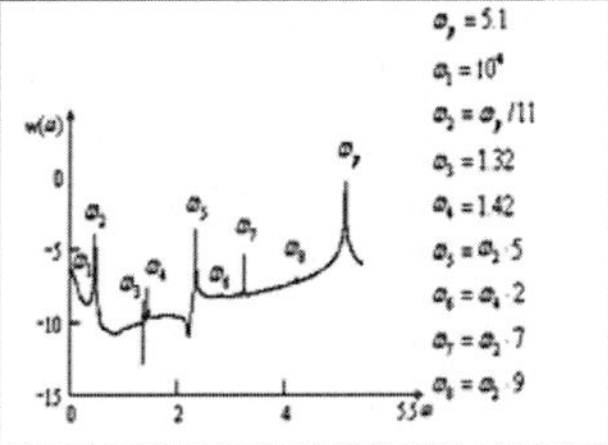

$\varepsilon_1 = 5,\ q_0 = 1.5 \cdot 10^5$

$x = 0.25$	$x = 0.5$	$x = 0.75$
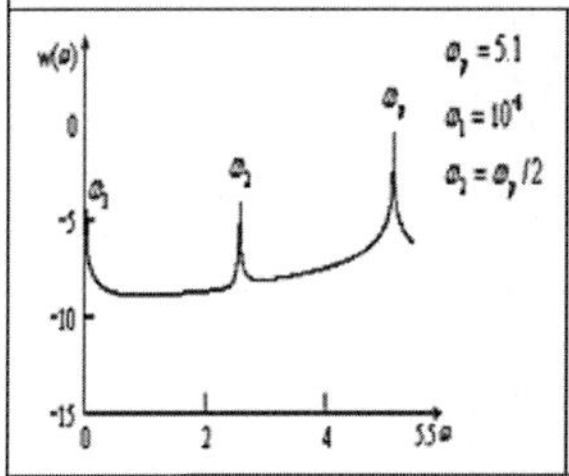	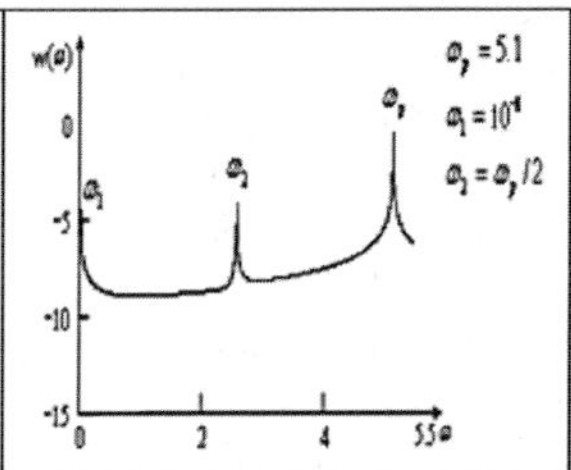	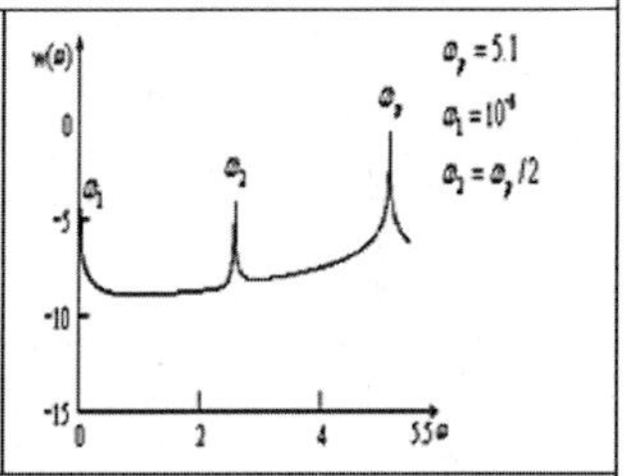

$\varepsilon_1 = 1,\ q_0 = 0.7 \cdot 10^5$

$x = 0.25$	$x = 0.5$	$x = 0.75$
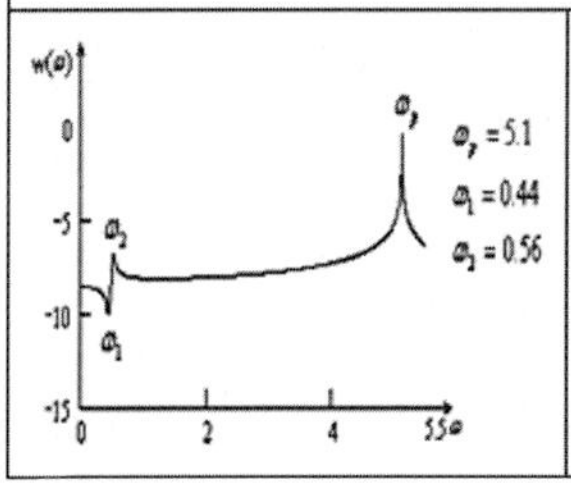	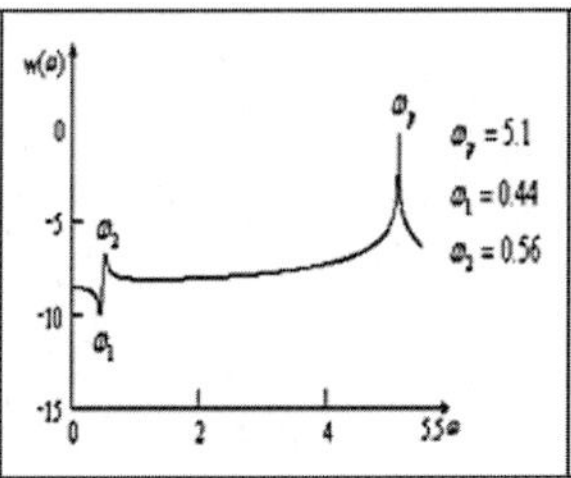	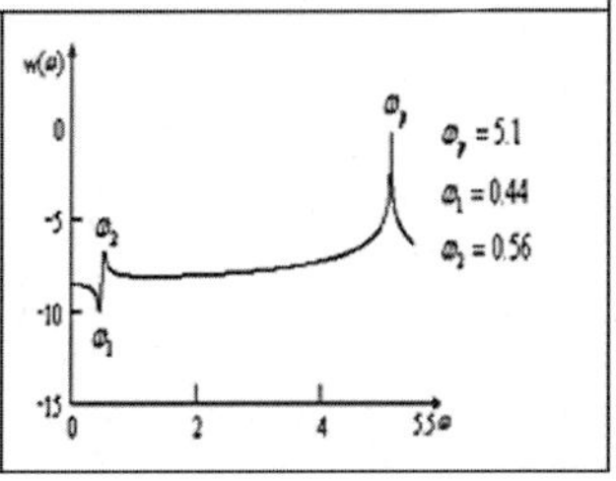

studied point and the clamping yield an increase of the frequencies occurring in the spectrum. On the contrary, on the free beam edge a number of frequencies is small in comparison to other beam points.

For $\varepsilon_1 = 20,\ \varepsilon_2 = 0$ on the beam free edge only three frequencies appear (all of them are linearly dependent). In point $x = 0.75$ the number of frequencies increases

Table 15.2 Power spectra for $\lambda = 50$ (a)

$\varepsilon_1 = 20,\ \varepsilon_2 = 0,\ q_0 = 2.8 \cdot 10^5$			
$x = 1$	$x = 0.75$	$x = 0.5$	$x = 0.25$
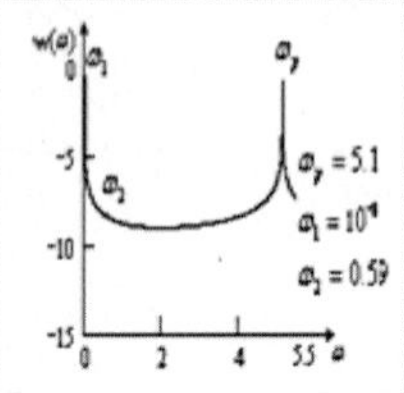	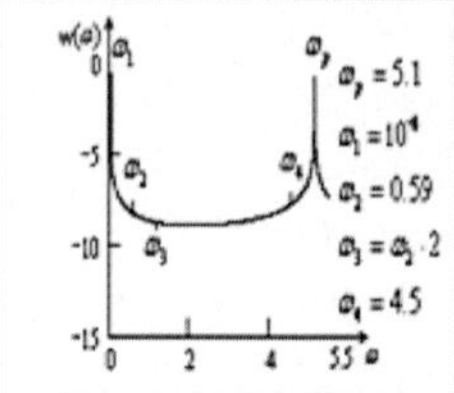	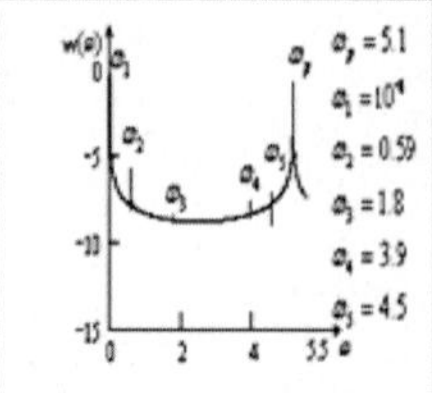	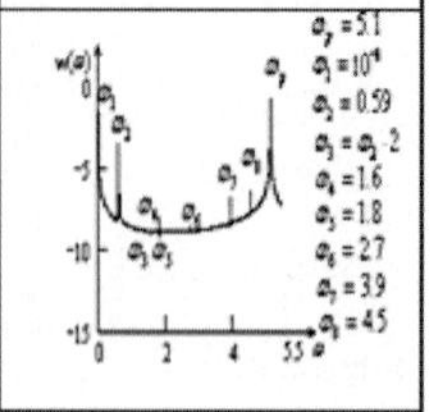
$\varepsilon_1 = 20,\ \varepsilon_2 = 0.05,\ q_0 = 2.8 \cdot 10^5$			
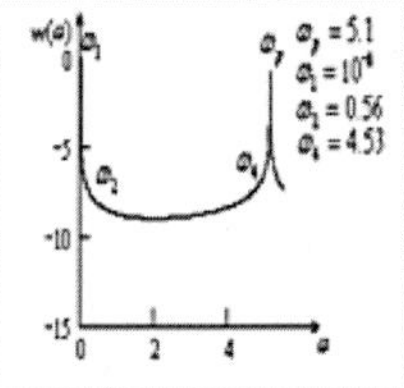	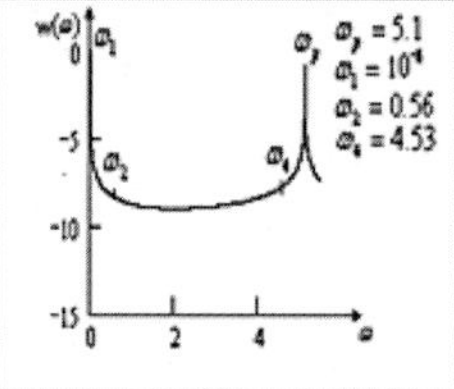	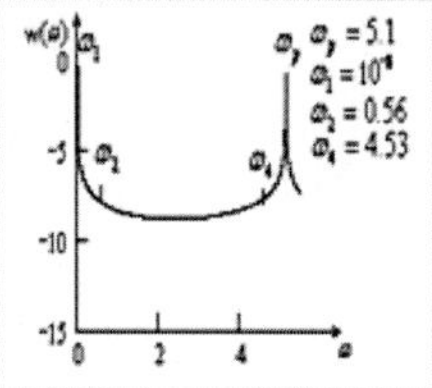	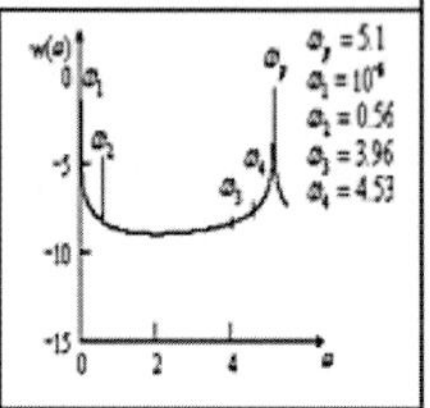
$\varepsilon_1 = 20,\ \varepsilon_2 = 0.1,\ q_0 = 2.8 \cdot 10^5$			
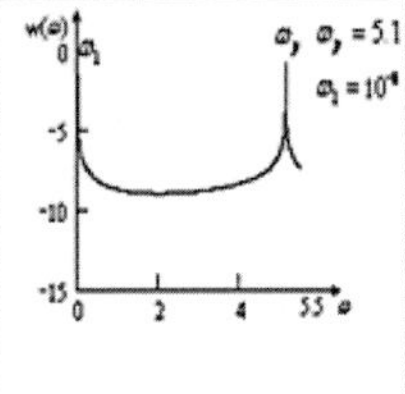	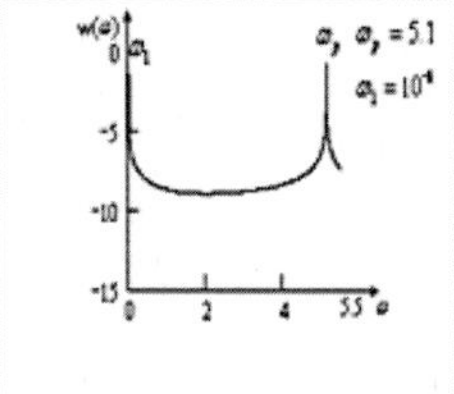	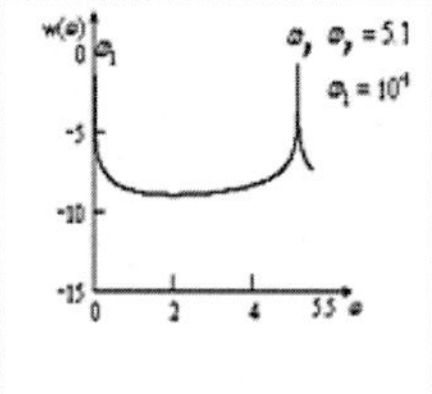	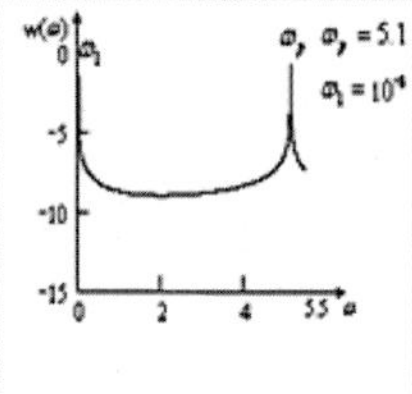
$\varepsilon_1 = 10,\ \varepsilon_2 = 0,\ q_0 = 1.65 \cdot 10^5$			
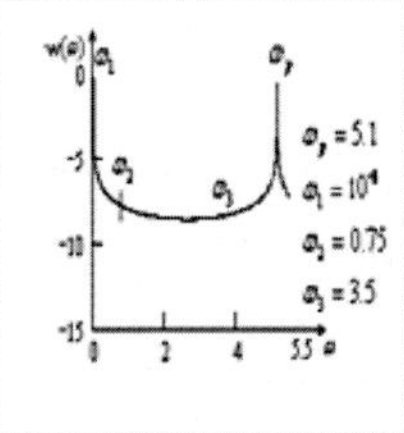	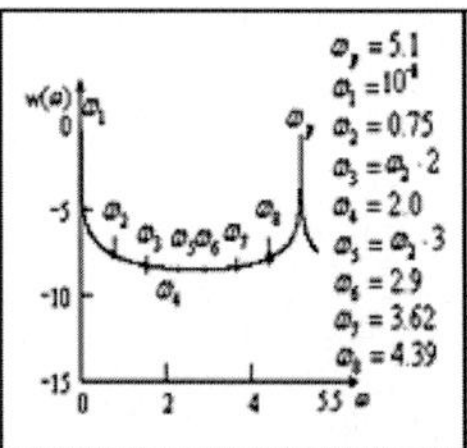	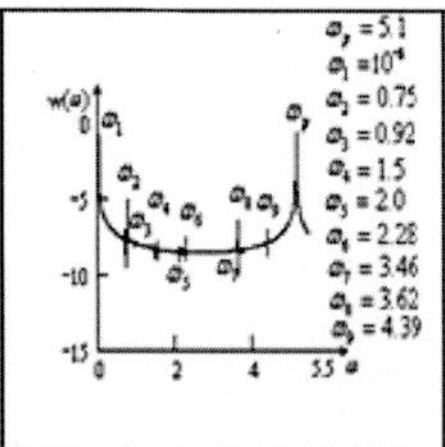	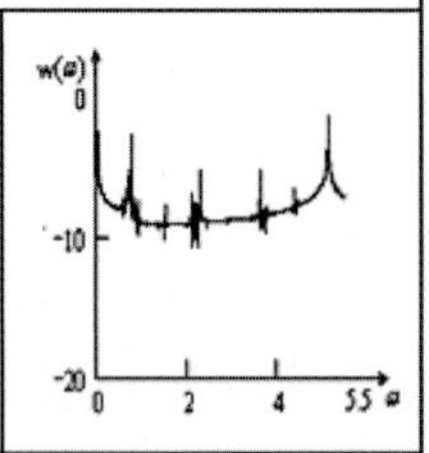
$\varepsilon_1 = 10,\ \varepsilon_2 = 0.05,\ q_0 = 1.65 \cdot 10^5$			
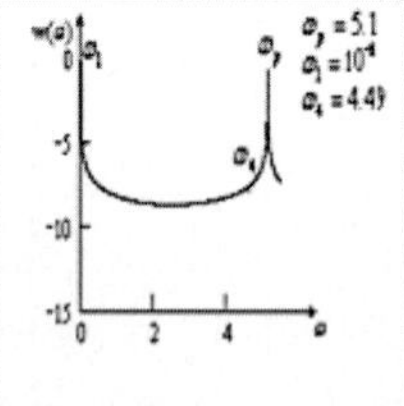	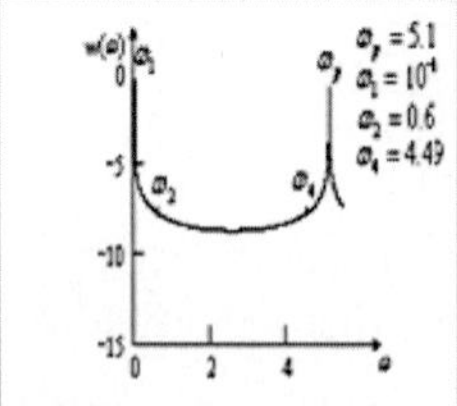	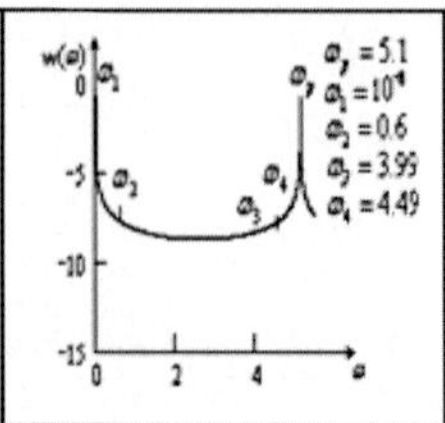	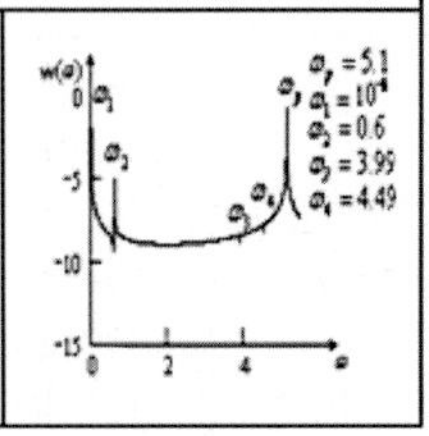

Table 15.3 Power spectra for $\lambda = 50$ (b)

$\varepsilon_1 = 10,\ \varepsilon_2 = 0.1,\ q_0 = 1.65 \cdot 10^5$

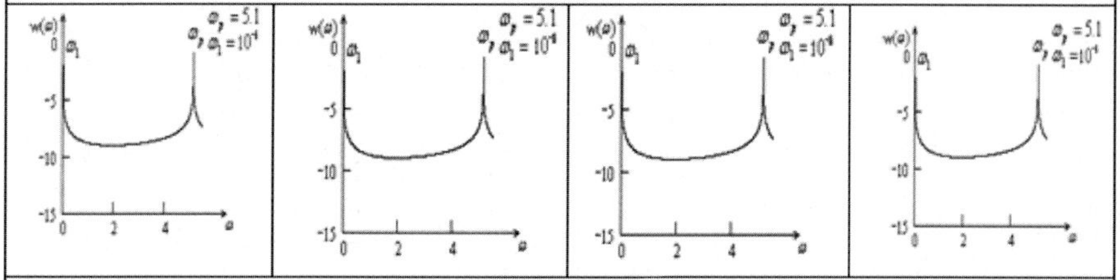

$\varepsilon_1 = 5,\ \varepsilon_2 = 0,\ q_0 = 0.95 \cdot 10^5$

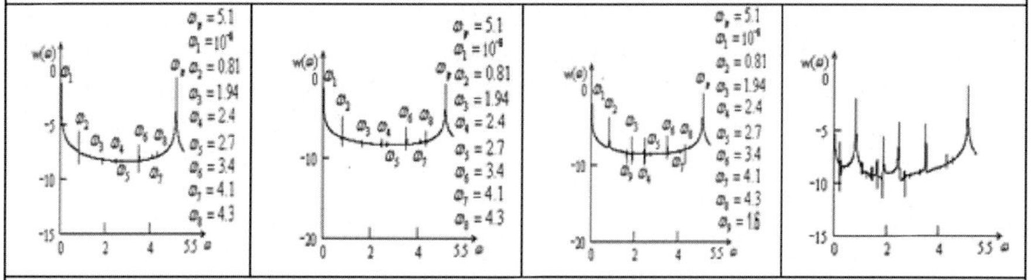

$\varepsilon_1 = 5,\ \varepsilon_2 = 0.04,\ q_0 = 0.95 \cdot 10^5$

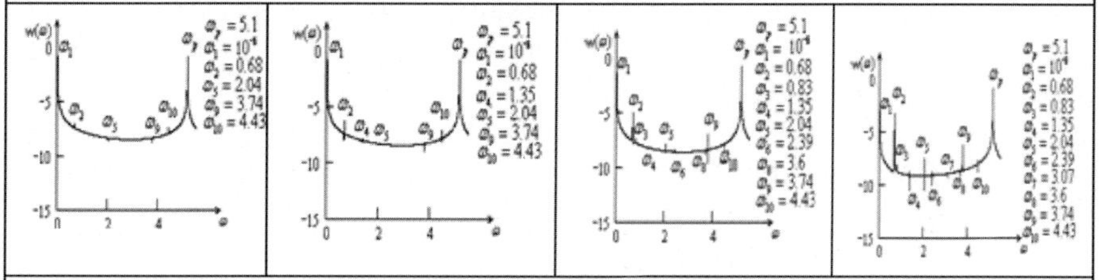

$\varepsilon_1 = 5,\ \varepsilon_2 = 0.07,\ q_0 = 0.95 \cdot 10^5$

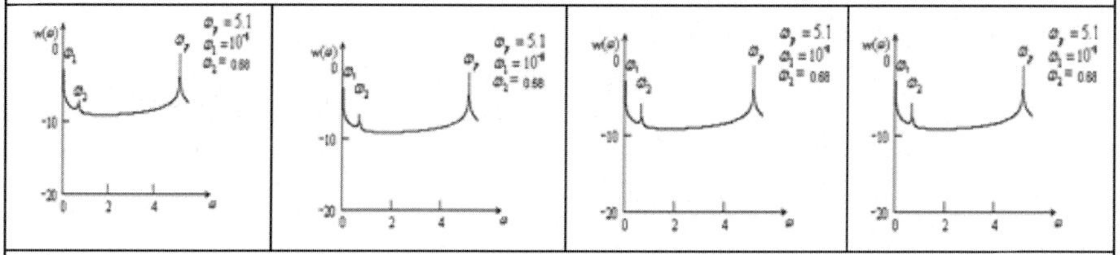

$\varepsilon_1 = 1,\ \varepsilon_2 = 0,\ q_0 = 0.32 \cdot 10^5$

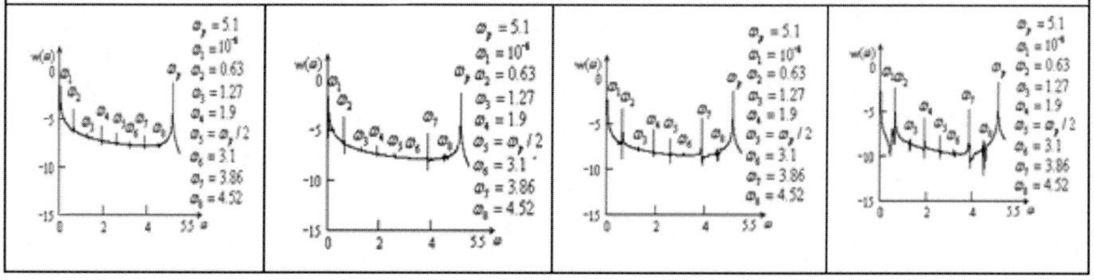

to five, and there is one linear combination; in the center ($x = 0.5$) we already have six frequencies, all of them linearly independent. For $x = 0.25$ nine frequencies occur and only one linear combination of the frequencies exists (the same as for point $x = 0.75$). In the case of $\varepsilon_1 = 20$ and $\varepsilon_2 = 0.05$ one observes a decrease of the number of frequencies along the whole beam length, but their number is different. Namely, for $\varepsilon_1 = 20$ and $\varepsilon_2 = 0.1$ the number of frequencies (two) is the same along the beam length, whereas for $\varepsilon_1 = 20,\ \varepsilon_2 = 0$ the number of frequencies varies from nine to three.

For $\varepsilon_1 = 10$ and $\varepsilon_2 = 0$ a picture analogous to the case of $\varepsilon_1 = 20,\ \varepsilon_2 = 0$ occurs, but more frequencies increase in comparison to the previous case. The number of linear combinations is minimal. For $x = 0.25$ the number of frequencies is so high that they are almost indistinguishable and the system moves into a chaotic state. An increase of the parameter ε_2 up to 0.05 yields a rapid decrease of the number of frequencies in the power spectrum, but their distribution along the beam length is different. They fully coincide only for $\varepsilon_2 = 0.1$.

Although for $\varepsilon_1 = 5$ and $\varepsilon_2 = 0$ the number of frequencies and their values for $x = 0.75$ and $x = 1$ fully coincide, in point $x = 0.5$ an increase in the number of frequencies is observed, and for $x = 0.25$ they increase again. Increasing the parameter ε_2 in a way analogous to that described so far, one may observe for $\varepsilon_2 = 0.07$ full coincidence of the number of frequencies.

In Table 15.4, taking into account results obtained so far, the graphs of convergence of the number of frequencies along the beam length for increasing of ε_2 are reported. The vertical axis corresponds to the beam length, whereas the horizontal axis corresponds to the number of frequencies. One corresponds to the free beam edge, whereas zero corresponds to the clamped beam edge. Each curve is supplemented by the associated damping coefficient ε_2.

Analysis of the reported results allows us to conclude that the influence of coefficient ε_2 is important, since its gradual increase results in observation of coincidences of the number of frequencies along the beam length.

If one takes graphs for an arbitrary ε_1, then a gradual decrease of the number of frequencies may be observed while approaching the studied beam point to the free beam edge ($x = 1$).

For $\varepsilon_2 = 0.1$ in majority of the studied cases, including $\varepsilon_1 = 20,\ 10$ full coincidences of the number of frequencies occurs. It occurs for $\varepsilon_1 = 5$ taking $\varepsilon_2 = 0.07$. This means that considering the values of $\varepsilon_1 = 5,\ 10,\ 20$ it is necessary to take into account the damping value in the beam material (ε_2).

For $\varepsilon_1 = 1$ and $\varepsilon_2 = 0$ the number and frequency values along the beam length coincide, i.e., they are optimal in the case of engineering computations. In other words, neglecting of the damping term ε_2 for $\varepsilon_1 > 1$ may lead to the occurrence of a new system state with completely different number of frequencies along the beam length.

Investigation of problem 3 ($\lambda = 100$) Next, we study problem 3 regarding the character of beam vibration variation depending on an increase of the relative beam length ($\lambda = 100$). Note that in the previous cases the beam of length $\lambda = 50$ has

Table 15.4 Graphs of frequency convergence

ε_1	
$\varepsilon_1 = 20$	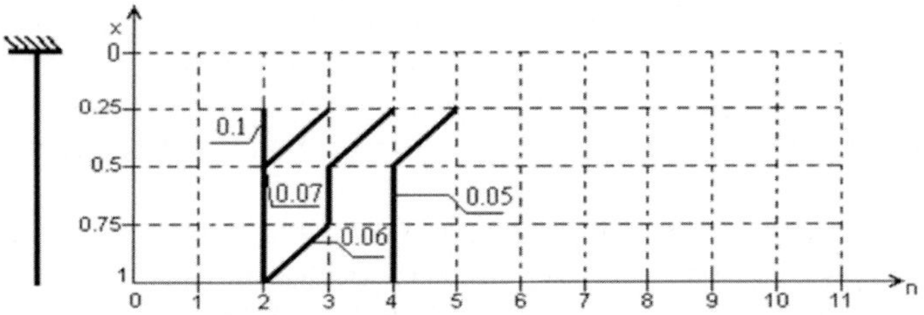
$\varepsilon_1 = 10$	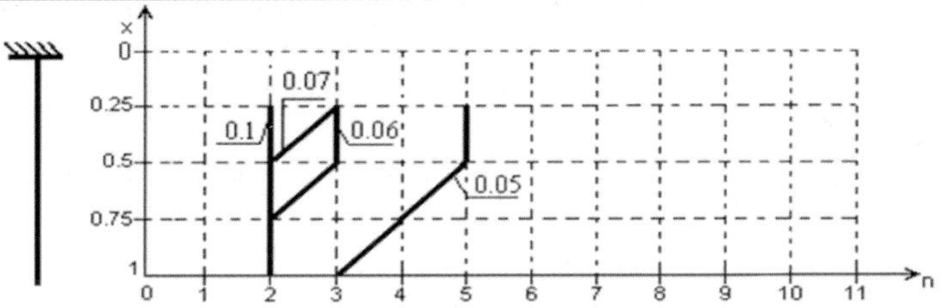
$\varepsilon_1 = 5$	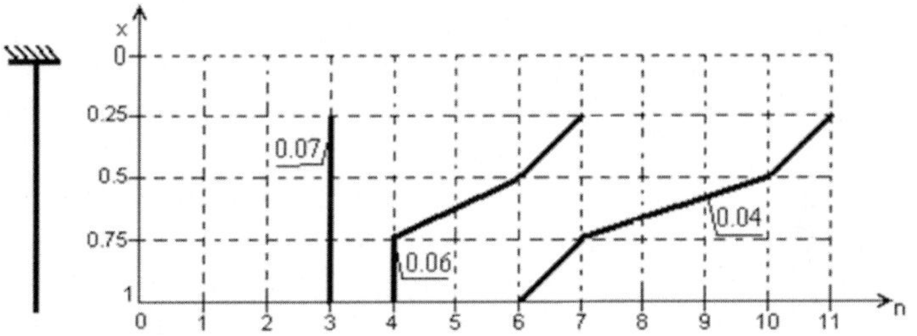

been considered. Having in mind that frequency of free beam vibrations $\omega_p = 10.2$ for $\lambda = 100$, we consider only the influence of coefficient $\varepsilon_1 (\varepsilon_2 = 0)$.

Comparing the results for $\lambda = 50$ (Tables 15.2 and 15.3) and for $\lambda = 100$ (Table 15.5) one may conclude that the number of frequencies is essentially higher for the first case, although the way in which new frequencies occur remains unchanged. In this case the occurrence of chaotic dynamics is not observed. In addition, linear combinations of any arbitrary frequencies for arbitrary values of ε_1 do not appear. Full coincidence of both the number of frequencies and values is observed only for $\varepsilon_1 = 1$. In the case of $\varepsilon_1 = 20$ only two frequencies occur on the free beam edge, whereas already for point $x = 0.75$ there are five frequencies; for $x = 0.5$ and for $x = 0.25$ there are six frequencies. However, the corresponding frequencies values for $x = 0.5$ and 0.25 are different.

Although the number of frequencies appearing for $\varepsilon_1 = 20$ is different depending on the considered beam point, it coincides for $\lambda = 50,\ 100$. In the free edge this

Table 15.5 Power spectra for $\lambda = 100$

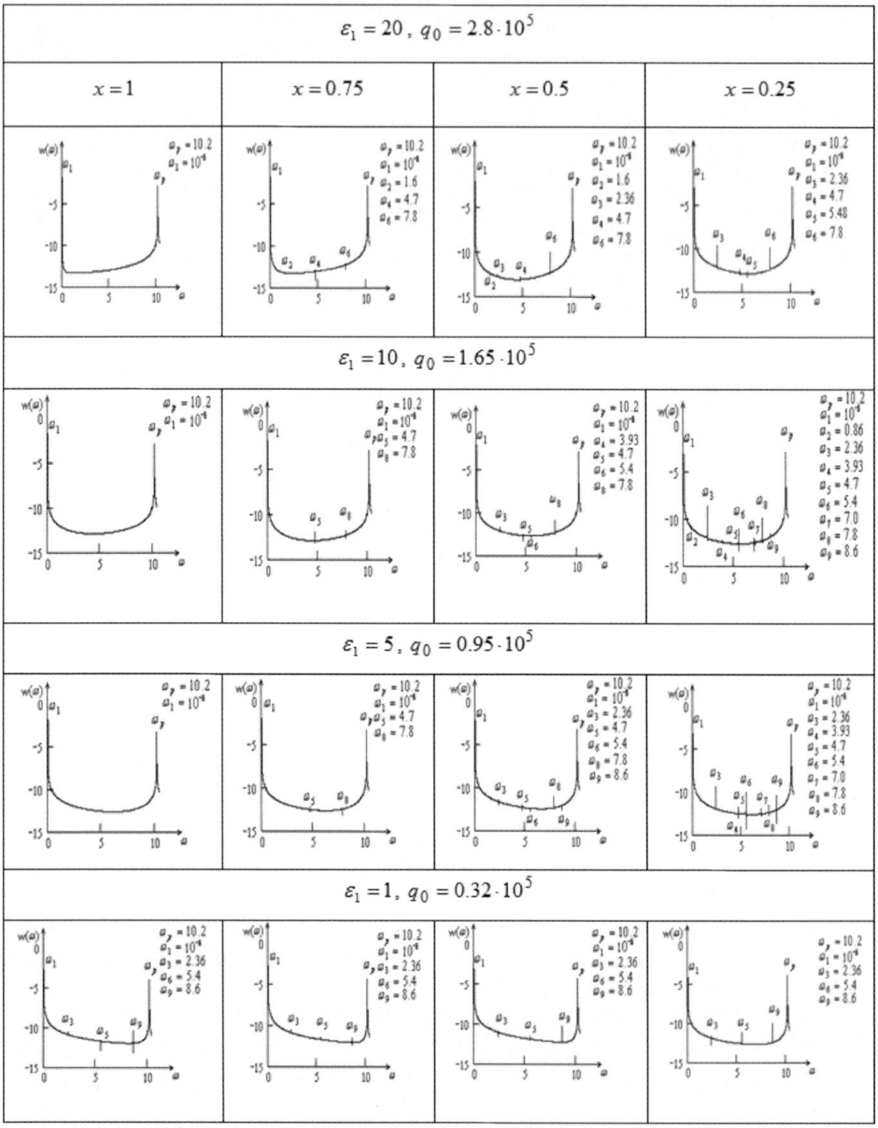

difference is a minimal one. For $\varepsilon_1 = 10$ a difference in the number of frequencies (for different λ) decreases. Beginning from $\varepsilon_1 = 5$ one may observe convergence of the number of frequencies depending on the considered point, whereas the full convergence appears for $\varepsilon_1 = 1$. To conclude this section, we have illustrated the convergence of the number of frequencies for the values $\varepsilon_1 = 1$, $\varepsilon_2 = 0$ independently of the geometrical beam parameters.

Table 15.6 Time histories and deformation curves

$w(t)$ for $\lambda = 50$	Deformation curves for $\lambda = 50$
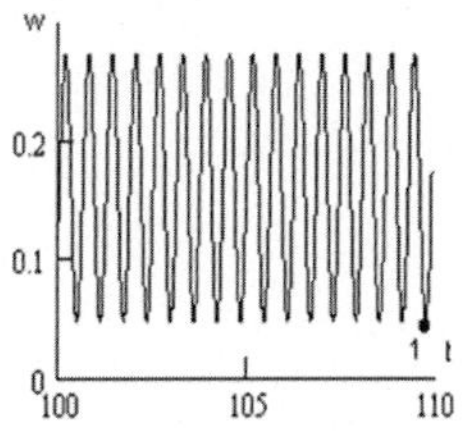	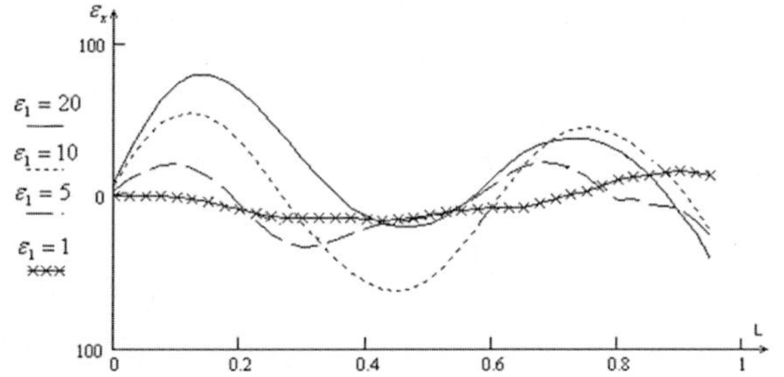
$w(t)$ for $\lambda = 100$	Deformation curves for $\lambda = 100$
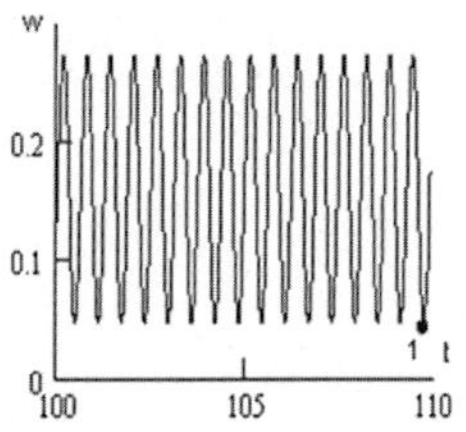	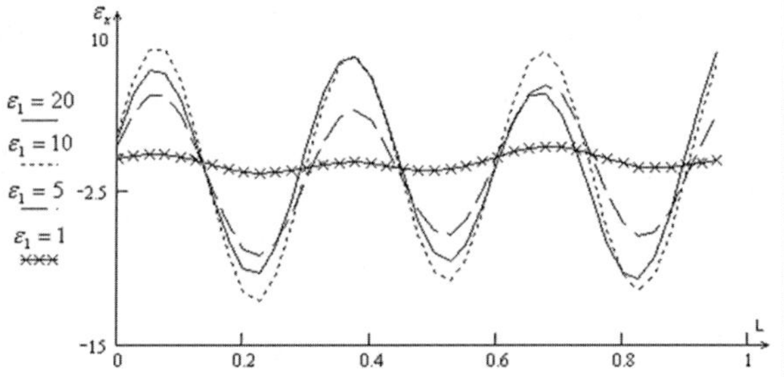

Investigation of deformation (problem 3) In Table 15.6 the graphs of middle beam surface deformation for problem 3 along the beam length $\lambda = 50,\ 100$ for series of ε_1 values in the time instant $t = 109.72$ (in time histories $w(t)$ the sign "1" is marked) have been presented.

Analysis of the obtained results shows that the number of the observed periods for $\lambda = 50$ and 100 is different. It may be explained through the difference of the chosen excitation frequencies $\omega_p = 5.1$ and 10.2, respectively.

In addition, for the case of $\lambda = 50$ asymmetry of the deflection graphs with respect to *OX* axis is observed, whereas for $\lambda = 100$ the mentioned graphs are symmetrically distributed with respect to the *OX* axis for all values of ε_1 (except of $\varepsilon_1 = 20$). Among the others, a period doubling for a doubled parameter λ has been observed.

15.5 Waves Generated by a Longitudinal Impact

Next, we consider the case of an impact on the free beam edge generated by a lumped body moving longitudinally (see Fig. 15.2).

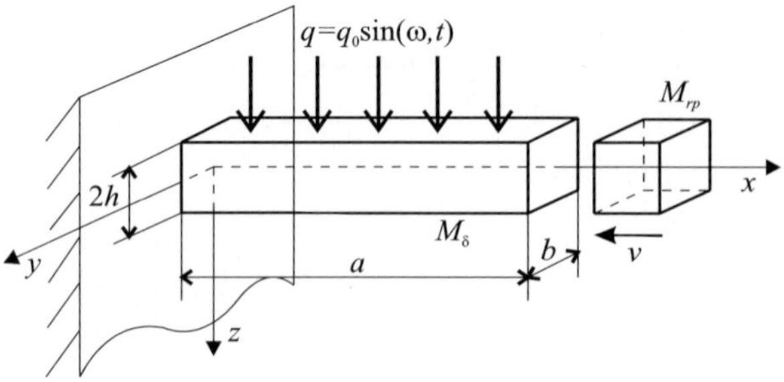

Fig. 15.2 Model of the studied problem

After an impact the mass remains attached to the beam and then dynamics of the system "beam-impacting mass" is studied.

We need to attach to Eqs. (15.3) the following dimensionless boundary conditions:

$$w(0) = u(0) = 0, \quad \frac{1}{\chi} N_x = \frac{\partial^2 u}{\partial t^2}, \quad M_x = 0, \tag{15.13}$$

where $\chi = \frac{M_{rp}}{M_\sigma}$, M_{rp} is the impacting body mass, M_σ is the beam mass.

Initial conditions are as follows:

$$\begin{gathered} w(x)_{|t=0} = u(x)_{|t=0} = 0, \\ \dot{w}(x)_{|t=0} = \dot{u}(x)_{|t=0} = 0, \quad x \neq 0, \quad \dot{u}(x)_{|t=0} = V \quad \text{for } x = 0, \end{gathered} \tag{15.14}$$

where V is the impacting body velocity in the impact time instant.

The defined infinite problem is reduced to a finite dimensional one using finite differences with approximation $O(h^2)$. In each mesh point a system of ODEs similar type to that of (15.3) is obtained, and the following additional equations on the border (15.15) are attached:

$$w_0 = u_0 = 0, \quad \frac{1}{\chi} N_x = \ddot{u}, \quad M_x = 0. \tag{15.15}$$

Initial conditions in this case take the form:

$$\begin{aligned} & w(x_i)_{|t=0} = u(x_i)_{|t=0} = 0, \\ & \dot{w}(x_i)_{|t=0} = \dot{u}(x_i)_{|t=0} = 0 \quad \text{for } x \neq 0, \\ & \dot{u}(x_i)_{|t=0} = N_x \frac{1}{\chi} \quad \text{for } x = 0, \quad i = 1, \ldots, n. \end{aligned} \tag{15.16}$$

Now we take the external load governed by (15.12), where $q_0 = 10^{-6}$, and the damping coefficients are $\varepsilon_1 = 1$, $\varepsilon_2 = 0$. We consider the cases for the following control parameter values $\chi = 1$, 1.5, 2 and the computed frequency spectra are reported in Table 15.7.

Table 15.7 Power spectra in the case of impact action

	Power spectrum
$\chi = 1$ $V = 10^{-3}$	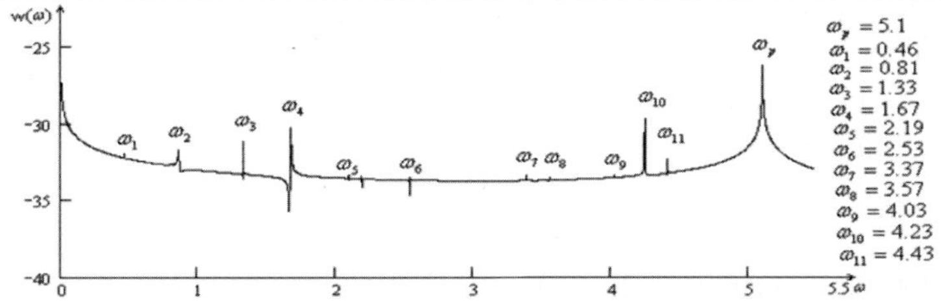
$\chi = 1.5$ $V = 10^{-3}$	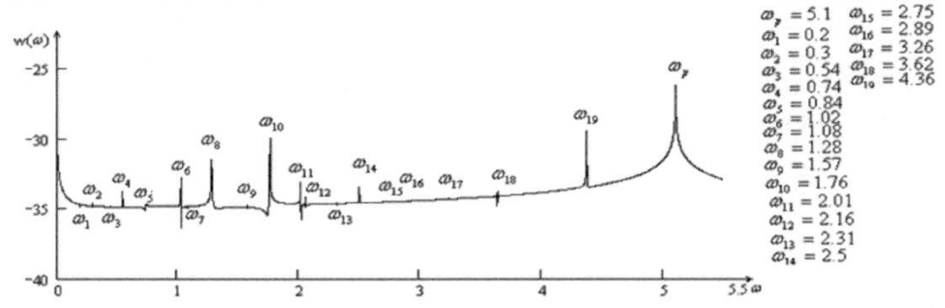
$\chi = 2$ $V = 10^{-3}$	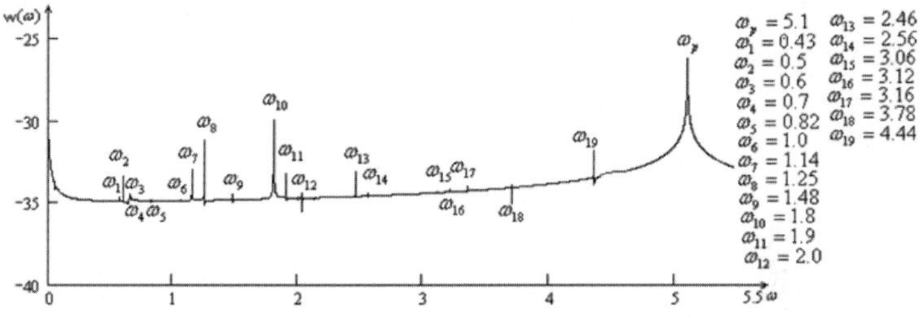

One may observe in the reported spectra the frequency number increase corresponding to increase of χ up to 1.5. For $\chi = 1$ only one Hopf bifurcation at ω_6 is observed, and the series of the dependent frequencies ω_p, ω_2, ω_4, ω_6, ω_7, ω_{10} differ from each other by the value of 0.87. The pairs of frequencies: ω_1–ω_6 differ from each other by the values of 0.35. The frequencies ω_7–ω_{11} differ from each other by the value of 0.2.

For $\chi = 1.5$ two Hopf bifurcations associated with ω_8, ω_{14} appear, and the linear combination of ω_3 and frequencies ω_7, ω_{12} appear. Frequencies ω_1, ω_2 and ω_4, ω_5 differ from each other by the value of 0.1. The remaining detected frequencies are linearly independent.

Two Hopf bifurcations are also detected for $\chi = 2$ for ω_8, ω_{14} and the frequencies ω_p, ω_7, ω_{10}, ω_{13}, ω_{16}, ω_{19} differ by the value of 0.66.

Then, for the parameters fixed so far, various χ and $V = 10^{-14}$, the wave movement along the beam length is studied. Below, graphs of the displacement waves for different time instants are analyzed. Among the others, time instants corresponding to qualitative changes in the time courses for $\chi = 1, 1.5, 2$ have been defined.

Table 15.8 Beam displacement generated by impact

Displacements	Parameters
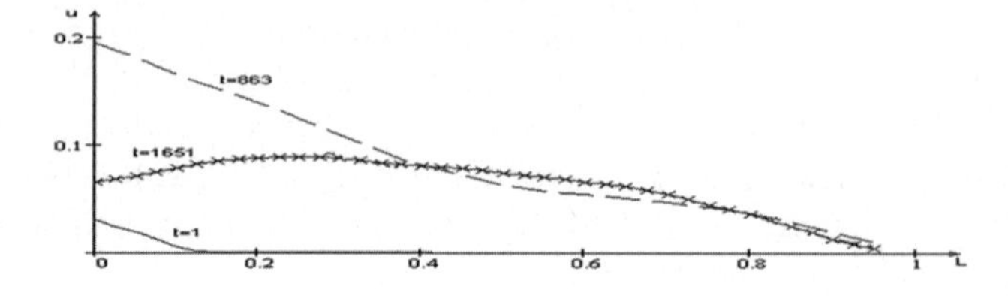	$\chi = 1$ $t = 1$ - initial time instant $t = 863$ - stationary time instant $t = 1651$-critical time
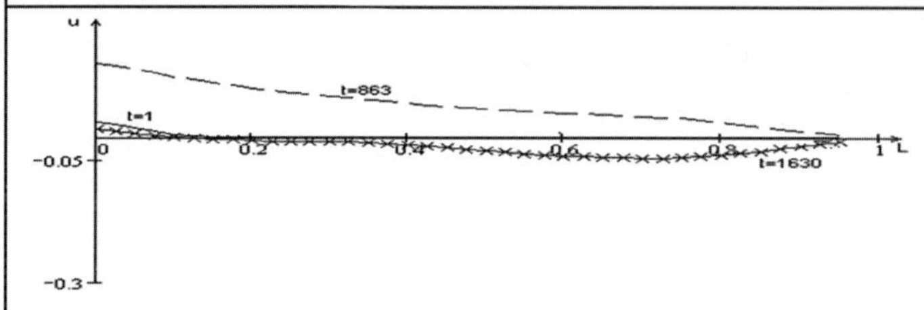	$\chi = 1.5$ $t = 1$ - initial time instant $t = 863$ - stationary time instant $t = 1630$-critical time
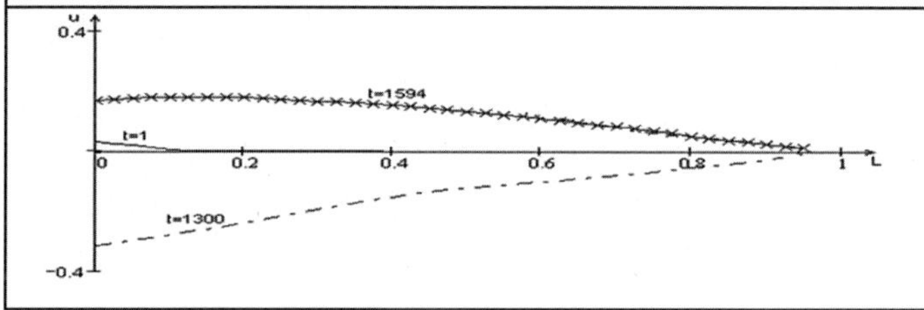	$\chi = 2$ $t = 1$ - initial time instant $t = 1300$ stationary time instant $t = 1594$ - critical time

For the initial time instant the reported displacements generated by impact (see Table 15.8) remain in the neighbourhood of the point $u = 0$ along the whole beam length with negligible differences occurring on the free beam edge.

A stationary time instant of the steady-state time histories occurs when the beam deflection becomes symmetric with respect to the beam axis, whereas it remains asymmetric up to this instant.

A critical time instant defines the time when interaction of both longitudinal and transversal beam vibrations appears. The occurrence of a critical time instant is characterized either by a convex graph or by the intersection of the *OX* axis. We have found that the critical time instant is minimal for $\chi = 2$, $t_{cr} = 1594$, maximal for $\chi = 1$, $t_{cr} = 1651$, and the critical time instant for $\chi = 1.5$ is 1594.

15.6 Conclusions

We have shown how important role a proper identification of damping parameters ε_1 and ε_2 plays in the analysis of dissipative vibrations of a flexible beam.

A choice of parameters ε_1 and ε_2 is essential, since their improper values may require a principally new computational scheme. In the case of $\varepsilon_1 > 1$ and $\varepsilon_2 = 0$

the beam vibrations may occur as the vibrations of uncoupled nonlinear oscillators, each of them exhibiting vibrations with its own spectrum of frequencies. A criterion of the choice of ε_1 and ε_2 is the observation that nonlinear beam vibrations should be considered as those generated by the system of coupled nonlinear oscillators. Modeling the beam as an elastic object and after application of the finite difference method along the spatial coordinates, one obtains the lumped system with 40 degrees of freedom being viewed as the system of 40 coupled oscillators. The proposed criterion of a reliable choice of damping parameters ε_1 and ε_2 remains valid also in the case of nonlinear beam vibrations.

Bibliography

1. Abolghasemi M., Jalali M.A., Attractors of a rotating viscoelastic beam. *International Journal of Non-Linear Mechanics*, 2003, Vol. 38, No. 5, 739–751.
2. Agladze K.I., Krinsky V.I., Pertsov A.M., Chaos in the non-stirred Belousov-Zhabotinsky reaction is induced by interaction of waves and stationary dissipative structures. Nature, 1984, 308, 834–835.
3. Aldraihem O.J., Baz A., Dynamic stability of stepped beams under moving loads. *Journal of Sound and Vibration*, 2002, Vol. 250, No. 5, 835–848.
4. Alfutov N.A., *Fundamentals on Computation of Stability of Elastic Systems*. Moscow, Mashinostroyeniye, 1991, in Russian.
5. Amabili M., Non-linear vibrations of doubly curved shallow shells. *International Journal of Non-Linear Mechanics*, 2005, Vol. 40, No. 5, 683–710.
6. Amabili M., Pellicano F., Paidoussis M.P., Non-linear dynamics and stability of circular cylindrical shells containing flowing fluid. Part III: Truncation effect without flow and experiments. *Journal of Sound and Vibration*, 2000, Vol. 237, No. 4, 617–640.
7. Amabili M., Pellicano F., Paidoussis M.P., Non-linear dynamic and stability of circular cylindrical shells containing flowing fluid. Part IV: Large-amplitude vibrations with flow. *Journal of Sound and Vibration*, 2000, Vol. 237, No. 4, 641–666.
8. Amabili M., Pellicano F., Vakakis A.F., Nonlinear vibrations and multiple resonances of fluid-filled, circular shells. Part I: Equations of motion and numerical results. *Journal of Vibration and Acoustics*, 2000, Vol. 122, No. 4, 346–354.
9. Amabili M., Pellicano F., Vakakis A.F., Non-linear vibrations and multiple resonances of fluid-filled circular shells. Part II: Perturbation analysis. *Journal of Vibration and Acoustics*, 2000, Vol. 122, No. 4, 355–364.
10. Amelchenko V.V., Krysko V.A., Investigation of flexible rectangular plates in post-critical state using computers. *Differential Equations and Computational Mathematics*. Saratov State University, Saratov, 1972, in Russian.
11. Andreyev L.V., Obodan N.I., Lebedev A.G., *Stability of Shells Under Non-axial Deformation*. Moscow, Nauka, 1988, in Russian.
12. Andronov A.A., Pontriagin L.S., Theory of "robust" systems. *DAN SSSR*, 1937, Vol. 14, 1–42, in Russian.
13. Andronov A.A., Witt A.A., Khaikin S.E., *Theory of Vibrations*. Nauka, Moscow, 1981, in Russian.
14. Arnold V.I., *Mathematical Methods of Classical Mechanics*, Springer-Verlag, New York, 1978.
15. Awrejcewicz J., *Bifurcation and Chaos in Coupled Oscillators*. World Scientific, Singapore, 1991.
16. Awrejcewicz J., *Bifurcation and Chaos in Simple Dynamical Systems*. World Scientific, Singapore, 1989.

17. Awrejcewicz J., *Nonlinear Dynamics of Machines*. Łódź TU Press, Łódź, 1994, in Polish.
18. Awrejcewicz J., *Oscillations of Discrete Deterministic Systems*. WNT, Warsaw, 1996, in Polish.
19. Awrejcewicz J., *Secrets of Nonlinear Dynamics*. Łódź TU Press, Łódź, 1997, in Polish.
20. Awrejcewicz J., *Synchronization and Chaos*. Łódź TU Press, Łódź, 1995, in Polish.
21. Awrejcewicz J., Krysko A.V., Analysis of complex parametric vibrations of plates and shells using Bubnov-Galerkin approach. *Archive of Applied Mechanics*, 73, 2003, 495–504.
22. Awrejcewicz J., Krysko A.V., Some problems of analysis and optimization of plates and shells. *Journal of Sound and Vibration*, 264(2), 2003, 343–376.
23. Awrejcewicz J., Krysko V.A., *Dynamics and Stability of Flexible Sectorial Shells Thermodynamically Loaded*. WNT, Warsaw, 2002, in Polish.
24. Awrejcewicz J., Krysko V.A., *Dynamics and Stability of Shells with Thermal Excitations*. WNT, Warsaw, 1999, in Polish.
25. Awrejcewicz J., Krysko V.A., *Nonclassical Thermoelastic Problems in Nonlinear Dynamics of Shells*. Springer-Verlag, Berlin, 2003.
26. Awrejcewicz J., Krysko V.A., *Numerical Analysis of Shells Oscillations with Thermal Load*. WNT, Warsaw, 1998, in Polish.
27. Awrejcewicz J., Krysko V.A., *Oscillations of Continuous Systems* WNT, Warsaw, 2000, in Polish.
28. Awrejcewicz J., Krysko V.A., *Techniques and Methods of Plates and Shells Analysis*. Łódź TU Press, Łódź, 1996, in Polish.
29. Awrejcewicz J., Krysko V.A., 3D theory versus 2D approximate theory of the free orthotropic (isotropic) plates and shells vibrations. Part 1: Derivation of governing equations. *Journal of Sound and Vibration*, 1999, Vol. 226, No. 5, 807–829.
30. Awrejcewicz J., Krysko V.A., 3D theory versus 2D approximate theory of the free orthotropic (isotropic) plates and shells vibrations. Part 2: Numerical algorithms and analysis. *Journal of Sound and Vibration*, 1999, Vol. 226, No. 5, 831–871.
31. Awrejcewicz J., Krysko V.A., Feigenbaum scenario exhibited by thin plate dynamics. *Nonlinear Dynamics*, 2001, Vol. 24, 373–398.
32. Awrejcewicz J., Krysko V. A., Nonlinear coupled problems in dynamics of shells. *International Journal of Engineering Science*, Vol. 41, No. 6, 2003, 587–607.
33. Awrejcewicz J., Krysko V.A., Krysko A.V., Complex parametric vibrations of flexible rectangular plates. *Meccanica*, Vol. 39, No. 3, 2004, 221–244.
34. Awrejcewicz J., Krysko V.A., Krysko A.V., On the economical solution method for a system of linear algebraic equations. *Mathematical Problems in Engineering*, 2004, Vol 4, 377–410.
35. Awrejcewicz J., Krysko V.A., Krysko A.V., Regular and chaotic behavior of flexible plates, In: *Proceedings of the Third International Conference on Thin-Walled Structures*, June 5–7, 2001, Krakow, Poland, Eds. J. Zaraś, K. Kowal-Michalska and J. Rhodes, Elsevier Science, 2001, 349–356.
36. Awrejcewicz J., Krysko V.A., Krysko A.V., Spatial-temporal chaos and solitons exhibited by von Karmán model. *International Journal of Bifurcation and Chaos*, 2002, Vol. 12, No. 7, 1465–1513.
37. Awrejcewicz J., Krysko V.A, Krysko A.V., Vakakis A., Solitons and chaos exhibited by flexible plates sinusoidally excited, In: J.M. Balthazar, P.B. Goncalves, R.M. Brasil, I.L. Caldas, F.B. Rizatto (Eds.), Nonlinear Dynamics, Chaos, Control and Their Applications to Engineering Sciences, Vol. 5: *Chaos Control and Time Series*, 2002, 258–267.
38. Awrejcewicz J., Krysko V.A., Kutsemako A.N., Oscillations of non-uniform shells. *Journal of Sound and Vibration*, 1999, Vol. 10, No. 3, 116–141.
39. Awrejcewicz J., Krysko V.A., Misnik P., *Three Dimensional Problems of Plates Theory in a Temperature Field*. WNT, Warsaw, 2003, in Polish.
40. Awrejcewicz J., Krysko V.A., Narkaitis G.G., Bifurcations of a thin plate-strip excited transversally and axially. *Nonlinear Dynamics*, 2003, Vol. 32, 187–209.
41. Awrejcewicz J., Krysko V.A., Nazar'iantz V., Chaotic vibrations of flexible infinite length cylindrical panels using the Kirchhoff-Love model. *Communications in Nonlinear Science and Numerical Simulation*, 2007, Vol. 12, No. 4, 519–542.

42. Awrejcewicz J., Krysko V.A., On the vibration of the Euler-Bernoulli beam with clamped ends deflection constraints. *Special Issue of International Journal of Bifurcation and Chaos*, 2005, Vol. 15, No. 3, 1864–1878.
43. Awrejcewicz J., Krysko V.A., Ovsiannikova O., Novel procedure to compute a contact zone magnitude vibrations of two-layered uncoupled plates. *Special Issue of Mathematical Problems in Engineering*, 2005, Vol. 4, 425–435.
44. Awrejcewicz J., Krysko V.A., Period doubling bifurcation and chaos exhibited by an isotropic plate. Zeitschrift für Angewandte Mathematik und Mechanik, 2000, Vol. 80, 267–268.
45. Awrejcewicz J., Krysko V.A., Wavelets-based analysis of parametric vibrations of flexible plates. *International Journal of Applied Mechanics*, 2003, Vol. 39, No. 10, 2003, 3–43.
46. Awrejcewicz J., Krysko V.A., Vakakis A.F., *Nonlinear Dynamics of Continuous Elastic Systems*. Springer-Verlag, Berlin, 2004.
47. Awrejcewicz J., Lamarque C.-H., *Bifurcation and Chaos in Nonsmooth Mechanical Systems*. World Scientific, Singapore, 2003.
48. Awrejcewicz J., Mosdorf R., *Numerical Analysis of Some Problems of Chaotic Dynamics*. WNT, Warsaw, 2003, in Polish.
49. Bacinov C.D., Nonlinear vibrations of plates with simultaneous action of static and vibrational loads. *Priklandnaya Mekhanika*, 1971, Vol. 7, No. 10, 126–130, in Russian.
50. Balamurugan V., Narayanan S., Finite element formulation and active vibration control study on beams using smart constrained layer damping (SCLD) treatment. *Journal of Sound and Vibration*, 2002, Vol. 249, No. 2, 227–250.
51. Battelli F., Fečkan M., Chaos in the beam equation. *Journal of Differential Equations*, 2005, Vol. 209, No. 1, 172–227.
52. Benamar R., Bennouna M.M., White R.G., The effects of large vibration amplitudes on the mode shapes and natural frequencies of thin elastic structures. Part III: Fully clamped rectangular isotropic plates measurements of the mode shape amplitude dependence and the spatial distribution of harmonic distortion. *Journal of Sound and Vibration*, 1999, Vol. 175, 377–395.
53. Bennouna M.M., White R.G., The effects of large vibration amplitudes on the fundamental mode shape of a clamped-clamped uniform beam. *Journal of Sound and Vibration*, 1984, Vol. 96, No. 3, 309–331.
54. Bernstein S.A., *Strength of Materials*. Vyshaia Shkola, Moscow, 1961, in Russian.
55. Bhat B.R., Wagner H., Natural frequencies of a uniform cantilever with a tip mass slender in the axial direction. *Journal of Sound and Vibration*, 1976, Vol. 45, No. 2, 304–307.
56. Bhat B.R., Kulkarni M.A., Natural frequencies of a cantilever with slender tip mass. *AIAA Journal*, 1976, Vol. 14, 536–537.
57. Birger I.A., General methods of plasticity. *Prikladnaya Mathematika i Mekhanika*, 1951, Vol. 15, 6–20, in Russian.
58. Bishop S.R., Thompson M.G., Foale S., Prediction of period-1 impacts in a driven beam. *Proceedings of Mathematical, Physical and Engineering Sciences*, 1996, Vol. 452, 2579–2592.
59. Bodenschatz F., Pesch W., Ahlers G., Recent developments in Rayleigh – Bénard convection. Annual Review of Fluid Mechanics, 32, 709–778.
60. Bogarian K.O., On convergence of the Bubnov-Galerkin and Ritz methods. *Doklady AN SSSR*, 1961, Vol. 191, No. 2, 267–269, in Russian.
61. Bolotin W.W., The problem of vibrations of bridges under moving loading. *AN SSSR, Department of Technical Science. Mechanics of Machines Construction*, 1961, No. 4, 109–115, in Russian.
62. Bubnov I.G., Review of S.P. Timoshenko's work, "On the Stability of Elastic Systems." *Collected Works*, 1956, Leningrad, 136–139, in Russian.
63. Budiansky B., Theory of buckling and post-buckling behavior of elastic structures. *Advances in Applied Mechanics*, 1974, Vol. 14, Academic Press, New York.
64. Budiansky B.L., A Reassessed of deformation theories of plasticity. *Journal of Applied Mechanics*, 1959, Vol. 26, No. 2, 160–172.

65. Butenin N.V., Neilmark Yu.I., Fufayev N.F., *Introduction into Theory of Nonlinear Vibrations*. Moscow, Nauka, 1976, in Russian.
66. Chan K.-T., Wang X.-Q., Leung T.-P., Free vibration of a Timoshenko beam partially loaded with distributed mass. *Journal of Sound and Vibration*, 1997, Vol. 206, No. 3, 353–369.
67. Chan K.-T., Wang X.-Q., Leung T.-P., Free vibration of beams with two sections of distributed mass. *Journal of Vibration and Acoustics*, 1998, Vol. 120, 944–948.
68. Chang S.I., Lee J.M., Bajaj A.K., Krousgrill C.M., Subharmonic responses in harmonically excited rectangular plates with one-to-one internal resonance. *Chaos, Solitons and Fractals*, 1997, Vol. 8, No. 4, 479–498.
69. Chen Y., On the vibration of beams or rods carrying a concentrated mass. *American Society of Mechanical Engineers Journal of Applied Mechanics*, 1963, Vol. 30, No. 2, 310–312.
70. Chen J.C., Babcock C.D., Nonlinear vibration of cylindrical shells. *American Institute of Aeronautics and Astronautics Journal*, 1975, Vol. 13, 868–876.
71. Chen L.-W., Hwang J.-R., Axisymmetric dynamic stability of transversely isotropic Mindlin circular plates. *Journal of Sound and Vibration*, 1988, Vol. 121, No. 2, 307–315.
72. Chen L.-Q., Yang X.-D., Transverse nonlinear dynamics of axially accelerating viscoelastic beams based on 4-term Galerkin truncation. *Chaos, Solitons and Fractals*, 2006, Vol. 27, No. 3, 748–757.
73. Cheng C.-J., Fan X.-J., Nonlinear mathematical theory of perforated viscoelastic thin plates with its applications. *International Journal of Solids and Structures*, 2001, Vol. 38, No. 36–37, 6627–6641.
74. Chetayev N.G., *Stability of Motion*. Nauka, Moscow, 1955, in Russian.
75. Cusumano J.P., Sharkady M.T., Kimble B.W., Experimental measurements of dimensionality and spatial coherence in the dynamics of a flexible-beam impact oscillator. *Philosophical Transactions: Physical Sciences and Engineering*, 1994, Vol. 347, No. 1683, 421–438.
76. Dediukin I.Yu., Krysko V.A., On criterions of dynamical instabilities of shells. *Prikladnaya Mekhanika*, 1954, Kiev, Vol. 30, No. 10, 56–60, in Russian.
77. Donnell L.H., A New theory for the buckling of thin cylinders under axial compression and bending. *Transactions of ASME*, 1934, Vol. E, No. 56, 795–806.
78. Donnell L.H., Wan C.C., Effect of imperfections on buckling of thin cylinders and columns under axial compression. *Journal of Applied Mechanics*, 1950, Vol.17, No. 1, 73–83; Discussion on the paper in *Journal of Applied Mechanics*, 1950, Vol. 17, No. 3, 340–342.
79. Douglas B.E., Yang J.C.S., Transverse compressional damping in the vibratory response of elastic-viscoelastic-elastic beams. *AIAA Journal*, 1978, Vol. 16, 925–930.
80. Douglas B.E., Compressional damping three-layer beams incorporating nearly incompressible viscoelastic cores. *Journal of Sound and Vibration*, 1986, Vol. 104, 343–347.
81. Doughty T.A., Davies P., Bajaj A.K., A comparison of three techniques using steady state data to identify non-linear modal behavior of an externally excited cantilever beam. *Journal of Sound and Vibration*, 2002, Vol. 249, No. 4, 785–813.
82. Dowell E.H., Ventres C.S., Modal equations for the nonlinear flexural vibrations of a cylindrical shell. *International Journal of Solids and Structures*, 1968, Vol. 4, 975–991.
83. Drucker D.C., A more fundamental dynamic buckling of clamped shallow spherical shells. *NASA, TN*, 1962, Vol. D-1510.
84. Duncan W.J. - *ARC RJ. M.* 1848, 1938.
85. Dwivedy S.K., Kar R.C., Dynamics of a slender beam with an attached mass under combination parametric and internal resonances, Part II: Periodic and chaotic responses. *Journal of Sound and Vibration*, 1999, Vol. 222, No. 2, 281–305.
86. Dwivedy S.K., Kar R.C., Non-linear dynamics of a slender beam carrying a lumped mass under principal parametric resonance with three-mode interactions. *International Journal of Non-Linear Mechanics*, 2001, Vol. 36, No. 6, 927–945.
87. Dwivedy S.K., Kar R.C., Simultaneous combination and 1:3:5 internal resonances in a parametrically excited beam-mass system. *International Journal of Non-Linear Mechanics*, 2003, Vol. 38, No. 4, 585–596.
88. Dzishkariani A.V., On approximating Ritz method for one non-linear equation. *Scientific Works of Tblissi Mathematical Institute of AN SSSR*, 1970, Vol. 36, 29–46, in Russian.

89. Dzishkariani A.V., On convergence velocity of the Bubnov-Galerkin method. *Journal of Computational Mathematics and Mathematical Physics*, 1964, Vol. 4, No. 2, 343–348, in Russian.
90. El-Bassiouny A.F., Single-mode control and chaos of cantilever beam under primary and principal parametric excitations. *Chaos, Solitons and Fractals*, 2006, Vol. 30, No. 5, 1098–1121.
91. Fegelman K.J.L., Grosch K., Dynamics of a flexible beam contacting a linear spring at low frequency excitation: Experiment and analysis. *Journal of Vibration and Acoustics*, 2002, Vol. 124, No. 2, 237–249.
92. Feigenbaum M.J., Quantitative universality for a class of nonlinear transformations., *Journal of Statistical Physics*, 1978, Vol. 19, No. 1, 25–52.
93. Feigenbaum M.J., The universal metric properties of nonlinear transformations. *Journal of Statistical Physics*, 1979, Vol. 21, No. 6, 669–706.
94. Feigenbaum M.J., Kadanoff L.P., Shenker S.J., Quasiperiodicity in dissipative systems. A renormalization group analysis. Physica D5, 1982, 370–386.
95. Feodosev V.F., On the method of solutions of stability of deformable bodies. *Prikladnaya Matematika i Mekhanika*, 1963, Vol. 27, No. 2, 265–275, in Russian.
96. Filippov A.P., *Vibrations of Continuous Systems*. Mashinostoyenye, Moscow, 1970, in Russian.
97. Filippov A.P., Kochmaniuk S.S., *Dynamical Influence of Moving Loading on Rods*. Naukowa Dumka, Kiev, 1968, in Russian.
98. Finlfyson B.A., *The Method of Weighted Residuals and Variational Principles*. Academic Press, New York, 1972.
99. Flügge W., Die stabilität der Kreiszylinderschale. *Ingènieur Archiu.*, 1932, Vol. 3, No. 5, 463–506.
100. Fletcher C., *Computational Galerkin Methods*. Springer-Verlag, Berlin, 1984.
101. Galerkin B. G., *Vestnik Inzhinerov*, 1915, No. 19, 897–908, in Russian.
102. Galerkin B.G., Rods and plates. Application of series in some problems of equilibriums of rods and plates. *Vestnik Inzhinerov*, Vol. 1, No. 19, 897–908, in Russian.
103. Galerkin B.G., Rods and Plates. *AN SSSR*, Moscow, 1952, Vol. 1, 168–195, in Russian.
104. Galinov K.Z., On general theory of plates and shells with finite displacements and deformations. *PMM*, 1951, Vol. 5, No. 6, 723–742, in Russian.
105. Ganapathi M., Varadan T.K., Large amplitude vibrations of circular cylindrical shells. *Journal of Sound and Vibration*, 1996, Vol. 192, No. 1, 1–14.
106. Gawinecki J., Existence, uniqueness and regularity of the first boundary - initial value problem for thermal stresses equations of classical and generalized thermomechanics. *Journal of Technical Physics*, 1983, Vol. 24, No. 4, 467–479.
107. Gawinecki J., Faedo-Galerkin method in the classical theory of heat stresses. *Bulletin WAT*, 1984, Vol. 33, No. 2, 17–34, in Polish.
108. Gawinecki J., On the first initial-boundary value problem for the equations of thermal stresses. *Bulletin of the Academy of Polish Science Series of Technical Sciences*, 1981, Vol. 29, No. 7–8, 400–404.
109. Gawinecki J., On the first initial-boundary value problem for thermal stresses equations of generalized thermomechanics. *Bulletin of the Academy of Polish Science Series of Technical Sciences*, 1981, Vol. 29, No. 7–8, 405–411.
110. Gelos R., Domingues H., Laura P., Application of the optimized Galerkin method to the determination of the fundamental frequency of a vibrating circular plate subjected to non-uniform in plane loading. *Journal of Sound and Vibration*, Vol. 114, No. 3, 598–900.
111. Goel R.P., Vibrations of a beam carrying a concentrated mass. *Journal of Applied Mechanics*, 1973, Vol. 40, No. 3, 821–822.
112. Gollub J.P., Benson S.V., Chaotic response to periodic perturbation of a convecting fluid. Physical Review Letters, 1978, 41(14), 948–951.
113. Grebogi C., Ott E., Yorke J.A., Chaotic attractors in crisis. Physical Review Letters, 1982, 48, 1507–1510.

114. Grigoluk E.I. Kabanov V.V., *Stability of Shells*. Nauka, Moscow, 1978, in Russian.
115. Grigoluk E.I., Kabanov V.V., Stability of circled cylindrical shells. *Mechanics of Solid Elastic Bodies*, VINITI, Moscow, 1969, in Russian.
116. Grigoluk E.I., On the Bubnov-Galerkin method. *Investigation of Theory of Plates and Shells*. University Press, 1975, Vol. XI, 3–41, in Russian.
117. Guo X., Mei C., Application of aeroelastic modes on nonlinear supersonic panel flutter at elevated temperatures. *Computers and Structures*, 2006, Vol. 84, No. 24–25, 1619–1628.
118. Han W., Petyt M., Geometrical nonlinear vibration analysis of thin, rectangular plates using the hierarchical finite element method-I: The fundamental mode of isotropic plates. *Computers and Structures*, 1997, Vol. 63, No. 2, 295–308.
119. Hai W., Duan Y., Pan L., An analytical study for controlling unstable periodic motion in magneto-elastic chaos. *Physics Letters A*, 1997, Vol. 234, No. 3, 198–204.
120. Heckl M.A., Coupled waves on a periodically supported Timoshenko beam. *Journal of Sound and Vibration*, 2002, Vol. 252, No. 5, 849–882.
121. Higuchi K., Dowell E.H., Effect of constant transverse force on chaotic oscillations of sinusoidally excited buckled beam. *International Journal of Non-Linear Mechanics*, 1991, Vol. 26, No. 3–4, 419–426.
122. Hopf E., Über die Anfangswertaufgabe fur die hydrodynamischen Grundgleichungen. *Mathematische Nachrichten*, 1951, No. 4, 213–231.
123. Ibrahim R.A., Afaneh A.A., Lee B.H., Structural modal multiblefurcation with internal resonance. I-Deterministic approach. II-Stochastic approach. *Journal of Vibration and Acoustics*, 1993, Vol. 115, 182–201.
124. Ilyushin A.A., *Plasticity*. Gostekhizdat, Moscow, 1948, in Russian.
125. Izwieshchenye od Instituta Inzhynerov Putiej Soobshchenya Imperatora Alexandra I. *Vestnik Putey Soobshcheniya*, No. 26, June 25th, 1911, in Russian.
126. Jackson E.A., On the control of complex dynamic systems. *Physica D*, 1991, Vol. 50, No. 3, 341–366.
127. Jackson E.A., The entertainment and migration controls of multiple-attractor systems. *Physicas Letters A*, 1990, Vol. 151, No. 9, 478–484.
128. Jenkins G., Watts D., *Spectral Analysis and Its Applications*. Holden Day, Merrifield, 1968.
129. Ji J.-C., Yu L., Chen Y.-S., Bifurcation and amplitude modulated motions in a parametrically excited two-degree-of-freedom non-linear system. *Journal of Sound and Vibration*, 1999, Vol. 228, No. 5, 1125–1144.
130. Kantor B.Ya., *Nonlinear Problems of Theory of Non-Homogeneous Shallow Shells*. Naukova Dumka, Kiev, 1971, in Russian.
131. Kantor B., Ya., *Contact Problems of Nonlinear Theory of Rotational Shells*. Kiev, Naukova Dumka, 1990, in Russian.
132. Kantorovich L.V., Krylov V.I., *Approximate Methods of Higher Analysis*. Fizmatgiz, Moscow, 1962, in Russian.
133. Kantorov L.V., Krylov V.I., *Approximate Methods of Higher Analysis*. Fizmatgiz, Moscow, 1962, in Russian.
134. Keldysh M.V., On the Galerkin method to solve boundary value problems. *Izviestia AN SSSR*, Mathematical Series, 1942, Vol. 6, No. 6, 309–330, in Russian.
135. Kirichenko V.F., Krysko V.A., Khametova N.A., On influence on thermo-elastic coupling of temperature and deformation fields on stability of shallow shells. *Prikladnaya Mekhanika*, 1988, Vol. 24, No. 1, 46–50, in Russian.
136. Kirichenko V.F., Krysko V.A., On solution existence of one nonlinear coupled problem of thermoelasticity. *Differential Equations*, 1984, Vol. 20, No. 9, 1583–1588, in Russian.
137. Koiter W.T., *Over de Stabiliteit Van Het Elastisch Evenwicht*. Ph.D. Thesis. Amsterdam, 1945.
138. Kolomec A.A., Kutsemako A.N., *Nonlinear Free Vibrations of a Closed Cylindrical Shell. Strength of Constructions in External Conditions*. Saratov State University, Saratov, 1992, 115–119, in Russian.

139. Kolomec A.A., Kutsemako A.N., Dynamical stability loss of a cylindrical shell subject to non-uniform external pressure. In: *Proceedings of the Conference Numerical Method Devoted to Analysis of Elasticity and Plasticity Problems*, Saratov State University, Saratov, 1986, 166–169, in Russian.
140. Kolomoec A.A., Kutsemako A.N., *Dynamical Stability Loss of Closed Cylindrical Shell Subject to Action of Non-uniform External Pressure. Problems of Optimal Design of Plates and Shells*. Saratov State University, Saratov, 1981, 54–56.
141. Kopmaz O., Telli S., On the eigenfrequencies of a two-part beam-mass system. *Journal of Sound and Vibration*, 2002, Vol. 252, No. 2, 370–384.
142. Korn G., Korn T., *Handbook of Mathematics*, Nauka, Moscow, 1973, in Russian.
143. Kornishin M.S., *Nonlinear Problems of the Theory of Plates and Shells and Methods of Their Solutions*. Nauka, Moscow, 1964, in Russian.
144. Kornishin M.S., Isanbaeva F.S., *Flexible Plates and Panels*. Nauka, Moscow, 1968, in Russian.
145. Kornishin M.S., Serazutdinov M.P., Dynamics of inclined shells during a solid body horizontal impacts. *Materials of Seminar in Statics and Dynamics of Shells*, Kazań, 1979, Vol. 12, 190–220, in Russian.
146. Kovtunov V.V., Chaotic vibrations of parametrically stabilized beam. *Strength of Materials and Theory of Constructions*, 1991, Vol. 59, 83–87, in Russian.
147. Kowalski T., Litewska K., Piskorek A., Uniqueness and regularity of the solution of the first-initial-boundary value problem in linear thermoelasticity. *Bulletin of the Academy of Polish Sciences Series of Technical Sciences*, 1982, Vol. 30, No. 3–4, 171–175.
148. Kowalski T., Piskorek A., Existenz der Lösung einer Anfangsrandwertaufgabe in der linearen Thermoelastizitatstheorie. *Zeitschrift für Angewandte Mathematik und Mechanik*, 1981, Vol. 61, No. 5, 250–252.
149. Kransoselskiy M.A., Vaynikko G.M., Zabreiko P.P., *Approximating Solution of Operator Equations*. Nauka, Moscow, 1969, in Russian.
150. Krasnoselskiy M.A., Convergence of the Galerkin method for nonlinear equations. *Doklady AN SSSR*, 1950, Vol. 26, No. 6, 1121–1124, in Russian.
151. Krotkova L.V., Vibrations of ellipse plates made from almost elastic material. Izviestia VUZ. *Series of Civil Engineering and Architecture*, 1970, Vol. 4, 62–65, in Russian.
152. Knudsen J., Massih A.R., Vibro-impact dynamics of a periodically forced beam. *Journal of Pressure Vessel Technology*, 2000, Vol. 122, No. 2, 210–221.
153. Krylov N.M., *On Approximated Solutions of the Mathematical Problems of Physics. Collection of Works in Three Volumes*. Kiev, 1961, Vol. 2, 150–204, in Russian.
154. Krylov N.M., *On Approximating Integration of Partial Differential Equations of Mathematical Physics. Collection of Works in Three Volumes*. 1961, Kiev, Vol. 2, 7–8, in Russian.
155. Krylov N.M., *On Error Allowed Influence During Application of the Ritz Method for Approximating Integration of Difference Equations. Collection of Works in Three Volumes*. Kiev, 1961, Vol. 1, 257–258, in Russian.
156. Krylov N.M., *On Various Generalization of the Ritz Method and Some Related Problems. Collection of Works in Three Volumes*. Kiev, 1961, Vol. 1, 195–243, in Russian.
157. Krylov N.M., Bogolubov N.N., On the theorems devoted to existence of integrals of partial hyperbolic differential equations. Izviestia AN SSSR. *Branch of Mathematical and Natural Sciences*, 1931, No. 3, 323–344, in Russian.
158. Krylov N.M., Tamarkin Ya.D., On the Ritz method to approximated solution of mathematical problems in physics. *Collected Works in Three Volumes*. Kiev, 1961, Vol. 1, 132–145, in Russian.
159. Krysko V.A., *Nonlinear Statics and Dynamics of Non-Homogeneous Shells*. Saratov State University, Saratov, 1976, in Russian.
160. Krysko V.A., Awrejcewicz J., Bruk V.M., On the solution of a coupled thermo-mechanical problem for non-homogeneous Timoshenko-type shells. *Journal of Mathematical Analysis and Applications*, 2002, Vol. 273, No. 2, 2002, 409–416.

161. Krysko V.A., Awrejcewicz J., Bruk V.M., The existence and uniqueness of solution of one coupled plate thermomechanics problems, *Journal of Applied Analysis*, 2002, Vol. 8, No. 1, 2002, 129–139.
162. Krysko V.A., Awrejcewicz J., Komarov S.A., Nonlinear deformations of spherical panels subjected to transversal load action, *Computer Methods in Applied Mechanics and Engineering*, Vol. 194, No. 27–29, 2005, 3108–3126.
163. Krysko V.A., Awrejcewicz J., Kutsemako A.N., Broughan K., Interaction between flexible shells (plates) and a moving lumped body, *Communications in Nonlinear Science and Numerical Simulation*, 2006, Vol. 11, No. 1, 13–43.
164. Krysko V.A., Awrejcewicz J., Narkaitis G.G., Nonlinear vibration and characteristics of flexible plate-strips with non-symmetric boundary conditions. *Communications in Nonlinear Science and Numerical Simulation*, 2006, Vol. 11, No. 1, 95–124.
165. Krysko V.A., Awrejcewicz J., Shchekaturova T.V., Chaotic vibrations of spherical and conical axially-symmetric shells. *Archive of Applied Mechanics*, 2005, Vol. 74, No. 5–6, 338–358.
166. Krysko V.A., Guba G.M., Fomin V.G., Dynamics of a shallow spherical shell subject to a heat impact. *Prikladnaya Mekhanika*, 1986, Vol. 22, No. 5, 21–27, in Russian.
167. Krysko V.A., Kravtsova I.V., Stochastic vibrations of flexible flat axisymmetric shells exposed to inhomogeneous loading. *International Conference, Dynamical Systems-Theory and Applications*, Łódź, Poland, 2003, 189–197.
168. Krysko V.A., Kutsemako A.N., *Stability and Vibrations of Non-Homogeneous Shells*. Saratov State University, Saratov, 1999, in Russian.
169. Krysko V.A., Savielyeva N.E., Statics and dynamics of closed cylindrical shells under nonhomogeneous loading. In: *Proceedings of the International Conference on Problems of Strength in Transport of Materials and Constructions*, Sankt-Petrsburg, 2004, 210–221, in Russian.
170. Krysko V.A., Shchekutorova T.W., Chaotic vibrations of cone shells. *Izviestia RAN MTT*, 2004, No. 4, 140–150, in Russian.
171. Kutsemako A.N., *Some Problems of Flexible Shells with Transversal Shears*. Ph.D. Thesis, Saratov, 1975, in Russian.
172. Kudriavtsev L.D., *Lectures on Mathematical Analysis*. Moscow, Vysshaia Shkola, 1989, Vol. 4, in Russian.
173. Kuznetsov Yu.A., Elements of Applied Bifurcation Theory. Springer Verlag, New York, 1998.
174. Kwuimy C.A., Nbendjo B.R.N., Woafo P., Optimization of electromechanical control of beam dynamics: Analytical method and finite differences simulation. *Journal of Sound and Vibration*, 2006, Vol. 298, No. 1–2, 180–193.
175. Ladyzhenskaya O.A., *Boundary Value Problems of Mathematical Physics*. Nauka, Moscow, 1973, in Russian.
176. Ladyzhenskaya O.A., On integral convergence of the approximating methods and solutions in functionals of linear elliptic operators. *Vestnik Leningradskago Universiteta*, 1958, Vol. 7, 60–69, in Russian.
177. Lai H.-Y., Chen C.-K., Yeh Y.-L., Double-mode modeling of chaotic and bifurcation dynamics for a simply supported rectangular plate in large deflection. *International Journal of Non-Linear Mechanics*, 2002, Vol. 37, No. 2, 331–343.
178. Lancosh K., *Practical Methods of Applied Analysis*. Moscow, Fizmatgiz, 1961, in Russian.
179. Landau D., Lectures of Academy of Science, *AS SSSR Press*, 1944, Vol. 7, 203–217, in Russian.
180. Landau D., On the problem of turbulence. *DAN SSSR*, 1944, Vol. 44, No. 8, in Russian.
181. Lavrentiev M.A., Ishlinskiy A.Yu., Dynamical forms of stability loss of elastic systems. *DAN SSSR*, 1949, Vol. 64, No. 6, 779–789, in Russian.
182. Lebedev L.P., On equilibrium state of free nonlinear plate. *PMM*, 1980, Vol. 44, No. 1, 161–165, in Russian.
183. Lebedev L.P., On existence of solutions in nonlinear theory of shallow shells. *PMM*, 1972, Vol. 36, No. 4, 691–704, in Russian.

184. Leibenson L.S., Variational methods of solution of theory of elasticity. Works collection. *AS SSSR Press*, 1951, Vol. 1, 177–463, in Russian.
185. Leibenson L.S., On application of potential displacements to approximate estimation of elastic equilibrium state. Collected Works. *AN SSSR Press*, 1957, Vol. 1, 39–49, in Russian.
186. Lenci S., Tarantino A.M., Chaotic dynamics of an elastic beam resting on a Winkler-type soil. *Chaos, Solitons and Fractals*, 1996, Vol. 7, No. 10, 1601–1614.
187. Li Q.S., Free vibration analysis of non-uniform beams with an arbitrary number of cracks and concentrated masses. *Journal of Sound and Vibration*, 2002, Vol. 252, No. 3, 509–525.
188. Li T.Y., Yorke J.A., Period three implies chaos. American Mathematical Monthly, 1975, 82(10), 985–992.
189. Liashko A.D., On convergence of the Galerkin method. *Doklady AN SSSR*, 1958, Vol. 120, No. 2, 242–244, in Russian.
190. Liashko A.D., On some variants of the Galerkin-Krylov method. *Doklady AN SSSR*, 1959, Vol. 128, No. 3, 468–470, in Russian.
191. Liashko A.D., On the variational method for nonlinear operator equations. *Uchiobnyje Zapiski Kazańskiego Universtiteta*, 1966, Vol. 125, 95–101, in Russian.
192. Liew K.M., Lim C.W., Ong L.S., Flexural vibration of doubly-tapered cylindrical shallow shells. *International Journal of Mechanical Science*, 1994, Vol. 36, No. 6, 547–565.
193. Lin W., Qiao N., Yuying H., Bifurcations and chaos in a forced cantilever system with impacts. *Journal of Sound and Vibration*, 2006, Vol. 296, No. 4–5, 1068–1078.
194. Lions J.L., Magenes E., *Non-homogeneous Boundary value Problems and Applications*. Springer-Verlag, Berlin, 1972.
195. Mandebrot B., The Fractal Geometry of Nature. New York, Freeman and Company, 1982.
196. Manneville P., Romeau Y., Intermittency and the Lorenz model. Physics Letters A, 1979, 75(1–2), 1–2.
197. Manneville P., Romeau Y., Different ways to turbulence in dissipative dynamical systems. *Physical*, 1980, Vol. 1, No. 1, 219–228.
198. Martin S., Leber H., Martienssen W., Oscillatory and diaotic states of the electrical conduction in Barium sodium miobate crystals. Physical Review Letters, 1984, 53, 303–306.
199. Martyniuk A.E., On generalization of a variational method. *Doklady AN SSSR*, 1957, Vol. 117, No. 2, 374–377, in Russian.
200. Martyniuk A.E., On the Galerkin momentous method. *Uchiobnyie zapiski Kazańskiego Universiteta*, 1957, Vol. 117, No. 2, 70–74, in Russian.
201. Matsunaga H., Vibration and buckling of multilayered composite beams according to higher order deformation theories. *Journal of Sound and Vibration*, 2001, Vol. 246, No. 1, 47–62.
202. Mead D.J., Comparison of some equations for the flexural vibration of damped sandwich beams. *Journal of Sound and Vibration*, 1982, Vol. 83, 363–377.
203. Mead D.J., Free vibrations of self-strained assemblies of beams. *Journal of Sound and Vibration*, 2002, Vol. 249, No. 1, 101–127.
204. Mead D.J., Markus S., The forced vibration of three-layer, damped sandwich beam with arbitrary boundary conditions. *Journal of Sound and Vibration*, 1969, Vol. 10, No. 2, 163–175.
205. Mead D.J., Markus S., Loss factors and resonant frequencies of encastré damped sandwich beams. *Journal of Sound and Vibration*, 1970, Vol. 12, No. 1, 99–112.
206. Michlin S.G., *Numerical Realization of Variational Methods*. Nauka, Moscow, 1966, in Russian.
207. Michlin S.G., *Problems of Minimum of Squared Functional*. Gostekhizdat, Moscow-Leningrad, 1952, in Russian.
208. Michlin S.G., *Some Problems of Theory of Errors*. Leningrad University Press, Leningrad, 1988, in Russian.
209. Michlin S.G., *Variational Methods in Mathematical Physics*. Nauka, Moscow, 1970, in Russian.
210. Michlin S.G., On convergence of the Galerkin method. *Doklady AN SSSR*, 1948, Vol. 61, No. 2, 197–199, in Russian.

211. Michlin S.G., On the Ritz method. *AN SSSR Press*, 1956, Vol. 106, No. 3, 391–394, in Russian.
212. Mikulas M.M., McElman J.A., On free vibrations of eccentrically stiffened cylindrical shells and flat plates. *NASA*, 1965, TN, D-3010.
213. Mioduchovski A., Pielorz A., Nadolski W., Haddow J.B., Finite oscillation of viscoelastic cantilever strip. *Acta Mechanica*, 1983, Vol. 50, No. 1–2, 39–48.
214. Morozov N.F., Investigation of vibrations of a prismatic rod subject to transversal load. Izviestia VUZ, *Series of Mathematics*, 1965, No. 3, 121–125, in Russian.
215. Morozov N.F., Nonlinear problems of thin plates. *Vestnik Leningradskego Instituta*, 1958, Vol. 4, No. 19, 100–124, in Russian.
216. Moorthy R.I.K., Kakodkar A., Srirangarajan H.R., Suryanarayan S., Finite element simulation of chaotic vibrations of a beam with non-linear boundary conditions. *Computers and Structures*, 1993, Vol. 49., No. 4, 589–596.
217. Moorthy R.I.K., Kakodkar A., Srirangarajan H.R., The significance of higher modes for evolution of chaos in structural mechanics, systems. *Journal of Sound and Vibration*, 1996, Vol. 198, No. 3, 267–277.
218. Mukhopadhyay M., Free vibration of rectangular plates with edges having different degrees of rotational restraint. *Journal of Sound and Vibration*, 1979, Vol. 67, No. 4, 459–468.
219. Mushtari Ch.M., Galimov K.Z., *Nonlinear Theory of Elastic Thin Shells*. Kazań, Tatknigizdat, 1957, in Russian.
220. Moussaoui F., Benamar R., White R.G., The effect of large vibration amplitudes on the mode shapes and natural frequencies of thin elastic shells. Part I: Coupled transverse-circumferential mode shapes of isotropic circular cylindrical shells of infinite length. *Journal of Sound and Vibration*, 2000, Vol. 232, No. 5, 917–943.
221. Murphy K.D., Virgin L.N., Rizzi S.A., Characterizing the dynamic response of a thermally loaded, acoustically excited plate. *Journal of Sound and Vibration*, 1996, Vol. 196, No. 5, 635–658.
222. Nagai K., Maruyama S., Oya M., Yamaguchi T., Chaotic oscillations of a shallow cylindrical shell with a concentrated mass under periodic excitation. *Computers and Structures*, 2004, Vol. 82, No. 31–32, 2607–2619.
223. Nagai K., Yamaguchi T., Chaotic vibrations of post-buckled beam with a variable cross section under periodic excitation. *Transactions of the Japanese Society of Mechanical Engineering C*, 1995, Vol. 61, No. 586, 2202–2209.
224. Nagai K., Yamaguchi T., Chaotic vibrations of post-buckled beam carrying a concentrated mass – experiment. 1st report. *Transactions of the Japanese Society of Mechanical Engineering C*, 1994, Vol. 61, No. 579, 3733–3740.
225. Nagai K., Yamaguchi T., Chaotic vibrations of post-buckled beam carrying a concentrated mass – theoretical analysis. 2nd report. *Transactions of the Japanese Society of Mechanical Engineering C*, 1994, Vol. 60, No. 579, 3741–3748.
226. Nagai K., Kasuga K., Kamada M., Yamaguchi T., Tanifuji K., Experiment on chaotic oscillations of a post-buckled reinforced beam constrained by an axial spring. *Transactions of the Japanese Society of Mechanical Engineering C*, 1998, Vol. 41, No. 3, 563–569.
227. Nashed M.Z., The convergence of the method of steepest descents for nonlinear equations with variational or quasi-variational operators. *Journal of Mathematical Mechanics*, 1964, Vol. 13, 765–774.
228. Nayfeh A.H., Mook D.T., *Non-Linear Oscillations*. John Wiley & Sons, New York, 1979.
229. Nazarov A.G., Impulse functions and their application in civil engineering. *Investigation of Theory of Constructions*. Gostekhizdat, Moscow-Leningrad, Vol. 4, 216–226, in Russian.
230. Nbendjo B.R.N., Woafo P., Active control with delay of horseshoes chaos using piezoelectric absorber on a buckled beam under parametric excitation. *Chaos, Solitons and Fractals*, 2007, Vol. 32, No. 1, 73–79.
231. Nemeth M.P., Young R.D., Collins T.J., Starnes J.H., Nonlinear analysis of the space shuttle super-lightweight LO2 tank. In: *Proceedings of the Structures, Structural Dynamics, and Material Conference*, 1998.

232. Newhouse S., Ruelle D., Takens F., Occurence of strange Axioms A attractor near quasi-periodic flow on Communications in Mathematical Physics, 1978, 64(1), 35–40.
233. Novitskiy V.V., Delta-function and its application in civil engineering. *Computation of 3D Constructions*. Gosstroizdat, Moscow-Leningrad, 1961, Vol. 8, 207–244, in Russian.
234. Novozhilov V.V., *Theory of Elasticity*. Sudpromgiz, Moscow, 1958, in Russian.
235. Orszag S.A., Vells L.C., Transition to turbulence in plane poiseville and plane couette flow. Journal of Fluid Mechanics Digital Archive, 1980, 96, 159–205.
236. Ott E., Grebogi C., Yorke J.A., Controlling chaos. *Physical Review Letters*, 1990, Vol. 64, 1196–1199.
237. Paidoussis M.P., Li G.X., Cross-flow-induced chaotic vibrations of heat-exchanger tubes impacting on loose supports. *Journal of Sound and Vibration*, 1992, Vol. 152, No. 2, 305–326.
238. Panovko Ya.G., Gubanova I.I., *Stability and Vibration of Elastic Systems*. Nauka, Moscow, 1987, in Russian.
239. Papkovich P.F., *Mechanics of a Ship*. Sudpromgiz, Leningrad, Vol. 2, 1942, in Russian.
240. Papkovich P.F., *Works on a Ship Constructions*. Sudpromgiz, Leningrad, Vol. 3, 1962, in Russian.
241. Papkovich P.F., *Mechanics of Constructions*, Part II. Sudpromgiz, Leningrad, 1939, in Russian.
242. Patash A.L., Convergence of the Bubnov-Galerkin method in problems of transversal deflection of three layered plates. Izviestia VUZ, *Series of Civil Engineering and Architecture*, 1982, Vol. 3, 42–45, in Russian.
243. Pellicano F., Amabili M., Dynamic instability and chaos of empty and fluid-filled circular cylindrical shells under periodic axial loads. *Journal of Sound and Vibration*, 2006, Vol. 293, No. 1–2, 227–252.
244. Petrov V., Gaspar V., Masere J., Showalter K., Controlling chaos in the Belousov-Zhabotinsky reaction. *Nature*, 1993. Vol. 361, 240–243.
245. Pezeshki C., Dowell E.H., An examination of initial condition maps for the sinusoidally excited buckled beam modeled by the Duffing's equation. *Journal of Sound and Vibration*, 1987, Vol. 117, No. 2, 219–232.
246. Ritzs W., Über eine neue Methode zur Lösung gewisser Variations probleme der mathematichen Physik. *Zeitschrift für Angewandte Mathematik und Mechanik*, 1908, Vol. 135, No. 1, 1–61.
247. Poincaré H., *Science et méthode*. Flam Marion, Paris, 1908.
248. Pomeau Y., Manneville P., Intermittent transition to turbulence in dissipative dynamical systems. *Communications in Mathematical Physics*, 1980, Vol. 74, No. 2, 189–197.
249. Poston T., Stewart I., *Catastrophe Theory and Its Application*. Dover Publications, New York, 1996.
250. Popov A.A., Thompson J.M.T., McRobie F.A., Chaotic energy exchange through autoparametric resonance in cylindrical shells. *Journal of Sound and Vibration*, 2001, Vol. 248, No. 3, 395–411.
251. Preobrazhenskiy I.N., *Stability and Vibrations of Plates and Shells with Holes*. Mashinostroyeniye, Moscow, 1981, in Russian.
252. Pyryev Yu., Dynamics of Contacting Systems with an Account of Heat Transfer, Friction and Wear. *Scientific Bulletin of Łódź Technical University*, Łódź, 2004, Vol. 936, in Polish.
253. Radstig Yu.A., Mirror functions and their applications in problems of civil engineering. *Scientific Works of Kazan's Air Institut*, Kazań, Tatgoizdat, 1948, 47–68, in Russian.
254. Reviews of Professors Bielelubski, Bielecki, Bubnov and Kolosov of the Timoshenko's work applying for D.I. Zhuravskiy's award. *Collection of Works of the Engineers Institute of Putey Soobshcheniya Imperatora Alexandra I*. Typography by Yu. Erlich. Sankt-Petersburg, 1913, Vol. 81, 1–40, in Russian.
255. Rabotnov Yu. N., *Creeps of Construction Elements*. Nauka, Moscow, 1966, in Russian.
256. Rand B., Ostlund S., Sethna J., Siggia E.D., Universal transition from quasi-periodicity to chaos in dissipative systems. Physical Review Letters, 1982, 49, 132–135.
257. Ribeiro P., Thermally induced transitions to chaos in plate vibrations. *Journal of Sound and Vibration*, 2007, Vol. 299, No. 1–2, 314–330.

258. Ribeiro P., Duarte R.P., From periodic to chaotic oscillations in composite laminated plates. *Computers and Structures*, 2006, Vol. 84, No. 24–25, 1629–1639.
259. Raouf R.A., Nayfeh A.H., Non-linear axisymmetric response of closed spherical shells to a radial harmonic excitation. *International Journal of Non-Linear Mechanics*, 1990, Vol. 25, No. 5, 475–492.
260. Ross C.T.F., Finite elements for the vibration of cones and cylinders. *International Journal of Numerical Methods Engineering*, 1975, Vol. 9, No. 4, 833–845.
261. Rössler O., Continuous Ehaos-From Prototype equations, Ann. New York Acad. Sci., 1879.
262. Ruelle D., Takens F., On the nature of turbulence. *Communications in Mathematical Physics*, 1971, Vol. 20, No. 3, 167–192.
263. Ramu A.S., Sankar T.S., Ganesan R., Bifurcations, catastrophes and chaos in a pre-buckled beam. *International Journal of Non-Linear Mechanics*, 1994, Vol. 29, No. 3, 449–462.
264. Shaw S.W., Pierre C., Normal modes for non-linear continuous systems. *Journal of Sound and Vibration*, 1994, Vol. 169, No. 3, 319–347.
265. Shaw S.W., Chaotic dynamics of a slender beam rotating about its longitudinal axis. *Journal of Sound and Vibration*, 1988, Vol. 124, No. 2, 329–343.
266. Schiff S.J., Jerger K., Duong D. H., Chang T., Spano M.L., Ditto W. L., Controlling chaos in the brain. *Nature*, 1994, Vol. 370, 615–620.
267. Schwartz L., *Mathematical Methods for Physical Sciences*. Mir, Moscow, in Russian.
268. Sharkovsky A.N., Existence of continuous mapping cycles into itself. *Ukrainian Journal of Mathematics*, 1964, Vol. 16, No. 1, 61–71.
269. Sheng D.-F., Chang C.-J., Dynamical behaviors of nonlinear viscoelastic thick plates with damage. *International Journal of Solids and Structures*, 2004, Vol. 41, No. 26, 7287–7308.
270. Shiau A.C., Soong T.T., Roth R.S., Dynamic buckling of conical shells with imperfection. *AIAA Journal*, 1974, Vol. 12, No. 6, 755–760.
271. Shinbrot T., Grebogi C., Ott E., Yorke J.A., Using small perturbations to control chaos. *Nature*, 1993, Vol. 363, 411–417.
272. Singer J., Wang Y.-Z., Bau H.-H., Controlling a chaotic systems. *Physical Review Letters*, 1991, Vol. 66, No.9 1123–1125.
273. Sisemore C.L, Smaili A.A., Darvennes C.M., Experimental measurement of compressional damping in an elastic-viscoelastic-elastic sandwich beam. In: *Proceedings of the American Society of Mechanical Engineers Noise Control and Acoustics Division*, 1995, 223–227.
274. Sisemore C.L, Smaili A.A., Houghton J.R., Passive damping of flexible mechanism systems: Experimental and finite element investigations. In: *Proceedings of the Tenth World Congress of the Theory of Machines and Mechanisms*, 1999, 2140–2155.
275. Sisemore C.L, Darvennes C.M., Transverse vibration of elastic-viscoelastic-elastic sandwich beams: Compression-experimental and analytical study. *Journal of Sound and Vibration*, 2002, Vol. 252, 155–167.
276. Skrypnik I.V., *Nonlinear Elliptic Equations of Higher Order*. Naukova Dumka, Kiev, 1973, in Russian.
277. Smelova-Reynolds T., Dowell E.H., The role of higher modes in the chaotic motion of the buckled beam – I. *International Journal of Non-Linear Mechanics*, 1996, Vol, 31, No. 6, 931–939.
278. Smelova-Reynolds T., Dowell E.H., The role of higher modes in the chaotic motion of the buckled beam – II. *International Journal of Non-Linear Mechanics*, 1996, Vol, 31, No. 6, 941–950.
279. Sobolev S.L., *On Application of Functional Analysis in Mathematical Physics*. Leningrad University Press, Leningrad, 1950, in Russian.
280. Sobolevskiy P.E., On equations with operators having sharp angle. *AN SSSR Press*, 1957, Vol. 116, No. 5, 752–757, in Russian.
281. Sobolevskiy P.E., On the Bubnov-Galerkin method for parabolic equations in Hilber Space. *AN SSSR Press*, 1968, Vol. 178, No. 3, 548–551, in Russian.
282. Soliman M.S., Gonçalves P.B., Chaotic behavior resulting in transient and steady state instabilities of pressure-loaded shallow spherical shells. *Journal of Sound and Vibration*, 2003, Vol. 259, No. 3, 497–512.

283. Sorokin S.V., Tcherniak D.M., Nonlinear dynamics and chaos of a finite-degree-of-freedom model of a buckled rod. *Meccanica*, 1996, Vol. 31, No. 3, 273–291.
284. Stanley A.J., Ganesan N., Free vibration characteristics of stiffened cylindrical shells. *Computers and Structures*, 1997, Vol. 65, No. 1, 33–45.
285. Sun Y.X., Zhang S.Y., Chaotic dynamic analysis of viscoelastic plates. *International Journal of Mechanical Sciences*, 2001, Vol. 43, No. 5, 1195–1208.
286. Svirsky I.V., *Methods of the Bubnov-Galerkin Type and a Sequence of Approximation*. Nauka, Moscow, 1968, in Russian.
287. Tcherniak D.M., *Computation of Flexible Plates Using the Bubnov-Galerkin Method on a Basis of Co-ordinate Functions. Theoretical and Experimental Investigation of Bridges*. Omsk, 1982, 105–109, in Russian.
288. Thom R., *Structural Stability and Morphogenesis*, Benjamin-Addison-Wesley, New York, 1975.
289. Thom R., Structural Stability and Morphogenesis: An outline of a General Theory of Models, Addison-Wesley, London, 1989.
290. Thompson J.M.T., *Instabilities and Catastrophes in Science and Engineering*. John Wiley, Chichester, 1982.
291. Thomson J.M.T., Bishop S.R., *Non-linearity and Chaos in Engineering Dynamics*. Center for Non-linear Dynamics, University College London, John Wiley & Sons, 1994.
292. Timoshenko S.P., Sur la stalilite des systemes elastiques. Application d'une nourelle me'thode dla recherch de stabilite de certaines parties constitutives des ponts. *Annales des Ponts et Chanssees*. L Partie, 9 serie, 1913, Vol. 15 No. 24, 496–566.
293. Timoshenko S.P., *Stability of Rods, Plates and Shells*. Nauka, Moscow, 1967, in Russian.
294. Timoshenko S.P., Stability of elastic systems. Application of a new method in analysis of stability of bridge constructions. *Kiev Polytechnik Institute Press*. Otdel Mechaniki Inzhyniernoy, 1910, Vol. 4, 375–560, in Russian.
295. Timoshenko S.P., On the longitudinal deflection of rods analysed in an elastic medium. *Sankt-Petersburg Polytechnik Institute Press*, Otdel Mekhaniki Inzhyniernoy, 1907, Vol. 7, No. 1, 145–157, in Russian.
296. Trindade M.A., Benjeddou A., Ohayon R., Piezoelectric active vibration control of damped sandwich beams. *Journal of Sound and Vibration*, 2001, Vol. 246, No. 4, 653–677.
297. Urabe M., Galerkin's Procedure for Nonlinear Periodic Systems. *Archive for Rational Mechanics and Analysis*, 1965, Vol. 20, No. 2, 120–152.
298. Urabe M., Reiter A., Numerical computation of nonlinear forced oscillators by Galerkin's procedure. *Journal of Mathematical Analysis and Applications*, 1966, Vol. 14, 107–140.
299. Urabe M., *Nonlinear Autonomous Oscillations* Analytical Theory Academic Press, 1967, New York.
300. Vainberg D.V., Roitfarb I.Z., Computation of plates and shells with discontinuous parameters. *Computation of 3D Constructions*. Moscow, Stroizdat, 1965, Vol. 10, 39–80, in Russian.
301. Vainikko G.M., On convergent operators. *AN SSSR Press*, 1968, Vol. 179, No. 5, 1029–1031, in Russian.
302. Vainikko G.M., On estimation of the Bubnov-Galerkin errors. *Scientific Bulletin of Tartyck University*, 1969, Vol. 150, 188–215, in Russian.
303. Volmir A.S., *Flexible Plates and Shells*. Gostekhizdat, Moscow, 1956, in Russian.
304. Volmir A.S., *Nonlinear Dynamics of Plates and Shells*. Nauka, Moscow, 1972, in Russian.
305. Volmir A.S., *Shells in Fluid Streams. Problems of Aeroelasticity*. Nauka, Moscow, 1976, in Russian.
306. Van de Vorst E.L.B., Van Campen D.H., de Kraker A., Foy R.H.B. Periodic solutions of a multi-DOF beam system with impact. *Journal of Sound and Vibration*, 1996, Vol. 192, No. 5, 913–925.
307. Van de Vorst E.L.B., Heertjes M.F. Van Campen D.H., de Kraker A., Fey R.H.B., Experimental analysis of the steady state behaviour of beam system with impact. *Journal of Sound and Vibration*, 1998, Vol. 212, No. 2, 321–336.

308. Vorovich I.I., *Mathematical Problems of Nonlinear Theory of Shallow Shells*. Moscow, Nauka, 1979, in Russian.
309. Vorovich I.I., Minakova N.I., Problem of stability and numerical methods in theory of spherical shells. *Achievements in Science and Technics VINITI*. Series of Mechanics of Deformable Body, Moscow, 1973, Vol. 7, 5–86, in Russian.
310. Vorovich I.I., On solutions existence in nonlinear theory of shells. *AN SSSR Press*, 1957, Vol. 117, No. 2, 203–206, in Russian.
311. Wenk H.U., On coupled thermoelastic vibration of geometrically nonlinear thin plates satisfying generalized mechanical and thermal conditions on the boundary and the surface. *Applied Mathematics*, 1982, Vol. 27, No. 6, 398–416.
312. White R.G., Effects of non-linearity due to large deflections in the resonance testing of structures. *Journal of Sound and Vibration*, 1971, Vol. 16, No. 2, 255–267.
313. White R.G., Developments in the acoustic fatigue design process for composite aircraft structures. *Computers and Structures*, 1990, Vol. 16, 171–192.
314. Wilkinson J.H., Reinsch C., *Linear Algebra*, Springer-Verlag, Berlin, 1971.
315. Volmir A.S., *Stability of Elastic Systems*. Fizmatgiz, Moscow, 1963, in Russian.
316. Vorovich I.I., *Mathematical Problems of Nonlinear Theory of Inclined Shells*. Nauka, Moscow, 1989, in Russian.
317. Woźniak Cz. (Ed.), *Mechanics of Elastic Plates and Shells*. Technical Mechanics. Vol. 8, PWN, Warszawa, 2001, in Polish.
318. Xiao Y.-G., Fu Y.-M., Zha X.-D., Bifurcation and chaos of rectangular moderately thick cracked plates on an elastic foundation subjected to periodic load. *Chaos, Solitons and Fractals*, 2008, Vol. 35, No. 3, 460–465.
319. Xie W.C., Lee H.P., Lim S.P., Normal modes of a non-linear clamped-clamped beam. *Journal of Sound and Vibration*, 2002, Vol. 250, No. 2, 339–349.
320. Yagasaki K., Bifurcations and chaos in quasi-periodically forced beam: Theory, simulation and experiment. *Journal of Sound and Vibration*, 1995, Vol. 183, No. 1, 1–31.
321. Yagasaki K., Homoclinic and heteroclinic behavior in an infinite-degree-of-freedom Hamiltonian system: Chaotic free vibrations of an undamped, buckled beam. *Physical Letters A*, 2001, Vol. 285, No. 1–2, 55–62.
322. Yakushev N.Z., Nonlinear Vibrations of Plates and Shells. *Investigation of Plates and Shells*, Kazań, 1978, No. 13, 203–216, in Russian.
323. Yamaguchi T., Nagai K., Chaotic vibrations of post-buckled beam carrying a concentrated mass. II-nd report. Theoretical analysis. *Transactions of the Japanese Society of Mechanical Engineering C*, 1994, Vol. 61, No. 579, 3741–3748.
324. Yang X.L., Sethna P.R., Non-linear phenomena in forced vibrations of a nearly square plate: Antisymmetric case. *Journal of Sound and Vibration*, 1992, Vol. 155, No. 3, 413–441.
325. Youngs D.L., Numerical simulation of turbulent mixing by Rayleigh-Taylor instability. Physica D, 1984, 12(1–3), 32–44.
326. Zarubin A.G., On velocity of convergence of Faedo-Galerkin method for linear nonstationary equations. *Differential Equations*, 1982, Vol. 8, No. 4, 639–645, in Russian.
327. Zarubin A.G., On velocity of convergence of projection methods for linear equations. *Journal of Computational Mathematics and Mathematical Physics*, 1979, Vol. 15, No. 4, 1048–1053, in Russian.
328. Zarubin A.G., On velocity of convergence of the Faedo-Galerkin method. Methods of algebraic and functional analysis during investigation of a family of operators. In: *Proceedings of the Mathematical Conference*, Tartu, 1978, 52–54, in Russian.
329. Zarubin A.G., Tiunchik M.F., On approximated solutions on one class of non-linear nonstationary equations. *Differential Equations*, 1973, Vol. 9, No. 11, 1966–1974, in Russian.
330. Zeeman E.C., *Applications of Catastrophe Theory*, Manifolds, Tokyo, 1973.
331. Zeeman E.C., Catastrophe theory in brain modeling. *International Journal of Neuroscience* 1973, Vol. 6, No. 1, 39–41.
332. Zeeman E.C., *Catastrophe Theory: Selected Papers 1972–1977*, Addison Wesley, London, 1977.

333. Zhelezovskiy S.E., Zhelozovskaya L.A., Kirichenko V.F., Krysko V.A., On velocity of convergence of the Bubnov-Galerkin method for hyperbolic equations. *Differential Equations*, 1990, Vol. 26, No. 2, 323–333, in Russian.
334. Zhelozovskiy S.E., Kirichenko V.F., Krysko V.A., On velocity convergence of the Bubnov-Galerkin method for one non-classical system of differential equations. *Differential Equations*, 1981, Vol. 23, No. 8, 1407–1416, in Russian.
335. Ziegler H., *Principles of Structural Stability*. Blaisidell Publishing Company, Waltham, 1968.
336. Zommerfeld, *Mechanics*. Foreign Literature Press, Moscow, 1947, in Russian.

Index

Printing: Krips bv, Meppel, The Netherlands
Binding: Stürtz, Würzburg, Germany